Dietary Reference Intakes (DRI)

The Dietary Reference Intakes (DRI) include two sets of values that serve as goals for nutrient intake—Recommended Dietary Allowances (RDA) and Adequate Intakes (AI). The RDA reflect the average daily amount of a nutrient considered adequate to meet the needs of most healthy people. If there is insufficient evidence to determine an RDA, an AI is set. AI are more tentative than RDA, but both may be used as goals for nutrient intakes. (Chapter 1 provides more details.)

In addition to the values that serve as goals for nutrient intakes (presented in the tables on these two pages), the DRI include a set of values called Tolerable Upper Intake Levels (UL). The UL represent the maximum amount of a nutrient that appears safe for most healthy people to consume on a regular basis. Turn the page for a listing of the UL for selected vitamins and minerals.

Estimated Energy Requirements (EER), Recommended Dietary Allowances (RDA), and Adequate Intakes (AI) for Water, Energy, and the Energy Nutrients

Age (yr)	Reference BMI (kg/m²)	Reference Height cm (in)	Reference Weight kg (lb)	Water[a] AI (L/day)	Energy EER[b] (kcal/day)	Carbohydrate RDA (g/day)	Total Fiber AI (g/day)	Total Fat AI (g/day)	Linoleic Acid AI (g/day)	Linolenic Acid[c] AI (g/day)	Protein RDA (g/day)[d]	Protein RDA (g/kg/day)
Males												
0–0.5	—	62 (24)	6 (13)	0.7[e]	570	60	—	31	4.4	0.5	9.1	1.52
0.5–1	—	71 (28)	9 (20)	0.8[f]	743	95	—	30	4.6	0.5	11	1.20
1–3[g]	—	86 (34)	12 (27)	1.3	1046	130	19	—	7	0.7	13	1.05
4–8[g]	15.3	115 (45)	20 (44)	1.7	1742	130	25	—	10	0.9	19	0.95
9–13	17.2	144 (57)	36 (79)	2.4	2279	130	31	—	12	1.2	34	0.95
14–18	20.5	174 (68)	61 (134)	3.3	3152	130	38	—	16	1.6	52	0.85
19–30	22.5	177 (70)	70 (154)	3.7	3067[h]	130	38	—	17	1.6	56	0.80
31–50	22.5[i]	177 (70)[i]	70 (154)[i]	3.7	3067[h]	130	38	—	17	1.6	56	0.80
>50	22.5[i]	177 (70)[i]	70 (154)[i]	3.7	3067[h]	130	30	—	14	1.6	56	0.80
Females												
0–0.5	—	62 (24)	6 (13)	0.7[e]	520	60	—	31	4.4	0.5	9.1	1.52
0.5–1	—	71 (28)	9 (20)	0.8[f]	676	95	—	30	4.6	0.5	11	1.20
1–3[g]	—	86 (34)	12 (27)	1.3	992	130	19	—	7	0.7	13	1.05
4–8[g]	15.3	115 (45)	20 (44)	1.7	1642	130	25	—	10	0.9	19	0.95
9–13	17.4	144 (57)	37 (81)	2.1	2071	130	26	—	10	1.0	34	0.95
14–18	20.4	163 (64)	54 (119)	2.3	2368	130	26	—	11	1.1	46	0.85
19–30	21.5	163 (64)	57 (126)	2.7	2403[j]	130	25	—	12	1.1	46	0.80
31–50	21.5[i]	163 (64)[i]	57 (126)[i]	2.7	2403[j]	130	25	—	12	1.1	46	0.80
>50	21.5[i]	163 (64)[i]	57 (126)[i]	2.7	2403[j]	130	21	—	11	1.1	46	0.80
Pregnancy												
1st trimester				3.0	+0	175	28	—	13	1.4	46	0.80
2nd trimester				3.0	+340	175	28	—	13	1.4	71	1.10
3rd trimester				3.0	+452	175	28	—	13	1.4	71	1.10
Lactation												
1st 6 months				3.8	+330	210	29	—	13	1.3	71	1.30
2nd 6 months				3.8	+400	210	29	—	13	1.3	71	1.30

NOTE: For all nutrients, values for infants are AI. Dashes indicate that values have not been determined.

[a] The water AI includes drinking water, water in beverages, and water in foods; in general, drinking water and other beverages contribute about 70 to 80 percent, and foods, the remainder. Conversion factors: 1 L = 33.8 fluid oz; 1 L = 1.06 qt; 1 cup = 8 fluid oz.

[b] The Estimated Energy Requirement (EER) represents the average dietary energy intake that will maintain energy balance in a healthy person of a given gender, age, weight, height, and physical activity level. The values listed are based on an "active" person at the reference height and weight and at the midpoint ages for each group until age 19. Chapter 8 and Appendix F provide equations and tables to determine estimated energy requirements.

[c] The linolenic acid referred to in this table and text is the omega-3 fatty acid known as alpha-linolenic acid.

[d] The values listed are based on reference body weights.

[e] Assumed to be from human milk.

[f] Assumed to be from human milk and complementary foods and beverages. This includes approximately 0.6 L (~2½ cups) as total fluid including formula, juices, and drinking water.

[g] For energy, the age groups for young children are 1–2 years and 3–8 years.

[h] For males, subtract 10 kcalories per day for each year of age above 19.

[i] Because weight need not change as adults age if activity is maintained, reference weights for adults 19 through 30 years are applied to all adult age groups.

[j] For females, subtract 7 kcalories per day for each year of age above 19.

SOURCE: Adapted from the *Dietary Reference Intakes* series, National Academies Press. Copyright 1997, 1998, 2000, 2001, 2002, 2004, 2005, 2011 by the National Academies of Sciences.

Recommended Dietary Allowances (RDA) and Adequate Intakes (AI) for Vitamins

Age (yr)	Thiamin RDA (mg/day)	Riboflavin RDA (mg/day)	Niacin RDA (mg/day)[a]	Biotin AI (µg/day)	Pantothenic acid AI (mg/day)	Vitamin B_6 RDA (mg/day)	Folate RDA (µg/day)[b]	Vitamin B_{12} RDA (µg/day)	Choline AI (mg/day)	Vitamin C RDA (mg/day)	Vitamin A RDA (µg/day)[c]	Vitamin D RDA (IU/day)[d]	Vitamin E RDA (mg/day)[e]	Vitamin K AI (µg/day)
Infants														
0–0.5	0.2	0.3	2	5	1.7	0.1	65	0.4	125	40	400	400 (10 µg)	4	2.0
0.5–1	0.3	0.4	4	6	1.8	0.3	80	0.5	150	50	500	400 (10 µg)	5	2.5
Children														
1–3	0.5	0.5	6	8	2	0.5	150	0.9	200	15	300	600 (15 µg)	6	30
4–8	0.6	0.6	8	12	3	0.6	200	1.2	250	25	400	600 (15 µg)	7	55
Males														
9–13	0.9	0.9	12	20	4	1.0	300	1.8	375	45	600	600 (15 µg)	11	60
14–18	1.2	1.3	16	25	5	1.3	400	2.4	550	75	900	600 (15 µg)	15	75
19–30	1.2	1.3	16	30	5	1.3	400	2.4	550	90	900	600 (15 µg)	15	120
31–50	1.2	1.3	16	30	5	1.3	400	2.4	550	90	900	600 (15 µg)	15	120
51–70	1.2	1.3	16	30	5	1.7	400	2.4	550	90	900	600 (15 µg)	15	120
>70	1.2	1.3	16	30	5	1.7	400	2.4	550	90	900	800 (20 µg)	15	120
Females														
9–13	0.9	0.9	12	20	4	1.0	300	1.8	375	45	600	600 (15 µg)	11	60
14–18	1.0	1.0	14	25	5	1.2	400	2.4	400	65	700	600 (15 µg)	15	75
19–30	1.1	1.1	14	30	5	1.3	400	2.4	425	75	700	600 (15 µg)	15	90
31–50	1.1	1.1	14	30	5	1.3	400	2.4	425	75	700	600 (15 µg)	15	90
51–70	1.1	1.1	14	30	5	1.5	400	2.4	425	75	700	600 (15 µg)	15	90
>70	1.1	1.1	14	30	5	1.5	400	2.4	425	75	700	800 (20 µg)	15	90
Pregnancy														
≤18	1.4	1.4	18	30	6	1.9	600	2.6	450	80	750	600 (15 µg)	15	75
19–30	1.4	1.4	18	30	6	1.9	600	2.6	450	85	770	600 (15 µg)	15	90
31–50	1.4	1.4	18	30	6	1.9	600	2.6	450	85	770	600 (15 µg)	15	90
Lactation														
≤18	1.4	1.6	17	35	7	2.0	500	2.8	550	115	1200	600 (15 µg)	19	75
19–30	1.4	1.6	17	35	7	2.0	500	2.8	550	120	1300	600 (15 µg)	19	90
31–50	1.4	1.6	17	35	7	2.0	500	2.8	550	120	1300	600 (15 µg)	19	90

NOTE: For all nutrients, values for infants are AI. The glossary on the inside back cover defines units of nutrient measure.

[a] Niacin recommendations are expressed as niacin equivalents (NE), except for recommendations for infants younger than 6 months, which are expressed as preformed niacin.

[b] Folate recommendations are expressed as dietary folate equivalents (DFE).

[c] Vitamin A recommendations are expressed as retinol activity equivalents (RAE).

[d] Vitamin D recommendations are expressed as cholecalciferol and assume an absence of adequate exposure to sunlight.

[e] Vitamin E recommendations are expressed as α-tocopherol.

Recommended Dietary Allowances (RDA) and Adequate Intakes (AI) for Minerals

Age (yr)	Sodium AI (mg/day)	Chloride AI (mg/day)	Potassium AI (mg/day)	Calcium RDA (mg/day)	Phosphorus RDA (mg/day)	Magnesium RDA (mg/day)	Iron RDA (mg/day)	Zinc RDA (mg/day)	Iodine RDA (µg/day)	Selenium RDA (µg/day)	Copper RDA (µg/day)	Manganese AI (mg/day)	Fluoride AI (mg/day)	Chromium AI (µg/day)	Molybdenum RDA (µg/day)
Infants															
0–0.5	120	180	400	200	100	30	0.27	2	110	15	200	0.003	0.01	0.2	2
0.5–1	370	570	700	260	275	75	11	3	130	20	220	0.6	0.5	5.5	3
Children															
1–3	1000	1500	3000	700	460	80	7	3	90	20	340	1.2	0.7	11	17
4–8	1200	1900	3800	1000	500	130	10	5	90	30	440	1.5	1.0	15	22
Males															
9–13	1500	2300	4500	1300	1250	240	8	8	120	40	700	1.9	2	25	34
14–18	1500	2300	4700	1300	1250	410	11	11	150	55	890	2.2	3	35	43
19–30	1500	2300	4700	1000	700	400	8	11	150	55	900	2.3	4	35	45
31–50	1500	2300	4700	1000	700	420	8	11	150	55	900	2.3	4	35	45
51–70	1300	2000	4700	1000	700	420	8	11	150	55	900	2.3	4	30	45
>70	1200	1800	4700	1200	700	420	8	11	150	55	900	2.3	4	30	45
Females															
9–13	1500	2300	4500	1300	1250	240	8	8	120	40	700	1.6	2	21	34
14–18	1500	2300	4700	1300	1250	360	15	9	150	55	890	1.6	3	24	43
19–30	1500	2300	4700	1000	700	310	18	8	150	55	900	1.8	3	25	45
31–50	1500	2300	4700	1000	700	320	18	8	150	55	900	1.8	3	25	45
51–70	1300	2000	4700	1200	700	320	8	8	150	55	900	1.8	3	20	45
>70	1200	1800	4700	1200	700	320	8	8	150	55	900	1.8	3	20	45
Pregnancy															
≤18	1500	2300	4700	1300	1250	400	27	12	220	60	1000	2.0	3	29	50
19–30	1500	2300	4700	1000	700	350	27	11	220	60	1000	2.0	3	30	50
31–50	1500	2300	4700	1000	700	360	27	11	220	60	1000	2.0	3	30	50
Lactation															
≤18	1500	2300	5100	1300	1250	360	10	13	290	70	1300	2.6	3	44	50
19–30	1500	2300	5100	1000	700	310	9	12	290	70	1300	2.6	3	45	50
31–50	1500	2300	5100	1000	700	320	9	12	290	70	1300	2.6	3	45	50

NOTE: For all nutrients, values for infants are AI. The glossary on the inside back cover defines units of nutrient measure.

Tolerable Upper Intake Levels (UL) for Vitamins

Age (yr)	Niacin (mg/day)[a]	Vitamin B$_6$ (mg/day)	Folate (μg/day)[a]	Choline (mg/day)	Vitamin C (mg/day)	Vitamin A (μg/day)[b]	Vitamin D (IU/day)	Vitamin E (mg/day)[c]
Infants								
0–0.5	—	—	—	—	—	600	1000 (25 μg)	—
0.5–1	—	—	—	—	—	600	1500 (38 μg)	—
Children								
1–3	10	30	300	1000	400	600	2500 (63 μg)	200
4–8	15	40	400	1000	650	900	3000 (75 μg)	300
9–13	20	60	600	2000	1200	1700	4000 (100 μg)	600
Adolescents								
14–18	30	80	800	3000	1800	2800	4000 (100 μg)	800
Adults								
19–70	35	100	1000	3500	2000	3000	4000 (100 μg)	1000
>70	35	100	1000	3500	2000	3000	4000 (100 μg)	1000
Pregnancy								
≤18	30	80	800	3000	1800	2800	4000 (100 μg)	800
19–50	35	100	1000	3500	2000	3000	4000 (100 μg)	1000
Lactation								
≤18	30	80	800	3000	1800	2800	4000 (100 μg)	800
19–50	35	100	1000	3500	2000	3000	4000 (100 μg)	1000

[a]The UL for niacin and folate apply to synthetic forms obtained from supplements, fortified foods, or a combination of the two.
[b]The UL for vitamin A applies to the preformed vitamin only.
[c]The UL for vitamin E applies to any form of supplemental α-tocopherol, fortified foods, or a combination of the two.

Tolerable Upper Intake Levels (UL) for Minerals

Age (yr)	Sodium (mg/day)	Chloride (mg/day)	Calcium (mg/day)	Phosphorus (mg/day)	Magnesium (mg/day)[d]	Iron (mg/day)	Zinc (mg/day)	Iodine (μg/day)	Selenium (μg/day)	Copper (μg/day)	Manganese (mg/day)	Fluoride (mg/day)	Molybdenum (μg/day)	Boron (mg/day)	Nickel (mg/day)	Vanadium (mg/day)
Infants																
0–0.5	—	—	1000	—	—	40	4	—	45	—	—	0.7	—	—	—	—
0.5–1	—	—	1500	—	—	40	5	—	60	—	—	0.9	—	—	—	—
Children																
1–3	1500	2300	2500	3000	65	40	7	200	90	1000	2	1.3	300	3	0.2	—
4–8	1900	2900	2500	3000	110	40	12	300	150	3000	3	2.2	600	6	0.3	—
9–13	2200	3400	3000	4000	350	40	23	600	280	5000	6	10	1100	11	0.6	—
Adolescents																
14–18	2300	3600	3000	4000	350	45	34	900	400	8000	9	10	1700	17	1.0	—
Adults																
19–50	2300	3600	2500	4000	350	45	40	1100	400	10,000	11	10	2000	20	1.0	1.8
51–70	2300	3600	2000	4000	350	45	40	1100	400	10,000	11	10	2000	20	1.0	1.8
>70	2300	3600	2000	3000	350	45	40	1100	400	10,000	11	10	2000	20	1.0	1.8
Pregnancy																
≤18	2300	3600	3000	3500	350	45	34	900	400	8000	9	10	1700	17	1.0	—
19–50	2300	3600	2500	3500	350	45	40	1100	400	10,000	11	10	2000	20	1.0	—
Lactation																
≤18	2300	3600	3000	4000	350	45	34	900	400	8000	9	10	1700	17	1.0	—
19–50	2300	3600	2500	4000	350	45	40	1100	400	10,000	11	10	2000	20	1.0	—

[d]The UL for magnesium applies to synthetic forms obtained from supplements or drugs only.
NOTE: An Upper Limit was not established for vitamins and minerals not listed and for those age groups listed with a dash (—) because of a lack of data, not because these nutrients are safe to consume at any level of intake. All nutrients can have adverse effects when intakes are excessive.

SOURCE: Adapted with permission from the *Dietary Reference Intakes* series, National Academies Press. Copyright 1997, 1998, 2000, 2001, 2002, 2005, 2011 by the National Academies of Sciences.

C

Nutrition Counseling and Education Skill Development

KATHLEEN D. BAUER

DOREEN LIOU

CAROL A. SOKOLIK
Montclair State University

WADSWORTH
CENGAGE Learning™

Australia • Brazil • Japan • Korea • Mexico • Singapore • Spain • United Kingdom • United States

Nutrition Counseling and Education Skill Development, Second Edition
Kathleen D. Bauer, Doreen Liou,
Carol A. Sokolik

Publisher/Executive Editor: Linda Schreiber-Ganster

Senior Acquisitions Editor: Peggy Williams

Assistant Editor: Shannon Holt

Senior Marketing Manager: Laura McGinn

Marketing Communications Manager: Linda Yip

Content Project Management: PreMediaGlobal

Art Director: John Walker

Print Buyer: Rebecca Cross

Rights Acquisition Specialist: Thomas McDonough

Production Service: PreMediaGlobal

Cover Designer: Riezebos Holzbaur/ Tae Hatayama

Cover Image:

Artichokes

Copyright: Minerva Studio

Bell Pepper

Copyright: Baloncici

Lettuce

Copyright: David Kay

Compositor: PreMediaGlobal

For product information and technology assistance, contact us at
Cengage Learning Customer & Sales Support, 1-800-354-9706
For permission to use material from this text or product,
submit all requests online at **www.cengage.com/permissions**
Further permissions questions can be emailed to
permissionrequest@cengage.com

Library of Congress Control Number: 2011929588

ISBN-13: 978-0-8400-6415-8

ISBN-10: 0-8400-6415-2

Wadsworth
20 Davis Drive
Belmont, CA 94002-3098
USA

Cengage Learning is a leading provider of customized learning solutions with office locations around the globe, including Singapore, the United Kingdom, Australia, Mexico, Brazil and Japan. Locate your local office at: **www.cengage.com/global**

Cengage Learning products are represented in Canada by Nelson Education, Ltd.

For your course and learning solutions, visit **www.cengage.com**

Purchase any of our products at your local college store or at our preferred online store **www.cengagebrain.com**

Printed in the United States of America
1 2 3 4 5 6 7 15 14 13 12 11

To my husband, Hank, and my children,
Emily so mee Rose and Kathryn sun hee Rose
and my grandchild, Kathleen hweng jae Rose
Thank you for patience, support, and love.
KDB

To my dear parents, Ming-Kung and Lihua Liou,
who are true educators and inspirational role models.

Thanks be to God for His wisdom and faithful guidance.
DL

I dedicate my contributions to the dietetics students
who always inspired me to do my best.
CAS

Contents

CHAPTER 11
Keys to Successful Nutrition Education Interventions   275

CHAPTER 12
Educational Strategies, Mass Media, and Evaluation   300

Preface

WELCOME TO THE SECOND EDITION OF NUTRITION COUNSELING AND EDUCATION

The second edition of this book continues to provide a step-by-step approach guiding entry-level practitioners through the basic components of changing food behavior and improving nutritional status. Behavior change is a complex process, and there is an array of strategies to influence client knowledge, skills, and attitudes. In order to be effective change agents, nutrition professionals need a solid foundation of counseling and education principles, opportunities to practice new skills, and knowledge of evaluation methodologies. This book meets all of these needs in an organized, accessible, and engaging approach. Because the goals and objectives of nutrition counselors and educators complement each other, the scope of this book has expanded to include elements unique to nutrition education.

INTENDED AUDIENCE

This book was developed to meet the needs of health professionals who have little or no previous counseling or education experience, but do have a solid knowledge of the disciplines of food and nutrition. Although the book addresses the requirements of nutrition professionals seeking to become registered dietitians, the approach focuses on skill development useful to all professionals who need to develop nutrition counseling and education skills. The goal of the book is to enable students to learn and use fundamental skills universal to counseling and education as a springboard on which to build and modify individual styles.

DISTINGUISHING FEATURES OF THE SECOND EDITION OF NUTRITION COUNSELING AND EDUCATION

- *Practical examples:* Recognizing that nutrition education and counseling takes place in a variety of settings, concrete examples, case studies, and first person accounts are presented representing a variety of wellness, private practice, and institutional settings.
- *Action based:* Exercises are integrated into the text to give students ample opportunity and encouragement to interact with the concepts covered in each chapter. Instructors can choose to assign the activities to be implemented individually at home or used as classroom activities. Students are encouraged to journal their responses to the exercises as a basis for classroom discussions, distance learning, or for documenting their own reflections. Instructors can assign journal entries and collect them for evaluation. This method provides a thorough understanding of how students are grasping concepts. Each chapter has a culminating assignment and a case study that integrates all or most of the major topics covered throughout the chapter.
- *Evidence-based:* Science-based approaches, grounded in behavior change models and theories, found to be effective for educational and counseling interventions are analyzed and integrated into skill development exercises.
- *Putting it all together—a four-week guided nutrition counseling program:* The text includes a step-by-step guide for students working with volunteer adult clients during four sessions. The objective of this section is to demonstrate how the

theoretical discussions, practice activities, and nutrition tools can be integrated for an effective intervention.

NEW EDITION HIGHLIGHTS

All chapters of the new edition have been updated to incorporate the latest professional standards, government guidelines, and research findings. Based on comments from students and instructors using this text, there have been a number of additions, thereby expanding the scope of the book and learning experiences. The following highlights new features of this text:

- **Keys to Nutrition Education Interventions.** Incorporating principles of nutrition education into Chapters 11 and 12 expands the scope of the second edition. This addition seemed to be a logical step because nutrition education and counseling share many commonalities.
- **Nutrition Care Process (NCP).** Chapter 5 provides an overview of the American Dietetic Association NCP. Emphasis is placed on developing PES (Problem-Etiology-Symptom) statements and ADIME (Assessment, Diagnosis, Intervention, Monitoring and Evaluation) documentation. Chapter 2 reviews the theoretical approaches covered in the NCP intervention guidelines.
- **Expansion of Motivational Interviewing (MI).** Using MI to influence health behaviors has become widely accepted. The review of the approach found in Chapter 2 is covered in more depth than the first edition. As in the first edition, the Nutrition Counseling Motivation Algorithm found in Chapter 4 uses the guiding tenets of MI.
- **Incorporation of Factors Influencing Food Behavior.** Chapter 1 explores the multitude of factors affecting food behaviors.
- **Expansion of Communication with Diverse Population Groups.** As in the first edition, multicultural issues have been integrated throughout the text. However, Chapter 9 explores the meaning of cultural competence

in the health care arena and examines special counseling, education, and communication issues related to selected populations groups.
- **Expansion of Group Facilitation and Group Counseling.** In order to accommodate learning experiences related to group work, a new chapter was created.

Chapter-by-Chapter Updates

The sequential flow of the chapters follows the needs of students to develop knowledge and skills during each step of the counseling and education process.

Chapter 1 Preparing to Meet Your Clients

- Nutrition education and counseling are defined.
- New to this edition is a section on factors affecting food behavior.
- A new chapter assignment has been added: Building a Collage.

Chapter 2 Frameworks for Understanding and Attaining Behavior Change

- Counseling philosophies, social psychological theories, behavior change models, and counseling approaches have been placed into a separate chapter.
- A number of new activities have been incorporated to enhance learning.
- Reviews of the Theory of Planned Behavior and Social Cognitive Theory have been added to this section.
- In order to better understand major factors influencing behavior change, a summary of effective behavior change attributes has been added.

Chapter 3 Communication Essentials

- Cultural influence on communication was enhanced.

Chapter 4 Meeting Your Client: The Counseling Interview

- The Cross-Cultural Nutrition Counseling Algorithm was moved to Chapter 9.

Chapter 5 Developing a Nutrition Care Plan – Putting It all Together

- A review of the Nutrition Care Process was incorporated into this chapter.
- Exercises were added to encourage understanding of ADIME documentation and development of PES statements.

Chapter 6 Promoting Change to Facilitate Self-Management

- Problem solving strategy was enhanced.
- Supporting self-management topics were added to this chapter.
- The Food Management Tool Assignment was updated to include MyPyramid.

Chapter 7 Making Behavior Change Last

- A review of influences of social network on behavior change was expanded.
- A review of mindful eating was added to this chapter.
- Ending a counseling relationship and evaluation of counseling were incorporated into this chapter.

Chapter 8 Physical Activity

- Recent national initiatives and Physical Activity Guidelines for Americans have been integrated into this chapter.

Chapter 9 Communication with Diverse Population Groups

- This is a new chapter in the second edition.
- Factors related to gaining cultural competence are reviewed.
- Cultural competence models are examined and applied to nutrition interventions.
- Communication essentials for working with individuals in various stages of the lifespan and people with disabilities, eating disorders, and experiencing weight bias are explored.

Chapter 10 Group Facilitation and Counseling

- This is a new chapter in the second edition.
- Nutrition interventions often require skills in group facilitation; therefore, this topic has been added to Chapter 10.
- Because evidence analysis indicates that group counseling is an effective strategy, this topic has been expanded.

Chapter 11 Keys to Successful Nutrition Education Interventions

- Chapter 11 is one of two chapters totally dedicated to nutrition education.
- Seven keys to successful nutrition education interventions are presented and the first four—needs assessment, educational philosophy, theory-based interventions, and goals and objectives—are covered in Chapter 11.
- Chapter 11 contains a continuous and interactive case study integrating key nutrition education processes.

Chapter 12 Educational Strategies, Mass Media, and Evaluation

- The last three keys to successful nutrition education interventions are presented. These include educational strategies, mass media, and evaluation.

Chapter 13 Professionalism and Final Issues

- The American Dietetic Association Three-Block Framework provides the guide for a discussion of professionalism.
- The section on Code of Ethics has been updated to reflect the 2009 American Dietetic Association Code.
- A new section on starting a private practice has been added.
- A new section on marketing has been added addressing emerging technology opportunities.

Chapter 14 Guided Counseling Experience

- The guided counseling experience has been updated to include the components of the Nutrition Care Process.

ACKNOWLEDGEMENTS

Thank you to all the reviewers and the individuals who shared their expertise and assisted in the development of the manuscript. Your insights and comments were invaluable to the second edition. We greatly appreciate the staff at Cengage Learning for their encouragement and tireless work shepherding us through the process. A special thanks to Shannon Holt, Assistant Editor of Biology and Nutrition, and Peggy Williams, Senior Acquisitions Editor of Life Sciences and Nutrition. In addition, we want to acknowledge the important contributions made by Dr. Nobuko Hongu, Assistant Professor and Nutrition Extension Specialist, The University of Arizona, Tucson, Arizona, for updating Chapter 8 on Physical Activity. We are also grateful for the guidance Jennifer Tomesko provided in the development of the Nutrition Care Process section of Chapter 5.

Reviewers:
Molly Kellogg
Psychotherapist, Nutrition Therapist, and Writer
Philadelphia, Pennsylvania

Anne B. Marietta
Southeast Missouri State University

Shaynee Roper
University of Houston

Virginia Bennett
Central Washington University

Jayne Byrne
College of St. Benedict/St. John's University

ABOUT THE AUTHORS

Kathleen D. Bauer, Ph.D., R.D., is the founder and has been the director of the Nutrition Counseling Clinic at Montclair State University for over fifteen years. She teaches both undergraduate and graduate nutrition counseling courses. Publications include book chapters and articles on cultural diversity and evaluation of nutrition counseling education methods. Her applied individual and group nutrition counseling experiences extend to faith-based and wellness programs, fitness centers, hospitals, nursing homes, and private practice.

Doreen Liou, Ed.D., R.D., has been the director of the Didactic Program in Dietetics at Montclair State University for the past ten years. She teaches both undergraduate and graduate courses in nutrition education and social marketing. Her research interests encompass qualitative and quantitative methods in addressing chronic disease risk and the applications of social psychological theories in minority population groups. Her nutrition education experiences extend to a variety of academic, community, and clinical settings.

Carol A. Sokolik, M.S., R.D., is a founder and former Co-Director of the Dietetic Internship at Montclair State University in New Jersey. She also taught in the University's Didactic Program in Dietetics for twenty eight years. In addition to her experience in hospital dietetics, she has worked as a dietetic consultant to several long term care facilities in New Jersey. At this time, she has relocated to southwestern Virginia and has continued her private practice in nutrition consulting.

Preparing to Meet Your Clients

Getty Images/Comstock/Jupiterimages

Not only is there an art in knowing something but also a certain art in teaching it.
—CICERO

Behavioral Objectives

- Define nutrition counseling and nutrition education.
- Identify and explain factors influencing food choices.
- Describe characteristics of an effective counselor.
- Identify factors affecting clients in a counseling relationship.
- Evaluate oneself for strengths and weaknesses in building a counseling relationship.
- Identify novice counselor issues.

Key Terms

- **Cultural Groups:** nonexclusive groups that have a set of values in common; an individual may be part of several cultural groups at the same time.
- **Culture:** learned patterns of thinking, feeling, and behaving that are shared by a group of people.
- **Cultural Values:** principles or standards of a cultural group.
- **Models:** generalized descriptions used to analyze or explain something.
- **Nutrition Counseling:** a supportive process guiding a client towards nutritional well-being.
- **Nutrition Education:** learning experiences aimed to promote voluntary adoption of health promoting dietary behaviors.
- **Worldview:** perception of the world that is biased by culture and personal experience.

INTRODUCTION

Nutrition counselors and educators provide guidance for helping individuals develop food practices consistent with the nutritional needs of their bodies. For clients, this may mean altering comfortable food patterns and long-standing beliefs and attitudes about food. Nutrition professionals work to increase knowledge, influence motivations, and guide development of skills required for dietary behavior change. This can be a challenging task. In order to be an effective change agent, nutrition counselors and educators need a solid understanding of the multitude of factors affecting food behaviors. We will begin this chapter by addressing these factors in order to enhance understanding of the forces influencing our clients. Then, we will explore the helping relationship and examine counselor and client concerns. Part of this examination will include cultural components. Nutrition professionals always need to be sensitive to the cultural context of their interventions from both their own cultural perspectives as well as their clients. Some of the activities in this chapter will provide opportunities for you to explore the cultural lenses that influence your view of the world.

FOUNDATION OF NUTRITION COUNSELING AND EDUCATION

Nutrition education has been defined as, "Any set of learning experiences designed to facilitate the voluntary adoption of eating and other nutrition-related behaviors conducive to health and well-being."[1] The needs of a target community are the focus of the nutrition education process. Nutrition counselors have similar goals, but interventions are guided by the needs of individual clients. In particular, **nutrition counseling** has been defined as the process of guiding a client toward a healthy nutrition lifestyle by meeting nutritional needs and solving problems that are barriers to change.[2] Haney and Leibsohn[3] designed a **model** of counseling to enable guidance to be effective, indicating that

> *counseling can be defined as an interaction in which the counselor focuses on client experience, client feeling, client thought, and client behavior with intentional responses to acknowledge, to explore, or to challenge. (p. 5)*

EXERCISE 1.1   DOVE Activity: Broadening Our Perspective (Awareness)

D—defer judgment
O—offbeat
V—vast
E—expand on other ideas

Divide into groups of three. Your instructor will select an object, such as a cup, and give you one minute to record all of the possible uses of the object. Draw a line under your list. Take about three minutes to share each other's ideas, and write the new ideas below the line. Discuss other possibilities for using the object with your group and record these in your journal. Use the DOVE technique to guide your thinking and behavior during this activity. Do not pass judgment on thoughts that cross your mind or on the suggestions of others. Allow your mind to think of a vast number of possibilities that may even be offbeat. How many more ideas occurred with sharing? Did you see possibilities from another perspective? One of the goals of counseling is to help clients see things using different lenses. What does this mean? How does this activity relate to a counseling experience? Write your thoughts in your journal and share them with your colleagues.

Source: Dairy, Food, and Nutrition Council, *Facilitating Food Choices: Leaders Manual* (Cedar Knolls, NJ: 1984).

FUNDAMENTALS OF FOOD BEHAVIOR

The heart of nutrition education and counseling is providing support and guidance for individuals to make appropriate food choices for their needs. Therefore, understanding the myriad influences affecting food choices is fundamental to designing an intervention. Influencing factors are often intertwined and may compete with each other leaving individuals feeling frustrated and overwhelmed when change is needed. Before we journey through methodologies for making change feel achievable, we will explore aspects of environmental, psychological, social, and physical factors affecting food choices as depicted in Figure 1.1.

- *Taste and Food Preference:* Taste is generally accepted as the most important determinant of food choices.[4] Biological taste preferences evolve from childhood based on availability and societal norms, but research shows that perferences can be altered by experiences and age.[5] Generally, young children prefer sweeter and saltier tastes than adults, and relocating to a new environment will often change eating patterns and even favorite foods.[6] Without consumers knowing, a number of food companies have been improving the nutritional quality of their foods by slowly changing recipes, such as, lowering sodium or sugar content or increasing fiber. For example, Ragu Old World Style Pasta Sauces stealthly reduced sodium by 25 percent from 2004 to 2007 with no loss of market share.[7] The fact that taste preferences can be modified should be reassuring for those who want to make dietary changes. Illness may also modify food preference. Individuals going through chemotherapy may find some of their favorite foods do not taste the same and lose the desire to eat them.[8]

- **Health Concerns:** Research has shown that health can be a driving force for food choice as illustrated by public campaigns to increase intake of fruits, vegetables, and whole grains.[9] In a national survey, sixty-nine percent of the participants indicated they are making an effort to improve the healthfulness of their diets by changing the types of food consumed.[10] Consumers are more likely to respond to healthful food messages if the advice stresses the good taste of wholesome foods and convenient ways to include them in the diet.[4] Health status of an individual, such as having loss of teeth or digestive disorders, can also affect the amount of food consumed and food choice.[8]

- **Nutrition Knowledge:** Traditionally, educators and nutrition counselors perceived their roles as disseminating information. After research indicated that many clients were not responsive to simple didactic approaches, their roles expanded to include a variety of behavior change strategies. However, the value of increasing knowledge should not be devalued. Those who have higher levels of knowledge are more likely to have better quality diets and lose more weight in weight loss programs.[11,12]

- **Convenience and Time:** For many individuals striving to make food choice changes, stressing convenient ways to prepare desired foods is imperative.[4] Our fast food **culture** has created a demand for easy to prepare and tasty food.[13] In a nationwide survey, busy housewives expressed a desire to spend less than fifteen minutes to prepare a meal.[14] Take out, value-added (precut, prewashed), and ready-made foods have become a cultural standard.

- **Culture and Religion:** Food is an intregral part of societal rituals influencing group identity.[15] Ritual meals solidify group membership and reaffirm our relationships to others. For example, all day eating at weekly family gatherings on Sundays or daily coffee breaks with sweet rolls are rituals that do much more than satisfy the appetite. If clients need to change participation in these rituals because of dietary restrictions, it is likely to create stress for clients, friends, and relatives. Culture also defines what is acceptable for consumption such as sweet red ants, scorpions, silk worms, or a glass of cow's milk. Culture also defines food patterns, and in the United States, snacking is common.[16] In addition, religions advocate food rituals and may also define food taboos such as restrictions against pork for Muslims, beef for Hindus, and shellfish for Othodox Jews. Due to increasing diversity, minorities make up one third of the people living in the United States. As a result, an array of ethnic foods are available in restaurants and grocery stores and have influenced the national palate.[13] For example, in the past, ketchup was considered a household staple; however, recent national sales of salsa now compete with ketchup and at times have surpassed ketchup sales.

- **Social Influences:** Food is often an intregral part of social experiences. Sharing a meal with friends after a football game or going out for ice cream to celebrate an academic

> A young man in his early twenties commenting about his food habits stated, "My friends do not say 'let's eat a salad together.' If you are a guy, it is a woozy thing to do. It is kind of looked down upon if you are a guy—weak. Eat the steak, eat the greasy stuff, be a man."*5

---

*Numerous first-person accounts from dietetic students or nutrition counselors working in the field are included throughout this book.

achievement help make special experiences festive. However, foods associated with sociability are often not the most nutritious. Social eating frequently encourages increased consumption of less nutritious foods and overconsumption.[17,18] Eating with friends and family increases energy intake by eighteen percent.[19] However, even though regular family meals have been shown to be a stabilizing influence in children's lives, an analysis of societal trends indicates that family meals at home are disappearing.[13]

> A female college student stated: "The whole society does not emphasize eating healthy. When you are eating, you have to think hard about what are the healthy foods to eat."

- *Media and Physical Environment:* North Americans are surrounded by media messages and most of them are encouraging consumption of high calorie foods that are nutritionally challenged. In 2004, food manufacturers spent nine billion dollars on advertising to persuade consumers. Commercials can have powerful influences on the quantity and quality of food consumed.[20,21] Not only do we encounter food messages repeatedly throughout the day, but we also have access to a continuous supply of unhealthy food and large portion sizes. Almost anywhere you go--drug stores, gas stations, movie rental stores, schools, for example--there are opportunities to purchase unhealthy food. Even laboratory animals put in this type of environment are likely to overeat the calorie dense food and gain excessive weight.[22]
- *Economics:* An individual's residence and socioeconomic status can influence a myriad of factors including accessibility to transportation, cooking facilities, refrigeration, grocery store options, and availability of healthful food choices. [23] For those who are economically disadvantaged, meeting nutritional guidelines is a challenge.[24] Low income households spend significantly less money on fruits and vegetables than high income households with nineteen percent buying none in any given week.[25]
- *Availability and Variety:* Individuals with increased numbers of food encounters, portion size, and variety of available choices tend to increase food intake.[26,27] Variety of food intake is important in meeting nutritional needs,

but when the assortment is excessive, such as making food selections from a buffet, overconsumption is probable.

- *Psychological:* Research has shown that individuals vary in their food response to stress. Some people increase consumption, whereas others claim feeling too stressed to eat. Certain foods have been associated with depression and mood alteration. Severely depressed individuals have been found to consume more chocolate (up to fifty-five percent) per month than others.[28]

An understanding of how all these factors influence our food behaviors is essential for nutrition educators and counselors. Since we are advocating lifestyle change of comfortable food patterns, we need to understand the discomfort that our clients are likely to feel as they anticipate and attempt the alterations. Our role is to acknowledge the challenge for our clients and to find and establish new patterns that provide a healthier lifestyle.

EXERCISE 1.2   Explore Influences of Food Behavior

Interview three people and ask them to recall the last meal they consumed. Inquire about the factors that influenced them to make their selections. Record your findings in your journal. Compare your findings to this section on influences of food choice.

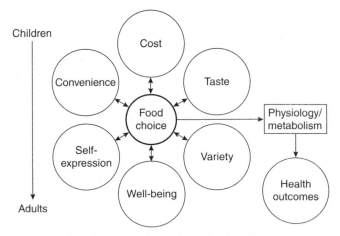

Figure 1.1   The Consumer Food Choice Model
Source: Adapted from A. Drewnowski, *Taste, Genetics, and Food Choice.* In Food Selection: From Genes to Culture, H. Anderson, J. Blundell, and M. Chiva, Eds. (Levallois-Perret, France: Danone Institute), 30. Copyright 2002.

UNDERSTANDING AN EFFECTIVE COUNSELING RELATIONSHIP

No matter what theory or behavior change model is providing the greatest influence, the relationship between counselor and client is the guiding force for change. The effect of this relationship is most often cited as the reason for success or failure of a counseling interaction.[2] Helm and Klawitter[28] report that successful clients identify their personal interaction with their therapist as the single most important part of treatment. To set the stage for understanding the basics of an effective counseling relationship, you will investigate the characteristics of effective nutrition counselors, explore your own personality and culture, examine the special needs and issues of a person seeking nutrition counseling, and review two phases of a helping relationship in the following sections.

Characteristics of Effective Nutrition Counselors

"Ideal helpers" have been described as possessing the following qualities:

> They respect their clients and express that respect by being available to them, working with them, not judging them, trusting the constructive forces found in them, and ultimately placing the expectation on them to do whatever is necessary to handle their problems in living more effectively. They genuinely care for those who have come for help. They are non-defensive, spontaneous, and always willing to say what they think and feel, provided it is in the best interest of their clients. Good helpers are concrete in their expressions, dealing with actual feelings and actual behavior rather than vague formulations, obscure psychodynamics, or generalities.[30] (p. 29)

EXERCISE 1.3   Helper Assessment

Think of a time someone helped you, such as a friend, family member, teacher, or counselor. In your journal, write down the behaviors or characteristics the person possessed that made the interaction so effective. After reading over the characteristics of effective counselors, compare their qualities to those identified by the leading authorities. Do they differ? Share your thoughts with your colleagues.

After thoroughly reviewing the literature in counseling, Okun[31] identified seven qualities of counselors considered to be the most influential in affecting the behaviors, attitudes, and feelings of helpees: knowledge, self-awareness, ethical integrity, congruence, honesty, ability to communicate, and gender and culture awareness. The following list describes these characteristics as well as those thought to be effective by nutrition counseling authorities:

- *Effective nutrition counselors are self-aware.* They are aware of their own beliefs, respond from an internal set of values, and as a result have a clear sense of priorities. However, they are not afraid to reexamine their values and goals.[32] This awareness aids counselors with being honest with themselves as to why they want to be a counselor and avoid using the helping relationship to fulfill their own needs.[33]

- *They have a solid foundation of knowledge.* Nutrition counselors need to be knowledgeable in a vast array of subjects in the biological and social sciences as well as have an ability to apply principles in the culinary arts. Because the science and art of nutrition is a dynamic field, the foundation of knowledge requires continuous updating. Clients particularly appreciate nutrition counselors who are experienced with the problems they face.[29]

- *They have ethical integrity.* Effective counselors value the dignity and worth of all people. Such clinicians work toward eliminating ways of thinking, speaking, and acting that reflect racism, sexism, ableism, ageism, homophobia, religious discrimination, and other negative ideologies.[34] Ethical integrity entails many facets that are addressed in the American Dietetic Association's Code of Ethics and a discussion of this topic can be found in Chapter 13.[35]

- *They have congruence.* This means the counselor is unified. There are no contradictions between who the counselor is and what the counselor says, and there is consistency in verbal and nonverbal behaviors as well. (For example, if a client shared some unusual behavior such as eating a whole cake covered with French dressing, the counselor's behavior would not be congruent if the nonverbal behavior indicated surprise but the verbal response did not.)

- **They can communicate clearly.** Clinicians must be able to communicate factual information and a sincere regard for their clients. Effective nutrition counselors are able to make sensitive comments and communicate an understanding about fears concerning food and weight.[28]

- **They have a sense of gender and cultural awareness.** This requires that counselors be aware of how their own gender and culture influence them. Effective counselors have a respect for a diversity of values that arise from their clients' cultural orientations.

- **They have a sense of humor.** Helping clients see the irony of their situation and laugh about their problems enriches counseling relationships. In addition, humor helps prevent clients from taking themselves and their problems too seriously.[32]

- **They are honest and genuine.** Such counselors appear authentic and sincere. They act human and do not live by pretenses hiding behind phony masks, defenses, and sterile roles.[32] Such counselors are honest and show

---

EXERCISE 1.4   People Skills Inventory

- Do you expect the best from people? Do you assume that others will be conscientious, trustworthy, friendly, and easy to work with until they prove you wrong?
- Are you appreciative of other people's physical, mental, and emotional attributes—and do you point them out frequently?
- Are you approachable? Do you make an effort to be outgoing? Do you usually wear a pleasant expression on your face?
- Do you make the effort to remember people's names?
- Are you interested in other people—all kinds of people? Do you spend far less time talking about yourself than encouraging others to talk about themselves?
- Do you readily communicate to others your interest in their life stories?
- When someone is talking, do you give him or her 100 percent of your attention—without daydreaming, interrupting, or planning what you are going to say next?
- Are you accepting and nonjudgmental of others' choices, decisions, and behavior?
- Do you wholeheartedly rejoice in other people's good fortune as easily as you sympathize with their troubles?
- Do you refuse to become childish, temperamental, moody, inconsistent, hostile, condescending, or aggressive in your dealings with other people—even if they do?
- Are you humble? Not to be confused with false modesty, being humble is the opposite of being arrogant and egotistical.
- Do you make it a rule never to resort to put-downs, sexist or ethnic jokes, sexual innuendoes, or ridicule for the sake of a laugh?
- Are you dependable? If you make commitments, do you keep them—no matter what? If you are entrusted with a secret, do you keep it confidential—no matter what?
- Are you open-minded? Are you willing to listen to opposing points of view without becoming angry, impatient, or defensive?
- Are you able to hold onto the people and things in your life that cause you joy and let go of the people and things in your life that cause you sadness, anger, and resentment?
- Can you handle a reasonable amount of pressure and stress without losing control or falling apart?
- Are you reflective? Are you able to analyze your own feelings? If you make a mistake, are you willing to acknowledge and correct it without excuses or blaming others?
- Do you like and approve of yourself most of the time?

Affirmative answers indicate skills you possess that enhance your ability to relate to others.

Source: Adapted from Scott N, "Success Often Lies in Relating to Other People," Dallas Morning News, April 20, 1995, p. 14C.

spontaneity, congruence, openness, and willingness to disclose information about themselves when appropriate. Honest counselors are able to give effective feedback to their clients.

- *They are flexible.* This means not being a perfectionist. Such counselors do not have unrealistic expectations and are willing to work at a pace their clients can handle.[29]

- *They are optimistic and hopeful.* Clients want to believe that lifestyle changes are possible, and they appreciate reassurance that solutions will be found.[29]

- *They respect, value, care, and trust others.* This enables counselors to show warmth and caring authentically through nonjudgmental verbal and nonverbal behavior, listening attentively, and behaving responsibly, such as returning phone calls and showing up on time. This behavior conveys the message that clients are valued and respected.

- *They can accurately understand what people feel from their frame of reference (empathy).* It is important for counselors to be aware of their own struggles and pain to have a frame of reference for identifying with others.[32]

It is one of the most beautiful compensations of this life that no man can sincerely try to help another without helping himself.

—Ralph Waldo Emerson

Understanding Yourself — Personality and Culture

According to Brammer,[33] our personalities are one of the principle tools of the helping process. By taking an inventory of your personality characteristics, you can have a better understanding of the ones you wish to modify.

Intertwined with a personality evaluation is a self-examination of why you want to be a counselor. What you expect out of a counseling relationship, the way you view yourself, and the personal attitudes and values you possess can affect the direction of the counseling process. You should be aware that as a helper, your self-image

---

EXERCISE 1.5   How Do You Rate?

Ask a close friend or family member who you supported at one time to describe what it was about your behavior that was helpful. Write these reactions down in your journal. Review the desirable characteristics for an effective counselor described in the previous section. Complete the personality inventory in Exercise 1.4, and then identify what characteristics you possess that will make you a good helper. What behaviors need improvement?

Write in your journal specific ways that you need to change to improve your helping skills.

---

is strengthened from the awareness that "I must be OK if I can help others in need." Also, because you are put into the perceptual world of others, you remove yourself from your own issues, diminishing concern for your own problems.[33]

Sometimes counselors seek to fulfill their own needs through the counseling relationship. Practitioners who have a need to express power and influence over others tend to be dictatorial, and less likely to be open to listening to their clients. This type of counselor expects clients to obey suggestions without questions. A counselor who is particularly needy for approval and acceptance will fear rejection. Belkin[36] warns that sometimes counselors try too hard to communicate the message "I want you to like me," rather than a more effective "I am here to help you." As a result, such counselors may be anxious to please their clients by trying to do everything for them, perhaps even doing favors. The tendency will be to gloss over and hide difficult issues because the focus is on eliciting only positive feelings from their clients. Consequently, clients will not learn new management skills, and dietary changes will not take place.

EXERCISE 1.6   Why Do You Want to Be a Helper?

Describe in your journal what it means to be a helper and why you want to be a helper. How does it feel when you help someone? Is it possible that you have issues related to dominance or neediness that could overshadow interactions with your clients?

Another important component to understanding yourself so as to become a culturally competent nutrition counselor and educator is to know what constitutes your **worldview** (cultural outlook). Each culture has a unique outlook on life, what people believe and value within their group. Our worldview provides basic assumptions about the nature of reality and has both conscious and unconscious influences. An understanding of this concept becomes clearer when we explore assumptions regarding supernatural forces, individual and nature, science and technology, and materialism. See Table 1.1. Kittler and Sucher[37] relate this unique outlook to its special meaning in the health community:

> . . . *expectations about personal and public conduct, assumptions regarding social interaction, and assessments of individual behavior are determined by this cultural outlook, or worldview. This perspective influences perceptions about health and illness as well as the role of each within the structure of society. (p. 37)*

Your worldview is determined by your culture and life experiences. Culture is shared history, consisting of "the thoughts, communication, actions, customs, beliefs, values, and institutions of racial, ethnic, religious or societal groups."[38] Possible societal groups include gender, age, sexual orientation, physical or mental ability, health, occupation, and socioeconomic status. Any individual will belong to several societal groups and acquire cultural characteristics and beliefs from each based on education and experiences within those groups. Because the experiences are unique, no two people acquire exactly the same cultural attributes. In addition, we are likely to migrate to and away from various cultures throughout our lives. For example, we may change jobs, religions, residence, or health status and as a result, cultural attributes will also alter. However, there are attributes that prevail and will affect the way we perceive ourselves and others.

> My aunt died of high blood pressure. Her religious belief was that her illness was God's will and should not be interferred with by taking medicine or changing her diet.

We share a commonality with those who are most like us. For example, many North Americans appreciate a friendly, open health care professional. People from other cultures, however, may feel uncomfortable interacting with a professional on such terms and may even view this behavior as a sign of incompetence. Your food habits can also be an important component of your culture. For example, Hindus find eating beef to be abhorrent – much the way many Westerners feel about Asians consuming dog meat.

Understanding the role of **cultural values** in your life as well as in lives of clients from cultures other than your own, provides a foundation for developing cultural sensitivity. Our cultural values

Table 1.1 Worldview Assumptions

Category	Assumption
Supernatural Assumptions	Supernatural assumptions include beliefs regarding God, malevolent spirits, ancestors, fate or luck being the cause of illness. The concept of soul loss causing depression or listlessness is prevalent in many societies. In order to alleviate supernatural problems, societies have devised ceremonies or rituals.
Individual and Nature	Not all societies make a clear distinction between human life and nature as in the United States. Some societies believe that we are subjugated by nature and need to show respect for natural forces and attempt to live in harmony with nature. The dominant culture in the United States sees human beings as having higher value than nature with a need to exploit or protect it.
Technology	The citizens of the United States put great fate in technology and the scientific method. Diseases are viewed as correctable mechanistic errors that can be fixed by manipulation. Americans tend to think science can help humanity—a view not as highly held in Europe.[39]
Materialism	Many people around the world believe that materialism dominates the worldview of Americans, that is, the need to acquire the latest and best possessions. This may have contributed to the popularity of "supersize food portions."

Source: Jandt F. An Introduction to Intercultural Communication: Identities in a Global Community. 6th ed. Thousand Oaks, CA: Sage Publications, Inc.; 2009.

are the "principles or standards that members of a cultural group share in common."[40] For example, in the United States, great value is placed on money, freedom, individualism, independence, privacy, biomedical medicine, and physical appearance. Cultural values are the grounding forces that provide meaning, structure, and organization in our lives. See Table 1.2. Individuals may hold onto to their values despite numerous obstacles or severe consequences. For example, Jung Chang describes in her family portrait, *Wild Swans: Three Daughters of China*, how her father actively supported Mao's Communist takeover of China and rose to be a prominent official in the party. His devotion to the party never wavered, even during the Cultural Revolution when he was denounced, publicly humiliated with a dunce hat, and sent to a rehabilitation camp.[41]

As nutrition counselors and educators advocate for change, there needs to be an appreciation of the

Table 1.2 Functions of Cultural Values

- Provide a set of rules by which to govern lives.
- Serve as a basis for attitudes, beliefs, and behaviors.
- Guide actions and decisions.
- Give direction to lives and help solve common problems.
- Influence how to perceive and react to others.
- Help determine basic attitudes regarding personal, social, and philosophical issues.
- Reflect a person's identity and provide a basis for self-evaluation.

Source: Adapted from Joan Luckmann, Transcultural Communication in Nursing. Belmont, CA: Delmar Cengage Learning, 1999.

high degree of importance placed on certain beliefs, values, and cultural practices. You can then empathize with individuals from nonwestern cultures

---

EXERCISE 1.7   What Is Your Worldview?

Indicate on the continuum the degree to which you share the following white North American cultural values; 1 indicates not at all, and 5 represents very much.

Not at All				Very Much	
1	2	3	4	5	Personal responsibility and self-help for preventing illness
1	2	3	4	5	Promptness, schedules, and rapid response-time dominates.
1	2	3	4	5	Future-oriented—willing to make sacrifices to obtain future goals.
1	2	3	4	5	Task-oriented—desire direct participation in your own health care.
1	2	3	4	5	Direct, honest, open dialogue is essential to effective communication.
1	2	3	4	5	Informal communication is a sign of friendliness.
1	2	3	4	5	Technology is of foremost importance in conquering illness.
1	2	3	4	5	Body and soul are separate entities.
1	2	3	4	5	Client confidentiality is of utmost importance; health care is for individuals, not families.
1	2	3	4	5	All patients deserve equal access to health care.
1	2	3	4	5	Desire to be youthful, thin, and fit.
1	2	3	4	5	Competition and independence.
1	2	3	4	5	Materialism.

Can you think of a time when your values and beliefs were in conflict with a person you were trying to associate with? What were the circumstances and results of that conflict? Write your response in your journal, and share your stories with your colleagues.

Source: Adapted from Kittler P and Sucher K, Food and Culture in America, 2d ed. (Belmont, CA: West/Wadsworth; 1998); and Keenan, Debra P. In the face of diversity: Modifying nutrition education delivery to meet the needs of an increasingly multicultural consumer base, *J Nutr Ed.* 1996;28:86–91.

EXERCISE 1.8   What Are Your Food Habits?

Record answers to the following questions in your journal; share them with your colleagues.

1. Who purchases and prepares most of the food consumed in your household?
2. What is your ethnic background and religious affiliation?
3. Are there foods you avoid eating for religious reasons?
4. List two foods you believe are high-status items.
5. What major holidays do you celebrate with your family?
6. List two rules you follow when eating a meal (for example, "Don't sing at the table").
7. Are there food habits that you find morally or ethically repugnant?
8. Are you aware of any of your own food habits that others would consider repugnant?

Source: Adapted from Kittler P and Sucher K, Food and Culture, 4th ed. (Belmont, CA: Wadsworth/Thomson; 2004), p 24–25.

---

who are experiencing confusion and problems as they try to participate in the North American health care system. Also, awareness can help prevent your personal biases, values, or problems from interfering with your ability to work with clients who are culturally different from you.

Conscious and unconscious prejudices unrelated to cultural issues that a counselor may possess could also interfere with emotional objectivity in a counseling situation. Individuals could have exaggerated dislikes of personal characteristics such as being obese, bald, aggressive, and poorly dressed. Awareness of these prejudices can help build tolerances and a commitment not to let them interfere with the counseling process through facial expressions and other nonverbal behavior.

Understanding Your Client

Just like counselors, clients come into nutrition counseling with unique personalities, cultural orientations, health care problems, and issues related to the counseling process. Each person's individual personality should be recognized and appreciated. Clients have their own set of needs, expectations,

concerns, and prejudices that will have an impact on the counseling relationship. In the rushed atmosphere of some institutional settings, health care workers can lose sight of the need to show respect, especially if a client has lost some of his or her physiological or mental functions due to illness.

From a cultural perspective, clients are diverse in many ways, belong to a number of societal groups, and have a set of unique life experiences contributing to a distinctive view of the world. Getting a fresh perspective from a counselor is one of the advantages of counseling. However, the farther away counselors are from their clients' cultural orientation, the more difficult it is to understand their worldview. If this is the case, then you will need to explore your clients' culture through books; newspapers; magazines; workshops; movies; and cultural encounters in markets, fairs, and restaurants. Learning your clients' beliefs about illness and the various functions and meanings of food are particularly important. While exploring **cultural groups**, you should remember that the characteristics of a group are simply generalities. You want to avoid stereotyping. Do not fall into the trap of believing that each characteristic applies to all people who appear to represent a particular group. Remember that the thoughts and behaviors of each individual develop over a lifetime and are shaped by membership in several cultural groups. For example, a homosexual male who grew up with a learning disability in Alabama with first generation parents from Italy and lives in Chicago as an adult would have a number of social groups and life experiences influencing his communication style, view of the world, and expectations. People totally, partially, or not at all embrace the standards of a culture they appear to represent.

The circumstances that bring clients to counseling can have a major impact on their readiness for nutrition counseling. Those who have been recently diagnosed with a serious illness may be experiencing shock or a great deal of physical discomfort to deal effectively with complex dietary guidelines—or any guidelines at all. They may display a tendency toward rebelliousness, a denial of the existence of the problems, anxiety, anger, or depression.[2,42] When counseling an individual with a life-threatening illness, nutrition counselors need to take into account a client's position on the continuum of treatment and recovery.[43]

An attitudinal investigation of young and well-educated patients with diabetes suggests a desire for a collaborative relationship with their health care providers helping them to explore options rather than simply being told what to do.[44] On the other hand, this same study identified a significant number of elderly with diabetes who did not desire an independent self-care role. Promoting self-sufficiency is often stated as a goal of nutrition counseling[45]; however, for some clients, that goal may need to be modified. This issue has also been addressed by the expert panel for the NIH report, *Identification, Evaluation, and Treatment of Overweight and Obesity*,[46] which states that a weight maintenance program consisting of diet therapy, behavior therapy, and physical activity may need to be continued indefinitely for some individuals.

> My client, a robust man in youth, was a World War II veteran who took part in the invasion of Normandy. But at age seventy-five, he suffered a stroke and went into a veterans' hospital for treatment. During his hospital stay, he asked a health care worker to help him get into bed because he wanted to go to sleep. The worker told him he would be able to go to sleep after he finished his lunch. My client became very angry and threw his lunch tray at the health care worker.

Some clients may regard the counseling process itself as an issue. The act of seeking and receiving help can create feelings of vulnerability and incompetence.[33] During counseling there is a presumed goal of doing something for the clients or changing them in some way. This implication of superiority can raise hostile feelings in the helpee because the act presumes that the helper is wiser, more competent, and more powerful than the helpee. This is illustrated in Helen Keller's account of her dreams about her teacher and lifelong friend, Annie Sullivan, who provided constant help for almost all aspects of Helen's existence:

> [T]here are some unaccountable contradictions in my dreams. For instance, although I have the strongest, deepest affection for my teacher, yet when she appears to me in my sleep, we quarrel and fling the wildest reproaches at each other. She seizes me by the hand and drags me by main force towards I can never decide what—an abyss, a perilous mountain pass or a rushing torrent, whatever in my terror I may imagine. (Herrman[47] pp. 165–166)

To help alleviate the negative impact of such issues on the counseling process, the motive for help

and the nature of the helping task as perceived by the counselor should be made clear to the receiver.[33]

Relationship Between Helper and Client

The helping relationship is often divided into two phases: building a relationship and facilitating positive action.[33] Building a relationship requires the development of rapport, an ability to show empathy, and the formation of a trusting relationship.[48] The goals of this phase are to learn about the nature of the problems from the client's viewpoint, explore strengths, and promote self-exploration.

The focus of the second phase of the counseling process is to help clients identify specific behaviors to alter and to design realistic behavior change strategies to facilitate positive action.[45] This means clients need to be open and honest about what they are willing and not willing to do. Lorenz et al.[48] state that in the successful Diabetes Control and Complications Trial, clients could better communicate their capabilities when health professionals articulated what problems could develop in attempting to improve blood glucose control. They found honesty more likely to occur in an environment in which clients do not feel they will be criticized when difficulties occur, but rather believe the caregivers will show understanding and work toward preparing for similar future circumstances. Nonjudgmental feedback was also an important component of the successful DASH (Dietary Approaches to Stop Hypertension) dietary trial for reducing hypertension.[50] Counselors must communicate their willingness to discover their clients' concerns and help them prioritize in a realistic manner.

EXERCISE 1.9   Exploring Food Habits of Others

Interview someone from a culture different than your own. Ask that person the questions in Exercise 1.8, and record his or her answers in your journal. What did you learn from this activity? How can you personally avoid ethnocentric judgments regarding food habits?

EXERCISE 1.10   Starting a Relationship

Lilly is forty-two years old, has three children, and is about twenty pounds overweight. She sought the help of a fitness and nutrition counselor, Joe, because she wants to increase her energy level and endurance. She tires quickly and feels that exercise will help her stamina.

JOE   Hello, Lilly. It's great you came a little early. Let's get you right on the scale. OK, at 163 pounds, it looks to me as if you need to shed about twenty pounds. You have a ways to go but worry not—we will get it off you. Everything will be fine.

LILLY   I really…

JOE   I am not kidding, Lilly—don't worry. We will start slowly. What you want to do is get your BMI down, your muscle tissue up, as well as get rid of the fat. If you follow me, I'll introduce you to everyone, sign you up for an aerobics class, and start you on your routine.

LILLY   Well, you see I only want…

JOE   Hey, Rick, this is Lilly. She is a newcomer.

RICK   Welcome, Lilly. Don't forget to take home some of our power bars—they are great for beginners who may not know how to eat right.

JOE   Yeah, and be sure to bring a sports drink in with you; you will get mighty thirsty. No pain, no gain!

In groups of three, brainstorm the concerns in this scenario. Why is this helping relationship off to a bad start? What questions or comments could Joe have made that may have been more helpful?

In summary, it would be futile to start designing behavior change strategies when an effective relationship has not developed and you do not have a clear understanding of your clients' problems or an appreciation of their strengths. According to Laquatra and Danish:[48]

> Attending to the second part of the counseling process without the strong foundation afforded by the first part results in dealing with the problem as being separate from the client, or worse yet, providing solutions to the wrong problems. Behavior-change strategies designed under these circumstances are not likely to succeed. (p. 352)

The scenario in Exercise 1.10 illustrates a common mistake helpers make—indicating that everything will be fine. Because it has no basis for reality, the comment belittles the client's feelings. If the client actually feels reassured by the comment, the benefit is temporary because no solution to the problem has been sought. Patronizing a client is self-defeating. It indicates superiority and can automatically create negative feelings. Effective counselors provide reassurance through clarifying their roles in the counseling process, identifying possible solutions, and explaining the counseling program.

Novice Counselor Issues

New counselors typically have concerns about their competency. A counselor who feels inadequate may be reluctant to handle controversial nutrition issues, sometimes giving only partial answers and ignoring critical questions. Confidence in your ability will increase with experience.

Client: *Are high-protein diets a good way to lose weight?*

Counselor: *Some people say they lose weight on them.*

In this example, the counselor is talking like a politician—not taking a stand, trying not to offend anyone. If you are not clear about an issue, you may want to tell your client that it is a topic you have not thoroughly investigated and you will review the matter. If after investigating the issue, you still do not have a clear answer, you should provide your client with what you have found out regarding the positives and negatives of the topic. The American Dietetic Association Code of Ethics[35] states, "The dietetics practitioner presents reliable and substantiated information and interprets controversial information without personal bias, recognizing that legitimate differences of opinion exist."

Another issue for novice nutrition counselors is assuming the role of expert or empathizer.[33] Combining the two roles can contribute to an effective intervention, but a single approach is likely to hamper progress. An authority figure is impressive and appears to have all the answers. Clients blindly accept the direction of the "guru," but little work is done to determine how to make the lifestyle changes work for them. As a result, clients revert to old eating patterns. On the other hand, the empathizer puts so much effort into focusing on client problems that the client receives little direction or information. With experience and determination, the two roles can be effectively combined.

REVIEW QUESTIONS

1. Define nutrition counseling and nutrition education.

2. What is generally considered the most important determinant of food choices?

3. Name and explain the seven qualities of counselors considered to be the most influential by leading authorities as identified by Okun.

4. Explain how taking on the role of helper improves the self-image of the helper.

5. Identify and explain how seeking to fulfill two basic needs of counselors through a counseling relationship can be detrimental to the relationship.

6. Why is it important for counselors to understand their worldviews to achieve cultural sensitivity?

7. Name and explain the two phases of the helping relationship.

8. Why is indicating to a client that everything will be fine unlikely to be productive? What is a more useful approach?

9. Identify three issues for novice counselors.

ASSIGNMENT—BUILD A COLLAGE

The purpose of this assignment is to reflect upon the aspects of your culture that have had the greatest impact on you. Part of becoming a culturally competent nutrition counselor is to understand your own beliefs, attitudes, and the forces that influenced them. This activity may help in the process of understanding the factors that have framed your values, views, and thinking patterns.

Culture is defined as "the thoughts, communication, actions, customs, beliefs, values, and institutions of racial, ethnic, religious or societal groups."[38] You are a member of several cultural groups. Select pictures from print media or use your own photographs that represent cultural forces that have influenced your worldview. Attach them to a poster board. Be prepared to discuss your collage with your colleagues.

REFERENCES

[1]Society for Nutrition Education. Joint position of Society for Nutrition Education (SNE), the American Dietetic Association (ADA), and American School Food Service Association (ASFSA): School-based nutrition programs and services. *J Nutr Educ.* 1995; 27:58–61.

[2]Curry KR, Jaffe A. *Nutrition Counseling & Communication Skills.* Philadelphia: Saunders; 1998.

[3]Haney JH, Leibsohn J. *Basic Counseling Responses.* Pacific Grove, CA: Brooks/Cole; 1999.

[4]Glanz K, Basil M, Maibach E, et al. Why Americans eat what they do: Taste, nutrition, cost, convenience, and weight control concerns as influences on food consumption. *J Am Diet Assoc.* 1998; 98:1118–1126.

[5]Drenowski A. Taste preferences and food intake. *Ann Rev Nutr.* 1997; 17:237–253.

[6]Sass C. Yummy! Yucky! Ick! Tasty! Know what your clients like (and hate) to eat. *ADA Times.* Jan–Feb 2007.

[7]Spittler L. Under the radar: Stealth nutrition in the food industry. *ADA Times.* March–April 2007.

[8]Hopkinson JB, Wright DNM, McDonald JW, Corner JL. The prevalence of concern about weight loss and change in eating habits in people with advanced cancer. *J Pain Symp Mngmt.* 2002; 32:322–331.

[9]Putnam J, Gerrior S. Chapter 7 Trends in the US food supply, 1970–1997. *American Eating Habits: Changes and Consequences, Agriculture Information Bulletin* No (AIB750) 1999, pp. 133–160. Ed. Elizabeth Frazao.

[10]International Food Information Council. *2009 Food & Health Survey: Consumer Attitudes toward Food, Nutrition & Health.* 2009. Available at: http://www.foodinsight.org/Resources/Detail.aspx?topic=2009_Food_Health_Survey_Consumer_Attitudes_toward_Food_Nutrition_and_Health. Accessed May 27, 2010.

[11]Klohe DM, Freeland-Graves JH, Anderson ER, et al. Nutrition knowledge is associated with greater weight

loss in obese and overweight low-income mothers. *J Am Diet Assoc.* 2006; 106:65–75.

[12]Variyam JN, Blaylock J, Smallwood D, Basiotis PP. USDA's Healthy Eating index and Nutrition Information. *Washington, DC: US Department of Agriculture: 1998. Technical Bulletin No. 1866.*

[13]Jarratt J, Mahaffie JB. The profession of dietetics at a critical juncture: A report on the 2006 environmental scan for the American Dietetic Association. *J Am Diet Assoc.* 2007; 107:S39–S57.

[14]Food Marketing Institute. *Trends in the United States. Consumer Attitudes & the Supermarket,* 1999. Washington, DC: Food Marketing Institute: 1999.

[15]Mintz SW, Bu Bois, CM. The anthropology of food and eating. *Ann R Anthropo.* 2002; 31:91–119.

[16]Piernas C, Poplin BM. Snacking Increased among U.S. Adults between 1977 and 2006. *J of Nutr.* 2010; 140:325–32.

[17]Liou D, Bauer K. Obesity Perceptions among Chinese Americans: the Interface of traditional Chinese and American values. *Fd Culture Soc.* 2010; 13:351–369.

[18]Salvy SJ, Howard M, Read M, Mele E. The presence of friends increases food intake in youth. *Am J Clin Nutr.* 2009; 90: 282–287.

[19]Hetherington, MM, Anderson AS, Norton BNM, et al. Situational effects on meal intake: A comparison of eating alone and eating with others. *Physiol Behav.* 2006; 88:498–505.

[20]Harris JL, Bargh JA, Brownell KD. Priming effects of television food advertising on eating behavior. *Health Psychology.* 2009; 28:404–413.

[21]Harris JL, Bargh JA, Brownell KD. Priming effects of television food advertising on eating behavior. *Health Psychology.* 2009; 28(4): 404–413.

[22]Tordoff MG. Obesity by choice: the powerful influence of nutrient availability on nutrient intake. *Am J Physiol Regul Integr Comp Physiol.* 2002; 282(5)RI536–539.

[23]Baker EA, Schootman M, Barnidge E, Kelly C. The role of race and poverty in access to foods that enable individuals to adhere to dietary guidelines. *Prev Chronic Dis* (serial online). Centers for Disease Control and Prevention Web site. July 2006. Available at: http://www.cdc.gov/pcd/issues/2006/jul/05_0217.htm. Accessed May 28, 2010.

[24]George GC, Milani TJ, Hanss-Nuss H, Greeland-Graves JH. Compliance with dietary guidelines and relationship to psychosocial factors in low-income women in late postpartum. *J Am Diet Assoc.* 2005; 105:916–926.

[25]Blisard N, Stewart H, Jolliffe D. Low-income households' expenditures on fruits and vegetables. *Agricultural Economic Report* No. 833. US Department of Agriculture Economic Research Service Web site. Available at http://www.ers.usda.gov/publications/aer833/aer833.pdf. Accessed May 30, 2010.

[26]Roll BJ, Liane SR, Meengs JS. Larger portion sizes lead to a sustained increase in energy intake over 2 days. *J Am Diet Assoc.* 2006; 106:543–549.

[27]Coulston AM. Limitations on the adage "eat a variety of foods"? *Am J Clin Nutr,* 1999; 199:69:350–351.

[28]Rose N, Koperski S, Golomb BA. Chocolate and depressive symptoms in a cross-sectional analysis. *Arch Intern Med.* 2010; 170(8):699–703.

[29]Helm KK, Klawitter B. *Nutrition Therapy: Advanced Counseling Skills.* Lake Dallas, TX: Helm Seminars; 1995.

[30]Egan G. *The Skilled Helper.* 9th ed. Pacific Grove, CA: Brooks/Cole; 2009.

[31]Okun B, Kantrowitz RE. *Effective Helping: Interviewing and Counseling Techniques.* Pacific Grove, CA: Brooks/Cole; 2007.

[32]Corey G. *Theory and Practice of Counseling and Psychotherapy.* 8th ed. Pacific Grove, CA: Brooks/Cole; 2008.

[33]Brammer LM. *The Helping Relationship Process and Skills.* 8th ed. Englewood Cliffs, NJ: Prentice Hall; 2002.

[34]Murphy BC, Dillon C. *Interviewing in Action: Process and Practice.* Pacific Grove, CA: Brooks/Cole; 1998.

[35]American Dietetic Association. American Dietetic Association/Commission on Dietetic Registration Code of Ethics for the Profession of Dietetics and Process for Consideration of Ethics Issues. *J Am Diet Assoc.* 2009; 109:1461–1467.

[36]Belkin GS. *Introduction to Counseling.* Dubuque, IA: Brown; 1984.

[37]Kittler PG, Sucher KP. *Food and Culture in America.* 5th ed. Belmont, CA: Thomson/Wadsworth; 2008.

[38]U.S. Department of Health and Human Services, OPHS Office of Minority Health. *National Standards for Culturally and Linguistically Appropriate Services in Health Care Final Report* (Washington, D.C.: U.S. Government Printing Office, March 2001).

[39]Jandt F. An Introduction to Intercultural Communication: Identities in a Global Community. 6th ed. Thousand Oaks, CA: Sage Publications, Inc.; 2009.

[40]Munoz C, Luckmann, J. *Transcultural Communication in Health Care.* Belmont, CA: Delmar Cengage Learning, 2004.

[41]Chang J. *Wild Swans: Three Daughters of China.* Simon & Schuster, 2003.

[42]Cohen-Cole SA. *The Medical Interview: The Three-Function Approach.* St. Louis, MO: Mosby Year-Book; 2000.

[43]Individualizing nutrition counseling for patients with cancer. *J Am Diet Assoc.* 1999; 99:1221.

[44]Anderson RM, Donnelly MB, Dedrick RF. Diabetes attitude scale. In: Redman BK, ed. *Measurement Tools in Patient Education.* New York: Springer; 2002:66–73.

[45]Berry M, Krummel D. Promoting dietary adherence. In: Kris-Etherton P, Burns JH, eds. *Cardiovascular Nutrition-Strategies and Tools for Disease Management and Prevention.* Chicago: American Dietetic Association; 1998:203–215.

[46]National Institutes of Health (NIH) Obesity Health Initiative. *Clinical Guidelines on the Identification, Evaluation, and Treatment of Overweight and Obesity in Adults,* NIH Publication No. 98-4083. Washington DC: US Department of Health and Human Services; 1998.

[47]Herrman D. *Helen Keller A Life.* New York: Knopf; 1998.

[48]Laquatra I, Danish SJ. Practitioner counseling skill in weight management. In: Dalton S ed. *Overweight and Weight Management: The Health Professional's Guide to Understanding and Practice.* Gaithersburg, MD: Aspen; 1997:348–371.

[49]Lorenz RA., Bubb J, Davis D, Jacobson A, Jannasch K, Kramer J, Lipps J, Schlundt D. Changing behavior: Practical lessons from the Diabetes Control and Complications Trial. *Diabetes Care.* 1996; 19:648–655.

[50]Windhauser MM, Evans MA, McCullough ML, Swain JF, Lin PH, Hobe KP, Plaisted CS, Karanja NM, Vollmer WM. Dietary adherence in the dietary approaches to stop hypertension trial. *J Am Diet Assoc.* 1999; 99:S76–S83.

2

Frameworks for Understanding and Attaining Behavior Change

majaiva/iStockphoto

Change and growth take place when a person has risked himself and dares to become involved with experimenting with his own life.

—HERBERT OTTO

Behavioral Objectives

- Explain the importance of behavior change models and theories for a nutrition practitioner.
- Describe and apply major concepts of selected behavior change theories and models.
- Describe major components of selected theoretical approaches to counseling.
- Differentiate counseling approaches for various durations of brief interventions.

Key Terms

- **Behavior Change:** conducting oneself differently in some particular manner.
- **Behavior Change Models:** a conceptual framework for analyzing and explaining behavior change.
- **Theories:** constructs to provide an explanation based on observation and reasoning of why phenomenon occurs.
- **Concepts:** the building blocks or major components of a theory.
- **Constructs:** concepts developed for use in a particular theory.
- **Models:** generalized descriptions used to analyze or explain something.
- **Motivation:** a state of readiness to change.
- **Self-Efficacy:** an individual's confidence to perform a specific behavior.
- **Self-Motivational Statements:** arguments for making a behavior change made by the client.

INTRODUCTION

Historically, nutrition counselors and educators overlooked many fundamental factors affecting food behavior and attempted to change food choices by simply dispensing facts and diets. The results were often disappointing. Eventually, nutrition professionals recognized a need for a new procedure and turned to established psychotherapy counseling approaches and theoretical models stemming from food-related research and social psychology to guide nutrition interventions.[1] During the 1980s, the focus was on behavior modification, giving way to goal setting and client-centered counseling in the 1990s. More recently, The Transtheoretical Model and Motivational Interviewing have provided guides for instituting behavior change in the health arena. An array of counseling philosophies, theories, behavior change models and counseling approaches are currently available to deal with the complex process of changing health behaviors. Table 2.1 summarizes the usefulness of using theories and models for formulating an intervention.

The following discussion summarizes the approaches most often identified as useful for designing interventions and guiding and appraising changes in dietary behavior. Note that some of the concepts overlap among the behavior change theories, therapies, models and approaches. We will start by discussing self-efficacy which is a construct of several behavior change theories and incorporated into some counseling approaches. Next, we will look at three theories that primarily focus on individual factors, such as knowledge, attitudes, beliefs and prior experience. These include Health Belief Model (HBM), The Transtheoretical Model (TTM), and Theory of Planned Behavior (TPB). The last theory to be addressed is Social Cognitive Theory (SCT) which does not look solely at individual traits for understanding behavior but incorporates a person's relationship with social groups and the environment. We will then turn our attention to counseling approaches frequently used to assist clients with making health behavior changes. Because Client-Centered Counseling provides guidance for establishing an effective counseling relationship, many practitioners utilize basic aspects of this approach. Then we will explore Solution Focused Therapy. This widely used counseling approach has not received much attention for changing dietary behavior, but offers some intriguing useful strategies in nutrition counseling. Next, we will review Cognitive Behavioral Therapy (CBT), which has repeatedly been shown to be effective for changing health behaviors, and finally Motivational Interviewing (MI), which is becoming widely used, especially with clients who are in the early stages of behavior change. You will observe a great deal of interplay among the theories, models and counseling approaches.

Table 2.1 Benefits of Theoretical Behavior Change Theories and Models

- Present a road map for understanding health behaviors
- Highlight variables (for example, knowledge, skills) to target in an intervention
- Supply rationale for designing nutrition interventions that will influence knowledge, attitudes, and behavior
- Guide process for eliciting behavior change
- Provide tools and strategies to facilitate behavior change
- Provide outcome measures to assess effectiveness of interventions

Source: Adapted from: American Dietetic Association. Nutrition Counseling Evidence Analysis Project. http://www.adaevidencelibrary.com.

SELF-EFFICACY

The concept of self-efficacy as a basic component of behavior change was developed by Albert Bandura.[2] Although sometimes considered a separate model, self-efficacy has been widely accepted and incorporated into numerous behavior change models. Bandura[3] defines self-efficacy as "the confidence to perform a specific behavior," such as a belief in ability to change food patterns. Attainment of health behavior changes has been found to correlate solidly with a strong self-efficacy,[4] probably because self-perception of efficacy affects individual choices, the amount of effort put into a task, views of barriers, and willingness to pursue goals when faced with obstacles. As a result, a person's confidence in his or her ability to accomplish a behavior change may be more important than actual skill.[2]

After the importance of change is acknowledged, counselors and educators can help clients to feel there is a "way out of this situation." Clients need to believe there are workable options that make change possible. If individuals perceive there is no solution, their discomfort may shift to defensive thinking: denial ("not really so bad"), rationalization ("didn't want anyway") or projection ("not my problem, but theirs").[5] The counselor's responsibility is to give clients hope by increasing awareness of options and assist in setting achievable goals. Successful experiences build confidence that more complex goals can be attained. Self-efficacy can also be strengthened by pointing out strengths, relating success stories, and expressing optimism for the future.

HEALTH BELIEF MODEL

The Health Belief Model (HBM) proposes that cognitive factors influence an individual's decision to make and maintain a specific health behavior change.[6] Central to making this decision, a person would need to (a) perceive personal susceptibility to a disease or condition; (b) perceive the disease or condition as having some degree of severity, such as physical or social consequences; (c) believe that there are particular benefits in taking actions which would effectively prevent or cure the disease or condition; and (d) perceive no major barriers which would impede the health action; (e) be exposed to a cue to take action; and (f) have confidence in personal ability to perform the specific behavior (self-efficacy).[7] See Table 2.2 for examples.

These beliefs interact with each other to determine a client's willingness to take action. For example, a woman who loves to eat sweets may believe that she is susceptible to getting dental cavities, but if she perceives the adverse effect (severity) on her life to be minimal, then she will not have an impetus to change. Studies have shown that a person with few overt symptoms has lower dietary

Table 2.2 Examples of Health Belief Model Constructs

Health Belief Construct	Sample Client Statements	Intervention Possibilities
Perceived susceptibility	"I am not sure if I should worry about having heart attack."	Educate on disease risk and link to diet, compare to an established standard. Example: "The American Heart Association recommends keeping total cholesterol below 200 mg/dl and yours is 250 mg/dl."
Perceived severity	"Well, I have high blood pressure, but I feel fine."	Discuss disease impact on client's physical, economic, social, and family life. Clarify consequences. Example: "High blood pressure increases risk of having a stroke."
Perceived benefits	"Eating breakfast will help me lose weight."	Specify action and benefits of the action. Example: "Also by eating breakfast you are likely to feel you have more energy throughout the day."
Perceived barriers	"The foods I need to eat to lower my cholesterol do not taste good."	Explore pros and cons; offer assistance, incentives, reassurance; correct misinformation; provide taste tests. Example: "There have been a number of new findings in recent years regarding foods that can lower cholesterol levels. Some of these you may find tasty."

(continued)

Table 2.2   Examples of Health Belief Model Constructs *(continued)*

Health Belief Construct	Sample Client Statements	Intervention Possibilities
Cues to action	"My mother always has sweet rolls on the counter for breakfast."	Link current symptoms to health problem, discuss media to promote health action, encourage social support, use reminder systems (sticky notes, automated cell phone messages, mailings). Example: "You could put the oatmeal box next to the stove at night as a reminder to make it in the morning."
Self-efficacy	"I am confident that I will eat fruit with lunch today."	Provide skill training and demonstrate behaviors, goal setting, provide verbal reinforcement. Example: "Yes, you are on the right track."

adherence.[8] Similarly, a man may believe that eating a plant-based diet will reduce his cholesterol level (benefits), but he may feel it is too inconvenient to change his food pattern (too many barriers) or feel incapable of taking the necessary steps to make the change (low self-efficacy). Cues to action to participate in a program or seek counseling can come from a number of sources, including physical symptoms, observation of a other person taking action, a media report, or advice of a physician. Counselors and clients can brainstorm together to design workable prompts to provide reminders to cue action, such as a note on the refrigerator.

Application of Health Belief Model

Using the HBM, a nutrition intervention in a community congregate food program was able to successfully increase consumption of whole-grains; improve knowledge regarding whole grains; and strengthen the belief that intake of whole-grain foods would reduce risk of disease.[9] The whole grain lesson plans for this intervention are available at http://noahnet.myweb.uga.edu. The following is an example of the application of the HBM constructs for changing whole-grain behavior in this study:

- **Perceived susceptibility and severity:** Personal risk was addressed by emphasizing increased risk for heart disease, cancer, type 2 diabetes and constipation.
- **Perceived benefits:** In order to encourage beliefs regarding benefits, nutritional superiority of whole grains over refined grains was highlighted.

EXERCISE 2.1   Health Belief Model Activity

Match the following descriptions with the appropriate Health Belief Model construct.

_____ 1. Perceived Benefits

_____ 2. Perceived Susceptibility

_____ 3. Perceived Barriers

_____ 4. Perceived Severity

_____ 5. Self-Efficacy

_____ 6. Cues to Action

a. Reading an article about heart disease prompts personal action in reducing dietary fat

b. Perception that heart disease can negatively affect a person's financial status

c. Individual's confidence in ability to engage in regular physical activity

d. Perception that eating fruits and vegetables may lower risk of developing colon cancer

e. Perception that eating healthfully will be costly and inconvenient

f. Personal belief in the chances of developing diabetes

- **Perceived barriers:** To overcome obstacles, taste tests and education regarding labeling of whole grains were provided.
- **Self-efficacy:** In order to increase confidence, lessons included demonstrations and opportunities to practice reading labels.
- **Cues to action:** Participants were given recipes, tip sheets, and educational materials to foster cues to action at home.

THE TRANSTHEORETICAL MODEL (STAGES OF CHANGE MODEL)

This model, developed by Prochaska and DiClemente, is often referred to as transtheoretical because it crosses over many behavior change models. This model provides a guide for explaining behavior change, supplies effective intervention designs and strategies, and evaluates dietary change interventions.[10-14]

Motivational Stages

The Transtheoretical Model (TTM) as depicted in Figure 2.1 describes behavior change as a process of passing through a sequence of distinct motivational stages (that is, five levels of readiness to take action). Implicit in this model, is that behavior change is a process that occurs over time. For an intended behavior change, an individual can begin at any one of the motivational levels or stages:

1. **Precontemplation:** A person in this stage has no intention of changing within the next six months and in fact resists any efforts to modify the problem behavior. The reasons for this include no awareness that a problem exists, denial of a problem, awareness of the problem but unwillingness to change, or feelings of hopelessness after attempting to change.

2. **Contemplation:** Contemplators recognize a need to change, but are in a state of ambivalence, alternating between reasons to change and reasons not to change. During an interview, a client may appear to be saying contradicting statements. For example, "I eat only good foods. I really enjoy the desserts in the lunch room at work." There is concern that the long-term health benefits of the change do not compensate for the short-term real or perceived costs.[14] Perceived barriers such as unacceptable tastes, economic constraints, or inconvenience are major obstacles. People can be stuck in this stage for years waiting for absolute certainty, the magic moment, or just wishing for different consequences without changing behavior. If asked, contemplators are likely to say they intend to change their undesired behavior in the next six months.

3. **Preparation:** Preparers believe the advantages outweigh the disadvantages of changing and are committed to take action in the near future (within the next thirty days). They may have taken small steps to prepare for a change, such as making an appointment with a nutrition counselor or inquiring about a walking club. A person in this stage would probably be willing to try a new recipe or to taste some new foods.

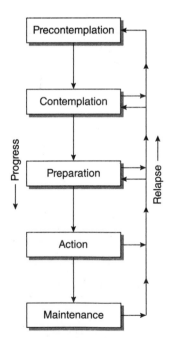

Figure 2.1 The Transtheoretical Stages of Change Model

Source: From BOYLE/HOLBEN. *Community Nutrition in Action,* 5E. © 2010 Brooks/Cole, a part of Cengage Learning, Inc. Reproduced by permission. www.cengage.com/permissions

4. **Action:** Clients are considered to be in this stage if they have altered the target behavior to an acceptable degree for one day or up to six months and continue to work at it. Although changes have been continuous in this stage, the new behaviors should not be viewed as permanent. The most common time for relapse to occur is the first three to six months of the action stage.[15]

5. **Maintenance:** A person in this stage has been engaging in the new behavior for more than six months and is consolidating the gains attained during previous stages.[16] However, the individual needs to work actively to modify the environment to maintain the changed behavior and prevent a relapse. Prochaska and Norcross[17] explain, "Perhaps most important is the sense that one is becoming more of the kind of person one wants to be."

A review of the various stages, see Figure 2.1, indicates that behavior change occurs in a linear order, in which people "graduate" from one stage to the next. However, it is normal for individuals to slip back one or more stages, even to have a relapse and then start to move forward again, progressing toward maintenance. (See Lifestyle Management Form 7.4 in Appendix D.) Figure 2.1 depicts the concept that although individuals move through a sequence of stages there is forward and backward movement in the various stages. Smoking research, for example, has shown that people commonly recycle four times through various stages before achieving long-term maintenance.[5] The fact that change is not perfectly maintained should not be viewed in a negative light.[16] By knowing from the onset that perfection is not realistic and lapses are to be expected, an intervention can be planned accordingly. Hopefully, by understanding that relapses are a normal occurrence in the change process, clients and counselors can maintain a realistic perspective and not become demoralized when they occur. In addition, individuals may be in different stages of change for various behaviors affecting a health outcome. For example, a person who would like to reduce cholesterol may be in an action stage for eating an ounce of nuts each day, but be only in contemplation stage for decreasing intake of high-fat cold cuts.

In this model, part of the decision to move from stage to stage is based on a client's view of the pros and cons of making a behavior change. Pros are considered an individual's beliefs about the anticipated benefits of changing (for example, eating vegetables will decrease my cancer risk.) On the other hand, cons are the costs of behavior change, which can include undesirable taste, inconvenience, and monetary, physical, or psychological costs. A shift in the balance of the two will contribute to advancing or backsliding.[17,18] In the precontemplation stage, cons clearly outweigh pros, resulting in a decision to not change an unhealthy food habit. In the contemplation stage, pros and cons tend to balance each other reflecting the ambivalence and confusion individuals experience at this stage. As individuals progress from preparation through maintenance, pros increase and cons decrease. For individuals in the precontemplation stage, pros need to increase twice as much as the cons for an individual to move to contemplation. Because clients at this stage are not interested in removing barriers, a nutrition intervention needs to emphasize benefits of change to increase pros.

Self-efficacy is also integrated into the TTM. Research indicates that self-efficacy tends to decrease between the precontemplation and contemplation stages, most likely due to an optimistic bias possessed by individuals in the precontemplation stage. Individuals in the contemplation stage may begin to realize the challenges of adopting a new behavior, which may be seen as daunting. As individuals progress through the action and maintenance stages, self-efficacy gradually increases.[19]

The Transtheoretical Model as a Behavior Change Guide

Besides helping understand and explain behavior change, TTM also serves as a guide to identify potentially effective messages and intervention strategies to facilitate movement through the stages to reach and remain at the maintenance stage. Because the strategies clients find useful at each stage differ,[11,20] the treatment intervention needs to be tailored to a client's stage of change. Traditionally, nutrition

---

EXERCISE 2.2   Determine Your Stage

The following is a list of health behaviors commonly accepted as desirable. Review the stages of change, and circle the corresponding number that indicates your stage.

1 = Precontemplation, 2 = Contemplation, 3 = Preparation, 4 = Action, 5 = Maintenance

• Floss teeth at least once a day.	1	2	3	4	5
• Exercise at least ninety minutes a week.	1	2	3	4	5
• Go to the dentist at least once a year.	1	2	3	4	5
• Eat at least five servings of fruits and vegetables a day.	1	2	3	4	5
• Always use a seat belt when driving.	1	2	3	4	5
• Refrain from smoking.	1	2	3	4	5
• Consume at least one thousand milligrams of calcium every day.	1	2	3	4	5
• Eat at least three servings of whole grains every day.	1	2	3	4	5
• Consistently use sunscreens.	1	2	3	4	5

In your journal, write what you learned about yourself. Describe what you learned about the stages of change construct.

Source: This activity was adapted from one developed by Mary Finckenor, Adjunct Professor, Montclair State University, Upper Montclair, New Jersey. Used with permission.

---

interventions have not taken readiness into consideration and treated all people as if they were actively searching for ways to make behavior changes (giving information, offering advice, and developing a diet plan). This approach has been counterproductive, because most individuals with dietary problems are in a preaction stage—precontemplation, contemplation, or preparation. In fact, giving advice to individuals who do not believe they have a problem could make them feel beleaguered and defensive, making change even less likely to occur.[5] In some cases, nutrition counselors may have erroneously assumed that an individual enrolled in a program is ready to take action.[16] The person may in fact have decided to participate because of pressure from a loved one, or serious consideration may have been given to the problem, but the person is not actually ready to make a behavior change. Authorities estimate that only 20 percent of the individuals who seek behavior change assistance are actually in the action stage.[13]

I walked into the hospital room of an obese teenage boy to give a discharge calorie-controlled, weight reduction diet. As soon as I introduced myself and explained the purpose of my visit, the boy said he didn't want another diet. He said he tried them all before, and none of them worked. He said he was fat, his whole family was fat, and that is the way it would always be. Although I was sympathetic to his plight, I proceeded to explain the diet. During the whole explanation, he rolled his eyes, and the rest of his body language indicated that he was annoyed with me. Even at the time I knew that the encounter was not productive. I just transmitted a bunch of facts, even though he obviously was not listening. I felt it was my responsibility to go over the diet with him and chart in his record that the diet order was accomplished. Now that I have had a counseling course, I believe I would have spent the limited time I had with him dealing with his frustration and would have told him to come see me as an outpatient after discharge if he had a change of heart. Now I wouldn't even attempt to go over the diet.*

Prochaska and Norcross[17] have identified effective intervention strategies to assist clients' progress from one stage to another. In general, cognitive (thinking-related) and affective (feeling-related) strategies are more effective in the early stages, whereas behavioral (action-oriented) strategies in the latter stages are more likely to meet client needs.[16] See Table 2.3. As individuals move through stages, intervention strategies need to be adjusted; therefore, counselors need to reassess their clients' stage periodically.

Using the Transtheoretical Model to Measure Outcomes

By tracking movement through various stages, the TTM has given nutrition counselors a new tool for measuring outcomes. For example, counselors should consider their intervention successful if a

---

*Numerous first person accounts from dietetic students or nutrition counselors working in the field are included throughout this book.

Table 2.3 Stages of Change Summary

Stage	Key Intervention Objectives	Intervention Strategies and Do's	Intervention Don'ts
Precontemplation			
No intention of changing within the next six months	Increase information and awareness, emotional acceptance.	Ask how life would be different if change is made. Provide personalized information. Emphasize benefits of change. Allow client to express emotions about the need to make dietary changes.	Do not assume client has knowledge or expect that providing information will automatically lead to behavior change. Do not ignore client's emotional adjustment to the need for dietary change, which overrides ability to process relevant information.
Contemplation			
Aware of problem, thinking about changing behavior within the next six months	Encourage self-reevaluation, increase confidence in ability to adopt recommended behaviors.	Discuss and resolve barriers to change. Encourage support networks. Give positive feedback about client's abilities. Help clarify ambivalence about adopting behavior, and emphasize expected benefits.	Do not ignore the potential impact of family members and others on client's ability to comply. Do not be alarmed or critical of a client's ambivalence.
Preparation			
Intends to change within the next thirty days, may have made small changes	Resolution of ambivalence, firm commitment, and development of a specific action plan.	Encourage client to set specific, achievable goals (for example, use 1% milk instead of whole milk). Remove cues for undesirable behavior. Reinforce small changes that client may have already achieved. Encourage client to make public the intended change.	Do not recommend general behavior changes (for example, "Use less salt"). Do not refer to small changes as "not good enough."
Action			
Actively engaged in behavior change for less than six months	Collaborative, tailored plans, behavioral skills training and social support.	Develop or refer to education program to include self-management skills. Cultivate social support. Consider reward possibilities. Remove cues for undesirable behaviors and add cues for desirable ones. Set realistic goals.	Do not refer client to information-only classes. Do not assume that initial action means permanent change.

(continued)

Table 2.3 Stages of Change Summary *(continued)*

Stage	Key Intervention Objectives	Intervention Strategies and Do's	Intervention Don'ts
Maintenance			
Engaged in the new behavior for at least six months	Collaborative, tailored revisions, problem solving skills and social and environmental support.	Identify and plan for potential difficulties (for example, maintaining dietary changes on vacation).	Do not be discouraged or judgmental about a lapse or relapse.
		Collect information about local resources (for example, support groups, shopping guides).	
		Encourage client to "recycle" if a lapse or relapse occurs.	
		Recommend more challenging dietary changes if client is motivated.	

Source: Adapted from the Journal of the American Dietetic Association, 99:683, Kristal A.R., Glanz K., Curry S.J., Patterson R.E., How can stages of change be best used in dietary interventions?, © 1999, with permission from Elsevier.

client has moved from "I do not need to make a change" to "Maybe I should give some thought to a change." This measure of success may provide encouragement to health professionals who become discouraged with the slow pace of change.[21]

Application of the Transtheoretical Model

The Diabetes Stages of Change (DiSC) was a program administered in Canada using the Transtheoretical Model as a guide to design and implement a 12-month intervention to improve self-care and improve diabetes control in 1,029 individuals with type 1 or type 2 diabetes.[22] Participants were in one of three levels of pre-action motivation groups: precontemplation, contemplation, or preparation for self-monitoring of blood glucose, healthy eating, or smoking cessation. Participants were given usual care or a tailored intervention based on their stage of change called Pathways to Change which included personalized assessment reports, self-help manuals, newsletters, and individual phone conversations using stage appropriate counseling strategies. Participants who received the Pathways to Change intervention as compared to usual care showed significant movement to action or maintenance stage for improving their diets by decreasing fat intake and increasing fruits and vegetables. They also had better control of their diabetes as indicated by blood glucose measures.

THEORY OF PLANNED BEHAVIOR

In the Theory of Planned Behavior (TPB) originally known as the Theory of Reasoned Action[23,24] an individual's health behavior is directly influenced by intention to engage in that behavior ("In the upcoming week, I intend to read labels for sodium content."). Three factors affecting behavioral

---

EXERCISE 2.3   Match Intervention Strategy with Stage of Change

You are hired by the corporate wellness director to design a nutrition intervention promoting intake of fresh fruit and vegetables with the goal of consuming at least 5 servings per day. A needs assessment revealed that employees were in precontemplation, contemplation, and action stages. Review the following groupings of behavior change strategies. For each group, indicate which approach best meets the needs for those in the precontemplation, contemplation, or action stage.

1. Provide coupons, recipes, cooking demonstrations.
2. Provide a self-assessment quiz to compare individual intake of fruits and vegetables against a standard. Provide free samples in lobby.
3. Provide posters and flyers about the importance of eating fruits and vegetables.

intention include attitude, subjective norm, and perceived behavioral control.

- *Attitudes* are favorable or unfavorable evaluations about a given behavior. They are strongly influenced by our beliefs about the outcomes of our actions (outcome beliefs) and how important these outcomes are to the client (evaluations of outcomes). For example, "eating whole grain foods will increase my energy levels" and "having high energy levels is extremely important to me."

- *Subjective norm* or perceived social pressure reflect beliefs about whether significant others approve or disapprove of the behavior. Subjective norms are determined by two factors. Normative beliefs are the strength of our beliefs that significant people approve or disapprove of the behavior. For example, significant family members may want a client to eat less salt. Motivation to comply is the strength of our desire to comply with the opinion of significant others. For example, how much does the client want to comply with family members' recommendations?

- *Perceived behavioral control* is an overall measure of an individual's perceived control over the behavior. Such as, "What is your overall perception of control in purchasing healthy food?" Control beliefs are influenced by presence or absence of resources supporting or impeding behavioral performance. For example, a supportive resource may include family members ("My wife always cooks without salt.") and barriers may include social or physical environmental factors ("My company provides lunch free of charge. If I want a low sodium lunch, I will not be able to eat most of the meals."). Control factors can be internal factors, such as skills and abilities or external factors, such as social or physical environmental factors. The impact of each resource to facilitate or impede the desired behavior is referred to as perceived power of the variable.

Application of the Theory of Planned Behavior

In a study to investigate the intention of dietitians to promote whole-grain foods, the TPB was used.[25]

Intention was measured assessing likelihood of encouraging consumption of whole-grain foods in the next month. Attitude was evaluated by the likelihood that intake of whole-grain foods would result in health benefits for clients. Subjective normative beliefs were based on the belief that other health professionals thought they should promote whole-grain foods and their motivation to comply with health professionals' opinions. Perceived behavioral control was evaluated by measuring barriers to promotion and assessing knowledge and self-efficacy for promotion of whole-grain foods. Results indicated that attitude for promotion of whole-grain foods was high, as well as the belief that other health professionals wanted them to promote these foods and a majority of study participants wanted to comply with this subjective normative belief. Perceived control (self-efficacy and barriers, including knowledge) was low indicating a need for continuing education for dietitians regarding promotion of whole-grain foods.

SOCIAL COGNITIVE THEORY

The Social Cognitive Theory (SCT)[3] formerly known as the Social Learning Theory provides a basis for understanding and predicting behavior, explaining the process of learning, and designing behavior change interventions. See Table 2.4 for a summary of the components of this theory. In this theory, there is a dynamic interaction of personal factors, behavior, and the environment with a change in one capable of influencing the others (known as reciprocal determinism). For example, a change in the environment (husband develops high blood pressure), produces a change in the individual (desire to learn about food choices to help husband), and a change in behavior (increase intake of fruits and vegetables). Key personal factors can include values and beliefs regarding outcomes of a behavior change and self-efficacy. Behavior change may occur by observing and modeling behaviors, and using self-regulating behavior change techniques such as journaling or goal setting. Environmental changes may include buying new cooking equipment or altering types of food available in the home.

EXERCISE 2.4 Evaluation of a Desired Behavior Change using the Theory of Planned Behavior

Think of a behavior you are trying to change and analyze it according to the Theory of Planned Behavior constructs. Describe the behavior you wish to change.

Circle your responses to the questionnaire and answer the following questions in your journal.

1. Why did you select the level you did for the two attitude questions?
2. How do significant others feel about your possible change?
3. Do people in your family and social circles perform the desired behavior themselves?
4. What factors could help you perform the new behavior?
5. Describe the internal and/or external barriers to adopting the new behavior.
6. Evaluate the three components affecting behavioral intention – (attitude, subjective norm, and perceived behavioral control) for your intended behavior change. Choose one of the three that is the most influential and explain why.

Intention: Indicate your level of intention (motivation) to change the behavior in the upcoming week.	Very unlikely	Unlikely	Unsure	Likely	Very likely
Attitude: What is your attitude toward the behavior change?	Extreme dislike	Dislike	Neutral	Enjoyable	Very enjoyable
Attitude: What do you feel about the outcomes of the new behavior?	Extreme dislike	Dislike	Neutral	Enjoyable	Very enjoyable
Normative Beliefs: Do significant others think you should change the behavior?	Highly unlikely	Unlikely	Unsure	Likely	Highly likely
Motivation to Comply: How likely are you to comply with significant others' opinions?	Highly unlikely	Unlikely	Unsure	Likely	Highly likely
Perceived Behavioral Control: What is your overall perception of control over the behavior?	Totally not under my control	Not under my control	Unsure	Under my control	Totally under my control

Table 2.4 Social Cognitive Theory Concepts and Intervention Strategies

Concept	Definition	Implications For Interventions
Reciprocal determinism	Dynamic interaction of the person, behavior, and the environment.	• Consider multiple behavior change strategies • Motivational interviewing • Social support • Behavioral therapy (for example, self-monitoring, stimulus control) • Change environment
Outcome expectations	Beliefs about the likelihood and value of the consequences of behavioral choices	• Provide taste tests • Educate about health implications of food behavior
Self-regulation (control)	Personal regulation of goal-directed behavior or performance	• Provide opportunities for decision making, self-monitoring, goal setting, problem solving, and self-reward. • Stimulus control

(continued)

Table 2.4 Social Cognitive Theory Concepts and Intervention Strategies *(continued)*

Concept	Definition	Implications For Interventions
Behavioral capacity	Knowledge and skill to perform a given behavior	• Provide comprehensive education, such as cooking classes
Expectations	A person's beliefs about the likely outcomes or results of a behavior	• Motivational interviewing • Model positive outcomes of diet and exercise
Self-Efficacy	Beliefs about personal ability to perform behaviors that lead to desired outcomes	• Skill development training and demonstrations • Small, incremental goals and behavioral contracting • Social modeling • Verbal persuasion, encouragement • Improving physical and emotional states
Observational learning	Behavior acquisition that occurs by watching the actions and outcomes of others' behavior, and media influences	• Demonstrations • Provide credible role models, such as teen celebrities who practice good health behaviors • Group problem solving session
Reinforcement	Responses to a person's behavior that increase the likelihood of its recurrence	• Affirm accomplishments • Encourage self-initiated rewards and incentives • Offer gift certificates or coupons
Facilitation	Providing tools, resources, or environmental changes that make new behaviors easier to perform	• Alter environment • Provide food, equipment, and transportation

Source: Adapted from Baranowski T., Parcel G.S. How Individuals, Environments, and Health Behavior Interact: Social Learning Theory, in Health Behavior and Health Education-Theory, Research, and Practice, 3rd ed., eds. K. Glanz, F. M. Lewis, and B. K. Rimer (San Franciso: Jossey-Bass, 2002) Copyright 2002 by Jossey-Bass, Inc., Publishers. Used with permission.

Application of the Social Cognitive Theory (SCT)

A guided goal setting intervention called EatFit using computer technology with middle school adolescents in various school and community settings used constructs of SCT to improve eating and fitness choices.[26,27] This program was developed by the Expanded Food and Nutrition Education Program administered by the University of California, Davis and received a Dannon Institute Award of Excellence in Community Nutrition. This intervention started with students selecting one of six possible dietary goals and one of four physical activity options. These goals were reinforced through nine experiential lessons that focused on a variety of healthy behaviors. EatFit curriculum can be found at: http://ucanr.org/sites/EFNEP_CA. Many of the SCT constructs were used in the intervention but the main three guiding constructs included the following:

- Self-efficacy was enhanced by many skill-building activities, such as reading food labels, verbal encouragement, and utilization of social modeling by interviewing their parents about goal setting experiences.

- Self-regulation was implemented by self-assessments.
- Outcome expectancies were addressed by matching goals with adolescent desired outcomes predetermined by focus group sessions with adolescents before the onset of the intervention. These outcome expectancies included improved appearance, increased energy, and increased independence.

CLIENT-CENTERED COUNSELING

Carl Rogers was the founder of client-centered counseling, also referred to as "nondirective" or "person-centered."[28] The basic assumption in this

---

EXERCISE 2.5   Using Social Cognitive Constructs

Interview an individual in your social circle regarding an experience with goal setting. How did the process work out for your friend or relative? What barriers and hurdles needed to be overcome? Write your answers in your journal.

theory of counseling is that humans are basically rational, socialized, and realistic and that there is an inherent tendency to strive toward growth, self-actualization, and self-direction. Clients actively participate in clarifying needs and exploring potential solutions.[29] They realize their potential for growth in an environment of unconditional positive self-regard. Counselors help develop this environment by totally accepting clients without passing judgments on their thoughts, behavior, or physique. This approach includes respecting clients, regardless of whether they have followed medical and counseling advice.

Total acceptance is extremely important for a level of trust to develop in which clients feel comfortable to express their thoughts freely. This portion of the theory has special meaning for nutrition counselors. A study of nutrition counselor perceptions and attitudes toward overweight clients indicates a need for training in sensitivity and empathy.[30] Another important component of this approach for a nutrition counselor is the underlying assumption that simply listening to knowledge cannot help a client. In client-centered therapy, clients discover within themselves the capacity to use the relationship to change and grow, thereby promoting wellness and independence. Listening to a client's story has been compared to the role of a pharmacologic agent meaning there is great value in developing an open and trusting relationship with a client.[31] Nutrition counselors should not lose sight of the fact that the educational component of dietary therapy has been shown to be extremely valuable.[32] However, person-centered theory of counseling can help guide nutrition counselors by stressing the importance of respect and acceptance for developing a counseling relationship.

> When I started working for the WIC Program, I worried that I might have trouble totally accepting an unmarried client who was pregnant or had a baby. However, my biggest problem was accepting the fact that the young women were very pleased with themselves and full of positive expectations about the upcoming births of their children.

COGNITIVE-BEHAVIORAL THERAPY

Cognitive-behavioral therapy (CBT) incorporates components of both cognitive therapies and behavior therapy and includes a wide range of treatment approaches.[33] Both are based on the assumption that behavior is learned and by altering the environment or internal factors, new behavior patterns develop. Many therapists use a combination of the two therapies and refer to themselves as cognitive-behavioral therapists, even if they rely more on one more than the other. An American Dietetic Association expert panel analysis of the usefulness of nutrition counseling theoretical approaches for changing health and food behavior, gave CBT high marks.[34] The following provides a discussion of each approach.

Cognitive Therapies Leaders in this field include Albert Ellis, who developed *rational emotive behavior therapy* (REBT);[35,36] Aaron T. Beck, who developed *cognitive therapy* (CT);[37,38] and Donald Meichenbaum,[39] who developed *cognitive behavior modification*. The premise of this approach is that negative self-talk and irrational ideas are self-defeating learned behaviors and the most frequent source of people's emotional problems. Clients learn to distinguish between thoughts and feelings, become aware of ways in which their thoughts influence feelings, critically analyze the trueness of their thoughts, and develop skills to interrupt and change harmful thinking.[40] Clients are taught that harmful self-monologues should be identified, eliminated, and replaced with productive self-talk. By influencing a person's pattern of thinking, feelings and actions are modified.[41] An example of an individual with a high cholesterol level using negative self-talk and creating an emotional turmoil for herself would be "I am a fool for eating that cheesecake. I have no self-control. I'll just die of a heart attack." This could be changed into better-coping self-talk: "I am learning how to handle these situations. Next time I will ask for a small taste. I am on the road to a healthier lifestyle."

Cognitive therapists have developed a number of techniques to improve positive feelings and help problem-solving ability. These include relaxation training and therapy, mental imagery, thought stopping, meditation, biofeedback, cognitive restructuring and systematic desensitization.

See Chapter 6 for elaboration on several of the strategies.

Behavioral Therapy Behavioral counseling evolved from behavioral theories developed by Ivan Pavlov, B. F. Skinner, Joseph Wolpe, Edward Thorndike, and Albert Bandura.[42,43] The premise of this type of counseling is that many behaviors are learned, so it is possible to learn new ones. The focus is not on maintaining will power but on creating an environment conducive to acquiring new behaviors. Three approaches to learning form the basis for behavior modification:

1. **Classical conditioning** focuses on antecedents (stimuli, cues) that affect food behavior. For example, seeing or smelling food, watching television, studying, or experiencing boredom may be a stimulus to eat. In nutrition counseling, clients may be encouraged to identify and eliminate cues, such as removing the cookie jar from the kitchen counter.

2. **Operant conditioning** is based on the law of effect, which states that behaviors can be changed by their positive or negative effect. In nutrition counseling, generally a positive approach to conditioning is applied, such as a reward for obtaining a goal. The change in diet itself can be the reward, as in the alleviation of constipation by an increased intake of fluids and fiber.

3. **Modeling** is observational learning, such as learning by watching a video or demonstration, observing an associate, or hearing a success story.

Application of Cognitive-Behavioral Therapy

Cognitive behavior strategies were used in a 12-week study with one hundred eight subjects who smoked and wanted to lose weight.[44] The strategies included self-monitoring, goal setting, stimulus control, cognitive restructuring, stress management, and social support. As compared to the control group, the intervention group decreased body weight, improved the quality of their diet, and increased self-efficacy for quitting smoking, and for controlling their weight.

SOLUTION-FOCUSED THERAPY

Insoo Kim Berg developed solution-focused therapy, and Steve de Shazer[45] brought the topic to international attention. Solution-focused therapists work with their clients to concentrate on solutions that have worked for them in the past and identify strengths to be expanded upon and used as resources. Focus of sessions is not on discovering and solving problems but may well be an exception to the normal course of action—that is, the one time the client was able to positively cope. By investigating the accomplishment, no matter how small, adaptive strategies are likely to emerge. For example, a middle-aged executive who complains that business lunches and dinners are a frequent difficulty would be asked to think of an occasion when healthy food was consumed at one of these meals. After identifying the skills the executive used to make the meal a healthy experience, the nutrition counselor would focus on helping replicate and expand those skills. The aim is for clients to use solution-oriented language—to speak about what they can do differently, what resources they possess, and what they have done in the past that worked. Language (solution-talk) provides the guide in solution-focused therapy. Examples of questions a solution-focused counselor may ask include the following:

- What can I do that would be helpful to you?
- Was there a time when you ate a whole-grain food?
- When was the last time you ate fruit?
- Has a family member or friend ever encouraged you to eat low-sodium foods?

> In the cardiac rehabilitation center where I worked, there was a client whose quality of life was severely affected by his weight. He was working as a security guard and had difficulty climbing steps or walking any reasonable distance due to his weight and his need to lug an oxygen tank. After several months of trying a variety of intervention strategies, I asked him whether he had ever been on a diet that worked. He said the only time he lost weight was when he cut bread out of his diet. We set "no more bread" as a goal, and that was the beginning of a successful weight loss program that allowed grains in other forms, such as cereal, pasta, and rice.

Table 2.5   Overview of What Is Motivational

1. Knowledge of consequences
2. Self-efficacy
3. A perception that a course of action has been chosen freely
4. Self-analysis (giving arguments for change)
5. Recognition of a discrepancy between present condition and desirable state of being
6. Social support
7. Feelings accepted

MOTIVATIONAL INTERVIEWING

A major factor for backsliding on the readiness continuum is lack of motivation (that is, eagerness to change). Motivational interviewing (MI) is an approach to counseling that complements the Transtheoretical Model because it entails a focus on strategies to help motivate clients to build commitment to make a behavior change. Miller and Rollnick, founders of MI provide the following definition: "a client-centered, directive method for enhancing intrinsic motivation to change by exploring and resolving ambivalence."[5] In this approach, motivation is not viewed as a personality trait or a defense mechanism, but considered a state of readiness to change that can alter and be influenced by others. Since counselors can impact motivation, to do so is considered an inherent part of their intervention responsibility. MI is particularly useful in the early stages of behavior change when there is a great deal of ambivalence about making a decision to change.[46] If a client has clearly indicated desired behavior change, spending precious counseling time exploring ambivalence would probably be frustrating and as a result counterproductive.

MI works to cultivate a client's own natural motivation for change (intrinsic).[5] Motivation can come from coerced external forces ("Lose weight or you can't be in my wedding.") or intrinsic (internal) due to specific values ("I want to be able to be a good role model for my children.").[47] Even if perceived self-efficacy and competence are the same, if motivation originates from internal beliefs and values, there will be enhanced performance, persistence and creativity to accomplish the task. An overview of factors usually found to be motivational can be found in Table 2.5.

Rosengren[48] provides a model to represent four interconnected elements of MI: MI spirit, MI principles, change talk, and OARS (acronym for counseling skills). See Figure 2.2.

Spirit of Motivational Interviewing   The guiding philosophy of MI has three components: collaboration, evocation and autonomy. A collaborative approach in the search for ways to achieve behavior change is essential for the motivational interviewing process. The expertise of both the counselor and the client are respected. The counselor brings

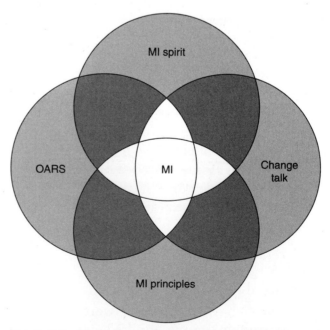

Figure 2.2   Elements of Motivational Interviewing
Source: D.B. Rosengren, *Guiding Motivational Interviewing Skills: A Practitioner Workbook* 1st ed., 2009, The Guildford Press (New York, NY: The Guildford Press, 2009) p. 9.

a wealth of knowledge and experience and the client is the expert on past experiences, influencing pressures, and personal beliefs and values. In MI there is a basic assumption that individuals have an intrinsic desire to do what is truly important to them, and the counselor's responsibility is to facilitate clients to evoke that motivation (evocation) and to bring about change. Autonomy recognizes that decisions to change always need to come from the client. The counselor creates an atmosphere where clients understand that they are not reacting to the force of any other person (such as counselor, parent, or doctor) but have chosen to make changes based on their own beliefs and values.

Motivational Interviewing Guiding Principles MI is a counseling approach described to be "like dancing rather than wrestling."[49] To achieve this outcome, there are four guiding principles with the acronym RULE: (1) resist the righting reflex, (2) understand and explore motivations, (3) listen with empathy, and (4) empower the client.

1. Resist the Righting Reflex Counselors may have become accustomed to trying to make things right because of a desire to help others lead healthier lives. If a client is ambivalent about change, he or she has a good argument for both changing and not changing. Your natural reaction may be to "right off the bat" set things straight and provide all the reasons for changing an established food pattern. For example, a counselor may tell an ambivalent client, "You should eat breakfast. You will have more energy throughout the day, be more focused in your work and have better control of your appetite all day. Successful dieters typically eat breakfast." This is good advice, and when and how to give advice will be reviewed in Chapter 3. However, an ambivalent client is likely to respond with all the reasons the good advice will not work. This scenario is not likely to produce a good outcome because it goes against two rules of human nature: we tend to believe what we hear ourselves say and there is an inherent resistance to persuasion. If your client is giving arguments for not changing, your interaction is building commitment to *not* change. More arguments for change on your part will likely produce more resistance

and defensive behaviors. You and your client will feel as if you are wrestling. Resistance is a natural survival response and is likely to occur if clients feel they are not in control, believe they do not have a choice, are confused, or feeling that a counselor is acting as if they are ready for change when they simply want to consider changing.[56] Additional signs of resistance include engaging in denial, putting up objections, changing the subject, interrupting, and showing reluctance to discuss a subject.[5] Kellogg[56] offers the following as signs of resistance:

- "yes, but . . ."
- "Well, I guess I could try."
- Agrees too quickly.
- Body language shows reluctance.

You may feel a need to push against the resistance because of knowledge of the long-term consequences of the client's present food behavior or your desire to be successful as a counselor. However, pushing for change by arguing, judging, persuading, or discounting feelings is likely to escalate the resistance. If you encounter resistance, you should acknowledge resistance and back off. Naming the resistance takes away some of the power of resistance. You can use the following guide when encountering resistance:[56]

- State what you see/hear. "You really do not like gyms."
- Acknowledge the resistance. "You believe that plan will not work for you."
- Shift back in your chair and breathe.
- Offer to let go. "How about we let that plan go." "Let's take a step back."
- Invite working together. "Would you like to brainstorm some other ideas?"

Additionally, you may wish to explore or revisit readiness to change. For example, you could say, "I can tell we've gotten off track here. Can you help me review what is most important to you right now?" Encourage the client to assume control. For example, "What would you like to work on next?" Another possibility is posing a change as an experiment. You could say, "Maybe we could design an experiment together to gather more information about all this. Are you curious what would happen?"[56]

2. Understand and Explore Motivations In MI you guide the counseling session to allow your clients to explore perceptions and see a discrepancy regarding their current behavior compared to their values, beliefs, and concerns. The guiding encourages your client to use change talk, that is, clarifying important goals, vocalizing reasons for change, and exploring the potential consequences of their present behavior. When the discrepancy overwhelms the need to keep the present behavior, there is likely to be a decision to start taking action to change. You know you are on the right track when your client is voicing concerns, giving reasons to change, and expressing an intention to change.

3. Listen with Empathy Acceptance facilitates change. The underlying assumption of expressing empathy is an acceptance and understanding of a client's perspective. This does not mean that a counselor has the same perspective or would have made similar choices. However, basic acceptance ("You are OK") creates an environment for change.[5] A message of "You are not OK" creates resistance to change. In MI clients are invited to explore conflicts. Unless a counselor communicates with empathy, clients are not likely to feel safe revealing discrepancies between their behavior and their beliefs and values.

4. Empower the Client Belief in the ability to change is an important motivator.[49] As previously discussed in this chapter, there are numerous methods of increasing self-efficacy. In the MI paradigm, supporting self-efficacy by stressing the importance of the client, not the counselor, as the one responsible for selecting and carrying out changes is essential. By doing so, you have indicated that you believe the client is capable of this task and thereby can increase self-efficacy.

Elicit Change Talk The objective of change talk is to resolve ambivalence by providing opportunities and encouragement for the client, rather than the counselor, to make arguments for change. When clients express the need for change or the reasons why change is necessary, the balance of indecision begins to shift toward taking action. As change talk strengthens, commitment increases as well as the likelihood of behavior change.[50] There are four categories of change talk statements:[51]

1. Cognitive. Problem recognition; for example, "I get headaches from my high blood pressure."
2. Cognitive. Optimism for change; for example, "Lots of people have to take insulin. I can do it, too."
3. Affective. Expression of concern; for example, "I'm so worried about my diabetes. I hope eating better and exercise brings down my blood sugar levels."
4. Behavioral. Intention to change; for example, "In the past, I always enjoyed eating fruit. I will eat a banana with breakfast and dried fruit with my lunch tomorrow."

Strategies to Elicit Change Talk

- **Evaluate Importance and Confidence** This technique usually involves two questions. First, clients are asked to rate on a scale of zero to ten (with ten being the highest) the importance of the behavior change (for example, increase intake of fruits and vegetables). Next they are asked to rate again on the same scale their confidence in making a change. Follow-up questions explore choices. For example, "Why did you choose the number four and not two?" What would you need to get to the number seven instead of four?" An individual may feel that a change is worthwhile and may even elicit change talk indicating the importance of change, but if that person has little confidence in the ability to make the change, then implementation of action strategies are not likely to be successful. For example, a woman may feel confident in her ability to increase her calcium intake, but if she does not consider the issue important enough, her degree of readiness to change is reduced. Likewise, a woman who feels an increase in calcium intake is important, but does not feel confident in her ability to make the increase, will be at a lower level of readiness to change. In general, lowest levels of readiness are often associated with low importance. Differences between the terms are illustrated in Table 2.6

Table 2.6   Three Topics in Talk about Behavior Change

Importance: Why?	Confidence: How? What?	Readiness: When?
Is it worthwhile?	Can I?	Should I do it now?
Why should I?	How will I do it?	What about other priorities?
How will I benefit?	How will I cope with x, y, and z?	
What will change?	Will I succeed if . . .	
At what cost?	What change . . .?	
Do I really want to?		
Will it make a difference?		

Source: Rollnick S, Mason P, Butler C. Health Behavior Change: A Guide for Practitioners. New York: Churchill Livingstone; © 1999; p. 21. Used with permission.

- **Values Clarification – Card Sort** This technique was used successfully in the Healthy Body Health Spirit Trial.[52] Clients are asked to sort cards, each having a personal core value (such as, being a good parent, competent, or attractive) according to how important the value is to them. Then clients are asked if there are any connections between the health behavior desires and their values. This strategy has been incorporated into Exercise 2.7.

- **Change Roles** Tell your client that you are going to change roles, and ask the client to convince you to make the contemplated behavior change. Gradually allow the client to persuade you.

EXERCISE 2.7   Values Clarification Card Sort

1. Obtain twenty-three index cards.
2. On one of the cards label IMPORTANT TO ME, on a second label VERY IMPORTANT TO ME, and on a third label NOT IMPORTANT TO ME. These are the anchor (title) cards, put aside.
3. On one card, label a behavior change you are contemplating, such as drink less coffee, and on a second card write another behavior change you are contemplating, such as exercising more.
4. On the remaining cards make 18 value cards, write one of the following values. (Note some are actually attributes or goals):

Good parent	Good community member	Competent
Good spouse or partner	Respected at home	Attractive
Wealth	Spiritual	Successful
Loved	On top of things	Independent
Health	Energetic	Responsible
Creativity	Considerate	Disciplined

5. Shuffle the eighteen value cards.
6. Team up with a partner and give your cards to your colleague.
7. Your partner will read the following script to you: "I am placing three title cards in front of you. Take the eighteen value cards and your two behavior change cards, look at each one, and place them under a title card. The only rule is that there can not be more than four cards in the VERY IMPORTANT TO ME pile."
8. Your partner will then ask you, "How may your desired behavior change desires relate to these goals or values?

Source: This activity is based on one used by the Healthy Body, Healthy Spirit Project; Resnicow K, Jackson A, Blissett D, Wang, et al. Results of the Healthy Body Healthy Spirit Trial. Health Psychology. 2005;24(4):339–348.

- **Typical Day Strategy** Ask your clients to take about five to ten minutes to describe a typical day and explain how their health issue (for example, diabetes) and their food needs are affecting their life. This strategy is discussed in more detail in Chapter 4.

Reinforce Change Talk When clients have made change talk statements, the counselor should take note and reinforce their meaning. The counselor's responsibility is to direct the intervention towards change talk and then to amplify clients' arguments for change.

The following are some methods for strengthening the statements:

- Request clarification (for example, how much, how many, and give an instance) on previous self-motivational statements.
- Reinforce change talk both non-verbally (for example, a nod) and verbally with a statement such as "I can understand why this has been so difficult for you."

Foundation Skills - OARS MI relies on basic counseling skills, such as those found in Table 2.7, to encourage clients to make a decision to change. Four skills found to be the most useful for MI and can be remembered with the acronym OARS, open-ended questions, affirmations, reflective listening and summaries.

- **Open-ended questions** Open-ended questions are used to explore and gather information from the client's perspective. They are questions that are not likely to be answered with a yes or no or a few words. To use these effectively, your approach must communicate curiosity, concern, and respect. You should not appear to be conducting an inquisition to gather information against your client. These types of questions are covered in more detail in Chapter 3, but the following have been found to be particularly useful for MI:
  - ❏ Ask about the pros and the cons of the client's present eating pattern and the contemplated change.

Table 2.7 General Motivational Interviewing Counseling Strategies

- Encourage clients to make their own appraisals of the benefits and losses of an intended change.
- Do not rush clients into decision making.
- Describe what other clients have done in similar situations.
- Give well-timed advice emphasizing that the client is the best judge of what can work.
- Provide information in a neutral, non-personal manner.
- Do not tell clients how they should feel about a medical or dietary assessment.
- Present choices.
- Clarify goals.
- Failure to reach a decision to change is not a failed consultation.
- Make sure clients understand that resolutions to change break down.
- Expect commitment to change to fluctuate, and empathize with the client's predicament.

  - ❏ Ask about extremes related to the problem. For example, "What worries you the most?"
  - ❏ Ask the client to envision the future after the change has been accomplished.
  - ❏ Ask about priorities in life (that is, what is most important to the client). Then ask how the contemplated behavior change fits into the hierarchy.

- **Affirmations** Affirmations recognize client efforts and strengths and provide another source of motivation. Pointing out a job well done or persistence in the face of numerous obstacles reminds clients that they possess inner qualities that make behavior change possible. Rosengren[48] suggests that affirmations should focus on specific behaviors, avoid use of the word "I", and highlight non-problem areas. For example, "You are providing a good food environment in your home." Rather than, "I am happy you decided not to buy soda anymore."

- **Summaries** Summaries are done periodically throughout a MI session to help organize thoughts, reinforce change talk, clarify discrepancies or links during the session, and transition to a new topic. The technique will be covered at greater length in Chapter 3.
- **Reflective listening** Reflective listening is a key skill in MI and entails using basic listening skills, interpreting the heart of your client's message, and reflecting the interpretation back to your client. By acting as a mirror and reflecting back your understanding of the intent or your interpretation of the underlying meaning, clients are encouraged to keep talking. This show of interest is an expression of empathy creating an environment for self-exploration about the challenges of making a behavior change. You also have the opportunity to select what you would like to reinforce. The following dialogue illustrates a nutrition counselor listening reflectively and attempting to identify the underlying meaning of a client's statements:

Client: *Everyone is getting on my back about my cholesterol level—my wife, my doctor, my brother. I guess I have to do something about my diet.*

Counselor: *You're feeling harassed that other people are pushing you to change the way you eat.*

Client: *I suppose they're right, but I feel fine.*

Counselor: *You're worried about the future.*

Client: *Yeah. I have a lot of responsibilities. I have two children and I want to be around to take care of them, see them grow up, and get married. But it doesn't thrill me to give up meatballs and pizza.*

Counselor: *You're wondering about what food habits you are willing to change.*

Client: *You know, I wouldn't mind eating more fish. I've heard that is a good food to eat to lower cholesterol levels. What do you think about oatmeal?*

Note that the formulation of a response is an active process. You must decide what to reflect and what to ignore. In this dialogue example, the counselor chose to respond to the client's statement

"I suppose they're right" rather than "I feel fine." The counselor guessed that if the client thought all those others were right, then he must be worried about his health. If the counselor had chosen to reflect on the feeling fine part of the client's second statement, what would have happened? Of course we can only "guess," but it doesn't seem likely that a client-initiated discussion of diet changes would have occurred so quickly. To respond reflectively is particularly useful after asking an open-ended question when you are trying to better understand your client's story.

The development of reflective listening skills can be a complex task for novice counselors.[49] If this is a skill you decide to develop, explore the motivational interviewing resources at the end of this chapter and consider attending motivational interviewing workshops.

Integrating Motivational Interviewing with Other Behavior Change Approaches

MI is a communication style, which can be integrated with other behavior change approaches. For example, MI may be used during an initial session with a client who is ambivalent about making dietary changes, and when the decisional balance shifts toward a commitment to change, the nutrition counselor could incorporate cognitive-behavioral techniques. In addition, a counselor may see a need to come back to a MI approach as a client begins to expand dietary changes. For example, someone who has high cholesterol and high blood pressure may begin working on making dietary changes by setting goals to eat fish three times a week and nuts or beans each day. After the food habits have been established, a client may have ambivalence about making other changes such as decreasing sodium or fried food and using a MI approach would again be helpful. In the PREMIER study to lower blood pressure, motivational interviewing integrated well with self-applied behavior modification techniques, Social Cognitive Theory, and The Transtheoretical Model to help individuals lower blood pressure and change dietary behaviors.[53]

BRIEF ENCOUNTERS USING MOTIVATIONAL INTERVIEWING

Health care practitioners are often involved in brief interventions that do not allow full development of the MI approach. However, using components of MI providing the "spirit" of motivational interviewing have met with success when time is limited.[46,49] Table 2.8 elucidates three kinds of interventions with suggested goals and skills according to time allotment. All of these approaches focus on encouraging behavior change. For brief encounters, the goal may be to encourage a client to think about changing health behaviors and to accept a referral. Many of the components for approaching health care counseling have been incorporated into the analysis and flow of a nutrition counseling session and is found in Chapter 4.

SUMMARY OF BEHAVIOR CHANGE ATTRIBUTES

Health behavior change models, theories and approaches provide a picture of what predisposes individuals towards making successful health behavior changes. Table 2.9 summarizes the attributes counseling and education interventions, which practitioners hope to cultivate with their clients. Not all six qualities need to present for change to occur, but they provide an overall view of desirability for practitioners.

The art of nutrition counseling and education is an evolving process for both the profession and the professional. Making a decision to change one's diet and implementing that decision is guided by a complex interaction of psychological factors.[7] No one orientation meets all

Table 2.8 Three Kinds of Behavior Change Interventions Based on Available Time

	Brief Advice (BA)	Behavior Change Counseling (BCC)	Motivational Interviewing
Context			
Session time	5-15 minutes	5-30 minutes	30-60 minutes
Setting	Mostly opportunistic	Opportunistic or help-seeking	Mostly help-seeking
Counseling Techniques			
	• Demonstrate respect • Communicate risk • Provide information	• BA goals • Establish rapport • Identify client goals • Assess importance and confidence • Exchange information • Choose strategies based on client readiness	• BA and BCC goals • Develop a relationship • Resolve ambivalence • Develop discrepancy
Goals			
	Initiate thinking about change in problem behavior	Build motivation for change	Elicit commitment to change
Style			
Practitioner-recipient	Active expert-passive recipient	Counselor-active participant	Leading partner-partner
Confrontational or challenging style	Sometimes	Seldom	Never

(continued)

Table 2.8   Three Kinds of Behavior Change Interventions Based on Available Time *(continued)*

	Brief Advice (BA)	Behavior Change Counseling (BCC)	Motivational Interviewing
Empathic style	Sometimes	Usually	Always
Information	Provided	Exchanged	Exchanged to develop discrepancy
Skills*			
Ask open-ended questions	**	**	***
Affirmations	**	**	***
Summaries	*	***	***
Ask permission	**	***	***
Encourage choice and responsibility in decision making	**	***	***
Provide advice	***	**	*
Reflective listening statements	*	**	***
Elicit change talk	*	**	***
Roll with resistance	*	***	***
Help client articulate deeply held values	*	**	***

*Skills range from non-essential to essential using a 3-point scale (one, two, or three asterisks).

Source: Rollnick S, Allison J, Ballasiotes S, Barth T, Butler CC, Rose BS, Rosengren DB. Variations on a Theme Motivational Interviewing and Its Adaptations, In: Motivational Interviewing Preparing People for Change 2nd Ed., New York: The Guilford Press, 2002, p. 274.

Table 2.9   Putting It All Together: Successful Behavior Change Attributes Based on Theories and Models

1. Strongly desires and intends to change for clear, personal reasons
2. Faces a minimum of obstacles (information processing, physical, logistical, or environmental barriers) to change
3. Has the requisite skill and self-confidence to make a change
4. Feels positively about the change and believes it will result in meaningful benefit(s)
5. Perceives the change is congruent with his or her self-image and social group(s) norms
6. Receives reminders, encouragement, and support to change at appropriate times and places from valued persons and community sources, and is in a largely supportive community or environment for the change.

Reprinted from American Journal of Preventive Medicine 22:267–284. Whitlock E.P., Orleans T., Pender N., Allan J., Evaluating Primary care behavioral counseling interventions. An evidence-based approach, 2002, with permission from Elsevier.

the needs of a complex, fluid society, nor can one methodology be a perfect fit for an individual nutrition counselor or educator. Nutrition professionals must use their professional judgment regarding selection of an intervention. We have a large array of theories and approaches to choose among. Sigman-Grant[54] states that sixty different behavior change models have been developed to explain changes in health behavior and to guide nutrition interventions. See Table 2.10 for an overview of the ones most often cited as a guide for explaining the nature and dynamics of food behavior. Addressing this issue, James et al.[55] suggest an eclectic approach merging the most useful ideas from various models.

The motivational nutrition counseling algorithm presented in a step-by-step manner in Chapter 4 has taken this approach and provided a flexible client-centered solution orientation.

In addition, a basic knowledge of a variety of counseling approaches is desirable to meet the needs of clients as they progress in a counseling program. Their readiness for particular interventions may change.

Table 2.10   Summary of Behavior Change Models and Approaches

Behavior Change Model or Approach	Focus	Key Concepts
Self-efficacy	A component of numerous behavior change models and approaches; confidence in ability to perform a behavior	• Positive self-efficacy increases probability of making a behavior change
Health Belief Model	Perception of the health problem and appraisal of benefits and barriers of adopting health behavior are central to a decision to change	• Perceived susceptibility • Perceived severity • Perceived benefits • Perceived barriers • Cues to action • Self-efficacy
The Transtheoretical Model	Behavior change is explained as a readiness to change	• Behavior change is described as a series of changes • Behavior change occurs over time • Specific behavior change strategies are identified for each stage
Theory of Planned Behavior	Intention influences behavior	• Intention is a result of attitude, subjective norm, perception of behavioral control
Social Cognitive Theory and Social Learning Theory	People learn by observing social interactions and media; personal factors, behavior, and the environment interact continuously, each influencing the other	• Self-efficacy • Enhancement of knowledge • Skill development • Social support • Observational learning • Reinforcements increase reoccurrence of behavior
Client-Centered Counseling	Clients actively participate in clarifying issues and exploring solutions	• Counselors develop an environment of unconditional positive self-regard
Cognitive Behavior Therapy	Behavior is learned so it can be unlearned; irrational ideas are self-defeating; focus is on changing the environment	• Distinguish between thoughts and feelings • Challenge pattern of thinking • Stress management • Self-monitoring • Identify and remove cues • Provide substitutions • Emphasize consequences • Modeling
Solution-Focused Therapy	Focus on identifying strengths and expanding on past successes	• Use solution talk • Do not solve problems
Motivational Interviewing	Explore and resolve ambivalence	• Client-centered, directive • Develop discrepancy • Reduce resistance • Support self-efficacy • Listen reflectively

CASE STUDY   Helping Relationships

John is a seventy-year-old white veteran of the Vietnam War who was admitted to the nursing home because he was no longer able to care for himself. His diagnoses on admission included cerebrovascular accident (stroke), angina, cancer of the prostate, and major depression. He is generally confined to a wheelchair, but he can ambulate eight to ten steps with the assistance of two people. He is unable to dress himself due to right-sided hemiparesis (paralysis); he is continent of bowel but at times is incontinent of bladder. All of his laboratory values are within normal limits. John is mentally alert and not at all confused but has clinical depression. He has no family or visitors.

John is able to feed himself and has an excellent appetite. He consistently consumes 100 percent of his meals. His weight on admission can be indicated only as over 300 pounds, as the scale cannot measure over 299 pounds. It is estimated that he weights about 320 pounds, which is approximately 100 pounds overweight. He has no difficulty chewing or swallowing and receives a regular diet of regular consistency. He loves to eat, and some staff members bring him food items from home, especially on the 3–11 shift. This helps calm him down during the evening hours, allowing the nurses to do their work.

At times John has outbursts of anger at the staff, particularly when given instructions on what he should do or when he is awakened from sleep. He calls the nurses "Babe" or "Sweetie" and can often be heard telling staff that they "look good today" and "you have a great set of gams." Most of the nurses, recreation staff, and social service staff are relatively young and find his comments to be offensive. Their attitude toward him is tolerant at best, and they do little for him beyond his basic care. The staff openly talk at the nurses' station about his repulsive attitude toward women. The nurses' aids complain about his weight because it is difficult to get him in and out of bed. Allowing him to ambulate as per doctor's orders is also a challenge because it takes two to three people to assist. John once fell, and the fire department had to come to get him off the floor because he was too heavy for staff members to lift.

In November, John had surgery for the removal of a cancerous prostate, and his prognosis continues to remain relatively poor considering his cancer and his heart disease. He will frequently comment on how he wants to lose weight; however, he will also say things like, "I could sure go for another one of those eclairs," or "That cook sure can make a great meat loaf—I could have eaten another whole lunch." At times he has even gotten angry if the staff does not meet his requests for seconds.

John spends his day sitting in the hallway watching the activities at the nurses' station and chain smoking. He enjoys some game shows and listens to country music. He does not attend recreational activities.

EXERCISE 2.8   Applying Theoretical Approaches for the Helping Relationships Case Study

Behavior Change Approach	Application of the Approach
Self-efficacy	In order to increase self-efficacy, a counsellor would point out John's strengths, give verbal encouragement, set goals, encourage the social work staff to find social support for him.
Client-centered	A client-centered counselor would have unconditional positive regard for John and clearly communicate understanding of his concerns. Nutrition facts would be kept to a minimum, and the nutritionist would take direction from John as to what nutrition goals should be formulated.
Motivational Interviewing	A counsellor using motivational interviewing would use an empathic style, explore his ambivalence about making food behavior changes, and encourage change talk.
Behavioral	A behavioral counselor would work on changing the environment to improve John's food management. This may involve offering rewards, making certain foods available, or showing a video.

(continued)

EXERCISE 2.8   Applying Theoretical Approaches for the Helping Relationships Case Study *(continued)*

Behavior Change Approach	Application of the Approach
Cognitive	A cognitive counselor would be concerned with John's irrational thought pattern. Intervention could focus on cognitive restructuring (changing thought patterns) or relaxation techniques. Because John is described as depressed, this could be discussed at a unit meeting with the staff psychologist.
Solution focused	A solution-oriented counselor could ask John whether he ever did exercise in his wheelchair or when he believes he is eating healthy foods at the nursing home. After identifying the resources that have worked for him in the past, opportunities to expand on those resources would be sought.

❏ Review the theoretical approaches for interacting with John. Record in your journal which theoretical approach or approaches you believe would provide success in dealing with this client. As a nutrition counselor, what do you believe would be your goals for working with John? Identify some specific actions you would take to achieve these goals.

EXERCISE 2.9   Helping Relationships

After reading the Helping Relationships Case Study, record in your journal five behaviors, characteristics, or physical concerns that a nutrition counselor needs to consider prior to any interventions, and indicate how they will impact the relationship. Also, take into consideration that this is an institution, and a nutrition counselor has a limited amount of time to spend with each client.

❏ What is preventing the professional staff from building a helping relationship?
❏ What issues regarding John could be brought up at a medical staff meeting?
❏ What steps could the nutritionist take to develop an effective helping relationship?

REVIEW QUESTIONS

1. What are the benefits of using theoretical behavior change theories and models?
2. Why does a high level of self-efficacy correlate positively with health behavior changes?
3. Identify and explain the six constructs of the Health Belief Model.
4. Identify and explain the five stages of change in The Transtheoretical Model.
5. Explain the components of attitude, subjective norm, and perceived behavioral control of the Theory of Planned Behavior.
6. Explain reciprocal determinism, a main principle of Social Cognitive Theory.
7. Explain why unconditional positive regard is essential for client-centered counseling.
8. Which type of therapy works at changing harmful thinking?
9. Which type of therapy focuses on changing the environment?
10. Which type of therapy encourages clients to work not on problem solving, but on when the client is able to cope?
11. How does a Motivational Interviewing counselor encourage a client to engage in change talk?

ASSIGNMENT—OBSERVATION OF A NUTRITION COUNSELOR

Observe a nutrition counselor in an inpatient, outpatient clinic, or private office setting for two hours.‡ Answer the following questions in your journal or in a typed, formal paper to be handed in to your instructor. Use the corresponding number or letter for each answer.

1. Identify the name of the setting, location, starting and ending time of the observation, date, and name of the counselor you observed.

2. Describe the physical setting where the nutrition counseling sessions took place.

3. Describe the counselor's attire and its appropriateness.

4. Select a client you observed, and give the following information to the best of your ability:

   a. What was the client's gender, age, and cultural (including ethnic) orientation?

   b. Was a helping relationship established? If not, why not? If yes, what did the counselor specifically do or say to encourage an effective relationship?

   c. Explain the nature of the client's problem.

   d. Was there evidence of collaboration between the counselor and the client to define dietary objectives? Explain.

   e. Were short- or long-term goals established? If yes, what were they?

   f. Describe any teaching or visual aids.

   g. Was there evidence of tailoring dietary objectives to address the client's lifestyle issues? Explain.

   h. Give your impression of the client's educational level and needs.

   i. Give your impression of the client's health belief and self-efficacy regarding his or her dietary objectives.

   j. What were the client's barriers to meeting the dietary objectives?

   k. Estimate and explain what you believe was the client's stage of change.

   l. Was there evidence of social support for the client to meet the dietary objectives?

   m. Complete a counseling observation checklist. See below.

5. Review Client-Centered Counseling, Solution Focused Therapy, Cognitive-Behavioral Therapy, and Motivational Interviewing

COUNSELING OBSERVATION CHECKLIST	RARELY	OCCASIONALLY	UNDECIDED	OFTEN	ALMOST ALWAYS	ALWAYS
Counselor Name: _____ Date of Session: _____ Length of observation: _____ minutes Estimated percentage of time counselor talked: _____%						
Did the nutrition counselor appear to be comfortable with the client and with the subject areas discussed?						
Did the counselor avoid imposing values on the client?						
Did the counselor remain objective?						
Did the counselor focus on the client, not just on the procedure of providing a diet instruction?						
Were the counselor's skills spontaneous and non-mechanical?						
How would you describe the likelihood that the client would return to this nutrition counselor again?						
Comments:						

approaches to counseling covered in this chapter. List each one, and indicate whether any components of the approaches were demonstrated in your observations. If yes, explain. If no, how could they have been incorporated?

6. Describe your general impressions of the counseling session. What did you learn from this experience?

If a counselor is not available, an alternative would be to use a video of a counseling session that could be critiqued individually or in groups. These are available on You Tube.

Nutrition Care Process (NCP) Connection*

Theoretical basis/approach for nutrition education and nutrition counseling in NCP include:
- Cognitive-Behavioral Theory
- Health Belief Model
- Social Learning Theory
- Transtheoetical/Stages of Change
- Motivational Interviewing – listed as a strategy in the NCP

---

*Throughout the book connections to the American Dietetic Association Nutrition Care Process will be highlighted when appropriate. The Nutrition Care process is a systematic approach to providing high quality nutrition care including nutrition assessment, diagnosis, intervention and monitoring evaluation. The process is elaborated upon in Chapter 5.

Answers to Exercises

Exercise 2.1: 1 = d, 2 = f, 3 = e, 4 = b, 5 = c, 6 = a
Exercise 2.3: 1 = action, 2 = contemplation, 3 = precontemplation

SUGGESTED READINGS, MATERIALS, AND INTERNET RESOURCES

Overview of Psychological Theories and Counseling Approaches

Nutrition Education: Linking Research, Theory, and Practice. 2nd Ed. Isobel Contento. Sudbury, MA: Jones and Bartlett Publishers; 2011.

Health Behavior and Health Education: Theory, Research, and Practice, 4th Ed. Karen Glanz, Barbara Krimer and K. Viswanath. San Francisco: Jossey-Bass Publishers; 2008.

Theory and Practice of Counseling and Psychotherapy, 5th Ed. Gerald Corey. New York: Brooks Cole; 2008.

Theory at a Glance. 2nd Ed. Barbara Rimer and Karen Glanz. http://www.cancer.gov/PDF/481f5d53-63df-41bc-bfaf-5aa48ee1da4d/TAAG3.pdf Karen Glanz. NIH Publication No. 05-3896; 2005.

Client-Centered Counseling

http://www.carlrogers.info/index.html Website: Carl Rogers Info interviews, videos of therapy sessions and presentations.

The Handbook of Person-Centred Psychotherapy and Counseling. Edited by: Mick Cooper, Maureen O'Hara, Peter F. Schmid and Gill Wyatt, Palgrave Macmillan, 2007. A comprehensive overview of theoretical and practical aspects of the person-centered approach. Highlights for nutrition counselors include using the approach for group work, families, medical settings, and counseling across differences.

Cognitive-Behavioral Therapy

http://www.abct.org/Home/ Association for Behavioral and Cognitive Therapies (BCT). This site includes information about BCT, factsheets, listing of educational opportunities, syllabi, PowerPoint presentations, and videos and podcasts.

Feeling Better, Getting Better, Staying Better: Profound Self-Help Therapy For Your Emotions, Albert Ellis, Impact Publishers, Inc., 2001. This book explains the basics of cognitive-behavioral therapy.

Transtheoretical Model

Changing for Good. James Prochaska, John Norcross, Carlo DiClemente. New York: Harper Paperbacks; 1995. Definitive guidelines written in a readable format to implement the stages of change model.

http://www.umbc.edu/psyc/habits/content/the_model/index.html The Habits Lab at UMBC Health and Addictive Behaviors: Investigating

Transtheoretical Solutions. This site, maintained by the University of Maryland Baltimore County, contains downloadable PowerPoint presentations, educational resources, and useful links.

Motivational Interviewing

http://www.motivationalinterview.org/ Motivational Interviewing Resources for Clinicians, Researchers, and Trainers. This website includes information about motivational interviewing, research, training manuals seminars, training manuals and materials, and videos and powerpoint presentations.

http://www.stephenrollnick.com/ Stephen Rollnick's Motivational Interviewing and Training Resources.

Motivational Interviewing: Preparing People for Change, 2nd ed. William R. Miller & Stephen Rollnick, New York: The Guilford Press, 2002. The landmark book on the motivational interviewing approach.

Building Motivational Interviewing Skills, a practitioner workbook. David B. Rosengren. New York: The Guilford Press, 2009. Written by a trainer of MI, the book is full of practice dialogues, concept quizzes, and excellent exercises.

Solution Focused Brief Therapy

Solution Focused Therapy: A Handbook for Health Care Professionals. Hawkes D, Marsh TI, Wilgosh R. Hanover, NH: Butterworth-Heinemann Medical, D. Hawkes, T.I. Marsh, & R. Wilgosh, 1998. This step-by-step guide supplies many case studies and strategies.

REFERENCES

[1]Gillespie AH, Grun JK. Trends and challenges in nutrition education research. *J Nutr Educ.* 1992;24:222–226.

[2]Bandura A. Self-efficacy: Toward a unifying theory of behavioral change. *Psychosoc Rev.* 1977;191–215.

[3]Bandura A. *Social Foundations of Thought and Action.* Englewood Cliffs, NJ: Prentice-Hall; 1986.

[4]Strecher VJ, DeVellis BM, Becker MH, et al. The role of self-efficacy in achieving health behavior change. *Health Educ Q.* 1986;13:73–91.

[5]Miller WR, Rollnick S. *Motivational Interviewing Preparing People for Change,* 2nd ed., New York: The Guilford Press, 2002.

[6]Rosenstock IM, Strecher VJ, Becker MH. Social learning theory and the health belief model. *Health Educ Quart.* 1988;15(2):175–183.

[7]Champion VL, Strecher VJ. The Health belief model. In: Glanz K, Rimer BK, Vswanath K. eds. *Health Behavior and Health Education: Theory, Research, and Practice.* 4th ed. San Francisco, CA: Jossey-Bass; 2008:45–66.

[8]Meichenbaum D, Turk DC. *Facilitating Treatment Adherence: A Practitioner's Guidebook.* New York: Plenum; 1987.

[9]Ellis J, Johnson MA, Fischer JB, Hargrove JL. Nutrition and health education intervention for whole grain foods in the Georgia Older Americans Nutrition Program. *J Nutr Elderly,* 2005;2467–83.

[10]Prochaska JO, DiClemente CC, Norcross JC. In search of how people change: Applications to addictive behaviors. *Am Psychol.* 1992;47:1102–1114.

[11]Greene GW, Rossi SR, Rossi JS, Velicer WF, Fava JL, Prochaska JO. Dietary applications of the Stages of Change Model. *J Am Diet Assoc.* 1999;99:67–678.

[12]Prochaska JO, Norcross JC, DiClemente V. *Changing for Good: A Revolutionary Six-Stage Program for Overcoming Bad Habits and Moving Your Life Positively Forward.* New York, NY: Avon Books, 1994.

[13]DiClemente CC, Prochaska J. Toward a comprehensive, transtheoretical model of change: Stages of change and addictive behaviors. In: Miller WR, Heather N. eds. *Treating Addictive Behaviors,* 2nd ed. New York: Phenum; 1998.

[14]Ruggiero L, Prochaska JO. Introduction. *Diabetes Spectrum.* 1993;6:22–24.

[15]Sandoval WM, Heller KE, Wiese WH, Childs DA. Stages of change: A model for nutrition counseling. *Top Clin Nutr.* 1994;9:64–69.

[16]Grommet JK. Weight management; framework for changing behavior. In: Dalton S. ed. *Overweight and Weight Management: The Health Professional's Guide to Understanding and Practice.* Gaithersburg, MD: Aspen; 1997:332–347.

[17]Prochaska JO, Norcross JC. *Systems of Psychotherapy: A Transtheoretical Analysis.* 3d ed. Pacific Grove, CA: Brooks/Cole; 1994.

[18]Prochaska, JO. Decision making in the transtheoretical model of behavior change. *Medical Decision Making.* 2008;28:845–849.

[19]Ma J, Betts NM, Horacek T, et al. The importance of decisional balance and self-efficacy in relation to stages of change for fruit and vegetable intakes by young adults. *Am J Health Prom.* 2002;16(3):157–166.

[20]Nothwehr F, Snetselaar L, Yang J, Wu H. Stage of change for healthful eating and use of behavioral strategies. *J Am Diet Assoc.* 2006;106:1035–1041.

[21]Sigman-Grant M. Stages of change: A framework for nutrition interventions. *Nutr Today.* 1996;31:162–170.

[22]Jones H, Edwards L, Vallis TM, et al. Changes in diabetes self-care behaviors make a difference in glycemic control: The Diabetes Stages of Change (DiSC) study. *Diabetes Care.* 2003;26(3):732–737.

[23]Ajzen I, Driver BL. Prediction of leisure participation from behavioral, normative, and control beliefs: An application of the theory of planned behavior. *Leisure Science.* 1991;13:185–204.

[24]Fishbein M, Ajzen I. Understanding Attitudes and Predicting Social Behavior. Englewood Cliffs, NJ: Prentice-Hall, Inc.; 1980.

[25]Chase K, Reicks M, Jones JM. Applying the theory of planned behavior to promotion of whole-grain foods by dietitians. *J Am Diet Assoc.* 2003;103:1639–1642.

[26]Shilts MK, Horowitz M, Townsend M. An innovative approach to goal setting for adolescents: Guided goal setting. *J Nutr Ed Behav.* 2004;36:155–156.

[27]Horowitz M, Shilts MK, Townsend M. EatFit: A goal oriented intervention that challenges middle school adolescents to improve their eating and fitness choices. *J Nutr Ed Behav.* 2004;36:43–44.

[28]Rogers CR. *Client-Centered Therapy.* Boston: Houghton Mifflin; 1951.

[29]Berry M, Krummel D. Promoting dietary adherence. In: Kris-Etherton P, Burns JH, eds. *Cardiovascular Nutrition-Strategies and Tools for Disease Management and Prevention.* Chicago: American Dietetic Association; 1998:203–215.

[30]McArthur LH, Ross JK. Attitudes of registered dietitians toward personal overweight and overweight clients. *J Am Diet Assoc.* 1997;97:63–66.

[31]Butler C, Rollnick S, Stott N. The practitioner, the patient and resistance to change: Recent ideas on compliance. *Can Med Assoc J.* 1996;154:1357–1362.

[32]National Institutes of Health (NIH) Obesity Health Initiative. *Clinical Guidelines on the Identification, Evaluation, and Treatment of Overweight and Obesity in Adults,* NIH Publication No. 98-4083. Washington DC: US Department of Health and Human Services; 1998.

[33]British Association for Behavioural and Cognitive Psychotherapies. http://www.babcp.com/ Accessed: May 30, 2010.

[34]Spahn JM, Reeves RS, Keim KS, et al. State of the evidence regarding behavior change theories and strategies in nutrition counseling to facilitate health and food behavior change. *J Am Diet Assoc.* 2010;110:879–891.

[35]Ellis A, Harper R. *A Guide to Rational Living.* North Hollywood, CA: Wilshire; 1997.

[36]Ellis A, Dryden W. Rational-emotive therapy. *Nurse Pract.*1984;12:16.

[37]Beck AT. *Cognitive Therapy and Emotional Disorders.* New York: International Universities Press; 1976.

[38]Beck AT. Cognitive therapy: past, present, and future. *J Consult Clinl Psychol.* 1993;61:194–198.

[39]Meichenbaum D. *Cognitive Behavior Modification: An Integrative Approach.* New York: Plenum; 1977.

[40]Association for Behavioral and Cognitive Therapies. What is Cognitive Therapy? http://www.abct.org/Professionals/?m=mPro&fa=WhatIsCBT Accessed June 15, 2010.

[41]Baldwin TT, Falcigia GA. Application of cognitive behavioral theories to dietary change in clients. *J Am Diet Assoc.* 1995;95:1315–1317.

[42]Skinner BF. *The Behavior of Organisms.* New York, NY: Appleton-Century-Crofts; 1938.

[43]Skinner BF. *Contingencies of Reinforcement: A Theoretical Analysis.* New York ,NY: Appleton-Century-Crofts; 1969.

[44]Sallit J, Ciccazzo M, Dixon Z. A cognitive-behavioral weight control program improves eating and smoking behaviors in weight-concerned female smokers. *J Am Diet Assoc.* 2009;109(8):1398–1405.

[45]de Shazer S. *Keys to Solutions in Brief Therapy.* New York, NY: Norton; 1985.

[46]Berg-Smith SM, Stevens VJ, Brown KM, et al. for the Dietary Intervention Study in Children (DISC) Research Group. A brief motivational intervention to improve dietary adherence in adolescents. *Health Educ Res.* 1999;14:101–112.

[47]Ryan RM, Deci EL. Self-determination theory and the facilitation of intrinsic motivation, social development, and well-being. *Am Psychol.* 2000;55(10): 68–78.

[48]Rosengren DB. *Building Motivational Interviewing Skills – a practitioner workbook.* New York, NY: The Guilford Press; 2009.

[49]Rollnick S, Miller WR, Butler C. *Motivational Interviewing in Health Care Helping Patients Change Behavior.* New York, NY: The Guilford Press; 2008.

[50]Amrhein PC, Miller WR, Yahne CE, Palmer M, Fulcher L. Client commitment language during motivational interviewing predicts drug use outcomes. *J Consul Clin Psych.* 2003;71(5): 862–878.

[51]Miller WR, Rollnick S. *Motivational Interviewing Preparing People for Change*, 1st ed., New York, NY: The Guilford Press, 1991.

[52]Resnicow K, Jackson A, Blissett D, Wang, et al. Results of the Healthy Body Healthy Spirit Trial. *Health Psychology.* 2005;24(4):339–348.

[53]Lin PH, Appel LJ, Funk K, et al. The PREMIER Intervention Helps Participants Follow the Dietary Approaches to Stop Hypertension Dietary Pattern and the Current Dietary Reference Intakes Recommendations. *J Am Diet Assoc.* 2007;107: 1541–1551.

[54]Sigman-Grant M. Change strategies for dietary behaviors in pregnancy and lactation. *University of Minnesota Educational Videos from the 1999 National Maternal Nutrition Intensive Course.* Minneapolis: University of Minnesota School of Public Health; 1999.

[55]James RK, Gilliland BE. *Theories and Strategies in Counseling and Psychotherapy.* 5th ed. Upper Saddle River, NJ: Allyn & Bacon; 2002.

3

Communication Essentials

An ounce of dialogue is worth a pound of monologue.

—ANONYMOUS

Behavioral Objectives

- Identify stages of skill development.
- Describe the impact of communication dynamics on nutrition interventions.
- Explain intercultural influence on communication.
- Identify three intents for formulating counseling responses.
- Evaluate effectiveness of counselor's nonverbal communication.
- Identify common messages from North American body language.
- Identify communication roadblocks.
- Demonstrate skills for building a relationship.
- Utilize basic counseling responses.

Key Terms

- **Communication Roadblocks:** obstacles that hamper self-exploration.
- **Counseling Focus:** placement of emphasis in a counseling response.
- **Counseling Intent:** rationale for selecting a particular intent.
- **Empathy:** true understanding of another's perspective.
- **Skill:** an acquired ability to perform a given task.
- **Synchrony:** harmony of body language.
- **Trait:** an inherent quality of mind or a personality characteristic.

INTRODUCTION

This chapter is devoted to exploring communication basics related to nutrition interventions. First, we will review basic nutrition counseling goals using the Haney and Leibsohn model of counseling, which includes facilitating awareness, supporting healthy lifestyle decision making, and encouraging clients to take appropriate actions. This chapter is devoted to the basic skills and counseling responses that aid in helping clients take these actions. In order to accomplish these goals, nutrition professionals are likely to need to learn new skills, so we will explore factors related to skill development. Then, we will examine a model of communication. Because we talk and interact with others every day, the need to review such a model may not seem obvious. However, interaction may not lead to communication, and by examining the model, we can see specific points where communication may break down. Cultural differences between two individuals can add another layer of complexity for communication, and this topic will be highlighted. Then, guidelines for effective communication in a counseling environment will be addressed. Finally, basic counseling responses will be covered stressing appropriate times to use them in an intervention.

NUTRITION COUNSELING GOALS

Using the Haney and Leibsohn model of counseling, we may identify three specific goals in nutrition counseling.[1] The first is to *facilitate lifestyle awareness*, which can be achieved by keeping the focus on your client; acknowledging feelings, experience, and behavior; and providing information. Exploring feelings, ambivalence, inner strengths, behavior, and alternative options can increase the likelihood of obtaining the second goal, *healthy lifestyle decision making*. The ultimate goal in nutrition counseling is for your client to *take appropriate action* to obtain a healthier lifestyle and become self-sufficient.[1*2] This is done by exploring issues and encouraging your client to view his or her situation differently. In particular, nutrition counselors help clients take appropriate actions by encouraging them to take risks, tolerate incongruities, and give new behaviors and thoughts a chance before discounting them.[3]

STAGES OF SKILL DEVELOPMENT

A student of nutrition counseling has the task of learning many new technical, social, and conceptual skills. For some individuals the job will be easier than for others. This is because we are all born with traits, which are a quality of mind or personality characteristics. The special traits that are part of our essence may or may not easily interface with the interpersonal skills needed for counseling. However, no matter what special abilities we possess, we can all learn counseling skills through patience and practice.

As with learning any new skill, we must pass through a sequence of steps before mastering the skill. While reviewing the following sequence, keep in mind a skill that took you some time to develop, such as learning to drive a car.

1. **Motivation.** The first important step for developing a skill is having a desire to learn. A motivated student will progress to a much higher level of expertise than an unmotivated one, no matter what special traits a person possesses. Motivation can be enhanced by learning in a supportive environment that encourages success by mastering skills in a sequential, stepwise manner.

2. **Learning.** Acquiring knowledge, skills, and attitudes necessary to become an effective nutrition counselor comes from reading, participating in learning activities, making observations, engaging in discussions, and listening to presentations.

[1*]Although self-sufficiency is often stated as the ultimate goal of nutrition counseling, it should be noted that in a study of older patients with diabetes, a significant number reported that they did not desire an independent self-care role (Anderson et al., 1998). Also, the guidelines developed by the expert panel for the Identification, Evaluation, and Treatment of Overweight and Obesity headed by Dr. F. Xavier Pi-Sunyer acknowledge that a weight maintenance program consisting of diet therapy, behavior therapy, and physical activity may need to be continued indefinitely for some individuals (NIH Publication, 98-4083, 1998).

3. **Awkwardness.** If possible, the initial attempts at using a new counseling skill should be undertaken with volunteers under supervision, such as a role-playing situation. A novice counselor must be willing to go through a period of discomfort in order to acquire effective counseling skills. A degree of awkwardness should also be expected when conducting your first counseling session.

4. **Conscious awareness.** As ability is gained, a counselor is likely to feel more comfortable using the specific skill, but will still be consciously aware of the process.

5. **Automatic response.** Eventually the skill will become an automatic reaction with little or no forethought or discomfort.

6. **Proficiency.** A high level of expertise will be obtained when a counselor can perform and modify the skill under varying conditions. As nutrition counselors gain proficiency, they are likely to feel free to experiment with new approaches and to modify and expand their skills.

MODEL OF COMMUNICATION

Understanding the dynamics of communication is essential for developing good counseling skills. Counselors in particular need to realize that a speaker's statements can be interpreted in several ways, as illustrated in Figure 3.1. In this model a speaker's intended meaning can be distorted at three main junctures: (1) The talker may not communicate clearly because of a faulty *encoding process*, or the ability to express a thought.

This happens when language skills are not adequate or a person uses abstractions or generalizations as a way of dealing with denial or anxiety. (2) Distortions also occur when words are not heard properly. (3) Lastly, a listener can distort a message during the *decoding process*, which refers to analysis of the thoughts expressed by others. We all interpret statements through mental filters created by past experiences. Because no two people have precisely the same life experiences, their filters will differ and interpretations simply become best guesses. Any remark contains multiple meanings. For example, the statement "I binged last Saturday" could mean eating two brownies, consuming half the food in the refrigerator, or getting drunk. If counselors silently equate their interpretations with exact meaning, communication will break down.

CULTURAL INFLUENCE ON COMMUNICATION

Cultural orientation has a major impact on the process of communication. The closer two individuals share a common culture, the greater the likelihood that distortions will be minimal and conversation will flow smoothly. Each society has a conscious and an unconscious series of expected reciprocal responses.[4] For example, in the United States a "You're welcome" is expected to follow a "Thank you." When expected responses do not happen, a feeling of discomfort ensues. Race, gender, age, and nationality have the greatest influence on cross-cultural communication, although equally influential can be degree of acculturation or

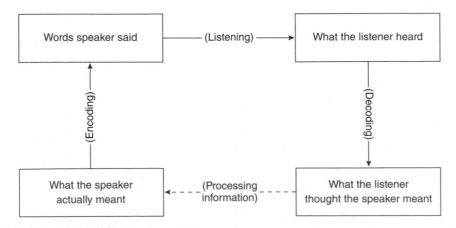

Figure 3.1   A Model of Communication
Source: Adapted from Gordon T, *Parent Effectiveness Training* (New York: Three Rivers; 2000).

EXERCISE 3.1   Generating Alternative Meanings

The purpose of this activity is to encourage you to practice generating alternative meanings. Work with colleagues in triads. Each person should complete this statement: "One thing that I like about myself is that I . . ." The statement should be relatively abstract and have a degree of ambiguity. Concrete statements such as physical attributes should be avoided, such as "One thing I like about myself is that I have blue eyes." Each person should take a turn making one of the statements. The two listeners consider various meanings and respond five times with "Do you mean that you . . .?" The volunteer can only answer yes or no. Here is an example:

SPEAKER   One thing I like about myself is that I am strong.
LISTENER   Do you mean that you can lift a lot of weight?
SPEAKER   No.
LISTENER   Do you mean that you are there to help people if there is a problem?
SPEAKER   Yes.
LISTENER   Do you mean that you can handle a lot of problems at one time?
SPEAKER   No.
LISTENER   Do you mean that you don't fall apart when a problem occurs?
SPEAKER   Yes.

In a counseling situation, you would not interrogate a client with a series of "What do you mean?" questions but rather you should listen closely and consider alternative meanings. In the following sections of this chapter, various counseling responses will be covered to help clients clarify their meanings to you and to themselves.

❏ When you were the speaker in this activity, were you frustrated by the limitation of only being able to answer yes or no? Generally as attempts are made to clarify meanings, a person undergoes a deeper self-evaluation and will feel the need to elaborate. How does this activity relate to the counseling process?

Source: Miller WR, Rollnick S. Motivational Interviewing. New York: Guilford Press; 1991:168.

---

assimilation, socioeconomic status, health condition, religion, educational background, group membership, sexual orientation, or political affiliation.[1]

Key differences among cultural groups often involve body language, variations in use of expressive language, degree of directness, use of eye contact, amount of personal space needed, and acceptable duration of silence. See Table 3.1 for specific examples of communication differences of various cultural groups. While reviewing this list, keep in mind that considerable individual variation exists within any particular cultural group.

GUIDELINES FOR ENHANCING COUNSELING COMMUNICATION EFFECTIVENESS

This section reviews selective skills for enhancing communication in a counseling session. These include an introduction to the use of focuses and intents for the formulation of responses, an overview of effective nonverbal behavior, an explanation of the value of harmonizing verbal and nonverbal behaviors, an analysis of nonverbal behavior, an examination of communication roadblocks, and a review of the importance of empathy for developing an effective counseling relationship.

Use Focuses and Intents when Formulating Responses

Flow of communication in a counseling setting for the most part is not like having a conversation with a friend. Counselors need to modify some previously learned behaviors such as talking about oneself, asking many questions, and avoiding lulls and silences.[1] Counselors use verbal and nonverbal counseling responses with a specific intent and focus to address counseling objectives. **Counseling intent** is a rationale for selecting a particular response, and **counseling focus** is the placement of

Table 3.1 Cultural Comparisons of Communication Styles

African Americans	Asians	Latinos	Middle Easterners	Native Americans	Majority Whites
Speak quickly, with affect and rhythm	Speak softly	Speak softly—may perceive normal white voice as yelling	Speak softly	Speak slowly and softly	Speak loudly, quickly—control of listener
Direct eye contact when speaking, may avert if prolonged—look away when listening	Avert eyes as sign of respect	Eye contact direct between members of same sex—may seem to stare—aversion an insult, though women may avert eyes with men	Direct gaze between members of same sex—women may avert eyes with men	Indirect gaze when speaking and listening	Direct eye contact when speaking and listening—prolonged contact rude
Interject often (taking turns)	Head nodding may indicate active listening—rarely interject	Seldom make responses to indicate active listening or to encourage continuation—rarely interject	Facial gestures express responses	Seldom make responses to indicate active listening or to encourage continuation—rarely interject	Head nodding, Murmuring
Very quick response	Delayed auditory (silence valued)	Mild auditory delay	Mild auditory delay	Delayed auditory (silence valued)	Quick response
Expressive, demonstrative	Polite, restrained, articulation of feelings considered immature	Men restrained, women expressive but not emotional	Expressive, emotional	Expression restrained	Task-oriented, focused
Respectful, direct approach	Indirect approach (Japanese); direct approach (Chinese, Koreans)	Indirect approach	Indirect approach	Indirect approach—stories about others may be metaphors for self	Direct approach, minimal small talk for urban whites—more indirect for rural whites
Assertive questioning	Rarely ask questions	Will ask questions when encouraged	Will ask polite questions	Rarely ask questions—Yes or no answer considered complete	Ask direct questions
Firm handshake, smile	May or may not exchange soft handshake	Firm handshake among men, soft handshake with women	Numerous greetings, salaam—may or may not exchange soft handshake, smile	Quick handshake, smile	Firm handshake, smile
Touching common—reluctance to touch may be interpreted as rejection—stand and sit closer than majority whites	Non-touching culture—Stand and sit farther away than majority whites	Touching common—Stand and sit closer than majority whites	Touching common between members of same gender—stand and sit closer than majority whites	Minimal touching	Moderate touching

(continued)

Table 3.1   Cultural Comparisons of Communication Styles *(continued)*

African Americans	Asians	Latinos	Middle Easterners	Native Americans	Majority Whites
High-context use of pictures, graphs, charts useful	Very high-context use of pictures, graphs, charts important	Moderately high context	High-context use of pictures, graphs, charts useful	Very high-context use of pictures, graphs, charts important	Low- to medium-high context
Polychronistic	Polychronistic—punctual	Polychronistic	Polychronistic	Polychronistic	Monochronistic

Counseling the culturally diverse: Theory and practice, Derald Wing Sue and David Sue. Copyright © 2002. Adapted with permission from John Wiley & Sons, Inc.

*Polychronistic: Circular, not defined by time, Monochronistic: Linear, time-oriented.

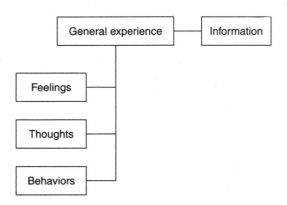

Figure 3.2   Nutrition Counseling Response Focuses
Source: Text for image is from Haney JH, Leibsohn J. *Basic Counseling Responses.* Pacific Grove, CA: Brooks/Cole, Wadsworth © 1999.

Figure 3.3   Counseling Response Intents
Source: Haney JH, Leibsohn J. *Basic Counseling Responses.* Pacific Grove, CA: Brooks/Cole, Wadsworth © 1999. Reprinted with permission.

the emphasis in a response. The focus of a response could be placed on information about a client or a client's general experiences. The experience response can be subdivided further into feelings, thoughts, or behaviors. See Figure 3.2.

The counseling model developed by Haney and Leibsohn[1] has three intents, or rationales for selecting a particular response. See Figure 3.3. By recognizing the hoped-for outcome of a response, a counselor is better able to formulate an effective response. The intents include the following:

1. **To acknowledge.** If a counselor's intent is to acknowledge, then responses would be selected that identify observations, affirm, show respect, or recognize the worthiness of a client. Relationship-building responses would fall into this category.

2. **To explore.** The objective of a response could be for a client to explore ambivalence, consider new information, or gain insight. If this is the case, counselors might ask questions, provide information, or make clarifying responses.

3. **To challenge.** If the intent is to help clients see their situation differently or to take a

Table 3.2   Summary of Possible Counseling Intents

	Possible Counseling Intents				
	Acknowledge			Explore	Challenge
Responses	**Relationship Building**	**Continue Talking**	**Counselor is Listening**	**Clarify Concern**	**Note a Discrepancy**
1. Attending	X	X	X		
2. Empathizing	X	X	X	X	
3. Legitimation	X				
4. Respect	X				
5. Personal support	X				
6. Partnership	X				
7. Mirroring			X		
8. Paraphrasing		X	X	X	
9. Giving feedback				X	X
10. Questioning				X	
11. Clarifying				X	
12. Noting discrepancy					X
13. Directing				X	X
14. Advice				X	X
15. Silence				X	
16. Self-disclosing	X	X			

different course of action, then a response that notes a discrepancy could be selected.

The following dialogue illustrates responses using various focuses and intents:

Counselor: *Your blood evaluation indicates that you have a cholesterol level of 330. Your dietary evaluation shows a high saturated fat intake and a low intake of fiber, fruits, and vegetables.* (information focus, intent to explore)

How did you handle the party last week? (experience focus[2*], intent to explore)

You have a right to feel angry about having to handle another dietary modification. (feeling focus, intent to acknowledge)

I am getting the impression that you are thinking that you are a bad person because you ate a lot of cheese at the party. (thought focus, intent to explore)

You set a goal to limit your intake of cheese to one ounce a day, but at the party you ate much more. (behavior focus, intent to challenge)

The communication analysis in the case study at the end of this chapter presents an examination of the focus and intent of responses made to a client. Note again that a particular interpretation of an intent or a focus can be debated because communication is influenced by a multitude of factors—cultural orientation, body language, and voice inflection to name a few. However, by studying intent and focus evaluations, student counselors can enhance their abilities to formulate counseling responses. See Table 3.2 for a list of possible counseling intents for common counseling responses.

[2*]Note that this is a general question allowing the client to choose a more specific focus (that is, feeling, thought, or behavior).

Table 3.3   Effective and Ineffective Counselor Nonverbal Behavior

Nonverbal Mode of Communication	Ineffective Nonverbal Counselor Behavior	Effective Nonverbal Counselor Behavior
Space	Distant or very close	Approximately arm's length
Posture	Slouching; rigid; seated leaning away	Relaxed but attentive; seated leaning slightly toward
Eye contact	Absent; defiant; jittery	Regular
Time	You continue with what you are doing before responding; in a hurry	Respond at first opportunity; share time with client
Feet and legs (in sitting)	Used to keep distance between the persons	Unobtrusive
Furniture	Used as a barrier	Used to draw persons together
Facial expression	Does not match feelings; scowl; blank look	Match your own or other's feelings; smile
Gestures	Compete for attention with your words	Highlight your words; unobtrusive; smooth
Mannerisms	Obvious; distracting	None or unobtrusive
Voice: volume	Very loud or very soft	Clearly audible
Voice: rate	Impatient or staccato; slow or hesitant	Average or a bit slower
Energy level	Apathetic; sleepy; jumpy; pushy	Alert; stay alert throughout a long conversation

Use Effective Nonverbal Behavior

A great deal of our communication, up to 85 percent, is based on body language.[5] Generally people learn to trust perceptions of nonverbal behavior over verbal remarks as a better indication of the meaning of messages. In other words, people inherently believe the adage "actions speak louder than words." This tendency probably occurs because much of our body language is under unconscious control, whereas verbal statements are more likely to be deliberate and subject to censorship.

Developing good nonverbal behavior is an extremely important skill for counselors to create an environment conducive to the development of a trusting relationship. Facilitative body behaviors have been shown to result in positive client ratings, even in the presence of ineffective and detracting verbal messages.[6] A counselor's joyful expression or attentive silence can communicate an understanding of a client's emotional state. Match the intensity of your own verbal and nonverbal messages with each other to create congruence. Communication will be hampered by unproductive

nonverbal behavior such as frequently looking at a watch, yawning, slouching, tapping or swinging feet, or playing with hair or a pencil.[5] These distracting behaviors indicate that the listener is not interested in what the speaker has to say.

Table 3.3 lists effective and less effective nonverbal behaviors; however, before condemning any behaviors,

The first client I ever counseled was in a nutrition counseling class. My first impression was that this person was not communicative, as she sat with her arms crossed in front of her. I thought she was putting up a barrier. Later she put her head in her hand and rested her elbow on the table. For some reason I instinctively followed both behaviors, even though they would not be found on a counseling etiquette list. This client gradually opened up, and I felt mimicking her behavior contributed to the harmony that developed between us.

EXERCISE 3.2   Identifying Effective Nonverbal Behavior

Work with a colleague and do this exercise twice, exchanging roles as speaker and listener. The speaker engages in a monologue for five minutes on what it was like growing up in his or her home. The listener facilitates the discussion using only body language. No verbal sounds, such as "Mm-hmm," are permitted. After completing this exercise, exchange information with your associate as to what was done that communicated listening and encouraged you to keep talking. What would you have said if you could have talked while you were the listener?

Source: Miller WR, Rollnick S. Motivational Interviewing. New York: Guilford; 1991, pp. 164–165.

the context of an encounter needs to be taken into consideration, including type of client, verbal content, timing in session, and the client's perceptual style.[7] The list should be considered as a guide, not as steadfast rules.

Harmonize Verbal and Nonverbal Behaviors

Your behaviors should also harmonize with your clients' expressive state. For example, a client who is animated and loud will have more trouble getting in synch with a counselor who has reserved body movements and a quiet voice. Body language harmony between two people is referred to as synchrony.

Mirroring and matching a client's body language have been advocated for business and sales personnel as a way to increase sensitivity and establish rapport.[8] Similarly, Magnus[9] suggests that a counselor mirror a client's silence behavior for those who are culturally accustomed to long periods of silence. In an investigation reported by Curry and Jaffe,[8] students who were able to calibrate their behavior to match their clients had more successful counseling interventions. These students matched behaviors, such as cocking of the head, or made responses incorporating words used by their clients. Care should be taken not to use this method to an extreme, otherwise, clients will feel as if they are being mocked. In order not to feel overwhelmed when learning this strategy, try to select only one aspect of your client's behavior to mimic.

Although this method is useful in most counseling situations, you may wish to modify the technique when a client's expressive state, such as one who is agitated or distracted, interferes with attending to the counselor and considering positive changes. Kellogg[10] suggests briefly harmonizing with your client and then shifting to a more attentive, focused manner which would encourage your client to become calmer and more focused.

For a nutrition counseling assignment, I visited an Indian Hindu temple for a ceremony, followed by a meal. The temple was extremely crowded, and in the beginning of my visit, people were busy preparing for the ceremony and meal. Several times I was physically pushed aside with no apology. I found myself getting angry about the whole experience until I discussed the situation with one of the women. I asked if maybe this was some kind of cultural thing, and the woman told me that no offense was meant. She explained that gentle pushing was common when their temple was crowded and there was a lot of work to be done. I still do not know if this is an Indian Hindu practice or just what happens at that particular temple, but I did feel much better after the conversation. In fact, I thoroughly enjoyed the experience, and I found the people at the temple to be warm and anxious to share and explain their culture to me.

Analyze Nonverbal Behavior of Your Client

Besides paying attention to your own nonverbal behavior, care should be taken to observe and interpret your client's body language. Clues regarding a client's feelings can come from body language, including expressions of autonomic nervous system reactivity (sweaty palms, flushed face, and so forth).[11]

Habit and culture complicate the overall task of interpreting nonverbal behavior, so counselors need to be wary of jumping to conclusions. Studies indicate that no single aspect of nonverbal communication can be universally translated across all cultural groups.[12] For example, nodding the head usually means yes in North America, but a single nod in the Middle East means no. More than seven thousand different gestures have been recognized,[13] thereby creating many opportunities for misunderstanding of particular cultural meanings. Magnus[9] suggests if you are unsure of a particular behavior, you should ask for clarification. For example, you could ask, "I notice that you are mostly looking down. Would you tell me what that means for you?"

EXERCISE 3.3   Video and Analyze Nonverbal Behavior

Video a five-minute conversation with a colleague or friend, during class or out of class. After the discussion, write down your feelings during the dialogue. Play the video twice—with the sound on and again with the sound off.

❑ In your journal, describe your nonverbal behavior each time. Was your nonverbal behavior congruent with your recorded feelings? Analyze your behavior. Do you have any distracting habits that communicate inattention (for example, biting lips, playing with hair, and so forth)? What did you learn from this experience?

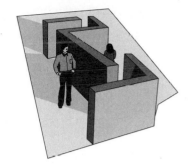

EXERCISE 3.4   Interpreting Common Nonverbal Cues Among North Americans

Select a person to act out the following behaviors. Write down your interpretation of the message portrayed by the behavior. The end of the chapter has a list of the common meanings of these behaviors for North Americans.

Behaviors

1 Hand over mouth
2 Finger wagging
3 Crossed arms
4 Clenched fists
5 Tugging at the collar
6 Hand over eyes
7 Hands on hips
8 Eyes wide, eyebrows raised
9 Smile
10 Shaking head
11 Scratching the head
12 Making eye contact
13 Avoiding eye contact
14 Wringing hands
15 Biting the lip
16 Tapping feet
17 Hunching over
18 Erect posture
19 Slouching in seat
20 Shifting in seat
21 Sitting on edge of seat

❏ How did you do? Do not feel too bad if you were not able to correctly identify all behaviors. Generally nonverbal behaviors are expressed in clusters, and we usually do not focus on one aspect of the cluster. We interpret nonverbal behavior based on a general impression.

Communication Roadblocks

Be aware of **communication roadblocks** and use them only when justified. Communication roadblocks are obstacles that counselors inadvertently put up that block self-exploration. They happen when counselors impose their own views, feelings, opinions, prejudices, and judgments. According to Miller and Jackson,[14] the underlying message of the counselor is, "Listen to me because I know better, I'm more important, or there is something wrong with you." Twelve types of responses that create roadblocks have been identified by Gordon[15] and elaborated on by Miller and Jackson.[14] See Table 3.4. A nutrition counselor using roadblocks in a dialogue with a client is illustrated in Table 3.5.

Roadblocks can be used effectively in the counseling process; however, many times they are employed too soon or too often. They are frequently

made with good intentions and not meant to block communication. Nevertheless, a counselor needs to be aware of their affect of blocking, stopping, diverting, or changing direction of communication. There are times in a counseling session that you do want to take a new direction and one of the responses in Table 3.4 would be appropriate. Generally, this would be after you believe you have listened carefully and understood your client's "story."

EMPATHY

Emphasis for developing an effective relationship is often placed on the first meeting and at the beginning of each session. However, a productive relationship needs to be continuously nurtured, for this process can in itself be an instrument of change. The core of such a relationship is the counselor's ability to experience and show **empathy**. Without this quality a therapeutic relationship cannot move forward.[16] The degree of empathy demonstrated by counselors has been shown to significantly affect outcomes in a study on drinking behavior.[17]

Empathy is a true understanding of another's unique perspective and experience without judging, criticizing, or blaming.[17,18] To allow empathic insight to enhance counseling, the counselor becomes immersed in another's experience without losing one's own sense of self. By maintaining a separate perspective, a counselor can gain insight for designing worthwhile interventions. To help explain the process of empathy, Murphy and Dillon[16] offer the following clarifications:

1. *Empathy is not sympathy. Sympathy is what I feel toward you; empathy is what I feel as you.*

Table 3.4   Examples of Roadblocks

Response	Examples
1. Ordering, directing, or commanding	"Don't say that." "Go right back and tell her . . ."
2. Warning or threatening	"You're really asking for trouble when you eat like that." "You better get your blood pressure down." "It is risky to carry around so much weight."
3. Giving advice, making suggestions, or providing solutions	"What you need to do . . ."
4. Persuading with logic, arguing, or lecturing	"Yes, but . . ." "Let's reason this through . . ."
5. Moralizing, preaching, or telling them their duty	"You should . . ." "You really ought to . . ."
6. Judging, criticizing, disagreeing, or blaming	"You're wrong. It is too bad that you can't . . ." "You did this to yourself." "You have only yourself to blame for this condition."
7. Agreeing, approving, or praising	"You did the right thing." "You're doing well at . . ." "You are absolutely right."
8. Shaming, ridiculing, or name-calling	"How foolish can you be!" "You are acting like a child." "You should be ashamed of yourself."
9. Interpreting or analyzing	"You know what your real problem is?" "I know what's troubling you." "You didn't really mean to do that."
10. Reassuring, sympathizing, or consoling	"Everything is going to be all right." "You will have your cholesterol down in no time." "Before you know it, it will all be over."
11. Questioning or probing	"Why did you say that?" "How did you come to that conclusion?"
12. Withdrawing, distracting, humoring, or changing the subject	"We can talk about that next week." "Let me tell you about what happened to me." "Look at how hard the rain is falling."

Source: Miller WR, Jackson KA. Practical Psychology for Pastors. 2d ed. Englewood Cliffs, NJ: Prentice Hall; 1995.

Table 3.5   Example of a Counselor Using Road Blocks

Roadblock	Speaker	Response
	CLIENT	*Everyone is getting on my back about my cholesterol levels—my wife, my doctor, my brother.*
Disagreeing	COUNSELOR	*They have your best interests in mind.*
	CLIENT	*I suppose they're right, but I feel fine.*
Warning	COUNSELOR	*You feel fine now but a cholesterol level of 300 is nothing to take lightly.*
	CLIENT	*I guess I have to do something about my diet.*
Agreeing; giving advice	COUNSELOR	*I think so, too. Eating the proper foods can really help bring down cholesterol levels.*
	CLIENT	*But it doesn't thrill me to give up meatballs and pizza.*
Reassuring; making suggestions	COUNSELOR	*It really isn't that bad. It's true that you probably can't have them as often as you have been eating them, but they could be worked into a meal plan. Now, have you ever tried eating soy foods, like tofu?*

Compare this dialogue with the reflective listening example in Chapter 2.

2. *Empathy is much more than just putting oneself in the other person's shoes. Empathy requires a shift of perspective. It's not what I would experience as me in your shoes; empathy is what I experience as you in your shoes.*

3. *Empathy requires a constant shifting between my experiencing as you what you feel, and my being able to think as me about your experience. (p. 88)*

Empathy will not have a meaningful impact on a counseling relationship unless effectively communicated both verbally and nonverbally. After a client perceives that she or he has been accurately seen and heard by another, a supportive environment is created conducive to growth and finding solutions. The skills reviewed in this section will aid in communicating this message. However, the greatest challenge of learning empathic skills lies in the integration of these skills into an interpersonal style that feels genuine to the counselor and is perceived as such by the client.[11]

Being able to empathize with another individual requires the ability to hear and sense the experiential world of that person.[5] This can be a challenge when a client's experiences are totally different from your own. However, empathy is a developmental process that can be consciously fostered and strengthened by expanding life experiences with people who are different from ourselves.[16] This process can also be supported through indirect encounters, through watching a movie, play, or interview, or reading a biography or novel. For example, movies such as *Philadelphia Story* or *Hoop Dreams* and books such as *Angela's Ashes* by Frank McCourt, *The Scalpel and the Silver Bear* by Lori Alvord and Elizabeth Van Pelt, or *When I Was Puerto Rican* by Susan Sheehan could be the conduit for such an experience. In this way the range of reactions that people have to various situations can be learned.

Many skills can be employed to foster a helpful empathetic counseling interaction. The next section describes some of these in detail.

BASIC COUNSELING RESPONSES

Nutrition counselors need a fundamental knowledge of counseling responses. The following list of basic responses is geared to accomplish three objectives: (1) to develop productive relationships, (2) to enhance listening and exploring to understand clients' messages—their needs and concerns, and (3) to provide the tools to utilize motivational strategies covered throughout this text. The following responses have been identified as being particularly useful in the health care arena (because a particular response may be known by several names, alternative terms are given in parentheses):[11]

1. Attending (active listening)
2. Reflection (empathizing)
3. Legitimation (affirmation, normalization)
4. Respect
5. Personal support
6. Partnership
7. Mirroring (parroting, echoing)
8. Paraphrasing (summarizing)
9. Giving feedback (immediacy)
10. Questioning
11. Clarifying (probing, prompting)
12. Noting a discrepancy (confrontation, challenging)
13. Directing (instructions)
14. Advice
15. Allowing silence
16. Self-referent (self-disclosing and self-involving)

EXERCISE 3.5 Entering the World of Others

Think of a time in your life when you were able to enter into the emotional world of a person whose life experiences were alien to you.

❏ Describe the experience in your journal; explain whether in any way the experience influenced your ability to empathize with others.

Relationship Building Responses

The following responses are particularly useful for building a relationship:

- Attending
- Reflection
- Legitimation
- Respect
- Partnership
- Personal support

Attending (Active Listening)

Attending is the most basic skill on which all other counseling skills are built. This involves giving undivided attention to your clients, listening for verbal messages, and observing nonverbal behavior. Your focus is on what you see and hear, not on what you know. This allows you to understand your clients' needs and concerns and how they view the world. Many attending behaviors are nonverbal but some nondescript verbal sounds need to be used in order to convey an impression of being engaged. Ivey et al.[5] identify four key components of attending behavior:

- **Eye contact.** Look at your client during dialogs. Refrain from staring, and permit natural breaks. Note, there are cultural variations in the acceptability of eye contact.
- **Attentive body language.** In North American culture, this generally means a slight forward trunk lean with a calm, flexible posture, and an empathetic facial expression. Gestures should be relaxed but kept to a minimum.
- **Vocal style.** Speech rate, volume, and tone should indicate concern.
- **Verbal following.** Give brief verbal and nonverbal responses, such as nods or an occasional "Hmm-hmm" or "Yes, I see," to indicate that a client's message has been received. Responses should relate to the topic.

Listening

Nature has given men one tongue and two ears, that we may hear twice as much as we speak
—**Epictetus**

Sometimes listening coupled with attending behavior is referred to as *active listening*. Actually, listening and attending are interrelated skills. One does not come without the other. Both are essential for the development of rapport and the communication of empathy. Murphy and Dillon[16] explain the interrelationship of listening, attending, and empathy:

> *It is important for clinicians to create an ambiance of focused attention in which meaningful communication can occur. Clinicians attend in order to listen; they listen in order to understand. Understanding contributes to empathy, and empathy engenders a readiness to respond. Thus, focused attending is an essential component of the therapeutic process. (pp. 55–56)*

Nutrition counselors have been criticized for controlling too much counseling time by doing most of the talking and spending too much time giving diet instructions and advice.[19] In one study, counselors who talked a great deal during sessions were described as unhelpful, inattentive, non-understanding, and disliked by the client.[20] A counselor who is doing most of the talking may have "missed the boat." Inattentive counselors solve problems and address issues important to themselves, which may not be of concern to their clients. Only by attending and listening can counselors accurately hear and understand their clients, respond appropriately, and find effective interventions. Good listening skills and attentive behavior indicate caring and concern, creating the impression that the counselor is capable and effective.

The problem for nutrition counselors, as for most people, is that listening skills are not well developed. Active listening is not simply a matter of hearing words but rather hard work requiring focused attention and concentration. Curry-Bartley[21] has identified three essential components of effective listening. See Table 3.6.

Reflection (Empathizing)

Reflection is labeling a client's expressed verbal and nonverbal emotion. When a counselor has accurately sensed an emotional state and has effectively employed reflection responses, clients feel understood, thereby facilitating self-acceptance and self-understanding. Teyber[22] explains the importance of understanding in a counseling relationship: "Clients begin to feel that they have been seen and are no longer invisible, alone, strange, or unimportant. At that moment, the client begins to perceive the therapist as someone who is different from most other people and possibly as someone who can help" (p. 49). No matter how empathic you feel, your client will not know if you do not verbally acknowledge your client's feelings.

Table 3.6 Essential Components of Effective Listening

Openness	Good listeners are willing to allow others to influence their perception of the world. Personal biases are put aside to hear viewpoints that could be in conflict with one's own belief system.
Concentration	Conscious attention needs to be focused on the speaker while tuning out everything else including fears, rational and irrational thoughts, and peripheral noises or activities.[21] If you have not been fully listening, your body language is likely to portray the fact unconsciously. This creates a barrier indicating to your client that you are not particularly interested in what he or she has to say. The most common reasons for interference with attention include the following:
	Lag time self-talk. An individual with an average intelligence can process information at speeds approximately five times faster than human speech, creating time for unproductive mental dialogue such as "Where did she buy those lovely earrings? They would go perfectly with the dress I bought for my cousin's wedding. Oh, why is my cousin marrying such a dingbat?"
	Rehearsing rebuttals. Using the extra mental capacity to rehearse rebuttals or questions will also break concentration.
	Assumptions. Assuming you know a solution and deciding that what the client has to say is uninteresting or irrelevant can interfere with communication as well.
Comprehension	By attending to the first two skills of listening—openness and concentration—the counselor increases the likelihood of comprehending the meaning and importance of what was said.

EXERCISE 3.6   Attending Success

Think of a time you were telling a story to someone and it was obvious that the person was engaged. In your journal describe this experience and explain what you felt during the encounter. Did this attentive behavior surprise you? What effect did this encounter have on your relationship?

Although your body language may convey empathy, nonverbal signals could be missed or misinterpreted, especially if a cultural difference exists.

Steps in Reflecting

The following steps will help you reflect more effectively as you communicate with clients.

1. Correctly Identify the Feeling Being Expressed. There are five major feeling categories: anger, fear, conflict, sadness, and happiness. Table 3.7 presents a list of commonly used feeling words at three levels of intensity.

Listening Guidelines for Counseling

- Remind yourself to focus and concentrate before each session.
- Listen for meaning, not just words.
- Use thinking-speaking lag time to examine and comprehend client's meaning.
- Avoid judgments. Be inquisitive and keep an open mind.
- Do not allow your mind to drift; bring your focus back to your client if your thoughts start to wander.
- Use verbal (uh huh, go on, I see) and nonverbal prompts (head nods, open face) to encourage talking.
- Maintain good eye contact, if culturally appropriate.

This step requires careful listening and close observation of nonverbal behavior and voice quality. Sometimes you will need to rely on your intuition.[23] You can also imagine how you would feel in a similar situation. For example, consider this scenario:

Client: *Now that I have diabetes, I have to think about what I eat all the time. I don't know if I will ever learn to cope with this. There are so many things I have to do in life already. The children need constant attention, and I have a stressful job. It just doesn't seem possible to think constantly about my blood sugar and insulin all day.*

Feeling: *Overwhelmed*

Emotions are easy to identify for those who are demonstrative by nature and vividly display their feelings. However, people who present themselves in a straightforward and businesslike manner are not so easily understood. Dubé et al.[24] suggest using the following question as a way to open the door to a discussion of feelings: "How has this whole illness (problem) been for you and your family—I mean, emotionally?"

EXERCISE 3.7   Listening Awareness

Over the next two days, choose three distinctly different listening encounters and represent each as an *X*, *O*, and *R*. For example, listening to your mother during dinner could be represented as an *X*, listening to your friend on the telephone could be an *O*, and listening to your psychology professor could be an *R*. These are illustrated on the first line. Put the symbols in the following continuum categories on the place that best fits your style of listening for each situation. I could describe myself as:

am alert.............X........R........O ....am bored

feel nonjudgmental ....feel judgmental

feel calm...feel volatile

listen to emotional messages ....listen only to facts

listen attentively ....give fake attention

think, then respond ....react before thinking

am in the here and now ....am occupied with past or future

❑ In your journal, describe the context and the participants of the listening encounter, and identify the corresponding symbol on the continuum. Compare and contrast the experiences. What implications does this experience have for your counseling endeavors?

Source: Adapted from Curry-Bartley KR, The art of science and listening. Topics in Clinical Nutrition. 1:18–19 © 1986 Aspen Publishers. Used with permission from Wolters Kluwer Health.

---

2. Reflect the Feeling You Have Identified to the Client. Drop the tone of your voice at the end of the statement; do not bring it up as if you are asking a question. Questions give a slight indication that you think the client should not feel that way and should reconsider his or her feelings.[17] A reflection statement rather than a question communicates understanding and acceptance. As an illustration say the following to yourself:

Less effective: *You're really angry that the burden of your father's care has been put on your shoulders?* (voice turns up at the end)

More effective: *You're really angry that the burden of your father's care has been put on your shoulders.* (voice turns down at the end)

You may also want to begin your sentence with a *stem tentative phrase.* However, care should be taken not to overdo such phrases because they can become annoying, especially if the same one is used repeatedly. Here are some examples:

"Perhaps you are feeling . . ."

"I imagine that you're feeling . . ."

"It appears that you are feeling . . ."

"It sounds as if . . ."

"It seems that . . ."

In response to the client who seems overwhelmed in the prior example, you might say, "It seems that you are feeling overwhelmed with trying to fit the care of diabetes into your life."

3. Match the Intensity of Your Response to the Level of Feeling Expressed by the Client. This can be done by choosing an appropriate word in Table 3.7 or by using modifying words such as a little, sort of, or somewhat to soften the response or really, very, or quite to make the feeling response stronger. Consider the following exchange:

Client: *I hate myself! I am such a jerk! I sit in front of the television eating junk food all night!*

Counselor (less effective): *It sounds as if you are slightly annoyed with yourself.*

Counselor (effective): *It sounds as if you are very angry with yourself.*

When in doubt, it is better to undershoot rather than overshoot your response. The effect of overstating a feeling can be a denial of the feeling

Table 3.7 Feeling Words

Relative Intensity of Words	Feeling Category				
	Anger	**Conflict**	**Fear**	**Happiness**	**Sadness**
Mild feeling	Annoyed Bothered Bugged Irked Irritated Peeved Ticked	Blocked Bound Caught Caught in a bind Pulled	Apprehensive Concerned Tense Tight Uneasy	Amused Anticipating Comfortable Confident Contented Glad Pleased Relieved	Apathetic Bored Confused Disappointed Discontented Mixed up Resigned Unsure
Moderate feeling	Disgusted Hacked Harassed Mad Provoked Put upon Resentful Set up Spiteful Used	Locked Pressured Torn	Afraid Alarmed Anxious Fearful Frightened Shook Threatened Worried	Delighted Eager Happy Hopeful Joyful Surprised Up	Abandoned Burdened Discouraged Distressed Down Drained Empty Hurt Lonely Lost Sad Unhappy Weighted
Intense feeling	Angry Boiled Burned Contemptuous Enraged Fuming Furious Hateful Hot Infuriated Pissed Smoldering Steamed	Ripped Wrenched	Desperate Overwhelmed Panicky Petrified Scared Terrified Terror-stricken Tortured	Bursting Ecstatic Elated Enthusiastic Enthralled Excited Free Fulfilled Moved Proud Terrific Thrilled Turned on	Anguished Crushed Deadened Depressed Despairing Helpless Hopeless Humiliated Miserable Overwhelmed Smothered Tortured

Source: From Helping Relationships and Strategies, 2nd ed. by D. Hutchins and C. Cole. Copyright © Brooks/Cole. Reprinted by permission of Wadsworth.

EXERCISE 3.8   Practice Reflection Responses

Over the next two days, make it a point to practice acknowledging feelings with friends, family members, coworkers, supermarket clerks, and others.

❑ Record in your journal three of the experiences describing when, where, and what happened. What is your impression of the effect of reflection statements?

and backing away from a feeling discussion.[14] Understating does not tend to have that effect; a client is likely to clarify the level of feeling—for example, "A 'little' happy! I'm elated!"

4. You Should Respond to the Feelings of Your Client, Not to the Feelings of Others.[23] Take a look at this sample dialogue:

Client: *I was upset last night at dinner when my sister kept talking about my weight.*

Counselor (ineffective): *Your sister is feeling uneasy about your weight.*

Counselor (effective): *You feel annoyed when someone nags you about your weight.*

Reflection responses have been presented here as a relationship-building skill; however, this response has other advantages, too. For example, Laquatra and Danish[19] emphasize the use of this response as a technique to encourage your clients to continue talking and to help clarify their problems. This clarification also helps counselors understand problems from the viewpoint of their clients.

Legitimation (Affirmation, Normalization)

Reflection responses involve identification and acknowledgment of a client's feelings; *legitimation* communicates the acceptance and validation of the client's emotional experience.[11] The counselor acknowledges that it is normal to have such feelings and reactions. Usually it is a good idea to receive verification that you have correctly identified your client's feelings before making a statement that the feelings are legitimate and make sense to you. For example:

Counselor: *It seems to me that you are feeling overwhelmed with the whole ordeal of this illness. (empathizing response)*

Client: *Yeah, it stinks.*

Counselor: *I can understand why you would feel like this. Anyone would under the circumstances. (legitimation statement)*

This statement could also be made without first identifying the feeling if your client is especially communicative about his or her feelings. For example:

Client: *This is terrific! I am so happy! Exercise and eating all that rabbit food have really paid off. I actually enjoy all those fruits, vegetables, and whole grains. My blood pressure is so good my doctor is taking me off medication. This is wonderful!*

Counselor: *You deserve to feel so happy after getting such good results and working so hard to make changes in your life.*

Respect

Respect for your clients and their coping abilities is implied by attentive listening and nonverbal behavior. However, explicit statements of respect show genuine appreciation for the worth of the client and can help build rapport, improve the client-counselor relationship, and help your client cope with difficult situations.[11] Respect responses include words of appreciation on the ability to overcome adversity and adjust to difficult situations. The fact that the client has come to a nutrition counseling session can in itself show positive coping behavior. The person can be complimented on a willingness to search for nutrition interventions to deal with the problem. Examples of statements a nutrition counselor could make include these:

"I am impressed that you are here searching for ways to lower your cholesterol levels through diet."

"Despite the fact that you have so many responsibilities, you have done a great job of making exercise a priority in your life."

"You have done such a terrific job of keeping a food journal."

Personal Support

You should make clear to your clients that strategies for solving their problems are available, and you are there to help them implement those strategies. Your clients should know you want to help. For statements of support to have a positive effect on building a relationship, they need to be honest. The following is an example of a supportive statement:

"There are a number of dietary options and strategies available to get your diabetes under control. I look forward to working with you to make that happen."

Partnership

Successful interventions begin with establishing a collaborative relationship with your client. This means that the client and counselor respect each other and work together to find solutions. The following is an example of a partnership statement:

"I want us to work together to find and implement strategies that will work for you. After we talk about your problems and strengths, we will look at some options for finding a solution."

Mirroring (Parroting, Echoing)

Parroting or *mirroring* responses repeat back to a client exactly what was said or with few words changed. This response lets a client know you are listening and encourages the person to keep talking and exploring. Care should be taken not to overdo this response, or else your client is likely to talk less. Here's an example of a mirroring exchange:

Client: *I had chocolate hidden under my bed.*

Counselor: *You had chocolate hidden under your bed.*

You may also consider echoing back a key word or key words, especially if your client has used the word repeatedly or if you want further clarification of the word. For example, your client may say to you, "My diet is a disaster." You could echo back, "a disaster" with turning up your voice or you could simply ask a question, "What do you mean by 'a disaster'?"

Paraphasing (Summarizing)

Paraphrasing responses are a rephrasing of the content of what the client said and meant. They can summarize prior statements or several statements of a conversation. Remember that the model of communication illustrated in Figure 3.1 indicates that alternative meanings are possible for any statement. These responses are a counselor's best guess as to what a client actually means. Paraphrasing responses let clients know you are listening, encourage clients to continue talking, and assist clients in clarifying concerns to themselves and the counselor. You could begin this response by simply rephrasing, or you may wish to use a lead-in such as the following:

"What I hear you saying is . . . ."
"Let me see if I understand this correctly . . . .."
"So what you are thinking is . . . ."

You should not be concerned if you have missed the actual meaning because the client's typical response will be to clarify back to the counselor the intended meaning. The following is an example of a counselor using paraphrasing responses:

Client: *I was really surprised to find that my cholesterol jumped to 300. It had always been around 190. I guess now that I am well into menopause it is going to be harder to control. I thought my diet was pretty good, so this is really annoying.*

Counselor: *It must seem unfair to you to have this happen.*

Client: *Yeah. When we go out to eat with my brother-in-law, he always orders an expensive steak, loads the butter on the baked potato, and never asks for salad dressing on the side. I don't know his cholesterol level, but he is alive and kicking. Well, these are the cards I was dealt in life, and I guess I can live with it.*

Counselor: *Even though you have some negative feelings about what has happened, you think that you can cope with what has to be done to get your cholesterol under control.*

Client: *That's right. I've been reading about good foods to eat to lower cholesterol levels. I am eating oatmeal and drinking soy milk just about every day. One problem I have is with the nuts. I buy double chocolate chips to mix with the nuts to make them tastier, but I find that when I feel stressed, I am going for the chips and eating too many of them. I have got to stop that.*

Counselor: *You're looking for a way to put an end to eating the chips.*

Client: *Not really. I am looking for a way to end eating them out of control, but I guess if I am not successful, I should stop buying them.*

Note that in the last statement the counselor did not fully pick up on the client's intended meaning, but that did not present a problem. The client simply clarified her meaning and even continued with a deeper self-exploration.

Paraphrasing an extended interaction is referred to as *summarizing* (Chapter 4 provides examples of counseling summaries). Periodic summaries of what has transpired in a counseling session can be used to transition to a new topic; integrate client behavior, thoughts, and feelings; provide closure at the end of a session; furnish a vehicle to elicit self-motivational statements; and allow checking for any misunderstandings.[17] They communicate a sense that the counselor is listening and trying to understand. If needed, this technique can provide a "therapeutic breathing space" for counselors to make a decision on what the next step should be in the session.[11] The following are possible lead-ins to summaries:

"Because our session is about to come to a close, I would like to review what we covered today so we can agree on where we are and where we are going."

"Let me summarize what we have covered so far and see whether we are in agreement."

Giving Feedback (Immediacy)

Giving feedback is telling clients what you have directly observed about their verbal and nonverbal behavior. Often this is not new information to your client, but by pointing out the behavior, you are inviting the client to examine the implications and increase self-awareness. Haney and Leibsohn[1] provide the following guidelines for giving feedback: be positive and specific; note behavior, not traits; and do not put the client on the defensive. Here are a couple examples:

> "When you said you wanted to give up drinking so much coffee, you looked sad."

> "I noticed that you started to wring your hands when you started to talk about your mother."

Questioning

Questions are effective responses for gathering information, encouraging exploration, and changing the focus of a discussion. See Exhibit 3.1 for a list of questions historically found effective for clinicians.[16] Although the use of questions appears to be an easily learned response, the challenge is asking appropriate questions and timing them well. Novice as well as seasoned nutrition counselors have been criticized for asking too many questions to fill silence or to satisfy curiosity.[25] Questions should be asked only if there is a particular therapeutic purpose in mind. They interrupt concentration and lead to a discussion of concerns that interest the counselor but not necessarily the client.

Useful Questions

Closed-ended questions elicit yes or no or short answers. They commonly begin with *is, are, was, were, have, had, do, does,* or *did.* For example: "Do you eat breakfast?" or "Have you ever counted calories?" These types of questions do not allow for expansion of ideas so their usefulness is limited. Additionally, they tend to influence answers because the questions hint at an expected response. For example, the question "How many fruits do you eat each day?" assumes fruits are eaten each day, and it would not be easy to give the answer "Zero." Closed-ended questions are useful for ending a lengthy discourse or for soliciting a specific answer.

Open-ended questions give a person a great deal of freedom to answer and encourage elaboration. They commonly begin with *what* or *how.* For example: "How did your goals for the week work out?" or "What problems do you have with the diet plan?" Clients are not likely to feel threatened by open-ended questions, and they communicate interest and trust. This type of question is generally preferable when possible. However, open-ended questions can lead to rambling, lengthy answers.

Funneling questions are a sequence of questions beginning with a broad topic and narrowing down to a specific item. For example: "Can you tell me what a typical day looks like regarding your food intake?" "Do you generally drink anything with your lunch?" "What kind of milk do you put in your coffee?"

Problematic Questions

Why questions are generally to be avoided because they often sound judgmental, put clients on the defensive, and seem to require an excuse. For example: "Why didn't you follow your plan and eat the orange when you came home from work?" Clients are likely to respond with an evasive answer, which provides no useful information, such as "I don't know." The client is likely to feel ashamed or unnatural. If you believe an investigation into motives for behaviors or feelings is warranted, Murphy and Dillon[16] recommend the following questions: "As you look back, what do you think was going on?" or "What do you think caused that to happen?" These types of questions set a state of curiosity and encourage clients to explore their concerns from various angles.

Multiple questions ask clients to respond to more than one question at a time. For example, "How did your goals to increase calcium work out? What did your family think of the changes?" Clients will become confused trying to decide which question to answer first.

Question-answer traps, termed by Miller and Rollnick,[17] are a series of questions causing clients to feel as if they are under interrogation. See the following example:

Counselor: *You're here to talk about a diet to lower your blood pressure. Is that right?*

Client: *Yes, that's right.*

Counselor: *Did your doctor give you any information about diet and blood pressure?*

Client: No.

Counselor: *Did your doctor talk to you about your weight?*

Client: *Yes, she said I should lose about fifteen pounds.*

Counselor: *Do you want to work on losing weight?*

Client: *Yes.*

Counselor: *Have you ever been on a diet before?*

Client: *Yes, once or twice.*

Counselor: *Did you lose weight on the diets?*

Client: *Some, but I gained it back.*

Counselor: *Did you use a selection of foods from food groups on the diets you followed?*

Client: *Yes.*

The question-answer trap is not effective for several reasons. It encourages a client to give short answers, does not allow for much in the way of self-exploration, does not elicit much information, and sets up the counselor as the expert who will provide the magical solution after enough questions have been asked. Even a series of open-ended questions can lead to a less obvious trap. Miller and Rollnick[17] suggest a general rule of no more than three questions in a row.

Clarifying (Probing, Prompting)

Clarifying responses encourage clients to continue talking about their concerns in order to be clear about their feelings and experiences. Stories that are not clear to you may also not be clear to your client. Clarifying responses can take the following forms:

- Communicate "Tell me more" through body language such as nods of the head or short comments such as "Uh-huh" or "Go on."
- Use trailing words such as "and . . ." or "and then . . ." or the last few words spoken by your client.

EXHIBIT 3.1   Tried-and-True Questions

1. What brings you here to see me? (reasons for coming)

2. What caused you to seek help now? (timing of request)

3. How did you think I might help? (anticipation of the experience)

4. What would you like to get done today? (client as the driver)

5. Where would you like to begin? (client as the driver)

6. Can you tell me more about your situation? (elaboration of person or situation)

7. Who else is available as a support or to help in this? (situation dynamics)

8. Who, if anyone, is making things more complicated just now? (situation dynamics)

9. Have you ever spoken to a nutrition counselor before? If so, how did it go? (vision of the work)

10. Are there other things you haven't mentioned yet that would be important for me to know? (elaboration)

11. What will we look for, to know that the changes you want have actually taken place? (concretizing desired outcome)

12. What is it like for you to be talking about these things with me? (relationship building, checking in)

13. How does the work we're doing compare with what you thought it would be like? (checking in)

14. Are there any other things that should be on our list of things to talk about? (double-checking)

15. Does what I'm saying make sense? (clarifying)

16. Could you put that in other words so I can understand it better? (not knowing)

17. Can you say more about that? (elaboration)

18. What is your response to this information? (checking in and encouraging integration)

EXERCISE 3.9 **Practice Using Questions**

Practice with a partner, each taking turns discussing an individual concern (for example, getting the children to eat breakfast, switching from whole milk to skim milk, and so forth). One person should take on the role of client and the other counselor. First the client should ask five closed questions and then five open questions.

❑ Explain what happened. How did you feel during each set of questions?

- Ask clarifying questions, such as the following:[26]

"Can you explain that in a slightly different way?"

"Can you think of an example in your life where that happens as well?"

"Let me make sure I understand what you have said because it seems to me that you are sharing something very, very important."

"Anything else?"

"Could you please clarify something you said earlier regarding . . ."

"I am very interested in something you said before about . . . . Could we talk a little more about it?"

Noting a Discrepancy (Confrontation, Challenging)

Individuals often experience a great deal of resistance to making lifestyle changes and giving up comfortable behavior patterns. This is likely to lead to denial or distortions. Commonly observed discrepancies occur between two statements, verbal and nonverbal communication, stated feelings and the way most others would feel, and what the client states as a value and his or her actual behavior.[25] A counselor who chooses to ignore discrepancies that lead to self-defeating or unreasonable behavior misses an important opportunity to help the client resolve ambivalence to change.[25] Contradictions and inconsistencies are often not obvious, but when they are brought to our attention, they can be illuminating. When counselors point out discrepancies, clients are better able to examine their ambivalent thoughts, feelings, and behaviors. As a result, clients are better prepared to make decisions and move forward in making lifestyle changes.

There are several ways of noting a discrepancy; some are more apt to bring up resistance than others. The intent should never be to criticize or attack. When the counselor comes across as caring and nonjudgmental, the client is most likely to be able to see the discrepancy and work to resolve it. The following illustrate some ways to note a discrepancy:

- State observation without *but* (a softer approach).

Counselor: *You say you want to increase your level of physical activity, and you feel that you do not have time to exercise.*

- State observation with *but* (a harder approach).

Counselor: *When we talked earlier you said that you do not eat fruits, but you ate some of the oranges served after the Chinese meal.*

- Use "on one hand . . .; on the other hand, . . ." expression.

Counselor: *On the one hand, you say that you would cut off your right arm to lose weight; on the other hand, you say that you do not want to exercise.*

- Tentatively name the discrepancy (with seems, appears, could there be, have a feeling).

Counselor: *I know you said you understood; I also get the feeling that there are aspects of these instructions you would like to go over again.*

- Directly name it ("I see an inconsistency"; "I'm hearing two things at once").

Counselor: *I see an inconsistency between what you say about your concern for yourself because of your father's heart attack and your willingness to cut back on your fat intake.*

No matter what wording is used to address the discrepancy, it is important to follow up with an offer to explore it further. The following provides two possible ways to make the offer:

Counselor: *How about we explore these two sides?*

Counselor: *Can you tell me more about your concern for your health? Now tell me what gets in the way of cutting back on your fat intake.*

Directing (Instructions)

The *directing* response is telling a client exactly what needs to be done. In nutrition counseling, directives are often important components of the educational portion of a session. For example, a counselor may explain to a client with a liver or kidney condition how to calculate fluid intake, or a diabetes educator may need to instruct an athlete in a new way of balancing food intake and insulin before a workout.

When giving directives, it is important to be clear and concise and to determine whether the instructions were completely understood. To be sure the message has been accurately communicated, consider asking a client to repeat back the instructions.

Advice

Advice is providing possible solutions for problems. Nutrition counselors have sometimes been criticized for giving too much advice; however, going to the other extreme and not giving any clear advice may leave your client confused and floundering.

Advice from physicians has been shown to be successful for changing smoking, exercise, and alcohol behavior.[27,28] To be effective, advice should (1) be given in a nonjudgmental manner, (2) identify the problem, (3) explain the need to change, (4) advocate an explicit plan of action, and (5) end with an open-ended question to elicit a response from the client.[17,29] Advice should only be given when there is a clear understanding of the problem, previous attempts to deal with the difficulty have been investigated, and the counselor has definite ideas for possible solutions.[19] If these criteria have not been met, advice giving is likely to lead to a "Yes, but . . ." scenario.[25] For example:

Counselor: *You might try walking to increase your physical activity.*

Client: *I have tried walking, but there just isn't enough time in the day.*

Counselor: *How about walking at lunchtime?*

Client: *There just isn't anywhere for me to walk at work.*

Counselor: *You could walk in the parking lot.*

Client: *Yes, but I wear high heels to work.*

Counselor: *You could bring sneakers to work and put them on for walking.*

Client: *Yes, but then I would get all sweaty, and I would not be comfortable at work for the rest of the day.*

This is an example of a counseling session going down the tubes. Obviously the criteria for giving advice have not been met. If you find yourself in a "yes, but" scenario, stop giving advice, back up, and spend more time exploring your client's issues.

Because giving advice is a roadblock to self-exploration, the timing of recommendations is important. Generally advice should be avoided early in a session when the goal is to understand and clarify issues. Ideally a client makes a request for advice; research indicates, however, that clients do not frequently make the request.[30]

Another possibility is for the counselor to ask permission prior to giving advice. This makes the "yes, but . . ." response less likely. Offering the advice and then asking for a response encourages clients to truly consider the idea and even come up with ideas of their own. For example:

Counselor: *I have some ideas for increasing your physical activity. Would you like to hear them?*

Client: *Well, OK. They need to fit in my schedule though.*

Counselor: *Many of my clients who are busy like you find a way to fit a walk in at lunch or during a break. What is your response to this idea?*

Client: *That wouldn't work for me because there is nowhere to walk at my job. My neighborhood is much nicer. You know, my dog would love more of a walk in the morning. Maybe I could go farther.*

Allowing Silence

At times in a counseling interaction, silence is a valuable tool. Sometimes clients need space for internal reflection and self-analysis, possibly after an open-ended question. For a novice counselor, the thought of allowing silence may seem intimidating when there may already be a fear that silence will occur because of not knowing what to say. Also, previous social conditioning can lead one to feel that talking is preferable to silence, no matter what the content.[31] Using silence effectively is an art and a skill that comes with practice. However, by attending to your client and using good listening skills, the point at which to use silence is likely to naturally flow.

In nutrition counseling, the need for silent contemplation could occur after a client has been given the results of an evaluation, during instructions of a complex dietary regimen, or after an emotional outburst due to the demands of coping with a newly diagnosed illness. If it appears that your client needs some space to process information, then it would be appropriate to divert your eyes for the moment and not maintain eye contact. Effective silent periods could be about thirty to sixty seconds.

One way to break the silence is to repeat the last sentence or phrase spoken by your client. For example, if your client's last words were "This all means making some big changes in my life," then your response could be "Some big changes in your life . . ."

Another possibility is asking your client what she or he was thinking about during the silence. A counselor should not always feel compelled to break a silence. There are times when you may want to challenge your client to formulate a response.

Self-Referent (Self-Disclosing and Self-Involving)

In a counseling relationship, there are likely to be times when shifting the focus of attention to yourself can be advantageous. The benefits of self-referent responses include increasing openness, building trust, providing a model to increase client level of self-disclosure, developing new perspectives, and creating a more impersonal atmosphere.[7,25] If these responses occur too frequently or too soon in the counseling relationship, the result could be a "chatty" session or a perspective that the counselor is self-absorbed. In addition, a client, possibly due to a cultural perspective, may perceive that the counselor lacks discretion. Counselors should also review their "need" to share. Could this be because you have unresolved issues that need airing or are you searching for a friend? Neither of these reasons would be appropriate for a self-disclosing statement. Before using a self-referent response, a counselor should always assess whether the intended sharing will be in the best interest of the client.

Self-disclosing and self-involving responses represent two types of self-referent responses. A *self-disclosing* response involves providing information about oneself; generally this is related to coping experiences. For example, a nutrition counselor who is also a kidney dialysis patient might disclose that fact and commiserate that sometimes the diet requirements are frustrating. In another case, a counselor could explain how she herself incorporates exercise into her busy schedule.

Self-involving responses actively incorporate a counselor's feelings and emotions into a session. Laquatra and Danish[25] provide the following format for a self-involving response: I (the counselor) feel (name feeling) about what you (the client) said or did. These responses can be used to provide feedback or to sensitively confront.

Here are examples of both self-involving and self-disclosing responses:

Client: *My diet has worked so well that my doctor lowered my blood pressure medication.*

Counselor: *I am delighted that you are doing so well. (self-involving)*

Client: *You are going to be so disappointed with me. I ate a ton of chocolate and didn't eat enough vegetables this week. You must wonder what I am doing here.*

> When I was working as a hospital dietitian, I was assigned to give dietary guidance to a middle-aged man newly diagnosed with diabetes. When I walked into his room and explained the purpose of my visit, he exploded with anger, berating just about everything concerning the hospital and diabetes. When he finished, we were silent for a while, and then I acknowledged that he was obviously upset and had a right to be after all he had gone through. I asked if he would like me to come back at a later time. To my surprise, he wanted to go over the diet. We actually had a good session. I believe both the silence as well as the legitimation statement changed his emotional state and allowed us to explore dietary modifications.

Counselor: *I'm concerned that you appear to be making changes for me rather than for yourself. (self-involving)*

Client: *I've always been afraid of shots, and now I need to give myself insulin injections.*

Counselor: *One of the reasons I became a diabetic counselor was because I also have diabetes. I was just like you. I wondered how I could possibly give myself shots. (self-disclosing)*

As described in the Chapter 2 Case Study, "Helping Relationships," many nursing home employees were offended by John's remarks. This greatly impacted on their relationship with him and the quality of care John received. John never had visitors, and it appeared that he often made scenes to receive attention in order to counteract his loneliness.

Keeping these factors in mind, Table 3.8 lists various types of responses that could have been made by a nurse to address the relationship issue and hopefully to improve the situation. A focus and an intent are identified for each response. In some cases, your interpretation of the underlying focus and intent may be different from those listed here. Also, actual reasons that a particular response is formulated will be influenced by the total context of the counseling intervention, including personalities of the counselor and counselee and the history of the relationship. In addition, your decoding of a response will be based on your life experiences, which gives you a unique perspective. As Haney and Leibsohn[1] emphasize, counseling is an art, and as a result a certain amount of ambiguity is to be expected. However, the process of evaluating why and how a certain response was phrased helps counselors in training grasp the basics of the counseling process.

CASE STUDY   Communication Analysis of John's Interactions

This case study delves deeper into the communication difficulties described in the Chapter 2 case study between John and nursing home staff. To prepare for this investigation, reread the "Helping Relationships" case study in Chapter 2, and review the model of communication illustrated in Figure 3.1. The following diagram illustrates distortions that occurred during the encoding and decoding process associated with a typical comment made by John to a member of the nursing staff.

Table 3.8 Helping Relationships Case Study Response Analysis

Type of Response	Example of Nurse Response	Focus of Response[1]	Intent for Selecting a Response[2]
Attending	The nurse leans forward and shows interest when John commented on something that appeared on television.	Experience	Acknowledge
Empathizing	"You must feel confused about what kinds of comments women find flattering today as compared to what you knew to be true in the 1960s."	Feeling	Acknowledge
Legitimation	"You have a right to feel angry after all you have gone through."	Feeling	Acknowledge
Respect	"It must have been some experience to have been part of the Vietnam War. You must be proud of yourself."	Feeling	Acknowledge
Personal support	"John, I want you to know that I want to help you get along better with the staff here."	Behavior	Acknowledge
Partnership	"I hope we can work together to make your experience here better."	Experience	Acknowledge
Mirroring	John says the nursing staff doesn't pay enough attention to him, and the nurse repeats the words back to John.	Experience	Acknowledge
Paraphrasing	"So what you are saying is that even though I am willing to talk to you, you don't think the rest of the staff will go along with our plan."	Thought	Explore
Feedback	"I noticed that you used the word *babe* four times this afternoon when talking to the staff."	Behavior	Explore
Question—closed	"Does it ever bother you that the nurses find some of your comments offensive?"	Feeling	Challenge
Question—open	"How do you feel about the nursing staff here?"	Feeling	Explore
Clarifying	"John, tell me more about what you want to say when you think someone is annoyed with you."	Thought	Explore
Noting a discrepancy	"On the one hand, you say the staff doesn't care about you; on the other hand, you say they bring you special treats from home."	Experience	Challenge
Directing	"Before I see you tomorrow I want you to think of two ways to compliment a staff member that you do not believe will be offensive."	Thought	Challenge
Advice	"To every action there is a reaction. If you act in a respectful manner to the staff, they will be respectful in return."	Behavior	Challenge
Silence	The nurse remains silent for thirty seconds after John bitterly complains about the staff following a question regarding how he feels about the staff at the nursing home.	Experience	Acknowledge
Self-referent–self-disclosing	"We all have times when we say things that others perceive differently. It happens with my husband and me, but we talk it through and it works out."	Experience	Challenge

[1]Focus can be information or experience. Experience can be subdivided into feeling, thought, or behavior.

[2]Intent can be to acknowledge, to explore, or to challenge.

EXERCISE 3.10   Evaluating Focuses and Intentions

Continuing with the Helping Relationships Case Study communication analysis, you will explore a comment made by John about food and identify the focus and intent of responses made to him by a dietitian. Keep in mind John's complicated medical condition and a weight problem that is seriously affecting the quality of his life. Refer to your readings and the communication analysis illustrated in Table 3.8 to provide guidance for completing the response table in this exercise.

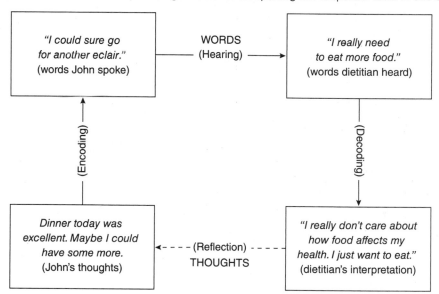

Type of Response	Example of Dietitian's Response	Focus of Response[1]	Intent for Selecting a Response[2]
1. Attending	The dietitian gently touches his arm, sits in a chair next to his wheelchair, and looks directly at him as he is speaking.		
2. Empathizing	"It's frustrating for you when you have food that you really enjoy, and the staff tells you that you can't have more of it."		
3. Legitimation	"You have a right to feel upset. Eating is the highlight of your day."		
4. Respect	"I think you have made some very difficult decisions in the past, and I will respect how you want to handle this serious weight issue."		
5. Personal support	"I want you to know that I am here to help you, even if that means making only baby steps toward your goals."		
6. Partnership	"If your weight is something you want to work on, you and I will work together to find a solution."		
7. Mirroring	John complains that all his weight is causing him a lot of problems, and the dietitian repeats his words.		
8. Paraphrasing	"So you are saying that you are willing to diet but that I shouldn't expect miracles."		
9. Feedback	"I noticed your voice became very soft when we talked about your dessert goal."		
10. Question–closed	"Are you willing to eat only one snack after dinner?"		

(continued)

EXERCISE 3.10 Evaluating Focuses and Intentions *(continued)*

Type of Response	Example of Dietitian's Response	Focus of Response[1]	Intent for Selecting a Response[2]
11. Question–open	"What do you think would work for you to help you lose weight?"		
12. Clarifying	"John, can we go back to talking about what happened at lunch yesterday? It seemed to go well and maybe you could tell me more about why it worked."		
13. Noting a discrepancy	"You say you would probably feel better if you lost some weight, but eating appears to be the best part of your day."		
14. Directing	"Before we meet for lunch tomorrow, I want you to tell me two things you think you can do to help you lose weight."		
15. Advice	"Losing weight will allow you to walk again and feel better."		
16. Silence	The dietitian asked John what he thought she could do to help him to lose weight. She waited during the silence until he was ready to answer.		
17. Self-referent– self-disclosing	"I've had times when I've had to lose weight and for me getting started is the hardest part of the process."		

[1]Focus can be information or experience. Experience can be subdivided into feeling, thought, or behavior.

[2]Intent can be to acknowledge, to explore, or to challenge.

REVIEW QUESTIONS

1. List six stages of skill development. Define *trait* and *skill.*

2. Explain three reasons why there could be distortions of a speaker's intended meaning.

3. Explain the use of focuses and intents for formulating counseling responses.

4. What is the value of harmonizing verbal and non-verbal behaviors with a client?

5. Give three examples of effective counseling non-verbal behavior.

6. Why do roadblocks impede self-exploration?

7. Describe empathy.

8. Identify the six relationship-building responses.

9. List Ivey's four key components of attending behavior.

10. Explain the three essential components of effective listening.

11. Give one example of each of the three relative intensity levels for each feeling category: anger, conflict, fear, happiness, and sadness.

Exercise 3.4 Answers

1. Hand over mouth	should not have spoken, regret
2. Finger wagging	judging
3. Crossed arms	angry, disapproving, disagreeing, defensive, aggressive
4. Clenched fists	anger, hostility
5. Tugging at the collar	discomfort, cornered
6. Hand over eyes	wish to hide, often from self
7. Hands on hips	anger, superiority
8. Eyes wide, eyebrows raised	surprise, guilt
9. Smile	happiness

10. Shaking head — disagreeing, shocked, disbelieving

11. Scratching the head — bewildered, disbelieving

12. Making eye contact — friendly, sincere, self-confident, assertive

13. Avoiding eye contact — cold, evasive, indifferent, insecure, passive, frightened, nervous, concealment

14. Wringing hands — nervous, anxious, fearful

15. Biting the lip — nervous, anxious, fearful

16. Tapping feet — nervous

17. Hunching over — insecure, passive

18. Erect posture — self-confident, assertive

19. Slouching in seat — bored, relaxed

20. Shifting in seat — restless, bored, nervous, apprehensive

21. Sitting on edge of seat — anxious, nervous, apprehensive

Sources: Arthur D. The importance of body language. *HR Focus.* 1995; 72:22–23 and Curry KR, Himburg SP. © 1988, The American Dietetic Association. *Establishing an Effective Nutrition Education/Counseling Program.* Used with permission.

Exercise 3.10 Answers

The following provides possible focuses and intents of the dietitian's responses.
1. Experience, acknowledge; 2. Feelings, acknowledge; 3. Feelings, acknowledge; 4. Experience, acknowledge; 5. Experience, acknowledge 6. Experience, acknowledge; 7. Experience, explore; 8. Thoughts, explore; 9. Behavior, explore; 10. Behavior, explore; 11. Thoughts, explore; 12. Behavior, explore; 13. Thoughts, challenge; 14. Experience, challenge; 15. Experience, challenge; 16. Thoughts, explore; 17. Experience, acknowledge.

ASSIGNMENT—OBSERVATION AND ANALYSIS OF A TELEVISION INTERVIEW

Observe a one-hour television interview. Record the interview so that the program can be reviewed for analysis. Your report should be typed. Use complete sentences to answer the following questions, and number each of your answers:

1. Record the name, date, and time of show observed; note who did the interviewing and who was interviewed.

2. Identify the purpose of the interview.

3. Explain how the interviewer handled the opening part of the interview: How did the interviewer address the interviewee (that is, Mr., Miss, first name, and so forth)? Was a rapport established? What statements were made or questions asked by the interviewer, and what body language of the interviewer facilitated or hampered the development of a rapport? Did the interviewee appear comfortable and willing to disclose information about him- or herself?

4. Explain how the interviewer handled the exploration phase. Did it appear that the interviewer had pre-planned and prepared an "interview guide"? Did the interviewees talk 60 to 70 percent of the time in response to questions?

5. The following list contains names of responses that the interviewer could have made. Give an example for each of the following, identify whether they were effective, and give an evaluation as to the effect of the response on the course of the interview. State whether there were no examples of a specific response in the interview.

- Attending
- Reflection
- Legitimation
- Respect
- Mirroring
- Paraphrasing
- Summarizing
- Giving feedback
- Open questions
- Closed questions
- Why questions
- Clarifying
- Noting a discrepancy
- Directing
- Advice
- Self-disclosing
- Self-involving

6. Explain how the closing was handled. In your opinion, was this an effective way to end the interview? Why?

7. Play the video for ten minutes without sound. Describe the body language of the interviewee and the interviewer. Was their body language congruent with what you heard verbally?

8. Identify three things you learned from this activity.

9. Are there things you observed regarding the manner in which the interviewer handled the session that you would definitely not do? What would you like to emulate in your work as a nutrition counselor?

SUGGESTED READINGS, MATERIALS, AND INTERNET RESOURCES

Books:

Haney JH, Leibsohn J. *Basic Counseling Responses: A Multimedia Learning System for the Helping Professions.* Pacific Grove, CA: Brooks/Cole Wadsworth; 1999. This book contains an effective interactive learning CD-ROM to promote understanding of basic counseling responses.

Motivational Interviewing in Health Care: Helping Patients Change Behavior. Stephen Rollnick, William R. Miller, and Christopher C. Butler, New York: The Guilford Press, 2008. Includes dialogues and vignettes illustrating core skills of MI and shows how to incorporate this brief evidence-based approach into any health care setting.

Website:

Molly Kellogg Website, www.mollykellogg.com Counseling Tips for Nutrition Therapists e-newsletter, webinars.

REFERENCES

1. Haney JH, Leibsohn J. *Basic Counseling Responses: A Multimedia Learning System for the Helping Professions.* Pacific Grove, CA: Brooks/Cole Wadsworth; 1999.
2. Lawn S, Schoo A. Supporting self-management of chronic health conditions: Common approaches. *Pt Educ Coun.* 2010; 80: 205–211.
3. Helm KK, Klawitter B. *Nutrition Therapy: Advanced Counseling Skills.* Lake Dallas, TX: Helm Seminars; 1995.
4. Kittler PG, Sucher KP. *Food and Culture.* 5th ed. Pacific Grove, CA: West/Wadsworth; 2008.
5. Ivey AE, Packard N, Ivey MB. *Basic Attending Skills.* 4th ed. North Amherst, MA: Microtraining Associates; 2007.
6. Fretz BR, Corn R, Tuemmier JM, et al. Counselor nonverbal behaviors and client evaluations. *J Counsel Psych.* 1979;26:304–311.
7. Cormier S, Cormier B. *Interviewing and Change Strategies for Helpers: Fundamental Skills and Cognitive Behavioral Interventions.* 6th ed. Pacific Grove, CA: Brooks/Cole; 2008.
8. Curry KR, Jaffe A. *Nutrition Counseling and Communication Skills.* Philadelphia: Saunders; 1998.
9. Magnus M. What's your IQ on cross-cultural nutrition counseling? *The Diabetes Educator.* 1996;96:57–62.
10. Kellogg M. Personal Communication, September 11, 2010.
11. Cohen-Cole SA, Bird J. *The Medical Interview: The Three-Function Approach.* 2d ed. St Louis, MO: Mosby; 2002.
12. Arthur D. The importance of body language. *HR Focus.* 1995;72:22–23.
13. Axtell R.E. *Gestures.* New York: Wiley; 1991.
14. Miller WR, Jackson KA. *Practical Psychology for Pastors.* 2d ed. Eugene, OR: Wipf & Stock Publishers; 2010.
15. Gordon T. *Parent Effectiveness Training.* New York: Three Rivers Press; 2000.
16. Murphy BC, Dillon C. *Interviewing in Action: Process and Practice.* Pacific Grove, CA: Brooks/Cole; 1998.
17. Miller WR, Rollnick S. *Motivational Interviewing Preparing People to Change.* 2d ed. New York: Guilford; 2002.
18. Maher L. Motivational interviewing: what, when, and why. *Patient Care.* 1998;32:55–64.
19. Laquatra I, Danish SJ. Practitioner counseling skill in weight management. In: Dalton S, ed. *Overweight and Weight Management: The Health Professional's Guide to Understanding and Practice.* Gaithersburg, MD: Aspen; 1997:348–371.
20. Kleinke DL, Tully TB. Influence of talking level on perceptions of counselors. *J Counsel Psych.* 1979;26:23–29.
21. Curry-Bartley K. The art and science of listening. *Top Clin Nutr.* 1986;1:14–24.
22. Teyber E, McClure F. *Interpersonal Processes in Psychotherapy.* 6th ed. Pacific Grove, CA: Brooks/Cole; 2010.
23. Danish S, D'Augelli AR, Hauer AL. *Helping Skills: A Basic Training Program.* 2d ed. New York: Human Sciences Press, Inc.; 1980.

[24]Dubé C, Novack D, Goldstein M. *Faculty Syllabus & Guide: Medical Interviewing*. Providence, RI: Brown University School of Medicine; 1999.

[25]Laquatra I, Danish SJ. Counseling skills for behavior change. In: Helm KK, Klawitter B, eds. *Nutrition Therapy Advanced Counseling Skills*. Lake Dallas, TX: Helm Seminars; 1995.

[26]King NL. *Counseling for Health & Fitness*. Eureka, CA: Nutrition Dimension; 1999.

[27]Long B, Woolen W, Patrick K, Calfas K, Sharpe D, Sallis J. *Project PACE Physician Manual*. Atlanta, GA: Centers for Disease Control; 1992.

[28]Russell MA, Wilson C, Taylor C, Baker CD. Effect of general practitioners' advice against smoking. *Br Med J* 1979;2:231–235.29.

[29]Berg-Smith SM, Stevens VJ, Brown KM, et al. A brief motivational intervention to improve dietary adherence in adolescents. *Health Educ Res*. 1999;14:399–410.

[30]Rollnick S, Miller W, Butler C. *Motivational Interviewing in Health Care: Helping Patients Change Behavior*. New York: Guildford Press; 2008.

[31]Dyer WW, Vriend J. *Counseling Techniques That Work*. Alexandria, VA: American Counseling Association; 1988.

4

Meeting Your Client: The Counseling Interview

PhotoDisc/Getty Images

There are two ways of spreading light;
To be the candle or the mirror that reflects it.
—EDITH WHARTON

Behavioral Objectives

- Explain the usefulness of counseling models.
- Describe the motivational nutrition counseling algorithm.
- Use a variety of readiness-to-change assessment tools.
- Demonstrate selected counseling strategies.
- Depict parts of a counseling interview.

Key Terms

- **Algorithm:** a step-by-step procedure for accomplishing a particular end.
- **FRAMES:** an acronym for the steps of a brief counseling intervention.
- **Models:** generalized descriptions used to analyze or explain something.
- **Motivation:** a state of readiness to change.
- **Motivational Nutrition Counseling Algorithm:** a step-by-step guide to direct the flow of a nutrition counseling session.

INTRODUCTION

Chapter 2 provided a review of a variety of behavior change and counseling theories and **models**. This chapter builds upon those approaches to formalize a concept of the nutrition counseling process. We begin by looking at a model that depicts an overview of the complete nutrition counseling procedure. Then, we will examine a **motivational nutrition counseling algorithm** for an individual counseling session that draws upon evidence based best practices. Several approaches for assessing **motivation** to change are offered. In addition, special considerations for acute care counseling and brief interventions are addressed.

NUTRITION COUNSELING MODELS

Models and algorithms can provide structure for conceptualizing the counseling process. These are aids for planning, implementing, and evaluating a counseling intervention because the session can be broken into component parts and addressed individually. The actual flow of a counseling session will adjust to the skills of the counselor and to the needs of the client. However, some structure can help a counselor visualize the direction of the counseling experience as well as the expected end point.

Figure 4.1 provides a model for nutrition counseling that can be used to visualize the major components of the total interaction a counselor has with a client. This model addresses the need for counselors to assume several roles to accomplish counseling tasks. As a diagnostician, a nutrition counselor reviews medical records, food patterns, medication intake, health history, socioeconomic conditions, and other factors. This *preparation* occurs before the first counseling session, during the first session, and periodically thereafter to better understand problems, skills, and resources related to food intake and readiness to take action. Nutrition counseling also requires an educational component that entails an *explanation* of the counseling process. In addition, nutrition counselors repeatedly assume the role of educator when communicating pertinent nutrition information

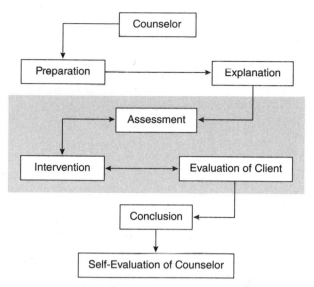

Figure 4.1   Model for a Nutrition Counseling Program
Source: L.G. Snetselaar, Nutrition Counseling Skills for Medical Nutrition Therapy, © 1997 Aspen Publishers. Adapted with permission of Jones and Bartlett Learning, Sudbury, MA. www.jblearning.com.

or providing hands-on educational experiences. During the *intervention* or treatment component of counseling, nutrition counselors take on the role of problem solver and expert using a variety of intervention strategies to help implement dietary goals.[1] Counselors assume the role of diagnostician to assess intervention strategies and evaluate client progress. The *assessment, intervention,* and *evaluation* components are part of each counseling session until the decision is made to conclude the program. At the *conclusion* of each session as well as the total program, the nutrition counselor resumes the role of expert when reviewing major issues and goals. The counselor becomes a learner in the last component involving *self-evaluation*. Here the objective is to learn from specific counseling experiences for the purpose of improving helping skills.

> *You cannot teach a man anything; you can only help him find it within himself.*
> **—Galileo**

EXERCISE 4.1   Explore Counseling Models

Review the counseling model presented in Figure 4.1. List two activities you might do to address the function depicted in each box.

NUTRITION COUNSELING MOTIVATIONAL ALGORITHM

A motivational nutrition counseling algorithm is presented in Figure 4.2 to direct the flow of a nutrition counseling session. This algorithm is based on one developed by Berg-Smith et al.[2] It takes into consideration that motivation is the underlying force for behavior change and that clients come into counseling at varying levels of readiness to take action. This algorithm incorporates concepts

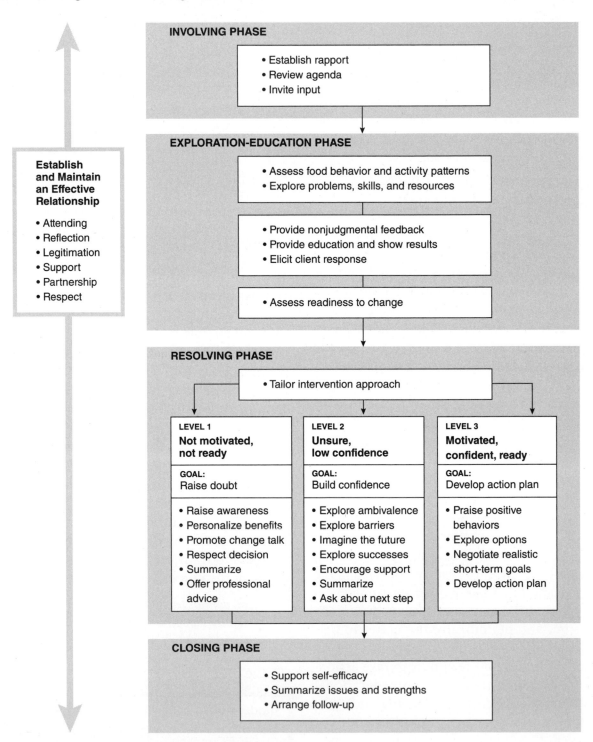

Figure 4.2 Motivational Nutrition Counseling Algorithm

Source: S. M. Berg-Smith, V. J. Stevens, K. M. Brown, et al., A brief motivational intervention to improve dietary adherence in adolescents. *Health Education Research,* 1999;14(3):399–410. Reprinted by permission of Oxford University Press.

from several intervention models and behavior change theories covered in Chapter 2 to provide the guiding force to influence motivation and change lifestyle behavior, including:[3-9]

- the transtheoretical model
- motivational interviewing
- solution-focused brief therapy
- self-efficacy

This algorithm provides the direction for the counseling interview and counseling protocol found in this chapter and for each session of the step-by-step guided counseling intervention outlined in Chapter 14 of this book.

ASSESSING READINESS TO CHANGE

The motivational nutrition counseling algorithm presented in Figure 4.2 provides for three levels of motivation requiring a need to make an assessment of readiness to change. Professionals have offered a variety of strategies to assess motivational level. The strategies generally attempt to identify a particular stage of the transtheoretical model, or a person identifies a position on a scale. Not all counseling situations lend themselves to a formal type of assessment, particularly when monitoring on a regular basis. A debate has emerged over whether to use an abbreviated continuum of readiness to change in a therapeutic setting as an alternative to the strictly defined stages of change.[11] Several looser assessments have been developed. The following describes four procedures for conducting readiness assessments:

1. ***Stage of Change Algorithm.*** Figure 4.3 presents an algorithm containing questions to determine stage of change for adopting a low-fat diet. It can be easily modified to assess stages for other dietary factors. This algorithm has been used in several nutritional studies.

2. ***Readiness-to-change open ended questions.*** Simple open ended questions can be effective to assess readiness to change.[7]
    - How do you feel about making a change now?
    - People differ in their desire to make changes. How do you feel?

   - When thinking about changing food habits, some people may not feel ready, others may feel they need time to think it over, and some people feel ready to start making changes. What are you feeling?

3. ***Readiness-to-change scale question.*** Windhauser et al.[10] have described a similar method, but without a tool, by simply asking this question: "On a scale from 1 to 10, with 1 being 'not at all' ready and 10 being 'totally ready,' what number would you pick that would represent how ready you are to make this change?"

4. ***Readiness-to-change graphic.*** Another possibility to assess readiness is showing a graduated picture of a thermometer, ruler or chart. See Figure 4.4 and Lifestyle Management Form 4.1 in Appendix D for a picture of a readiness graphic using ten numbers. You begin by asking your client to look at the picture and identify a spot on the graphic that indicates how ready he or she is to make a change. Some people have an easier time responding to a visual representation of their readiness level. This method has been incorporated into the step-by-step guide for four counseling sessions found in Chapter 14. We have found that a physical tool for this assessment is useful for beginning counselors.

The previously mentioned readiness scales can be modified for other assessments:

- ***Dietary adherence assessment.*** Similar assessment scales can be used to appraise adherence to a dietary protocol. In that case, *not ready* is equivalent to "never follow the dietary guidelines," and *ready* corresponds to "always follow the guidelines."

- ***Assessment of importance and confidence.*** The same types of scales can be used to measure importance and confidence, two components of readiness.[7] In these cases, the ends of the continuum are "very important" or "very confident" and "not important" or "not confident." These assessments provide a clearer picture of how someone feels about change and helps the counselor to know what to stress during an intervention.

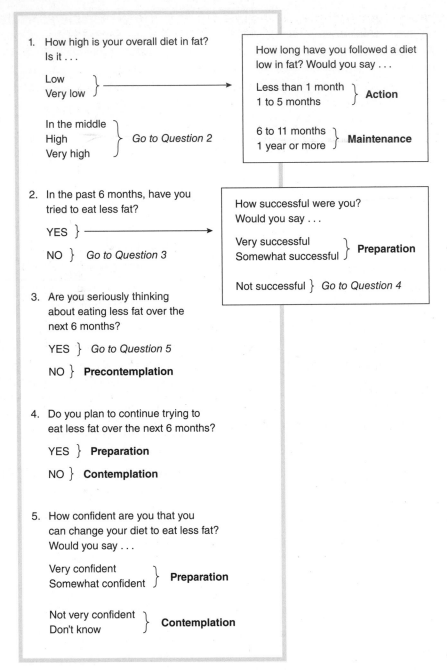

Figure 4.3 Questions and Algorithm Used to Assign Stages of Change for a Low-Fat Diet
Source: Reprinted from the *Journal of the American Dietetic Association*, 99:680, Kristal A.R., Glanz K., Curry S.J., Patterson R.E., How can stages of change be best used in dietary interventions?, © 1999, with permission from Elsevier.

NOT READY				NOT SURE	NOT SURE				READY
1	2	3	4	5	6	7	8	9	10

Figure 4.4 Readiness-To-Change Graphic
Source: Stott, et al. Innovation in clinical method: Diabetes care and negotiating skills. Family Practice. 1995;12:413–418. Adapted with permission from Oxford University Press.

Find a friend or relative willing to allow you to assess his or her motivation level regarding a needed behavior change. Use the algorithm assessment tool in Figure 4.3, one of the readiness-to-change open-ended questions, the readiness-to-change graphic (Lifestyle Management Form 4.1 in Appendix D), and the readiness-to change scale question.

❏ In your journal compare and contrast the techniques, and indicate which one gave you the clearest picture of the individual's readiness to make a change.

EXERCISE 4.3   Assess Importance, Confidence, and Readiness

Select five lifestyle behaviors you would like to evaluate and list them in your journal. Use the following readiness-to-change scale questions to assess importance, confidence, and readiness for making a change:

With the number one indicating not ready and ten representing totally ready, how do you feel right now about your readiness to make the change?

With the number one indicating not important and ten representing very important, what number represents how important it is for you to make a change right now?

If you decide to change, how confident are you that you will succeed on a scale of one to ten, with one signifying no confidence and ten indicating very confident?

Put the scores in columns with the change categories as headings. For example:

	Readiness	Importance	Confidence
Use my seat belt for every car trip.	7	10	8

NUTRITION COUNSELING PROTOCOLS: ANALYSIS AND FLOW OF A COUNSELING INTERVIEW AND COUNSELING SESSION

The following nutrition counseling protocols are based on the motivational algorithm for a counseling intervention presented in Figure 4.2. Table 4.1 presents an overview of the tasks and objectives covered in each phase of the algorithm. The word

interview refers to the collection of valid and accurate data, whereas the word *counseling* implies that a counselor is assisting an individual in making life change decisions. Because nutrition counselors address both tasks in their interactions with clients, the terms have been combined.

The following narrative offers a step-by-step guide for conducting a nutrition counseling session. The guide is organized into the four phases identified in Table 4.1 and Figure 4.2. The flow of the tasks and strategies follows the motivational nutrition counseling algorithm illustrated in Figure 4.2. The guide is provided to (1) give direction to a novice nutrition counselor, (2) furnish a counseling framework which can be molded to fit individual talents and needs, and (3) supply a springboard on which to build skills. There is no perfect method for all individuals, but the guide includes what is generally considered standard in nutrition counseling.

INVOLVING PHASE

The first stage of the counseling session is the involving phase, which includes such relationship-building activities as greeting the client and establishing comfort by making small talk, opening the session by identifying the client's goals and long-term behavioral objectives, explaining the counseling process, and making the transition to the next phase.

Greeting

The greeting sets the stage for the development of a trusting, helping relationship. At the beginning of each session, your greeting should indicate a sense of warmth and caring. The tone of your voice and your body language should convey the message that you are happy to meet your client. This is especially important for the initial session because a first impression is a lasting impression. As Will Rogers said, "You never get a second chance to make a first impression." The manner in which you and your client will address each other should be established. In many institutional settings and among some ethnic groups the custom will be to use a formal Mr., Mrs., or Miss. In less formal settings such as a health center or rehab program, calling clients by their first name may be more appropriate. If in doubt, start

Table 4.1   Overview of the Counseling and Interview Process

Phase	Possible Tasks[1]	Objectives
Involving	• Begin with greetings and introductions. • Identify client's long-term behavior change objectives. • Explain rational for recommended diet. • Explain counseling process. • Set agenda.	• Establish rapport, trust, and comfort. • Communicate an ability to help. • Interact in a curious and nonjudgmental manner.
Exploration-education	• Offer educational activities. • Assess food behavior, activity patterns, and past behavior change attempts. • Explore problems, skills, and resources. • Give nonjudgmental feedback. • Elicit client response. • Assess readiness to change.	• Provide information. • Show acceptance. • Learn nature of problems and strengths. • Promote self-exploration by the client. • Clarify problems and identify strengths. • Help the client to evaluate the situation.
Resolving	• Tailor intervention to the client's motivational level.	• Help the client make decisions about behavior change. • Indicate that the client is the best judge of what will work.
Closing	• Support self-efficacy. • Review issues and strengths. • Restate goal(s). • Express appreciation. • Arrange follow-up.	• Provide support. • Provide closure.

[1]The specific tasks to be addressed are dependent on motivational level and needs of your client as well as previous interactions with your client (for example, first session or fourth session).

out using a formal address or both names, such as "Mrs. Jones" or "Sally Jones." Inquire how the client should be addressed, such as "Do you prefer to be called Mrs. Jones or Sally?" Generally you should not use the first name unless given permission. Be sure to introduce anyone who is accompanying you, such as a colleague or an assistant. If there are friends or relatives with your client, be sure to also acknowledge and warmly greet them.

Establish Comfort

Aim to create a private and quiet environment. When the session has begun, pay attention to your own comfort level as well as that of your client. For example, sun in the eyes can be extremely distracting, as can a loud radio in an adjacent room.

A review of the intake records of a ninety-year old, new resident in the nursing home where I consulted indicated that this person had grown up on a farm close to where the nursing home was located. When I walked into this woman's room to introduce myself and do a diet consultation, I found the woman sitting in a corner with a scowl on her face. As I attempted to talk to her, her body language and grunts did not change until she realized what I was saying. I was telling her she had come home. I explained to her what I meant and her facial expression and body language changed completely. She then had no trouble hearing me, and we talked a little while about what it was like growing up on the farm. After that experience, I always tried to find something special about new residents from the records. It was a great way to develop rapport.

Small Talk

After your greeting, it may be appropriate to engage in some small talk, depending on the setting and the amount of time you have for the session. Generally this should be limited to a question or a comment about the office or building where the meeting is taking place or the weather. This verbal exchange can aid in the development of a comfortable atmosphere, but if carried on for more than a few comments, the counseling experience may be hampered by creating a superficial atmosphere that can permeate into the rest of the session. Particularly stay away from comments unrelated to the client or counseling experience, such as a current story in the news.[11]

Opening–First Session

A common opening after the greeting and small talk is to ask an open-ended question in a curious manner, such as "When we talked on the phone, you said the doctor told you that your blood pressure is elevated. What are you hoping to achieve in counseling?" or "What brings you here today?" or "How can I help?" This begins the process of attempting to understand your client's needs, expectations, concerns, and coping strengths. As clients clarify their needs, the direction to pursue intervention strategies becomes clearer. Some clients need time to feel truly comfortable expressing their thoughts and feelings. King[12] notes that some clients could take up to three sessions before they are relaxed enough to communicate openly. Examination of your client's issues will be more fully developed in the next phase (the exploring phase) and in subsequent sessions as well. For now, emphasis should be placed on the following specific counseling approaches during this part of the encounter:

- **Relationship-building responses.** There will probably be opportunities to use several of the relationship-building statements, particularly empathizing, legitimation, and respect.
- **Reflective listening.** This skill can aid in understanding client issues.
- **Responses that indicate attentiveness and help clarify meanings.** These include paraphrasing, summarizing, clarifying, and asking open-ended questions.

Opening–Subsequent Sessions

In subsequent sessions, an opening question is used to invite input, such as "Where would you like to begin?" or "How have things been going since we last talked?" This gives a client an opportunity to address any burning issues before getting into your agenda or you may decide to alter your agenda based on your client's immediate concerns.

Identifying Client's Long-Term Behavior Change Objectives (General Goals)

While discussing your client's needs and expectations, a long-term goal or goals should be established.

This topic needs to be covered in the first session and periodically reviewed thereafter if there is long-term involvement. Goals can be general in nature such as to feel better or to improve the nutritional quality of food intake. If possible set at least one measurable long-term goal, such as reducing cholesterol to 190 milligrams/deciliter or keeping blood sugar levels below 160 milligrams/deciliter. Specific goals are needed for measuring outcomes. If your client comes in with a vague goal, such as improving diet quality, you can collaborate with your client after an assessment to establish more specific goals that can be evaluated. For clients who wish to lose weight, efforts should be made to encourage clients to set realistic and maintainable goal weights and to focus attention on setting goals related to healthful eating and increased activity. Health benefits are often seen with a modest weight loss of 10%.[13] Be sure to listen carefully what your client says is important to him or her. If your client received a diet prescription from a physician, provide a rationale for the diet.

Explain Program and Counseling Process–First Session

At the beginning of the first meeting, your client should receive a description of what will happen in the course of the counseling program. This would include a description of the assessment tools, a general statement regarding the issues that will be discussed, and a survey of possible intervention strategies and activities. Review frequency of meetings and correspondence by telephone, e-mail, or text messaging.

Discuss confidentiality. A partnership statement is appropriate, clarifying your role as a source of expertise, support, and inspiration and your intention to work collaboratively with the client to make decisions about lifestyle changes. Your comments should indicate that ultimately the client will be making decisions, for he or she will be the one to implement changes and is also the best judge of what will work. This is particularly important in cases of acute illness, such as diabetes, where dietary practices are an integral component of care and in reality it is a self-managed disease.[14]

Many programs use a counseling agreement form to verify that many of these topics have been addressed. Lifestyle Management Form 14.2 in Appendix D is an example of such a form. Be sure to also review what your client is hoping to receive from a nutrition counselor. Here is an example of this part of the involving phase:

Counselor: *As we get into the counseling process, I'd like to share with you my hope of how we will be working together. I see myself as a source of information, support, and inspiration for you as we work together to find solutions for your issues. I hope to assist you in making an informed choice about what behaviors to change and whether or not to change the behavior at all. I can help you learn about healthy options and possible strategies, but only you know best what will fit into your life and what you are willing to tackle. I would like us to work together to build on skills that you already have for dealing with your nutrition issues. How does this all sound to you? Do you see things differently?*

Discuss Weight Monitoring, If Appropriate–First Session

If you have a client who would like to lose weight, discuss how to handle weight monitoring. Authorities do not agree how often dieters should weigh themselves. Weights can be taken once a day, once a week, or once a month. Also, nutrition counselors are not in agreement on whether counselors should be doing the actual weighing for uncomplicated outpatient clients.[15] Of course you want the focus of your sessions to be on making changes that allow weight loss to occur, not on numbers on a scale. Because several physiological factors can affect the reading, a one-or two-pound weight loss may not be readily seen. If a counselor is weighing a client, and the scale does not show a loss, this can have a negative impact on the whole counseling session. Discuss with your clients if they would like you to weigh them and how often. Sometimes clients prefer that you do the weighing and want the added pressure of someone else knowing their weight to help them maintain their food goals. If you set a long-term goal with your client of losing a specified amount of weight, measuring that outcome would be difficult if you never personally took your client's weight. Let your clients be the guides as to how they would like to handle the matter of taking weights. See Exhibit 4.1 for an alternative opinion regarding weight monitoring for clients with uncomplicated health issues. Also, consider using other parameters of health to monitor changes in health status, such as waist circumference, percent body fat, and cholesterol levels. In some individuals, these values may be more responsive to lifestyle changes than numbers on a weight scale.

Setting the Agenda–First Session

Establish and agree with your client as to what will be covered in the counseling session. For your first session this probably means going over the flow of the session—explaining the assessment process, reviewing preliminary results, selecting a food habit to address, assessing readiness to take action, and then setting a goal and plan of action (if ready), or exploring ambivalence and providing information (if not ready).

EXHIBIT 4.1   Debate over Dieting

Authorities do not agree on how much emphasis should be placed on body weights and the concept of a "diet."[16] Some believe the scale ought to be thrown in the trash, the concept of diet abolished, and emphasis placed on healthy eating, exercise, and body size acceptance. Some are concerned that too much importance on dieting contributes to body dissatisfaction among girls and women, which may lead to bingeing behavior, bulimia, and a host of other physiological and psychological problems.[17] Advocates of the non-diet approach maintain that healthful changes in eating and other lifestyle behaviors without "dieting" can have a positive impact on health. In the successful Kentucky Diabetes Endocrinology Center, many patients gained good control over diabetes, blood pressure, and blood lipids without placing primary emphasis on body weight.[18]

Setting the Agenda–Subsequent Sessions

Before your meeting, review your notes regarding previous sessions, and prepare educational experiences or materials as indicated. Ask your client for any issues he or she would like to address during your meeting, explain your intentions for the session, and then come to an agreement for an agenda. You may have additional assessment data to share with your client, and this may result in new behavior change options to address.

> The first time I ever counseled a client was in a nutrition counseling class in college. The person assigned to me volunteered to participate because his doctor told him his cholesterol was slightly elevated. The first day I saw him he made it clear that he was looking for dietary information related to this issue. I didn't actually address this concern until our fourth session. Until that time I thought he was uncooperative and disinterested, and then I saw a new client. At that point I understood what I should have been doing from the first meeting.

Transitioning to the Next Phase

Before entering into the exploration-education phase, make a statement indicating a new direction:

Counselor: *Now that we have gone over the basics of the program, we can explore your needs in greater detail.*

EXPLORATION-EDUCATION PHASE

During this phase, a nutrition counselor and client work together to understand a client's nutrition and lifestyle problem, search for strengths to help address difficulties, assess readiness to take action, and provide educational experiences. Counselors need to provide a nonjudgmental environment so clients feel free to elaborate on pertinent issues. A counselor's verbal and nonverbal behavior should be viewed as curious rather than investigative. Responses covered in Chapter 3 that can be especially useful to advance exploration include open questions, paraphrasing, reflection, probing, and directives.

Educational Activities

During your first session, assessment activities are likely to be time-consuming, resulting in too little time for involved educational activities. The main educational task should be to address health risks associated with your client's eating pattern, although there are likely to be opportunities to provide sound bites of information as a client expresses concerns and asks questions. The educational experiences of subsequent sessions should be geared to your client's needs and desires as determined during the assessment process. See Chapter 6 for a discussion of integration of information giving in a counseling session.

Assessment–First Session

Assessment is an important component of the counseling process to tailor an intervention to the needs of a client. Basic dietary and physical assessment procedures and commonly used forms in nutrition counseling are addressed in Chapter 5. While collecting information or reviewing completed forms, do not react with advice, criticism, or judgment, as this could inhibit disclosure. If at all possible, counselors should refrain from firing a series of questions to gain information. A more satisfying and valuable method would be to encourage discussions at certain points to allow clients to provide insight about their life experiences.

Often counselors attempt to have clients complete assessment or screening forms before the first session. This approach saves valuable counseling time and allows clients to focus ahead of time on issues and counseling expectations. If that is the case, when you do meet with your client, avoid rehashing exactly what is on the forms, but encourage open discussion of what the experience was about. Consider the following example:

Counselor: *Thank you for completing these forms. The information in them will be helpful as we work together to search for solutions for your food problems. I am wondering, what came to your mind as you were filling out these papers? What topics covered in these forms do you think have particular importance for your food issues? Did you feel a need to expand or clarify any of your answers? Which ones? Did it prompt you to think about what you would like to cover in our sessions together?*

Here are some appropriate topics to explore with your clients:

- Concerns about health risks associated with current eating behavior
- Concerns about changing food patterns
- Past lifestyle change successes
- Past experiences trying to change food habits by themselves, with the aid of a nutrition counselor or in an organized program
- Difficulties with making food habit changes in the past
- Strategies that worked or did not work when attempting to make a lifestyle change
- Selection of education topics to address in future sessions

If you have enough time in a counseling session (six to eight minutes), consider using "a typical day" strategy, which is similar to the diet history interview described in Chapter 5. This method encourages clients to drive the assessment discussion and to tell their "stories." If the story goes on for more than eight minutes, however, the activity becomes tiring for both parties, and the counselor should intercede to speed up delivery. See Exhibit 4.2 for guidelines for using this strategy.

Some authorities suggest an alternative approach of having at least one counseling session with your client before using assessment instruments to build motivation and understanding of the assessment process. They feel that the laborious task of completing the forms begins the counseling intervention with an impersonal tone and may be an obstacle to receiving treatment at all.[9] However, in some facilities where the number of nutrition counseling sessions is limited, this procedure may not be practical.

EXHIBIT 4.2   Guidelines for Using a Typical Day Strategy

1. Introduce the task carefully

Sit back and relax! Ask the client a question such as: "Can you take me through a typical day in your life, so that I can understand in more detail what happens? Then you can also tell me where your eating fits in. Can you think of a recent typical day? Take me through this day from beginning to end. You got up . . . "

2. Follow the story

- Allow the client to paint a picture with as little interruption as possible. Listen carefully. Simple open questions are usually all you need—for example, "What happened then? How did you feel? What exactly made you feel that way?"
- Avoid imposing any of your hypotheses, ideas, or interesting questions on the story you are being told. Hold them back for a later time. This is the biggest mistake made when first using this strategy. Don't investigate problems!
- Watch the pacing. If it is a bit slow, speed things up: "Can you take us forward a bit more quickly? What happened when . . . ?" If it is a bit too fast, slow things down: "Hold on! You are going too fast. Take me back to . . . . What happened . . . ?"
- If you are uncertain about details, and you are satisfied that you are being curious rather than investigative, ask the client to fill them in for you.
- You know you have got it right when you are doing 10 to 15 percent of the talking, the client seems engaged in the process, and lots of interesting information about the person is emerging.

3. Review and summarize

A useful question at the end of your client's story is "Is there anything else at all about this picture you have painted that you would like to tell me?" Now is the time to ask probing questions to clarify any descriptions, such as "Can you tell me what kind of bread you usually use to make your lunch sandwich?" This is also a good opportunity to be honest with the client about your reaction and to provide legitimation responses wherever possible. Having listened so carefully to the client, you will now be able to explore other topics quite easily. Often this leads into an investigation for the need for general information—for example, "Is there anything about . . . that you would like to know?"

Source: Rollnick S, Mason P, Butler C, Health Behavior Change: A Guide for Practitioners (New York: Churchill Livingstone; 1999); © 1999, pp. 113–114. Used with permission.

EXERCISE 4.4   Practice Using "A Typical Day" Strategy

Review the guidelines in Exhibit 4.2 for utilizing "a typical day" strategy. Work with a colleague, taking turns role-playing counselor or client.

❏ Write your reactions to this activity in your journal. What were your thoughts and feelings while you were the counselor? The client? How would you use this activity in an actual counseling interaction? What did you learn from this activity?

Assessments–Subsequent Sessions

Generally assessment activities are the most intense during the first session, but assessments should be made periodically to assist in setting new goals and to monitor progress. The assessment graphic (Figure 4.4) discussed earlier can be used to appraise how closely a client has been following a goal or a dietary protocol. Follow up the adherence question with simple open-ended questions to gain a deeper understanding of a client's progress.

Counselor: *Please look at the picture of this graphic. What square would you pick to describe how closely you have been following your food plan?*

Tell me something about the square you chose.

Why did you choose the square you did?

Giving Nonjudgmental Feedback

Assessment results should not be simply handed to a client but reviewed point by point in a neutral manner. Counselors need to provide clear norms for comparison, such as a therapeutic dietary protocol or the Dietary Reference Intakes. Some assessment forms have a standard on the form. (See Lifestyle Management Form 5.3 in Appendix D.) Give only the facts, allowing the client to make the initial interpretation. If you have a great deal of feedback to provide, pause regularly to allow the client to process the information and to check for comprehension:

Counselor: *Your assessment indicates an intake of ½ cup of vegetables a day. As you see, the standard*

recommendation based on the MyPyramid guidelines is 2½ cups a day.

Eliciting Client Thoughts about the Comparison of the Assessment to the Standard

Curiously ask simple open-ended questions to encourage a client to explore the meaning of the results through questions and personal reflection. The ideal response from a client would be something like "I see, I didn't really give much thought to this before," or "I'm wondering if . . ."[7] This approach provides an opportunity for people to discover discrepancies between their condition and a standard and to make self-motivational statements as discussed in the section on motivational interviewing in Chapter 2.

Counselor: *What do you think about this information?*

Do these numbers surprise you?

I have given you a lot of information. How do you feel about what we have gone over?

Did you expect the evaluation to look different?

If your client does ask you for clarification or meaning, present the information in a nonthreatening manner by avoiding the word *you*.[7] For example, "People who have a low intake of vegetables are at a higher risk for developing several types of cancers"—not "You have an increased risk for developing several types of cancers." Nor should a counselor tell a client how he or she should feel about the feedback—for example, "You should really be concerned about these numbers. They increase your risk for developing cancer." An alarming explanation interferes with the decoding process, causing explanations to be misunderstood and increasing resistance to change.

Determining What's Next

After an assessment, it is generally a good idea to summarize; include what the client is doing well, problems identified in the assessment, any self-motivational statements made by the client, and ask the client whether the summary needs any additions or corrections. You could include any ideas that the client picked up on. For example, "you like the idea of switching to skim milk and plan to do that right away." After confirming that your summary was accurate, your next task is to ask your client how he or she would like to proceed. Knowing there are acceptable options may assist a client in deciding to make a change. If your client asks your advice, give your impressions, provide options, and indicate that the client would know best what would work for him or her.

Assessing Readiness to Make a Change

At this point your client may have clearly indicated a desire to make a behavior change, and you will need to use your judgment about making a formal assessment for readiness. If you are using a continuum scale such as the assessment graphic in Lifestyle Management Form 4.1 in Appendix D, you may decide to check for importance and confidence as well as readiness.

Counselor: *To get a better idea of how ready you are to make a food behavior change, we will use this picture of a ruler. If 1 represents not ready and 10 means totally ready, where would you place yourself?*

There are actually two parts to readiness, importance and confidence. I think it might be useful to look at them separately. Using the same scale, how do you feel right now about how important this change is for you? The number 1 represents not important, and 10 is very important.

The other part of readiness is confidence. If you decided to change right now, how confident do you feel about succeeding? The number 1 indicates not confident at all, and 10 represents very confident.

RESOLVING PHASE

In the involving and exploration phases, the major objective was to assist clients in clarifying problems and identifying strengths to themselves and to the counselor. The direction of the remaining time of the counseling session will be determined by a client's motivation category. In the motivational interviewing algorithm described by Berg-Smith et al.[6] and incorporated into Figure 4.2, three pre-action motivational levels are illustrated

to address the needs of the majority of individuals with a dietary problem.[19] In the first and possibly the second level, the major issue related to motivation is likely to feel the behavior change as important. Those who pick a higher stage or higher number on a continuum are likely to feel the behavior change is important, but are struggling with confidence in ability to make a successful change.[7] Therefore, in the following analysis of the resolving phase, Level 1 counseling approaches will deal with importance, Level 2 with confidence issues, and Level 3 with selection of a goal and design of an action plan. Although the motivational levels are represented as three distinct entities, counselors need to be flexible in their approach to accommodate fluctuations in motivation level that can occur during an intervention.[5] In such cases there may need to be a cross-over in selection of counseling approaches among the three motivational levels. A summary of the counseling approaches for each level of readiness to change is presented in Table 4.2.

LEVEL 1: NOT READY TO CHANGE (PRECONTEMPLATIVE)

Level 1 clients have clearly indicated that they are not ready to change their behavior or are not doing well at attempting to change. Individuals have a right to decide their destiny, and a decision not to change should be respected. However, health care providers have an obligation to make clear the probable consequences of clients exercising their prerogative. The major goal of working with clients who fit into this category is to raise doubt about present dietary behavior; the major tasks are to raise awareness of the health and diet problems related to their dietary pattern. Often precontemplators have come to counseling because of the urgings of others, or they are "sitting ducks" in a hospital room as a dietitian walks in to give a consultation because of a diet order. They do not need solutions; they need to know they have a problem. The following discussion provides some strategies recommended for clients at this level of readiness to change.[2,7,20]

Table 4.2   Resolving Phase Summary of Tailored Intervention Approach

Readiness to Change		Counseling Approach	

Precontemplation	• **Level 1**
	
	Not ready
	
	Ruler = 1–3
		Goal:
		
		Major task:
		
		Approach:
			Raise doubt about present dietary behavior.
			
			Inform and facilitate contemplation of change.
			
			• Raise awareness of the health problem/diet options.
			• Personalize benefits.
			• Ask open-ended questions to explore importance of change and to promote change talk.
			  Elicit self-motivational statements regarding importance.
			  Elicit identification of motivating factors.
			• Summarize.
			• Offer professional advice, if appropriate.
			• Express support.

Contemplation	• **Level 2**
	
	Unsure
	
	Ruler = 4–7
		Goal:
		
		Major task:
		
		Approach:
			Build confidence and increase motivation to change diet.
			
			Explore and resolve ambivalence.
			
			• Raise awareness of the benefits of changing and diet options.
			• Ask open-ended questions to explore confidence and promote change talk.
			  Elicit self-motivational statements regarding confidence.
			  Elicit identification of barriers.
			• Explore ambivalence by examining the pros and cons.
			  Client identifies pros and cons of not changing.
			  Client identifies pros and cons of changing.
			• Imagine the future.
			• Explore past successes.
			• Encourage support networks.
			• Summarize ambivalence.
			• Ask about next step.

Preparation
 	
Action
	• **Level 3**
	
	Ready
	
	Ruler = 8–10
		Goal:
		
		Major task:
		
		Approach:
			Negotiate a specific plan of action.
			
			Facilitate decision making.
			
			• Praise positive behaviors.
			• Explore change options.
			  Elicit client's ideas for change.
			  Look to the past.
			  Review options that have worked for others.
			• Client selects an appropriate goal.
			• Develop action plan.

Raise Awareness of the Health Problem and Diet Options

Sometimes clients are not aware of the benefits of behavior change or the risks and consequences of their present dietary behavior. Others have misconceptions about the type of dietary changes that are needed. During an awareness discussion, emphasize anything positive that your client is doing that could be built on if a decision to change is made.

Counselor: *There is a lot of information in the news about dietary fat and cholesterol levels, but you may not be aware of all the other dietary factors associated with elevated cholesterol levels. I see that you enjoy eating salad and that you have soup with beans for*

lunch sometimes. Both of those choices could be built on to help lower your cholesterol level.

Personalize Benefits

Clients often know that improving their diet would probably be better for them. However, they may not have given thought to how they would benefit personally or how they may feel better.

Counselor: *Increasing fruit and vegetable intake could be particularly beneficial to you to help lower your blood pressure and aid in your efforts to lose weight. Focusing on these foods is likely to have a positive impact on your occasional constipation problem, too.*

Ask Key Open-Ended Questions to Explore Importance of Change and Promote Change Talk

Thinking and talking about changing behavior can help elicit self-motivational statements and aid in the development of motivation to change.[7] Change talk can be elicited by using key open-ended questions. The most effective open-ended questions for people who are pre-contemplative deal with the need to change. Counselors should listen carefully to the answers and concentrate on the exact meaning of what is being said. Follow up your client's answers with paraphrasing, reflective listening statements, or other open-ended questions. If you observe resistance in your follow-up, back up and use a different approach.[7] However, because the client has already indicated that there is little desire to change, it is generally best to begin this discussion with a tentative approach by requesting permission to discuss the issue.[7] See Table 4.3 for examples of questions appropriate for people at Level 1.

Summarize

Summaries help reinforce what has been said, tying together various aspects of a discussion and encouraging clients to rethink their position. Give a summary of reasons not to change before giving a summary of reasons to change. Be sure to end your summary with any self-motivational statements your client may have made. Finally, ask your client whether the summary was fair and whether he or she would like to make any additions.

Counselor: *Now that the session is coming to a close, I would like to review what we covered so we can agree on where we are and where we are going. You said you came today because of pressure from your doctor and your wife. Your cholesterol readings have been high. The last one was 320. You know that people are concerned about you, but you feel fine and wish people would get off your back. When we went over the types of foods that have been found to help lower cholesterol, you were surprised that there was more you could focus on than just fat. In fact, there were some foods that you enjoy eating that were on the review list. You thought if you did change the way you eat, that some of the people close to you wouldn't be so worried about you. Lastly, you said that beans would appeal to you when*

Table 4.3 Key Open-Ended Questions to Explore Importance of Change

Category	Examples
Ask Permission	*Would you be willing to continue our discussion and talk about the possibility of a change in your food habits?*
Explore Importance	*What do you believe will happen if you do not change the way you eat?*
	What is the worst thing that could happen if you continue to eat the way you have been eating?
	When we used the assessment questions to evaluate how important it was to you to change your food habits, you indicated that it was somewhat important. Why did you pick the number 4 instead of 1?
Explore Motivating Factors	*What would have to be different for you to believe that it is important to change your diet?*
	You indicated that changing your diet was somewhat important by choosing the number 4 on the ruler. What would cause you to view things differently and move up to the number 8?

cooked in several ways. Was that a fair summary? Did I leave anything out? Where does this leave us now?

Offer Professional Advice, If Appropriate

Well-timed and compassionate advice can aid in motivating behavior change.[5] In the ideal situation, a client asks for advice, but if that is not the case, then the counselor can ask permission to give advice. Review the guidelines in Chapter 3 for offering advice. Be sure the advice you give emphasizes that clients know best what will work and the choice is up to them. This would be a good time to offer educational materials.

Counselor: *It is really up to you and you know best what would work for you. There are some simple things you could do that might make a difference in your cholesterol level. For example, you enjoy bean soup and oatmeal. A good place to begin could be to start having soup for lunch or dinner or oatmeal for breakfast. I have some information that you could take home to read about foods to emphasize in your meals to lower cholesterol levels. How are you feeling at this point about making a change?*

Express Support

Relationship-building skills may be ignored, and a counselor could be tempted to argue with a client, especially in the case of a serious medical condition. However, this tactic is not likely to encourage a client to move toward the action stage, and may result in a stronger resistance to change. Letting the client know that you are there to offer guidance and support is likely to have a greater impact. For clients at this level of motivation, the objective is to create a doubt about their present behavior pattern, so preparing an action plan is not useful. Letting your clients know what others have done in their situation can have an impact. The fact that you and the client are not working toward making a behavior change at this time should be acknowledged, and the door should be left open for future contact.

Counselor: *I respect your decision not to change your diet. It is really up to you. I don't want to push you. I do want to be sure you know what could happen as a result of not changing. You probably need some time to think about this. Maybe you will feel differently about this in* *the future. I want you to know that I will always be here to work with you to find solutions. I have met others with your problem. Most individuals do decide to work on changing their diet. However, some do not. You are the best judge of what would work for you. Would you like me to call you next week to discuss how you are feeling about the diet prescription? If you have any questions or need clarification about anything, do not hesitate to call me.*

LEVEL 2: UNSURE, LOW CONFIDENCE

During the motivation assessment, clients in this category indicated that a diet change is possible. They know the problem exists, but something is needed to push the decisional balance in favor of making a change. The objective of working with people at this level is to build confidence in their ability to make a diet change, and the major task is to explore and resolve ambivalence. The following is a review of some of the approaches advocated for people who have low confidence in their ability to change.[2,7,20]

Raise Awareness of the Benefits of Changing and Diet Options

Clients at this level know they need a solution, but may not have all the facts regarding the benefits of changing. They may not really know what dietary changes would have to be made. Simple facts may be all that is needed to progress to a higher level of readiness to change.

Ask Key Open-Ended Questions to Explore Confidence and Promote Change Talk

The formats of these questions are similar to those posed for Level 1 clients; however, the objective switches from exploring importance to focusing on confidence and barriers. See examples in Table 4.4. For people at this motivational level, little thought may have been given to exactly what is keeping them from making dietary changes. By discussing their barriers, possible ways of dealing with them may be identified, and confidence in the ability to change may increase. Again, answers to these questions should be followed up with

Table 4.4   Key Open-Ended Questions to Explore Importance of Change

Category	Examples
Explore Confidence	*You have indicated that you are somewhat confident that you would be able to change you diet. When we did the ruler evaluation, you picked the number 6. What makes it a 6 rather than a 1?*
Explore Barriers	*You picked the number 4 on the picture of the ruler when we were evaluating how confident you were in your ability to change your diet. What is keeping you from moving up to the number 8? How could I help you get there?*
	What are your barriers to making the recommended dietary changes?
	What would need to be different for you to feel you are able to make diet changes?

Table 4.5   Balance Sheet for Someone Contemplating a Diet Change for High Blood Pressure

No Change	Change
Likes (Pros) I get to eat all foods I really like. I am comfortable with my food pattern.	**Likes (Pros)** I think I will feel better. Maybe I will lose weight. Maybe I could reduce the amount of medicine I take for my blood pressure.
Dislike (Cons) I am not a good role model for my children. I dislike taking medicine for my blood pressure.	**Dislike (Cons)** I don't think I will like the foods as much. I have to get used to eating new foods. I think the new diet will be more expensive. I will have to think about what I will eat all the time.

Source: Adapted from Rollnick S, Mason P, Butler C, Health Behavior Change: A Guide for Practitioners. (New York: Churchill Livingstone; 1999), p. 82.

paraphrasing, reflective listening statements, or more open-ended questions.[7] The objective is to elicit self-motivational statements regarding confidence in ability to change dietary habits. Table 4.4 illustrates key open-ended questions that promote change talk to make a good behavior change.

Explore Ambivalence by Examining the Pros and Cons

The objective of this exercise is for the client, not the counselor, to identify the pros and cons related to a contemplated change. Rollnick et al.[7] suggest using the words *like* and *dislike* or *pros* and *cons* rather than *advantages* and *disadvantages* or *costs* and *benefits*. The latter words may be confusing for some individuals. You could use a balance sheet as illustrated in Table 4.5 as a visual aid and even fill in the categories. However, remember the objective of the activity is for the client to fully explain his or her thoughts and feelings. Do not let the focus of the interaction be completion of the form and thereby interfere with the flow of conversation.

- Inquire if your client would like to examine the pros and cons.

    Counselor: *Many people find it useful to explore their likes and dislikes about this issue. Would you like to do that?*

- Guide your client to examine the pros and cons of the present diet. A comfortable beginning for this strategy is generally to start with what the client likes about his or her present diet. Any follow-up questions should be asked for clarification and not divert focus away from the primary subject. Listen carefully and remember key words used by your client. Before progressing to the next set of questions, summarize both sides of the position, interjecting words used by your client.[7]

    Counselor: *What do you like about your present eating habits?*

    What do you dislike about the way you are eating?

- Guide your client to identify pros and cons of new or additional change. Likes and dislikes of the present diet often mirror those of making a change to a new diet pattern. For example, a client may like the fact that all foods he or she enjoys can be eaten if no change is made, and one of the cons of changing is limiting some of the enjoyable foods. As a result, the conversation may have naturally flowed to the pros and cons of making a change. If not, questions can be asked to arrive at the topic and then provide a summary of the responses.

Counselor: *What do you think you would like about this new way of eating?*

What do you think you would not like about making these changes?

Imagine the Future

Use imagery to create a picture of a successful future assisting clients in identifying goals and hoped-for benefits. A variation of this question and one of the mainstays of solution-focused therapy is asking a client to suppose a miracle happened.[21] The final question, asking whether any part of the picture is presently happening, gives clues as to resources and skills already available that can be expanded on to produce hoped-for outcomes.

Counselor: *Let us create a picture of the future. Imagine that you made all the changes necessary to lower your cholesterol level. What is the first thing you notice that is different? What else is different? How do you feel? What does your brother, wife, or husband see you doing? Who notices that this happened? Are any small parts of this picture happening now? How would you like your diet to be in the future?*

Explore Past Successes and Provide Feedback about Positive Behaviors and Abilities

By exploring successes and identifying abilities, the counselor and client can lay a foundation of existing skills that can be built on to make needed changes. One strategy for identifying successes would be to ask whether the client was ever able to accomplish the desired task or another goal. After a success is identified, ask the client to elaborate by asking for details. Ask what the client did to make the success happen. Probe for identifying obstacles and how they were overcome. Clients should be complimented on any past or present coping abilities that have been identified. This will encourage clients to continue to make similar choices in the future.

Counselor: *What strategies have you used in the past to overcome barriers?*

Have you ever been able to go to a party and eat only one dessert when there were many available to choose among?

You are already drinking soy milk and that is a great substitution for milk to reduce your casein intake.

What permanent changes have you made in the past? Tell me what helped you do this so successfully.

Encourage Support Networks

Confidence in ability to make a behavior change, or *self-efficacy,* increases when we watch and interact with others who have made the same or similar changes. Support groups can provide excellent resources for modeling. Clients should also be encouraged to share their intentions to change with others. It often brings support and assistance from associates, friends, and family.

Summarize Ambivalence

The importance of periodic summaries has already been discussed. An effective time for providing a summary is after using a variety of motivational strategies. Summarize your client's ambivalence, and ask your client how she or he would like to continue.

Counselor: *What are your options?*

How would you like to continue?

Choose a Goal, If Appropriate

If your client would like to set a goal, follow the guidelines for goal setting for Level 3 clients, and review the goal-setting process described in Chapter 5. The objective will be to specify a goal to meet the client's motivational level. This may mean buying skim milk instead of low-fat milk or taking some active steps to increase awareness, such as reading informational literature.

LEVEL 3–MOTIVATED, CONFIDENT, READY

Level 3 clients have indicated that they are ready to make a lifestyle change. For these clients, the nutrition counselor serves as a resource person increasing awareness of possible alternatives for solving problems. The counselor and client collaborate to select lifestyle changes to alter, clarify goals, and tailor intervention strategies to achieve goals. If possible, past successes should be used to find viable solutions. These strategies are outlined in Chapter 5 and Chapter 6.

Praise Positive Behaviors

To reinforce desirable behavior patterns, counselors should point them out and offer praise. Also explore what skills your client is using to accomplish the desired outcome.

Counselor: *It is so good that you use skim milk on your cereal in the morning. Has that always been the case? How did you make the switch from whole milk to skim milk?*

Closing Phase

In this phase, review with your client what has occurred during the session, including a summary of the issues, identification of strengths, and a clear restatement of goals. In addition, an expression of optimism about the future and a statement of appreciation for any obstacles overcome should be made to support self-efficacy. Plan for the next counseling encounter, which could be a phone call, e-mail, text message, or counseling session. This is also a good time to use a partnership statement.

Counselor: *It was a pleasure to meet with you today. You came here because you wanted to know more about what you could do with your diet to lower your blood pressure. We reviewed your food pattern and identified several beneficial food habits, such as your use of skim milk in coffee, several servings of whole grains each day, and an adequate intake of water each day. I believe we did a good job setting a goal to eat a fruit each day for a snack at work. I am optimistic because you enjoy eating fruit and have a well-defined plan to put seven pieces of fruit in a bowl in the front of the refrigerator on Sunday, your shopping day. I am really impressed with all the research you have done about blood pressure and food and exercise. I look forward to working with you to make additional changes. Would you like me to send you a follow-up e-mail during the midweek?*

EXERCISE 4.5   Practice Using Counseling Strategies

Practice using the resolving strategies with your colleague, alternating the role of client and counselor. Each of you should choose a behavior change that you have been contemplating (such as following a walking program, increasing fiber, or flossing teeth). Select a behavior change that you feel somewhat ambivalent about implementing so that your assessment will fall into Level 1 or 2. When you are role-playing a counselor, conduct a readiness assessment and follow the suggested counseling strategies for your colleague's motivational level. Because the counselor is experimenting with a variety of new techniques, the experimenter is allowed to call time-out at any point to gather thoughts.

At the end of the designated period (usually eight to ten minutes), each person should share his or her feelings and reactions to the experience. First cover what went well, then what could have been done differently, and finally how each felt about what transpired.

❑ Write your reactions as a client and as a counselor to this activity in your journal. Which strategies did you find useful? What did you learn from this exercise?

Source: This activity was adapted from role-playing directions in the following sources: Rollnick S, Mason P, Butler C, Health Behavior Change: A Guide for Practitioners (New York: Churchill Livingstone; 1999), and Dubé C, Novack D, Goldstein M. Faculty Syllabus & Guide Medical Interviewing (Providence, RI: Brown University School of Medicine; 1999).

FRAMEWORK FOR BRIEF INTERVENTIONS

Motivational techniques can be employed even when time is limited. The framework for brief interventions can be remembered using the acronym **FRAMES**: feedback, responsibility, advice, menu, empathy, and self-efficacy.[5,22] This includes providing feedback in a non-judgmental manner as previously explained. In addition, the right to choose to change and client responsibility for change is emphasized. With permission from the client, the nutrition counselor offers clear advice and a menu of change strategies. Throughout the intervention, the counselor interacts with the client in an empathic style displaying warmth, active interest, respect, concern, and sympathetic understanding. Finally, the counselor offers hope for the future and enhances self-efficacy by expressing optimism for a client's ability to make a change. See Chapter 2 for elaboration on counseling options based time constraints.

CONSIDERATIONS FOR ACUTE CARE

Several factors need to be taken into account when counseling a patient in an acute care facility. For example, patients may have no idea that a registered dietitian was scheduled to see them. The diet consultation may be the result of an internal policy and a guideline of an accrediting agency, or the patient's doctor may have requested the meeting. Because of recent changes in health care, patients' counseling needs present some special challenges. The physical condition of patients is likely to be more distressed than in the past and hospital stays are shorter, limiting the number of possible inpatient counseling sessions.

Here are some tips for working with patients in acute care settings:

- Each time you visit a patient, introduce yourself, verify that you have the correct patient, and explain the purpose of your visit. For example, "Are you Mary Edwards?" If the answer is yes, introduce yourself and then proceed with the greeting: "I am glad to meet you, Mrs. Edwards." It is important for the professional to explain the reason for the contact and how the patient will benefit from the consultation.
- If you have an option to meet at another time, asking whether this is a good time for the meeting to take place is usually a good idea. Patients may be in too much pain or too tired to benefit from a consultation.
- Although time may be limited, do not disregard relationship-building skills and rush through the interview. A hurried atmosphere gives the impression that you do not view your discussion with the client as important or that you do not care enough to get all the facts.[22]

CASE STUDY   Nancy: Intervention at Three Levels of Motivation

Nancy is a twenty-six-year-old, overweight African American woman who was recently diagnosed with hypertension. She is five feet, four inches tall and weighs 174 pounds (BMI 30). While growing up in Columbia, South Carolina, Nancy enjoyed school sports including softball and volleyball. Weight was never an issue during her childhood; in fact, some of her relatives would tell her she was too skinny. After her marriage, Nancy moved with her mother and husband to northern New Jersey to live closer to her sister. Nancy has a sedentary job as the floor supervisor for an overnight mail delivery service and does not engage in any regular exercise program indicating that life is hectic and there is no time. After the birth of each child, Nancy found herself 15-20 pounds heavier than pre-pregnancy weight. She lives on the second story of a two-family house with her husband, mother, and three children ages, one, three, and five years. Nancy's physician prescribed a medication to lower her blood pressure and suggested that she consult with the clinic's nutritionist about her weight and diet.

Nancy's family history is a concern. Diabetes runs in the family, and there is a history of pica during Nancy's pregnancies. Her father suffered from several complications due to diabetes including poor eyesight, amputation of the right leg, and he died from a diabetic coma. Both her sister and mother have hypertension, and her mother has been having kidney problems. Nancy's husband does not work. He fell while working on a construction job and is receiving disability.

The following discussion contains three possible scenarios for Nancy, illustrating use of the motivational nutrition counseling algorithm at three different motivational levels. Examples of relationship-building responses are sporadically intertwined in the scenarios.

Level 1—Not Motivated, Not Ready

Nancy came into your office directly from her doctor's appointment after being diagnosed with hypertension. Nancy said her doctor told her to lose weight and go over her diet with the nutrition counselor. Nancy says she doesn't want to lose weight and she looks fine. Her husband likes her "with some meat on her." In fact, she says she

(continued)

CASE STUDY   Nancy: Intervention at Three Levels of Motivation *(continued)*

doesn't want to bother with her diet, either—she has enough problems. Her husband is on disability with a bad back, her mother has been having kidney problems, and she has three small children who take up a lot of time. You reflect and justify her feelings and ask whether she wants to talk about her diet and high blood pressure because she is here. Nancy says she doesn't really know why she is in your office, she really doesn't want to talk about her diet, and besides, she has the pills to take care of her blood pressure.

You say, "I respect your decision. It is obvious that any talk about diet at this time would not be useful. You may feel differently in the future. Sometimes when clients first get a diagnosis, they need some time to think about it and are ready to tackle food issues down the line. I want you to know that I will be here to assist you if you would like some help." You offer Nancy some literature about diet and blood pressure. She says she is willing to read the literature and will call you if she feels differently.

Level 2—Unsure, Low Confidence
Assessment

Nancy did not fill out assessment forms before her appointment. She says her doctor wants her to lose weight, but she doesn't believe she has a weight problem. Her husband likes her with "some meat on her." Nancy feels that taking care of three children and holding down a night job as a floor supervisor for a mail delivery service is stressful, and she doesn't need more problems in her life. Her husband is home on disability with a bad back, and the family needs the income and the benefits from her job. You say, "I hear what you are saying. You sound annoyed that you have been given another burden. You don't need any new problems. You certainly are entitled to feel this way with all the responsibilities you are shouldering. However, if you want to explore what you could do about your food intake to lower your blood pressure, we could work together to set goals that would fit into your schedule. Do you want to talk about your blood pressure?" Nancy says yes.

Dietary Evaluation

Although the family is on a tight budget, they have ample money for food. She is active in church activities, which often include food. Sunday dinner is served early—2 P.M.—and is a large meal often attended by her sister and her family and other relatives. On work days, Nancy wakes up at 9:30 in the morning. Her mother fixes her a large breakfast which often includes biscuits, eggs, sausage, juice, coffee, and grits. There will not be another formal meal until dinner which is usually around 4 P.M. because Nancy has to be at work at 6 P.M. Before dinner, she is likely to snack on chips, sweetened soda, or cookies. Dinner usually includes a vegetable, potatoes, and a meat which is often fried, possibly pork chops or chicken. Nancy's mother packs her a meal for work. This usually includes a sandwich made with cold cuts, white bread, mustard; a pickle, chips; cookies; and a candy bar. At work there is a snack room where coffee, tea and donuts, or other baked goods are always available.

You use a typical day strategy combined with a 24-hour recall and complete a short food frequency checklist. Nancy's typical calorie intake is approximately 3,200 calories, 142 grams total fat, 70 grams of saturated fat, and 3,600 mg of sodium per day.

Feedback

You show Nancy a list of lifestyle and food behaviors that can help lower blood pressure, compare Nancy's assessment to the list, and ask Nancy what she thinks. Nancy doesn't think it looks too good, but she is surprised that there is more that she can do than just losing weight. You explain the importance of weight loss and why her doctor emphasized the weight issue but point out that other important diet changes can be made. "By focusing on some of them, weight loss may even occur," you add. You also point out some positive aspects of her diet, such as the collard greens for dinner and the use of skim milk in her coffee.

Readiness

When you ask Nancy her readiness to consider any of the options, she says she is about a 6 on a scale of 1 to 10.

Exploring Ambivalence

You compliment Nancy on keeping the appointment despite her ambivalence and say, "There must be a part of you that believes you should make some diet changes." You ask Nancy why she might want to make some changes. Nancy's mother also has high blood pressure, and as a result, she knows some of the problems that occur with the disease, so she is concerned. However, Nancy feels fine and has the pills to control her blood pressure. You summarize her ambivalence about making changes and ask where that leaves her. Nancy says she would like to do something.

Goals and Action Plan

You point to the list of lifestyle and food behaviors for having a positive impact on blood pressure and Nancy's diet summary and ask what appeals to her. Nancy says she would like to eat more fruit. You and Nancy go through the goal-setting process building on past experiences. Nancy's goal is to eat a banana or an orange for a snack at work. She will buy the fruit on her way home and put a sticky note on the dashboard of her car to remind her to take the fruit to work. You tell Nancy that you would like to do what you can to support her in this endeavour and ask whether it would be all right to call her. Nancy says yes.

Follow-Up

When you call Nancy, she says that she took fruit to work every day for the past week. One of her co-workers has also started bringing fruit, so Nancy feels good that she has a positive influence on someone else. You congratulate Nancy for following through on her goal and tell her that you are there to support her if she wants assistance in making more changes. You express confidence in her ability to continue making dietary changes.

Level 3—Motivated, Confident, Ready

Assessment

Between her appointment with her physician and you, Nancy completed a client assessment questionnaire and a food frequency checklist. Nancy says her doctor suggested that she see the clinic nutritionist to talk about her food intake and losing weight. You used "a typical day strategy." Improving her food selections appeals to her, but the idea of losing weight surprised her. She never thought of herself as overweight, and her husband likes her "with some meat on her." You explain why the doctor suggested weight loss and say, "I will give you some literature to read about blood pressure and weight to help you make a decision. It is one of the best things you could do to help your blood pressure. However, there are still a lot of things we can work on to help lower your blood pressure that may also result in some weight loss." You share with her the results of an analysis of the assessment forms.

Feedback

After reviewing Nancy's assessments and comparing them to a list of lifestyle and food behaviors that can help lower blood pressure, Nancy says she would like to increase her intake of fruits and vegetables. She really likes fruit, but has never been in the habit of eating fruit. Eating vegetables will take some effort.

Readiness

When you ask her about her readiness to increase her fruits and vegetables, Nancy says she very much wants to make a change. On a scale of 1 to 10, she rates her confidence to succeed a 9 for fruits and a 7 for vegetables.

Exploring Ambivalence

Nancy's mother has started to experience some kidney problems from high blood pressure. If Nancy works on her diet, she thinks the whole family will benefit, especially her mother. Also, she wants to do what is best for the children. Nancy needs to be healthy. Right now her husband is home on disability with a bad back, and the family needs her to work for the income and the benefits. Nancy says she is not sure about trying to lose weight and she does have the pills.

Goals and Action Plan

You ask Nancy whether she has ideas about how to increase fruits and vegetables. She says her mother does all the cooking in the house and believes she would be willing to make soup with more vegetables and serve at least one vegetable with dinner. Nancy has noticed the cut-up veggie packs at the grocery and thinks they would be convenient to take to work. Nancy asks the counselor for ideas. You offer her reading material, recipes, and tips for increasing fruits and vegetables. You and Nancy go through the goal-setting process, building on past experiences. Nancy's goal is to

(continued)

CASE STUDY   Nancy: Intervention at Three Levels of Motivation *(continued)*

have homemade soup with vegetables and a four-ounce glass of grape juice on waking. For dinner she will have at least one serving of vegetables and will take a banana or orange and a veggie pack with her to work each day. You talk about self-monitoring and give her a food record designed to track fruit and vegetable intake. You also ask whether you can call her in a week. Nancy says she would like that.

Follow-Up

When you call Nancy, she says that she took fruit and veggie packs to work every day for the past week. Her mother has been making soups and more vegetables for dinner. The counselor reviews her goal of eating three vegetables and two fruits each day. Nancy says she is concerned that her mother is using too much salt when cooking and asks whether her mother can come with her for their next appointment. You tell her that sounds like a good idea. Nancy has noticed that everyone in the house is eating more fruits and vegetables and thinks that is good. You congratulate her on her success, adding that you believe she has the ability to continue making diet changes.

REVIEW QUESTIONS

1. Explain why the arrows for assessment, intervention, and evaluation of a client are reciprocal in the model for a nutrition counseling program in Figure 4.1.

2. Identify the two components of readiness to make a behavior change.

3. Describe four methods to assess readiness to change.

4. Describe each of the four phases of the counseling interview process.

5. Explain "a typical day strategy."

6. Why should a client be allowed to do the initial evaluation after receiving feedback on an assessment?

7. Identify goals and tasks for the resolving phase for each of the three motivational levels in the motivational nutrition counseling algorithm.

8. Explain the components of the FRAMES method for brief interventions.

ASSIGNMENT—CASE STUDY ANALYSIS

Refer to the case study in this chapter to answer the following:

1. Write three family or societal strengths that should be taken into consideration during a counseling intervention with Nancy.

2. The counseling algorithm in Figure 4.2 integrates several intervention models for behavior change. Review discussions of the transtheoretical model, motivational interviewing, self-efficacy and solution-oriented therapy discussed in Chapter 2. Give at least one example of how the methodology for each was illustrated in the case study scenarios about Nancy.

3. Review the resolving phase intervention strategies for motivational levels one and two discussed in this chapter. Select a strategy not illustrated in the case studies that you believe may have been useful in working with Nancy. Explain why you believe the method would be

useful. Write a statement a counselor would make when using the strategy.

4. Review the five relationship-building responses in Chapter 3, and underline examples of each in the three case studies about Nancy. Identify the type of response that was illustrated (that is, reflection, legitimation, support, partnership, or respect), and evaluate the effectiveness of each for the particular situation that was described.

SUGGESTED READINGS, MATERIALS, AND INTERNET RESOURCES

Nutrition Counseling for Medical Nutrition.
Snetselaar LG. *Nutrition Counseling Skills for the Nutrition Care Process.* 4th ed. Sudbury, MA: Jones and Bartlett Publishers; 2008. Provides an abundance of counseling strategies to address requirements for specific dietary modifications.

LIFESTEPS Weight Management Program.
http://www.lifestepsweight.com/ This program is designed to train registered dietitians and other health professionals to lead weight management programs; a five-week program offered on-line three times a year.

Facilitating Behavior Change: Key Strategies for Empowering Your Patients.
http://www.facilitatingbehaviorchange.org/ This is a continuing education program developed by the American Diabetes Association.

WIC Works. www.nal.usda.gov/ The WIC Online Learning section offers users free online registration to access 18 web modules including multicultural communication skills, motivational interviewing counseling skills, and outcomes-based nutrition assessment.

REFERENCES

[1]Snetselaar LG. *Nutrition Counseling Skills for the Nutrition Care Process.* 4th ed. Gaithersburg, MD: Aspen; 2009.

[2]Berg-Smith SM, Stevens VJ, Brown KM, et.al. A brief motivational intervention to improve dietary adherence in adolescents. *Health Educ Res.* 1999;14: 399–410.

[3]Prochaska J, DiClemente C. Transtheoretical therapy: Toward a more integrative model of change. *Psychother Theory Res Practice.* 1982;61:276–288.

[4]Prochaska J, DiClemente C. Toward a comprehensive model of change. In: Miller WR, Heather N. *Treating Addictive Behaviors: Processes of Change.* New York: Plenum; 1986:3–27.

[5]Miller WR, Rollnick S. *Motivational Interviewing Preparing People for Change,* 2nd ed., New York: The Guilford Press, 2002.

[6]Rollnick S, Heather N, Bell A. Negotiating behaviour change in medical settings: The development of brief motivational interviewing. *J Mental Health.* 1992;1:25–37.

[7]Rollnick S, Miller WR, Butler C. *Interviewing in Health Care Helping Patients Change Behavior.* New York, NY: The Guilford Press; 2008.

[8]Miller S, Hubble M, Duncan B., eds. *Handbook of Solution-Focused Brief Therapy.* San Francisco: Jossey-Bass; 1996.

[9]Bandura A. Towards a unifying theory of behavior change. *Psychol Rev.* 1977;84:191–215.

[10]Windhauser MM, Ernst DB, Karanja NM, et. al. Translating the dietary approaches to stop hypertension diet from research to practice: Dietary and behavior change techniques. *J Am Diet Assoc.*1999;99(suppl):S90–S95.

[11]Dyer WW, Vriend J. *Counseling Techniques That Work.* Alexandria, VA: American Counseling Association; 1988.

[12]King NL. *Counseling for Health & Fitness.* Eureka, CA: Nutrition Dimension; 1999.

[13]American Dietetic Association. Position of the American Dietetic Association: Weight management. *J Am Diet Assoc.* 2009;109:330–346.

[14]Funnell MM, Anderson RM. Putting Humpty Dumpty back together again: Reintegrating the clinical and behavioral components in diabetes care and education. *Diabetes Spectrum.* 1999;12:19–23.

[15]Kellogg, M. Tip #74 To Weigh Or Not to Weigh. http://www.mollykellogg.com/tipslist.html Accessed August 20, 2010.

[16]Flegal KM, Graubard BI, Williamson DF, et al. Excess deaths associated with underweight, overweight, and obesity. *J Am Med Assoc.* 2005;293(15):1861–1867.

[17]Herrin M, Parham E, Ikeda J, White A, Branen L. Alternative viewpoint on National Institutes of Health Clinical Guidelines. *J Nutr Ed.* 1999;31:116–118.

[18]Pohl SL. Facilitating lifestyle change in people with diabetes mellitus: Perspective from a private practice. *Diabetes Spectrum.* 1999;12:28–33.

[19]Snetselaar L. Counseling for Change. In: Mahan LK, Escott-Stump S, eds. *Krause's Food, Nutrition, & Diet Therapy,* 10th ed. Philadelphia: Saunders; 2007.

[20]Prochaska JO, Norcross JC, DiClemente CC. *Changing for Good.* New York: Avon; 1994.

[21]Hawkes D, Marsh TI, Wilgosh R. *Solution Focused Therapy a Handbook for Health Care Professionals.* Boston: Butterworth Heinemann; 1998.

[22]Britt E, Hudson SM, Blampied NM. Motivational interviewing in health settings: A review. *Pt Educ Counsel.* 2004;53:147–155.

Developing A Nutrition Care Plan: Putting It All Together

Digital Imagery/PhotoDisc, Inc.

*A life that hasn't a definite plan
is likely to become driftwood.*

—DAVID SARNOFF

Behavioral Objectives

- Develop goals that are specific, achievable, and measurable.
- Differentiate among anticipated results, broad goals, and specific goals.
- Design a plan of action for a goal.
- Evaluate dietary status utilizing standard assessment tools.
- Use common dietary assessment tools.
- Assess total energy expenditure.
- Use standard physical assessment methods to assess healthy weight.
- Define *overweight* and *obesity.*
- Describe functions of charting.
- Describe the four domains of the Nutrition Care Process.
- Use SOAP and Nutrition Care Process documentation formats.

Key Terms

- **ADIME:** four step process of the Nutrition Care Process; assessment, diagnosis, intervention, and monitoring and evaluation.
- **Android Fat Distribution:** waist and upper abdominal fat accumulation; apple shape.
- **Body Mass Index (BMI):** preferred weight-for-height standard; determinant of health risk; predictor of mortality.
- **Dietary Approaches to Stop Hypertension (DASH):** an eating plan focusing on whole foods emphasizing fruits, vegetables, nuts, and reduced sodium.
- **Dietary Assessment:** evaluation of nutrient intake and food patterns.
- **Dietary Reference Intakes (DRI):** four sets of nutrient recommendations for the United States and Canada: Estimated Average Requirements, Recommended Dietary Allowances, Adequate Intakes, and Tolerable Upper Intake Levels.

- **Evidence-Based Guidelines:** based on the scientific method, the best available research evidence for interventions to produce effective outcomes.
- **Gynoid Fat Distribution:** fat accumulation in hips and thighs; pear shape.
- **Joint Commission on Accreditation of Healthcare Organizations (JCAHO):** a non-profit organization that provides accreditation and ensures compliance with established minimum standards to subscriber hospitals and other health care organizations.
- **MyPyramid Food Guidance System:** pictorial representation of five major food groups indicating kinds and amounts of food to consume.
- **Nutritional Assessment:** a comprehensive analysis of an individual's dietary evaluation; medical, medication, and psychosocial history; anthropometric data; biochemical data; and physical examination.
- **Nutrition Care Process:** comprehensive model developed to standardize the process of nutrition care delivery.
- **Obesity:** a body mass index of 30 to 30.9.
- **Overweight:** a body mass index of 25 to 29.9.
- **SOAP Format:** a comprehensive documentation tool; subjective, objective, assessment, and plan.
- **Waist Circumference:** a method to assess upper abdominal fat distribution.

INTRODUCTION

In this chapter, we will review practical factors related to developing a care plan for a nutrition counseling or education intervention. Because interventions can take place in a variety of locations including clinical, commercial, community, and private practice settings, only the basics of developing a plan can be addressed. In addition, an employing facility may have specific guidelines that must be followed. This chapter will address universal best practice procedures, basic tools, and professional essentials for a care plan. We will start by reviewing and practicing the essentials of the goal setting process. Then we will explore various methods for

completing a dietary assessment including utilizing energy standards and commonly used physical assessment tools. The foundation of documentation and charting will be reviewed and demonstrated. Although the American Dietetic Association Nutrition Care Process has been integrated into this book at strategic places, a more thorough review of the process concludes the chapter.

An archer cannot hit the bull's-eye if he doesn't know where the target is.

—Anonymous

GOAL SETTING

In order to make successful behavior changes, individuals need to know what the target is and to clearly "see" it. The story of Florence Chadwick attempting to swim 26 miles from Catalina Island to Palos Verdes, California for the first time illustrates how important having a clearly visible goal helps attainment of an objective. After 15 hours of swimming in shark infested rough waters, a fog descended and she quit, just one-half mile from shore. Florence climbed into an escort boat, and after finding out how close she was to her goal, stated, "If I could have seen land, I know I could have made it."[1] Goal setting is a logical strategy for clients ready to make a behavior change—that is, those at Level 3 and possibly Level 2 in the motivational nutrition counseling algorithm (refer back to Figure 4.2 in Chapter 4). This process enables complex behavior changes to be divided into small achievable steps. Successful small changes improve self-efficacy and motivate clients to continue making lifestyle alterations. Goal setting is a key strategy in the **Nutrition Care Process,** Social Cognitive Theory, Motivational Interviewing, and Cognitive Behavioral Theory.

Counselors should be wary of entering into goal setting too quickly. In order to formulate achievable goals, the groundwork must be done. That means a counselor and client must have fully investigated the nutrition issues of concern. While exploring viable focus areas for making a food behavior change, Berg-Smith et al.[2] emphasize the need for a counselor to convey the following messages to a client:

1. A number of courses of action are available to you.
2. You are the best judge of what will work.
3. We will work together to review the options and select a course of action.

The sections that follow describe general guidelines for establishing goals.

Explain Goal Setting Basics

Before beginning the process of setting a specific goal, you may wish to explain the basics of the goal setting process to your client in order for there to be a clear understanding of the objectives. You can use the mnemonic SMART to remember the components of the process. See Table 5.1.

Explore Change Options

A major objective of this whole process is for you to work in partnership with your client to develop an action plan. Your job is not to be the one setting the goals, but to be sure the stated goal(s) meet the goal setting criteria. Clients must feel a sense of ownership over the plan for goal setting to be an instrument of change.

Elicit Client's Ideas for Change Your client may clearly state ideas about what behavior change he or

Table 5.1   Goal Setting Basics

Letter	Term	Description
S	Specific	Specific goals address the what, why, and how.
M	Measureable	Measureable goals are concrete and observable.
A	Attainable	Attainable goals usually mean small changes that are under the control of the client. They do not depend on another person.
R	Rewarding	Rewarding goals are stated positively.
T	Time bound	Putting an end point on a goal gives a clear target.

she would like to tackle during the assessment interview. The counselor should discuss these statements with the client to determine acceptable options.

Counselor: *When we went over the assessment data, several areas were identified that could be a focus area for making a food change. You know best what would work for you. Is there one particular area that appeals to you?*

There are probably a number of options, but what do you think will work for you?

Consider Using an Options Tool For those who appear to have difficulty selecting a specific area of focus, an options tool could be useful (see Figure 5.1).[2]

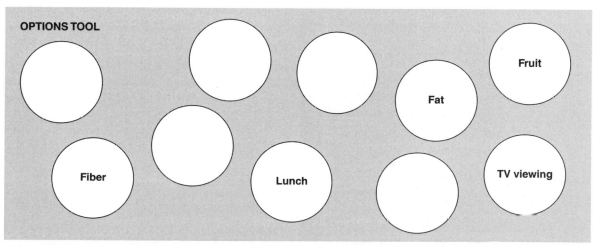

Figure 5.1   Options Tool
Source: Berg-Smith SM, Stevens VJ, Brown KM, et al, A brief motivational intervention to improve dietary adherence in adolescents. *Health Education Research* © 1999; 14:399–410. Adapted with permission of Oxford University Press.

This tool consists of a group of several circles. While reviewing the assessment with your client, you may have identified several areas that could be addressed for behavior change. Write each topic area in one of the circles. Be sure to leave some circles blank for your client. Ask your client what else could be addressed and add any suggestions to the circles.

Counselor: *This options tool may help in the decision process. As you can see, there are a number of circles on the paper. I'd like us to work together to brainstorm ideas of what you could focus on, and we will write them in the circles. As the weeks go by, we can use this tool to help us decide a new area for focusing a goal. What appeals to you the most?*

Explore Concerns Regarding a Selected Option

Probe to investigate any concerns you have about an option selected by your client. Further discussion may convince you to better understand a particular choice, or the process of clarifying could alter or modify a client's choice.

Counselor: *Help me understand why you feel this is the best choice.*

This seems to be the best choice, but will this work for you?

Identify a Specific Goal from a Broadly Stated Goal

Once a broadly stated goal, such as increasing fruit intake, has been selected, the task will be to narrow the focus area down to a specific goal. This specific goal will be one of the small steps that address an intended behavior change and lead to the anticipated outcome (for example, reduction of high blood pressure).

Table 5.2 lists the interrelationship of these concepts. Some highly motivated individuals will be capable of implementing substantial behavior changes; however, for most people, a gradual change in behavior is more likely to result in the desired outcome. Small successes lead to enhanced self-efficacy and more ambitious goal attainment in the future. Trying to accomplish too much too soon is likely to lead to disappointment.

Use Small Talk Botman[3] suggests focusing on identification of the smallest goal that is achievable and worthwhile rather than asking how much a person can accomplish. If a goal is too ambitious, it will not be attainable, and may leave a client feeling not capable of making changes.

Counselor: *What is the smallest specific goal you believe is worth pursuing?*

Explore Past Experiences If a client has not clearly stated a desired goal, then explore past experiences with the broadly stated goal. Successes of the past, no matter how small, are useful starting places for defining an achievable goal. For example, if the general goal was to increase fruit consumption, possible questions include the following:

Counselor: *When have you eaten fruit in the past?*

When did you last eat fruit you enjoyed?

Table 5.2 Anticipated Results, Broad Goals, and Specific Goals

Anticipated Results	Broad Goals—Not Specified Enough to be Achievable	Specific Goals—Concrete Achievable, Measurable, Positive
Decrease blood pressure.	Increase vegetables.	Eat two servings of vegetables for lunch and again at dinner.
Decrease cholesterol.	Increase fiber.	Eat oatmeal for breakfast five times this week.
Decrease cancer risk.	Increase fruit.	Take an apple or orange to work each day to eat during a break.
Maintain or increase bone density.	Increase calcium intake.	Prepare one kale recipe for dinner this week.

Build on Past Successes If a success was identified, try to build on that success. The objective is to identify strengths and skills the client already possesses that can be embellished to implement the new goal.

Counselor: *You said you remembered enjoying a clementine (orange) that was in a bowl at your mother's house. That's great because clementines are in season now and can be purchased in a handy box at the grocery store. Do you believe that placing a bag or bowl of these clementines in a convenient place would help you meet your goal?*

Define Goals

Sound goals are concrete, achievable (under the client's control), measurable, and positive. The goal should identify when, where, how often, and under what conditions the new behavior will occur.

Determine Achievable Goals To be achievable, goals need to be realistic, reasonable, and desirable. If goals are perceived as important, clients have greater determination to achieve them. Obstacles will be viewed as challenges to overcome. Do not overwhelm your client by setting numerous goals at one time. Set one to three goals depending on the needs and skills of the client. Clients are the best judges of what changes are workable. If a client appears averse to taking a risk of setting a goal for a needed behavior change, you may consider approaching the change as an experiment or trial run to evaluate what happens so as to better see what needs to be done in the future to make the change work.

Counselor: *Would you like to experiment with this goal for the next week?*

Define Measurable Goals When a client makes a specific goal, it should be clear when the goal is attained. Goal statements with vague terms such as *good, try, better, more,* or *less* cannot be measured, and clients will have trouble knowing when the goal has been reached. Also, as the need to produce outcome data for funding agencies increases, you will have a hard time reporting your results if you do not have well-defined outcomes. This requires measurable goals. You may consider working with your client to develop a goal attainment scale (GAS) with graduated levels of desired outcomes (see Chapter 7).

Set Goals over Which the Client Has Control Attainment of a goal should depend on the actions of the client, not another person. For example, "I want my husband to . . ." is not a good goal, because it focuses on someone other than the client. If a client can only accomplish the goal through the help of another person, then the goal should be abandoned and a new one sought. However, this does not mean that clients should not seek out the support of others. For instance, if the plan was for your client to walk after dinner three days next week, it would be appropriate to ask a friend or family member to participate. Your client, though, should be making plans that will occur despite another person's involvement.

Weight loss is not a behavioral goal and should never be a short-term goal, although it could be a long-term anticipated outcome. A number of physiological factors affect numbers on a weight scale over which clients have no control.

State Positive Goals What do you think and feel when you imagine a list of foods to avoid on a standard diet? Chances are that you created a depressing image of deprivation and despair. A goal that takes away a pleasurable pastime and leaves an empty void is doomed to failure. Positive images create a greater likelihood of success. For example, a common problem is eating high-calorie, low-nutrient dense foods while watching television. A goal of "I won't eat ice cream while watching television" is likely to leave a client feeling distressed. Better to state what will be done—for example, "Four days a week I will eat one cup of plain, low-fat yogurt with one teaspoon of fruit preserves while watching television. Three days a week I will eat one cup of low-fat ice cream." A non-eating activity, such as grooming the dog, can also be substituted if it appears workable.

EXERCISE 5.1 Analyze and Rewrite Goal Statements

In your journal, record the number of the following goal statements, identify the problem(s) with the statement, and write an alternative goal to meet standard criteria.

Example

Goal: Drink more water.

Problem(s): Not measurable, not specific.

Alternative goal: Take two sixteen-ounce bottles of water to school each day to drink throughout the day.

1. Cut back on salt.
2. Eat more fruit.
3. Try running at least twice this week.
4. Cut down on intake of bread and cheese.
5. Increase strength training.
6. Increase calories by consuming healthy foods.
7. Lose one pound this week.

A journey of a thousand miles must begin with a single step.

— Lao-Tzu

DESIGN A PLAN OF ACTION

Once a well-defined, achievable, and measurable goal has been identified, the next step is to design a plan to implement the goal. The importance of this component of the intervention cannot be over emphasized. Action planning is a significant predictor of both health risk and health protective behaviors beyond the influence of motivational factors and past behaviors.[4] This supports a common quote in weight loss programs, "You fail to plan, you plan to fail." The following reviews the factors that need to be considered for a well developed action plan.

Investigate the Physical Environment

Explore anything in the physical environment that could help or may hinder achievement of the goal.

Counselor: *Your idea of placing a bowl of clementines on the kitchen counter sounds great. Now let's* think about how you will get the clementines. It is usually not a good idea to rely on anyone else to achieve your goals. Is it possible for you to purchase the clementines? Where will you purchase them? When can you purchase them? Your idea of taking a clementine with you as you go out the door to eat for a mid-morning snack on Monday through Friday sounds like a good idea. Is there anything that could happen that would prevent that from occurring? Do you think you need any reminders, like a note in your car or bathroom?*

Examine Social Support

Explore whether there is anyone in your client's environment who can help or hinder achievement of the goal.

Counselor: *Talking about your goal with your co-worker who you usually eat a snack with at the office sounds like a good idea, especially since you said she wants to eat better, too. Maybe the two of you will be a role model for each other.*

Review the Cognitive Environment

Explore your client's cognitive environment regarding the planned goal.

Counselor: *What are you saying to yourself right now about this goal?*

Does it still seem achievable?

Are we being too ambitious to expect five days of fruit for a mid-morning snack?

What will you be saying to yourself if you miss a day?

Explain Positive Coping Talk, If Necessary

If your clients express negative judgments about themselves, suggest that they replace destructive self-talk with a positive coping talk. Explain that a problem is not a failure; it is simply part of the change process. The following is an example of an alternative self-dialogue:

Counselor: *If you find yourself berating yourself, you can substitute another dialogue, such as "I had trouble today. I learned that on days like today I am not likely* *to eat my orange because. . . . . Next time I will . . . . I am on the road to a healthier lifestyle."*

Modify Goal, If Necessary

If the goal appears to be too ambitious, be prepared to modify the goal or to completely put it aside.

Counselor: *It looks as if we got off track on this one. Let us put this one on hold until we meet next week.*

Select a Tracking Technique

Explore how the client would like to keep track of the goal—journal, chart on the refrigerator, empty fruit bowl, and so forth. Tracking procedures are covered in Chapter 6.

Verbalize the Goal

When you believe the goal is clearly defined, ask your client to verbalize the goal.

Counselor: *Just to be sure we are both clear about the food goal that we established for this week, could you please state the goal?*

Write Down the Goal

It has been said that goals that are not written down are just wishes. Write the goal on an index card and give it to your client. This is a good time to make a statement of support to enhance self-efficacy.

Counselor: *I feel we did a good job. I think you are ready to do this and will be able to achieve your goal this week. However, if there is any difficulty, do not despair—some problems are expected.*

The preceding discussion provides a clear map for implementing the goal setting process. For most clients who are ready to make a lifestyle change, the defined process will work well. However, counselors need to keep an open mind, listen carefully to their clients, and remain flexible.

EXERCISE 5.2   Write Your Own Short-Term Goal

Think about an issue, preferably a lifestyle concern, in your own life that you would like to address. In your journal, write a short-term goal that you intend to follow for the upcoming week. Then answer the following questions:

1. Explain why your goal meets the SMART guidelines in Table 5.1.
2. How will you tell whether you have reached your goal?
3. Do you have direct control over the achievement of your goal?
4. Use the assessment graphic, Lifestyle Management Form 4.1 in Appendix D, to identify your motivational level.
5. Design an action plan. Address environmental, social, and cognitive supports or hindrances.
6. How will you monitor progress of the goal?
7. At the end of the week, describe how successful you were. Analyze why or why not. How useful was the goal setting process for changing your behavior?

EXERCISE 5.3   Practice Goal Setting

Practice setting a goal with a colleague. Each of you should take turns playing the role of counselor and client. When you are the client, it would be best to choose a goal to work on that you feel motivated to implement (for example, write a letter to a friend or eat a piece of fruit for breakfast). If that is not possible, then role-play. The acting counselor should work with the acting client to define a goal and work through an action plan. Record your impressions of this experience in your journal.

Be sure you address all steps of the goal setting process:

❑ Explain goal setting process
❑ Explore change options: elicit client's ideas, use options tool, explore concerns
❑ Identify specific goal from broadly stated goal: small talk, explore past, build on successes
❑ Define goal: achievable, measurable, client control, positive

Design a Plan of Action:

❑ Physical environment
❑ Social environment
❑ Cognitive environment: explain coping talk
❑ Modify goal, if necessary
❑ Select a tracking technique
❑ Ask client to verbalize goal
❑ Write goal down

DIETARY ASSESSMENT

A **dietary assessment** is part of a comprehensive nutrition assessment. The components of a particular nutritional assessment will vary according the characteristics of the intervention. Most nutrition assessments contain the following features: dietary evaluation and nutrition related history; anthropometric parameters; biochemical data and medical test results; and medical and clinical history. In certain situations, an additional functional capacity assessment (for example, vision and dexterity) is needed to evaluate food preparation capability. After accurately gathering all of the assessment information and carefully analyzing the results and relationships between data, the assessor then compares findings to criteria, which could be a goal, nutrition prescription, or standard and makes a meaningful evaluation.[5] See Table 5.3 for a list of examples of nutrition assessment data using Nutrition Care Process assessment domains. A nutrition counselor may become involved in all aspects of a comprehensive nutritional assessment; however, the greatest impact will be on the dietary assessment.

Examination of data obtained from assessments can be used to accomplish the following functions:

- Furnish baseline parameters.
- Determine health and nutritional risks.
- Ascertain feasible alternatives for making dietary changes and planning interventions.
- Identify strengths and roadblocks for making lifestyle changes.
- Set priorities for making dietary changes.
- Monitor progress and success of intervention strategies.
- Anticipate appropriate outcomes.

> When goal setting with a client, I always followed the standard criteria of making sure client goals were specific, achievable, measurable, and positive and then developed an action plan. However, once I worked with a client who resisted dwelling on the specifics of implementing her goals. She asked if we could just set a rather clear goal, such as eating three vegetables a day, and not develop an action plan. Somehow this client felt a need to rebel against a specified plan when we had set one. So I followed her request, and she did well in the program. I understood the need to be flexible and to listen to clients' perception of their needs. I felt good that this client perceived that I was open to hear her concerns.

> I had a twenty-year-old male client whose food intake was a nutritionist's nightmare. I believe an eight-year old boy at a birthday party would have eaten better than my client. This person ate almost exclusively high-fat, high-sugar, and high-salt convenience foods. There were no servings of vegetables (except French fries), fruits, dairy, or whole grains. We were having trouble selecting a food goal until I brought up the concept of picturing a meal plate. I was surprised that he connected to this concept because it appeared that he rarely ate what most people would call a meal. Nonetheless, his goal was to eat one meal a day in which his plate consisted of an entrée, grain, and vegetable, each taking up a third of his plate. This goal worked well.

The objective of this section is not to present a comprehensive discussion of nutrition assessment, but to review some commonly used dietary assessment tools and procedures that can be used to practice nutrition counseling skills. Three steps are involved in completing a dietary evaluation: food intake data collection, data analysis, and interpretation of analysis.

Step 1: Food Intake Data Collection

Methods and tools used to collect data to aid in understanding the kinds and amounts of food consumed and the factors influencing choice include a client assessment questionnaire, food diary and daily food record, usual diet, diet history interview, food frequency, and 24-hour recall. Although these are the standard forms, a counselor working in the field of nutrition will find that a range of assessment tools are available varying in purpose, length, and complexity.[6] Selection of a particular instrument will depend on the objective of the interaction (the type of clientele), initial or follow-up sessions, number of planned visits, and available resources (computer, time, and so forth). Sometimes an instrument is needed to address a specific issue, such as hypertension,[7] readiness to lose weight,[8] or psychosocial variables. For example, the Accu-Check Interview is a computer software program used by diabetes educators to screen for depression or emotional distress while using a motivational interviewing approach.[9] At other times the assessment tool is employed as a general screening device and has a broader perspective, such as the "Determine Your Health" checklist or the Americans Mini Nutritional Assessment (MNA) developed for older adults.[10,11] A discussion of both forms can be found in Chapter 9.

Table 5.3 Nutrition Assessment Data Domains and Examples

Nutrition Care Process Domains (types of data)	Selected Examples
Food and Nutrition-Related History	• Food and nutrient intake • Food and nutrient administration • Medication and herbal supplement use • Knowledge, beliefs, attitudes, including readiness to change • Factors affecting access to food and food- and nutrition-related supplies • Physical activity and function • Nutrition-related patient- and client-centered measures including quality of life
Anthropometric Measurements	• Height and length • Weight • Body mass index • Waist circumference
Biochemical Data, Medical Tests, and Procedures	• Laboratory data such as hemoglobin, glycosylated hemoglobin (HbA_{1c}), lipid profile • Tests such as resting metabolic rate
Nutrition-Focused Physical Findings	• Appetite • Loss or excess of subcutaneous fat • Vital signs including blood pressure
Client History	• Medical and health history • Complementary and alternative medicine treatments • Social history (socioeconomic factors, housing situation, medical care support, etc.)

Source: American Dietetic Association. International Dietetics & Nutrition Terminology (IDNT) Reference Manual Standardized Language for the Nutrition Care Process. 3rd ed. Chicago, IL: American Dietetic Association; 2010.

While collecting data from your client, be careful not to give advice, "preach," or condemn because such remarks are likely to be interpreted as judgmental and inhibit the free flow of information. Even words of approval should be avoided because this could encourage a client to give information perceived as good answers. Some people who regularly conduct dietary assessments find it useful to tell clients that there are no wrong answers. Others tell clients not be afraid to give an answer because no one has a perfect diet, and the counselor may even give an example of a low nutrient-dense food he or she enjoys.

Each type of tool has advantages and pitfalls. A summary of the strengths and limitations can be found in Table 5.4. By using more than one instrument, the probability of obtaining a clear picture of your client's nutrition strengths and problems increases. Let's take a look at the features of several assessment tools.

Client Assessment Questionnaire (Historical Data Form) Sometimes referred to as an *intake form,* client assessment questionnaires generally contain several divisions addressing information about historical data. See Lifestyle Management Form 5.1 in Appendix D for an example. The top of this form has an administrative section and is usually followed by questions related to medical history. These questions are not asked for the purpose of making a diagnosis, but to ascertain any medical factors that could have a nutritional impact. The family health history portion of the form provides information about a possible tendency toward a particular health condition. This has nutritional implications for your client. For example, a family history of heart disease may warrant an emphasis on heart-healthy foods.

Drug history questions provide information about medications, herbal preparations, and nutrient supplements that may impact nutritional status. Care should be taken to check for any interactions

Table 5.4   Summary of Methods, Strengths, and Limitations of Selected Diet Assessment Tools/Procedures

Method	Strengths	Limitations
Client assessment questionnaire and historical data form: a preliminary nutritional assessment form usually divided into sections for administrative data, medical history, medication data, psychosocial history, and food patterns	• Provides clues to strengths and potential barriers	• May seem invasive • May not be culturally sensitive
Food diary and daily food record: a written record of an individual's food and beverages consumed over a period of time, usually three to seven days	• Does not depend on memory • Provides accurate intake data • Provides information about food habits	• Requires literacy • Requires a motivated client • Recording process may influence food intake • Requires ability to measure and judge portion sizes • Time-consuming
24-hour recall: a dietary assessment method in which an individual is requested to recall all food and beverages consumed in a 24-hour period	• Quick • Data can be directly entered into an analysis program • No burden for respondent • Does not influence usual diet • Literacy not required	• Relies on memory • May not represent usual diet • Requires ability to judge portion sizes • Under reporting and over reporting occurs
Food frequency: a method of analyzing a diet based on how often foods are consumed (that is, servings per day, week, month, or year)	• Furnishes overall picture of diet • Not affected by season • Useful for screening	• Requires ability to judge portion sizes • No meal pattern data
Usual diet: clients are led through a series of questions to describe the typical foods consumed in a day	• May be more of a typical representation than a 24-hour recall	• Not useful if diet pattern varies considerably
Diet history interview: a conversational assessment method in which clients are asked to review their normal day's eating pattern	• Provides clarification of issues	• Relies on memory • Requires interview training

of food and drugs that can alter the effectiveness of a drug and a client's nutritional status.

A section on socioeconomic history furnishes valuable information about your clients' support systems, family settings, or significant others—any of which can play a role in their ability to make successful diet changes. This information can be especially helpful during the goal setting process when exploring with your clients whether there are particular individuals in their lives who may interfere or help them achieve their goals.

The food pattern section, often referred to as *dietary history,* contains questions about food preferences and food selection variables that influence food intake. This knowledge will be particularly helpful for prioritizing goals and designing interventions. A final section requesting clients to identify nutrition issues they would like to explore helps in planning the educational component of future sessions.

These forms can be tailored to meet counseling needs for specific clientele. For example, if serving mainly low-income individuals or students living in a dormitory, you would have a greater need for questions about accessibility to refrigeration and cooking facilities. A form to be used with eating disorder clients could contain specific questions about laxative use or purging, and a form for weight control clients may have a request to detail weight history.

Portion Size Many of the following methods require clients to estimate or measure their portion sizes. Some short videos are available that can be used as an instruction tool for clients who will be keeping food records. (See the resources at the end of the chapter.) Several aids have been employed to help respondents recall portion sizes for retrospective data collection. These include two- and three-dimensional food models; various shapes of cardboard or plastic household cups, bowls, plates, glasses, and spoons; life-size photographs; graduated measuring spoons and cups for liquid and dry ingredients; and a ruler. Containers with two to three cups of dried beans, rice, or dry cereal can also be helpful for estimating portion sizes, as can premeasured plastic or net bags of beans in sizes equal to one cup, one-half cup, and one-quarter cup. Here are some other portion size equivalents, from the International Food Information Council:

Commonly Used Estimates of Portion Sizes
One-half cup fruit, vegetable, cooked cereal,
  pasta, or rice = a small fist
Three ounces cooked meat, poultry, or fish =
  a deck of cards
One tortilla (1 oz.) = a small (six inch) plate
Half bagel (1 oz.) = the width of a small
  drink lid
One teaspoon of margarine or butter =
  a thumb tip
One tablespoon of peanut butter = two checkers
A small baked potato = a computer mouse
One pancake or waffle (1 oz.) = a four-inch CD
One medium apple or orange (1 cup) =
  a baseball
1.5 ounces of cheese = six dice
1.5 cups soft drink or fruit drink (12 oz.) = 1 can

EXERCISE 5.4   *Estimating Portion Sizes*

MyPyramid no longer uses the term "servings" or portions for recommendations. All food group recommendations for MyPyramid are made in household units (cups for fruits, vegetables and milk, and ounce equivalents for grains and meat and beans). For clients to follow these recommendations, they still need to visualize a cup of vegetables or an ounce of meat to follow the recommendations. The **DASH** Food Plan and the Dietary Exchange List do use serving sizes. As nutrition counselors and educators, we need to be familiar with these resources so we need to work at visualizing amounts of food. Also, we need to help our clients visualize amounts eaten when completing dietary assessments. The following provides two activities to help with the process:

1. Go to the National Heart Lung and Blood Institute website, http://www.nhlbi.nih.gov and search for Portion Distortion. Download the serving size card. Complete both slide sets.
   ❑ Write two reactions to the activity in your journal, and explain what you learned that will help you as a nutrition professional.
2. Estimate the amount of liquid, cereal, and beans in various-sized cups, glasses, bowls, and plates set up by your instructor. In addition, estimate the serving sizes of each food item in both "standard" and "large portion" TV dinners. Measure the quantities and compare your findings to the NHLBI serving sizes indicated on the serving card.
   ❑ Record your reaction to this activity in your journal. Indicate how this experience relates to future counseling experiences.

Food Diary and Daily Food Record To employ this method, a client records food and liquid intake along with preparation method as it occurs for a specified period, generally three to seven days.[12] Sometimes additional information is recorded such as time, place, activities, social setting, degree of hunger, and emotional state. A limitation of this tool for assessment is the impact recording can have on food intake. The hassle of writing a food item in a journal could discourage consumption of some foods, and the activity of recording encourages a person to take time to evaluate the particular choice. As a result, food diaries can be used as an intervention technique to alter food habits as discussed under journaling in Chapter 6. Review of food records is especially useful for both the counselor and client to gain insight into the client's eating lifestyle. Identification of positive behaviors may help identify skills that merit expansion to help solve problem areas.

Clients need to be given directions for completing food record forms and guidelines for measuring, weighing, and estimating portion sizes. Because accuracy is thought to decline if weighing all food items is requested (clients may be less likely to eat some foods due to the burden of weighing), household measures are generally considered acceptable.[13] See Lifestyle Management Form 5.2 in Appendix D for an example of a food diary recording form.

Usual Intake Form The usual intake form gives a counselor an idea of a client's typical daily pattern of food intake. This form is simple and generally not time-consuming to complete. However, its usefulness will be limited for clients whose intake varies widely from day to day. In such cases, answering general questions would be difficult, and another assessment tool should be used.

The assessor begins by inquiring into the client's initial food or drink of the day. This line of questioning continues until a daily pattern has emerged. The counselor must refrain from asking leading questions that may influence answers. Probing questions should be asked to ascertain the nutrient characteristics of the items consumed. For example, if a sandwich is generally consumed for lunch, investigate type of bread, filling, and condiments. See Lifestyle Management Form 5.3 in Appendix D for an example of a usual intake form.

Diet History Interview A diet history interview is similar to asking clients about their usual diets; however, the emphasis is on minimizing questions and allowing clients to tell their "stories." Clients are invited to give an account of a normal day's eating pattern, with the counselor utilizing attending skills and interrupting as little as possible. In

this respect, the technique is similar to "a typical day" strategy covered in Chapter 3. After the narrative, the counselor selectively chooses follow-up questions to obtain only new and relevant information.

The conversational emphasis of this approach interfaces well with the motivational nutrition counseling protocol (Chapter 4) as well as the culturally sensitive respondent-driven interview (Chapter 9). During the process of conducting a diet history interview, a counselor could use the 24-Hour Recall and Usual Intake Form (Lifestyle Management Form 5.3 in Appendix D) to record a client's diet pattern. However, the act of completing the form should not be allowed to interfere with your clients' ability to relate to their stories. The conversational nature of this approach will be disrupted if clients are asked to repeat something, and attending skills will be less than adequate if a counselor is eying a piece of paper for most of the interview. Quickly jotting down notes during the story would probably work well. Alternatively, the usual diet form could be filled out after the client has related his or her story while probing questions are used for clarification. See Exhibit 5.1 for a protocol of the method.

Food Frequency Checklist The food frequency checklist is an assessment tool containing lists of food grouped according to similarity in nutrient quality and quantity. They are designed to be either read to clients by an interviewer or distributed in printed form for self-administration. The form contains a set of response options to be checked off that indicate how often certain foods are consumed (for example, by the day, week, month, and so forth). Some questionnaires are designed to consider one or two specific nutrients, such as calcium or fat and cholesterol, whereas others are comprehensive in scope such as the NHANES Food Frequency Questionnaire.[14] Food frequency questionnaires vary in the amount of detail requested regarding serving size and preparation methods. If they are too short, the knowledge gained is limited. If they are too long, the process of completing the form is tedious and accuracy can decline.

EXHIBIT 5.1   Protocol for Diet History Interview

Diet History

The purpose of the diet history is to obtain an account of a person's usual food intake. Structurally, it takes the form of a description of meals consumed throughout the day accompanied by a food frequency cross-check. One way of looking at the first component of the diet history is as a story with a beginning (usually breakfast) and end (usually supper or evening snack). Use of the narrative approach means that participants are given the opportunity to finish their story before they are asked any questions. In this way, the flow of participants' information is not interrupted (but what they say is acknowledged and supported by the interviewer). Additional comments, not necessarily on food per se, made during this description may provide the interviewer insights for questions or discussion addressed later. When introducing the diet history, the interviewer refers to the notion of usual, meaning within the past couple of months, and of a time sequence for the description, such as the duration of the day. Participants are asked to provide a general pattern and then point out variations.

Interview Protocol

- Explain the purpose of the interview. Advise the participant that you are seeking a description of usual eating patterns and suggest that she or he start with the beginning of the day.
- If the participant begins with the first meal of the day and uses time references or meal sequences of the day to progress with the description, do not interrupt the story; merely indicate that you are listening (nod, write, say "hmm" or "yes").
- If the participant stops at intervals along the way waiting for you to respond, provide narrative support to continue— for example, "Was that all for breakfast?" or "Do you have anything after that?"
- If the participant volunteers explanations for why or how she or he consumes certain foods, acknowledge the explanations in a supportive, nonjudgmental way, but keep the account on track.
- When the participant has reached the end of the day, look at what you have noted and identify areas for which you need more detail. This will depend on the purpose for taking the history. Ask specific, strategic questions.
- If the participant responds to a topic with "It depends," be sure to encourage all possible variations on that topic (usually a meal description).
- If the participant says "probably" in defining amounts of foods, use visual aids to assist in the estimation process.
- Summarize the overall pattern of the diet and ask whether there is a great deal of variation in this pattern. Note the variation.
- Proceed with a food frequency checklist and questions on food preparation.
- Ask the participant if there is anything else she (or he) would like to add to the diet story and if she or he thinks you have a true reflection of her or his usual eating pattern.

Source: Reprinted from Tapsell LC, Brenninger V, Barnard J, Applying conversation analysis to foster accurate reporting in the diet history interview. © 2000, The American Dietetic Association. Reprinted by permission from Elsevier.

In addition, the instrument may not list ethnic or child-appropriate foods. In a clinical setting, you may consider a general food frequency questionnaire and a 24-hour recall in order to better evaluate your client's food intake.

Overall, this method is easy for most people to use. This questionnaire helps counselors evaluate diet in terms of how often certain foods and food groups are eaten. Food groups not eaten often or omitted are indications of dietary imbalances. Close attention should also be given to the nutritional desirability of frequently consumed foods.

(See Figure 5.2 and Lifestyle Management Form 5.4 in Appendix D for examples.)

24-Hour Recall In this method, the interviewer asks the client to recall all foods, beverages, and nutritional supplements consumed, including amounts and preparation methods, over a twenty-four-hour period. Counselors can define the period of time from midnight to midnight of the previous day or the past twenty-four hours.[13] The starting point can be the most recent or the most distant of the twenty-four-hour period. The 24-Hour Recall

American Medical Association

Physicians dedicated to the health of America

Eating Pattern Questionnaire

Name _____ Date _____

Please answer the following questions and check the appropriate boxes that most closely describe your eating patterns.

1. Do you follow a special diet?
   - ☐ No   ☐ Diabetic   ☐ Low sodium
   - ☐ Low fat   ☐ Kosher   ☐ Vegetarian
   - ☐ Other

   Give examples of what guidelines or diets, if any, you follow: _____
   _____
   _____

2. Which meals do you regularly eat?
   - ☐ Breakfast   ☐ Lunch   ☐ Brunch   ☐ Dinner

3. When do you snack?
   - ☐ Morning   ☐ Afternoon   ☐ Evening
   - ☐ Late night   ☐ Throughout the day

   What are your favorite snack foods? _____
   _____
   _____

4. Do you eat out or order food in?
   - ☐ Yes   ☐ No

   How often?
   - ☐ Daily   ☐ Weekly   ☐ Monthly   ☐ Other
   _____

   What kind of restaurant(s)/eating facilities? _____
   _____

   What kinds of cuisine? _____
   _____
   _____

5. How is your food usually prepared? (check all that apply)
   - ☐ Baked   ☐ Broiled   ☐ Boiled   ☐ Fried
   - ☐ Steamed   ☐ Poached   ☐ Other

6. How many times each day do you have the following food items?
   a. Starch (bread, bagel, roll, cereal, pasta, noodles, rice, potato)
      ☐ Never   ☐ Less than 1   ☐ 1–2   ☐ 3–5   ☐ 6–8   ☐ 9–11
   b. Fruit
      ☐ Never   ☐ Less than 1   ☐ 1–2   ☐ 3–5   ☐ 6–8   ☐ 9–11
   c. Vegetables
      ☐ Never   ☐ Less than 1   ☐ 1–2   ☐ 3–5   ☐ 6–8   ☐ 9–11
   d. Dairy (milk, yogurt)
      ☐ Never   ☐ Less than 1   ☐ 1–2   ☐ 3–5   ☐ 6–8   ☐ 9–11
   e. Meat, fish, poultry, eggs, cheese
      ☐ Never   ☐ Less than 1   ☐ 1–2   ☐ 3–5   ☐ 6–8   ☐ 9–11
   f. Fat (butter, margarine, mayonnaise, oil, salad dressing, sour cream, cream cheese)
      ☐ Never   ☐ Less than 1   ☐ 1–2   ☐ 3–5   ☐ 6–8   9–11
   g. Sweets (candy, cake, regular soda, juice)
      ☐ Never   ☐ Less than 1   ☐ 1–2   ☐ 3–5   ☐ 6–8   ☐ 9–11

7. What beverages do you drink daily and how much?
   - ☐ Water   ___ times or glasses per day (8 oz)
   - ☐ Coffee   ___ times or cups per day
   - ☐ Tea   ___ times or cups per day
   - ☐ Soda   ___ times or glasses per day (12 oz)
   - ☐ Alcohol ___ times or glasses per day (12 oz)
   - ☐ Other   ___ times or glasses per day
   (Specify) _____

8. Would you like to change your eating habits?
   - ☐ Yes   ☐ No

   Which habits would you like to begin to change?
   _____
   _____
   _____
   _____

Figure 5.2   Eating Pattern Questionnaire
Source: Copyright 2003. American Medical Association. All Rights Reserved.

Form is similar to a Usual Diet Form, and these two tools have been combined in Lifestyle Management Form 5.3 in Appendix D.

An advantage of this method is that it is easy to administer and requires little effort on the part of the client. However, one day may not be representative of a person's usual intake. This difficulty can be overcome if the 24-hour recall is administered on several nonconsecutive days, including both weekdays and weekend days.[15] Although this form can be self-administered, the accuracy increases if counselors assist their clients in recalling their food

---

EXERCISE 5.5   Practice Gathering Information for a Dietary Assessment

Complete a client assessment questionnaire and a food frequency questionnaire (Lifestyle Management Forms 5.1 and 5.4 in Appendix D) based on your own diet history and food habits. Exchange forms with a colleague and take turns acting as a counselor. Gather information using the following interview guide. The collected data will be evaluated in Exercise 5.8.

❏ Ask your client whether she or he has any nutritional concerns.

❏ Review the completed Client Assessment Questionnaire, Lifestyle Management Form 5.1 in Appendix D—*What came to your mind as you were filling out this form? What topics covered in this form do you think have particular importance for your food issues?* Look over the form and ask for clarification where appropriate. Your client may have already covered relevant issues in response to your previous open questions.

❏ Conduct a diet history interview. Follow the protocol in Exhibit 5.1. While your client is telling you his or her story, fill in the 24-Hour Recall and Usual Diet Form, Lifestyle Management Form 5.3 in Appendix D.

❏ Summarize.

❏ Review the completed Food Frequency Questionnaire, Lifestyle Management Form 5.4 in Appendix D. Clarify portion sizes using food models or the serving size cards downloaded in Exercise 5.4, if needed—*Thank you for completing the Food Frequency Questionnaire. What came to your mind as you were filling out this form? Did you feel a need to clarify or expand on anything while you were completing this form?*

○ In your journal, write your impressions of this experience as a client and as a counselor.

---

consumption and portion sizes.[6] The following are some components of an effective 24-hour recall:

- Do not ask leading questions, such as those assuming a meal was eaten. Refrain from prompts, such as "What did you eat for breakfast?" Better to ask, "What liquids or foods were first consumed after waking up today?"

- Ask probing questions. For example, "You said you had a lot of butter on your toast. How much is a lot? What kind of bread was used to make the toast? Was anything else put on the toast besides butter?"

- Ask sequential questions about the day's activities, travels, and encounters with others to help clients recall foods consumed. Inquire if any foods were consumed during meal preparation or clean-up or during the middle of the night.

- Use portion size estimation tools to improve the accuracy of the answers.

- Research has shown that certain food items are frequently missed in twenty-four-hour recalls—crackers, breads, rolls, tortillas; hot or cold cereals; cheese added as a topping on vegetables or on a sandwich; chips, candy, nuts, seeds; fruit eaten with meals or as a snack; coffee, tea, soft drinks, juices; and beer, wine cocktails, brandy, any other drinks made with liquor. Sugerman et al.[6] suggest going over this list before completing the recall to be sure none of the items were missed.

- Another aid to increasing retrieval of memory is the *multiple-pass procedure*.[16] Depending on your need for accuracy, you may consider utilizing the validated USDA 5-Step multiple-pass method.[17,18] The five steps include: (a) a quick list, in which clients recall foods and beverages consumed in sequence during a twenty-four-hour period without interruption; (b) clients are queried on the forgotten food list, as described previously; (c) clients are probed to recall time and occasion at which foods were consumed; (d) the detail cycle, in which clients are asked to provide descriptions of foods and amounts with the aid of models and measuring guides; and (e) the final probe review, clients are questioned regarding type, amounts, additions and toppings, and preparation methods.

EXERCISE 5.6 Conduct a 24 Hour Recall

Using Lifestyle Management Form 5.3 in Appendix D, take turns with a colleague administering a twenty-four-hour recall. Use USDA 5-Step multiple-pass method and visual aids to help estimate portion sizes.

❏ In your journal, write your impressions of the experience.

Culturally Appropriate Assessment Instruments There is a critical need for the development of culturally specific techniques and tools to conduct nutritional assessments.[19] Depending on communication difficulties and cultural feelings about invasiveness, a counselor may find a qualitative rather than a quantitative approach to yield greater success.[20] To establish trust during the first session, consider using "a typical day technique" reviewed in Chapter 4 or the diet history interview data collection method covered in this chapter. These approaches eliminate the need to differentiate meals or categorize food items. A request for additional information such as frequency and portion size could be delayed until the next meeting.

Step 2: Data Analysis

After dietary information is collected, the data needs to be analyzed for food groups and components of food, such as energy, nutrients, or phytochemicals.

- **Food group evaluations** can generally be done quickly, making immediate feedback possible. Some forms have the standards on the collection form or as an attachment to the assessment form allowing for a speedy evaluation. See Lifestyle Management Form 5.3 in Appendix D.
- **Food component analysis** is rather time-consuming, so generally feedback cannot be given the same day data are collected. Nutrients can be analyzed from food composition tables, the U.S. Department of Agriculture (USDA) Nutrient Database (www.nal.usda.gov) or with the aid of a nutritional analysis software program.

Step 3: Interpretation of Analysis

Interpretation of analysis of dietary information is done by comparing the data analysis to a standard. Computer programs automatically execute both steps 2 and 3—that is, analyze and interpret, generally for food groups and nutrients. The following describes the most commonly used standards:

- The **MyPyramid Food Guidance System** MyPyramid is a pictorial representation of five major food groups and can be found at: http://www.mypyramid.gov/. Food intake data can be entered into the site and an analysis of food intake compared to individualized recommended food group amounts as well as for many Dietary Reference Intakes. The website provides a host of educational and motivational materials.
- **Dietary Reference Intakes (DRI)** DRI are commonly used standards when assessment of specific nutrients is desired and can be found on the inside cover. They are divided into four categories: Estimated Average Requirements (EARs), Recommended Dietary Allowances (RDAs), Adequate Intakes (AIs), and Tolerable Upper Intake Levels (ULs). For a description of these divisions, see Table 5.5 and the USDA Website (www.nal.usda.gov/).
- **Dietary Approaches to Stop Hypertension (DASH)** For clients who desire a more ambitious dietary regimen, comparisons could be made to the DASH eating plan, particularly if you are working with an individual who has elevated blood pressure. The National Heart Lung and Blood Institute has a booklet, *In Brief: Your Guide to Lowering Your Blood Pressure with DASH*, and can be downloaded from their website. The DASH Food Plan is a heart-healthy regimen rich in fruits, vegetables, fiber, and low-fat dairy foods and low in saturated and total fat. See Appendix A. Although the DASH food plan was developed to address high blood pressure, the plan has been found to be useful for everyone to guide healthful eating. After reviewing research findings of the DASH food plan, developers of MyPyramid incorporated components of

Table 5.5 Recommended Dietary Intake Terms

Term	Definition
Adequate Intake	Amount of a nutrient that maintains a function; used when recommended dietary allowance (RDA) cannot be determined
Recommended Dietary Allowances (RDAs)	The amount of a nutrient covering the needs of nearly all healthy individuals
Estimated Average Requirements (EARs)	Amount of a nutrient estimated to meet the requirement of half the healthy people in a given age and gender group
Tolerable Upper Intake Level (UL)	Maximum level of a nutrient that appears safe

EXERCISE 5.7 Explore Interactive DRI for Healthcare Professionals Website

Go to the USDA Food and Nutrition Information Center to explore the Interactive DRI for Healthcare Professionals at: http://fnic.nal.usda.gov/interactiveDRI/. Use this tool to calculate your BMI, daily calorie needs, and daily nutrient recommendations for dietary planning based on the Dietary Reference Intakes (DRIs).

❑ In your journal, describe your experience using the website. Explain how a nutrition educator or counselor could use this site for nutrition interventions.

EXERCISE 5.8 Data Analysis and Interpretation

To complete this activity, work with the same colleague you paired with for Exercise 5.5.

❑ Review the feedback form, point by point, in a nonjudgmental manner with your client. Compare the standards to your volunteer's food intake. *As you can see your usual vegetable intake is one cup a day, and the MyPyramid recommendation is about 2 ½ cups a day.* Continue in this vein until you have gone over all the findings.
❑ Ask your client his or her impression of the evaluation. *What do you think about this information?*
❑ If your client expresses interest in making a change, use the Assessment Graphic (Lifestyle Management Form 4.1 in Appendix to determine degree of motivation.
❑ Summarize.

○ Write your reactions to this exercise in your journal. What did you learn from this experience that you would like to incorporate or change when working with future clients?

the plan into MyPyramid recommendations. Some nutrition professionals like to use this plan to do a quick assessment of a client's food intake so immediate feedback can be provided.
- Canada's Food Guide to Healthy Eating for People Four Years and Older (http://www.phac-aspc.gc.ca/) can also be used to assess diet for general good health.

ENERGY DETERMINATIONS

Nutrition counselors may need to estimate *total energy expenditure* (TEE) of their clients for a variety of therapeutic reasons, including planning a weight loss program. There are three components making up TEE: *resting energy expenditure*, the *thermic effect of food*, and *energy expended in physical activity*. Generally only resting energy and physical activity energy are calculated in counseling interventions. The thermic effect of food is often omitted because the inherent error factor of the total equation is greater than the amount that would be added due to the thermic effect value.[21] The following sections outline the steps for calculating TEE.

Step 1: Determine Resting Energy Expenditure (REE)

REE is the energy needed to sustain life functions, such as respiration, beating of the heart, and kidney function over a 24-hour period. Nutrition counselors generally determine REE by using a formula or by indirect calorimetry as shown in Figure 5.3. See the end of the chapter for office-sized indirect calorimetry equipment resources. Several standard

Figure 5.3 Indirect Calorimetry
Source: Courtesy Korr Medical Technologies.

formulas can be used to estimate REE. See Table 5.6 for two commonly used formulas. The American Dietetic Association (ADA) Evidence Analysis Library recommends the use of indirect calorimetry to measure REE for overweight and obese individuals, and if that is not available they recommend using the Mifflin-St Jeor Equations for this population group.

Step 2: Select a Physical Activity (PA) Factor

Table 5.7 provides factors for physical activity level.

Step 3: Determine TEE

Multiply REE times PA to obtain the estimated TEE (kilocalories/day) to maintain weight.

$$REE \times PA = TEE$$

Step 4: Adjust for Weight Loss

If weight loss is desired, subtract 500 kilocalories/day to obtain adjusted caloric intake required to achieve weight loss of approximately one pound per week.

EXERCISE 5.9   Calculate Your Total Energy Expenditure (TEE)

Calculate your TEE using the Mifflin–St. Jeor and the Harris-Benedict equations for REE.

❏ Compare and contrast the two methods in your journal.

Table 5.6 Equations for Determining Resting Energy Expenditure for Adults*

Name	Equation
Mifflin-St Jeor Equations	Men: REE = [10 × weight (kilograms)] + [6.25 × height (centimeters)] − (5 × age) + 5
	Women: REE = [10 × weight (kilograms)] + [6.25 × height (centimeters)] − (5 × age) − 161
Harris Benedict Equation	Men: REE = [66.5 + 13.8 × weight (kilograms)] + 5.0 × height (centimeters) − 6.8 × age
	Women: REE = 655.1 + 9.6 × weight (kilograms) + 1.9 × height (centimeters) − 4.7 × age

To convert inches to centimeters, multiply inches times 2.54.
To convert pounds to kilograms, divide pounds by 2.2.

* Values rounded for simplicity.

Table 5.7 Physical Activity (PA) Level Factors

Activity Level	PA Men	PA Women	Typical Daily Living Activities
Sedentary	1.00	1.00	Only physical activities typical of daily living
Low active	1.11	1.12	30–60 minutes of moderate activity
Active	1.25	1.27	≥60 minutes of moderate activity
Very active	1.48	1.45	≥60 minutes of moderate activity plus 60 min. vigorous or 120 minutes of moderate activity

Note: Moderate activity is equivalent to walking at 3 to 4½ mph.

PHYSICAL ASSESSMENTS AND HEALTHY WEIGHT STANDARDS

Because weight issues are related to many of the major health problems in North America, nutrition counselors often need to address overweight or **obesity** concerns. The prevalence of these conditions is increasing globally as well as in the United States.[22] The 2009 Center for Disease Control Behavioral Risk Factor Surveillance data indicates about 63% of Americans are **overweight**, and more than 27% are obese.[23] See Figure 5.4. The following sections describe commonly used methods and standards for assessing weight.

Weight-for-Height Tables

In the past, height-weight tables were widely used to determine healthy weight. Authorities now discourage their use because the population used to establish the tables did not represent the current ethnic and racial distribution in the United States. Also, development of the data made several assumptions including weight of clothing, height of shoe heels, and frame size, which made the validity of guidelines questionable.

Body Mass Index

Body mass index (BMI) is the preferred weight-for-height standard and is used as a determinant of health risk and a predictor of mortality. It can be determined from existing tables, equations, or web sites. A BMI chart can be found in Appendix B. The following National Heart, Lung, and Blood Institute website provides downloadable BMI tables, an electronic calculation of BMI, and downloadable calculators for iPhone, Palm OS, or PocketPC 2003: http://www.nhlbi.nih.gov/index.htm. Clink on Health Professionals for resources. The standard calculation is based on metric units, but BMI can be estimated from another equation using pounds and inches:

BMI = weight (kilograms) ÷ height (meters) squared
(1 pound = 0.4536 kilogram)
(1 inch = 2.54 centimeters = 0.0254 meter)

BMI = (weight [pounds] ÷ height [inches]^2) × 703

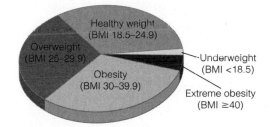

Figure 5.4 Distribution of Body Mass Index in U.S. Adults
Source: From WHITNEY/ROLFES. Understanding Nutrition, 12E. © 2011 Brooks/Cole, a part of Cengage Learning, Inc. Reproduced by permission. www.cengage.com/permissions

In general, a healthy weight-for-height is a BMI ranging from 18.5 to 24.9, with a midpoint of 22. It can be used to conveniently calculate an individual's goal weight. See Exhibit 5.2. The risk for developing associated morbidities or diseases such as hypertension, high blood cholesterol, type 2 diabetes, and coronary heart disease begins to climb above the desirable range (25).[22] Mortality risk increases for both underweight and overweight (Figure 5.5). In the United States, two-thirds of adults have a BMI greater than 25.[24] See Table 5.9 for weight classification according to body mass index.

The current BMI standards for increased risk for disease and mortality do not meet the needs of all population groups. The World Health Organization report suggests using 23 as the level indicating overweight and 27.5 as the cutoff for obesity for most Asians.[25,26] For Pacific Islanders, the cut off for overweight has been identified as 26. BMI cannot be

EXHIBIT 5.2   BMI Chart Used to Determine Healthy Weight

Find your client's height and run your finger along the corresponding horizontal line until you come to the weight that matches the desired BMI, such as 24. That weight will be the healthy body weight, which could be used as a goal weight. For example, look at the BMI chart in Appendix B. Suppose a man was five feet, six inches tall and wanted to know a healthy weight. You begin at the horizontal line corresponding to that height and run your finger along the line until you reach the weight number that vertically corresponds to 24. That weight is 148 pounds. If the person's weight is 158 pounds, that would mean he would need to lose 10 pounds to achieve a BMI of 24.

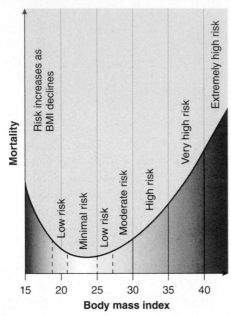

Figure 5.5 Body Mass Index and Mortality

EXERCISE 5.10   Use the BMI Chart to Determine Healthy Body Weight

Determine the healthy body weight for a male who is five feet, seven inches tall using the BMI chart in Appendix B. For guidance refer to Exhibit 5.2.

❏ Record your answers in your journal.

---

used as a standard to estimate body fat risk for pregnant and lactating women, highly muscular people, growing children, or adults over age sixty-five.

Waist Circumference

Waist circumference is an indicator of upper abdominal fat (stomach area) accumulation and is used to assess central or upper-body obesity. People with this type of obesity are sometimes referred to as "apples" or as having **android** (man-like) **fat distribution**, and they are at an increased risk for heart disease, stroke, diabetes, hypertension, and some types of cancer. See Figure 5.6. As a general rule, risk of central obesity-related health problems increases for women with a waist circumference greater than 35 inches (88 centimeters) and for men greater than 40 inches (102 centimeters).[22] To modify health risks for Asians, the cut off points should be modified to 31.5 inches (80 centimeters) in Asian women and 33.5 inches (85 centimeters) in Asian men.[27]

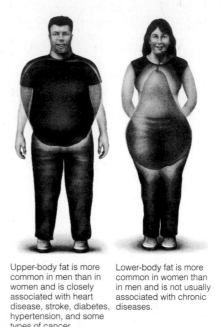

Upper-body fat is more common in men than in women and is closely associated with heart disease, stroke, diabetes, hypertension, and some types of cancer.

Lower-body fat is more common in women than in men and is not usually associated with chronic diseases.

Figure 5.6 Examples of "Apple" and "Pear" Body Shapes
Source: From WHITNEY/ROLFES. Understanding Nutrition, 12E. © 2011 Brooks/Cole, a part of Cengage Learning, Inc. Reproduced by permission. www.cengage.com/permissions

Smoking, high alcohol intake, and menopause tend to increase abdominal fat, and exercise tends to decrease it.[24] People who accumulate fat in their hips and thighs are not as susceptible to the obesity-related diseases, but their fat is more resistant to breaking down from calorie deprivation and exercise. These people are often referred to as having a pear shape or **gynoid** (woman-like) **fat distribution**. See Table 5.8 for the methodology to correctly measure waist circumference. For individuals who have very large waists and for whom health risks have already been determined by BMI and other risk

Table 5.8 Measuring Waist Circumference

To correctly measure waist circumference:

1. Locate the upper hip bone and the top of the right iliac crest.
2. Place a measuring tape in a horizontal plane around the abdomen at the level of the top of the iliac crest (waist).
3. Be sure that the tape is snug, but does not compress the skin, and is parallel to the floor.
4. Read the measurement at the end of a normal expiration of breath.

Source: Adapted from The Practical Guide to the Identification, Evaluation, and Treatment of Overweight and Obesity in Adults. National Heart, Lung, and Blood Institute and North American Association for the Study of Obesity. Bethesda, MD: National Institutes of Health; 2000. NIH Publication number 00-4084, October 2000.

Table 5.9   Health Risk Relative to Body Mass Index and Waist Circumference

Classification	BMI	Risk with Normal Waist	Risk with High Risk Waist
Underweight	<18.5	Increased	Increased
Normal	18.5–24.9	Not elevated	Increased
Overweight	25.0–29.9	Increased	High
Obesity class I	30.0–34.9	High	Very high
Obesity class II	35.0–39.9	Very high	Very high
Obesity class III	≥40	Extremely high	Extremely high

Normal waist: Men ≤ 40 inches, women ≤ 35 inches
High risk waist: Men > 40 inches, women > 35 inches

Source: Adapted from The Practical Guide to the Identification, Evaluation, and Treatment of Overweight and Obesity in Adults. National Heart, Lung, and Blood Institute and North American Association for the Study of Obesity. Bethesda, MD: National Institutes of Health; 2000. NIH Publication number 00-4084, October 2000.

factors, a waist measurement is not likely to provide additional useful data. See Table 5.9. The following is a list of conditions for which waist measurements can provide useful data for an intervention:[28,29]

- Individuals with a normal BMI, to determine disease risk.
- Individuals with a BMI <35, to determine disease risk (Those who have a BMI ≥35 are at risk for disease in spite of waist circumference.)
- Clients who have been making diet and exercise changes, to monitor progress.

EXERCISE 5.11 Assess Your Weight

Complete the chart for yourself and transfer the data to your journal.

MEASUREMENTS	STANDARD
Actual weight =	BMI healthy weight (Exhibit 5.2 and Appendix B) =
Body mass index =	Desirable = 18.5–24.9
Waist circumference =	High risk: males, >102 centimeters (40 inches), Asian males, >85 centimeters (33.5 inches); females, >88 centimeters (35 inches), Asian females, >80 centimeters (31.5 inches)

☐ Some authorities believe that North American health officials often assess health through a thinness lens. What is your impression of that statement in light of the various methods you just used to assess your healthy weight?

According to the National Institutes of Health (NIH) *Clinical Guidelines*,[22] BMI and waist circumference provide practitioners with the most accessible and accurate measurements of degree of overweight and obesity.

DOCUMENTATION AND CHARTING

"If it's not documented, it didn't happen."
—Anonymous

After completing a counseling session, the next step is to reflect, evaluate, document, and plan. The amount of time available for this step will depend on the setting of the counseling session. The Client Concerns and Strengths Log, Lifestyle Management Form 5.6 in Appendix D and Exhibit 5.3, can aid in that endeavor by guiding reflection on the concerns and strengths expressed by your client and identified by you. This activity is particularly useful to a novice counselor who may become bogged down with concerns over which the client has little control. This reflection could be shared with your client during the next session, or the activity may help you to form the framework for future counseling sessions. The specific criterion in the documentation depends on institutional standards or the setting of the intervention. Charting can provide the following benefits:

- Evidence of care
- Demonstration of accountability in meeting legal, regulatory and professional standards

- A basis for evaluation and planning to ensure quality care
- Documentation for legal protection of clients, practitioners and facility
- A tool for communication among health care team members
- Justification for third-party reimbursement

Because charting in medical records is considered a legal document, care must be taken to provide clear, well-written notes that address **JCAHO** standards. The development of automated computer systems streamlines charting and has become commonplace. However, there may still be facilities that require hand-written documentation, and if you are in a private or small practice, you may prefer hand written records. The following are some general guidelines for documentation of counseling sessions:

- Notes should be concise. Goals and plans should not be embodied in a lengthy narrative. Physicians are more likely to respond to dietitian recommendations when goals and plans are easily identified.[30]
- If hand-written, entries should be clear and legibly written in blue or black ink. Electronic charting offers numerous advantages, such as a reduction of duplication and repetition and an increase in care management tools, including alerts or reminders.
- Documentation should be accurate for ongoing referencing.
- Entries should be appropriate and pertinent. Personal opinions and criticisms should be avoided.
- Notes should be in chronological order, leaving no blank spaces.
- Entries should be made as soon as possible after the encounter.
- Date and sign all entries with full name and credentials.

There are a variety of documentation styles. Each institution defines a charting format for its facility. Many of the formats are similar in that they present objective data, provide an assessment, and end with a plan of action and expected outcomes. The **SOAP** (subjective, objective, assessment, and plan) **format** was almost universally used in medical facilities. Although this is no longer the case, it is still commonly used in a number of clinical settings, so we will review the process in this section. Also, the SOAP method is appropriate to use in a lifestyle management program and can be integrated with the American Dietetic Association Nutrition Care Process.

SOAP Format

Because of the comprehensive nature of this format, new practitioners find practice with this method particularly useful for developing charting skills. The case study addendum in this chapter contains three samples of SOAP notes for each of the three levels of motivation case studies presented in Chapter 4. Each segment of the SOAP format contains the following components:

S (Subjective)

- Information relayed to you from the client or the family.
- Citations do not need to be in complete sentences. Because the notes are under *S,* the assumption is that the information came from the client and there is no need to begin a statement with "Client says."
- Entries may include information about physical activity; weight patterns; appetite changes; socioeconomic conditions; work schedule; cultural information; significant nutritional history, such as usual eating pattern, cooking, and dining out.

O (Objective)

- Information generally comes from charts and laboratory reports and includes factual and scientific information that can be proven.
- Citations do not need to be in complete sentences.
- Examples of possible information include: age, gender, diagnosis, nutritionally pertinent medications, anthropometrics, laboratory data, clinical data (nausea, diarrhea), height, weight, healthy body weight, changes in weight, and diet order, estimation of nutritional needs.

A (Assessment)

- This is your interpretation of the client's status based on subjective and objective information including nutrition diagnosis.
- Information should be written in complete sentences as a paragraph.
- The following can be the format for this entry:
  - Begin with a statement summarizing the client's nutritional status and concerns.
  - Reflect on subjective and objective data and their impact on concerns including evaluation of nutritional history, possible problems and difficulties with self-management and the effect of medications on nutritional status.
  - Provide possible approaches and interventions.
  - Assess degree of readiness, comprehension of information provided, and previous goal achievement.

P (Plan)

- Notations are generally short, concise statements written in complete sentences that can include the following:
  - Long-term goals and specific, measurable short-term goals
  - Need for additional diagnostic data—assessments, lab work, consultations
  - Therapeutic plans—changes in nutrition care plan, diet prescription, supplement recommendations
  - Educational plans to address dietary issues

EXHIBIT 5.3   Client Concerns and Strengths Log

1. List all concerns identified by you or expressed by your client.

too little time	nonsupportive husband—NC
children responsibilities—NC	low intake of fiber, fruits and vegetables
too little exercise	little dairy, does not like, low intake of calcium
big family dinners every Sunday	frequent consumption of fast food constipation
eating while watching television	little knowledge about the role of nutrition in treating hypertension
no planning for meals	

2. Write "NC" next to of all concerns over which you or your client have no control.

3. Categorize the remaining concerns that you and your client can address to set realistic goals.

NUTRITIONAL*	BEHAVIORAL	EXERCISE
Low intake of fiber, fruits and vegetables	No planning for meals	Too little exercise
Frequent consumption of fast food	Eating while watching television	
Little dairy, low intake of calcium	Too little time	
	Big family dinners every Sunday	
	Frequent trips to fast-food establishments	

*Address food pattern, frequency, and variety concerns, if appropriate.

4. List strengths and skills that could be used to set goals that are applicable to the previously listed concerns (for example, organizational skills, knowledge of calories and food groups, cooking skills, regular activity).

- has an exercise bike; enjoyed it at one time
- a walking club in client's church
- tasted soy milk once; liked it
- enjoys oatmeal, whole-grain crackers
- enjoys apples, dried apricots, cherries, carrot and celery sticks
- good organizational skills
- cooking and good preparation knowledge

- good support system—Mom
- has taken children to high school track to play while client ran; children enjoyed themselves
- her mother has made vegetable platters for the family dinners; would probably do so more frequently if the request was made
- shredded carrots in canned spaghetti sauce once; children didn't seem to mind

(continued)

EXHIBIT 5.3   Client Concerns and Strengths Log *(continued)*

5. Categorize the strengths and skills in the following chart:

NUTRITIONAL	BEHAVIORAL	EXERCISE
Soy milk Oatmeal, whole-grain crackers Apples, dried apricots, cherries Shredded carrots in spaghetti sauce	Mother makes vegetable platters	Exercise bike High school track Walking club available

6. What strengths and skills can be used to address the concerns? List them in the following chart.

STRENGTHS AND SKILLS	CONCERNS	POSSIBLE INTERVENTION STRATEGIES
Likes oatmeal, whole-grain crackers Likes apples, dried apricots, cherries Likes carrots and celery sticks Interest in cooking Shredded carrots in spaghetti sauce Supportive mother Good organizational skills	Too little fiber Low fruit intake Low vegetable intake Limited variety High sodium intake No planning for meals	Prepare oatmeal with cut-up apples Take veggie packs to work for a snack Ask Mom to put vegetables in soup Heart-healthy cooking classes with Mom Cooking demonstrations Include mother in next counseling session Plan week's menus on one hour break

CASE STUDY   Nancy: Documentation at Three Levels of Motivation

The following contains a follow-up of the three scenarios of Nancy at different motivational levels in Chapter 4's case study. Included here are examples of SOAP notes for each of the scenarios. An example of using the **ADIME** format is found later in the chapter.

Level 1—Not Motivated, Not Ready

S Doctor wants me to lose weight, don't want to; husband likes me with "meat on my bones"; too many personal problems to worry about dieting.

O 5'4", 174#, BMI 30, 34# overweight

A Client referred by physician for diet interventions for weight loss to reduce hypertension. She expresses no interest in learning about diet options, as she feels overwhelmed by personal problems. Her feelings were acknowledged, and intervention was limited to provision of literature.

P Rx: Provision of business card for future referral; Provision of literature on diet and hypertension
Goal: Increased awareness of the role of diet and hypertension; Referral in next 3 months for further counseling

Level 2—Unsure, Low Confidence

S Doctor wants me to lose weight, don't want to; husband likes me with "meat on my bones"; too many personal problems to worry about dieting.

O 5'4", 174#, BMI 30, 34# overweight, hypertension

A Referred by physician for diet interventions for weight loss secondary to hypertension. Her ambivalence to commit to a counseling program due to stress and personal responsibilities was acknowledged. She exhibits several lifestyle concerns that negatively impact her health. Motivation level 6 on a scale of 1–10. Discussion of the role of

(continued)

CASE STUDY   Nancy: Documentation at Three Levels of Motivation

diet and lifestyle changes was well received. Focusing on positive actions, involvement in realistic goal setting, and stressing the relationship of her health to meeting her family responsibilities are expected to increase her degree of readiness.

P Rx: Client education
1. Diet and hypertension
2. DASH Food Plan
3. Seeking a support system—work and home

Will follow up with telephone call in 1 week

Goal: 1. Eat a banana or orange daily at work for a snack.

Level 3—Motivated, Confident, Ready

S Doctor wants me to lose weight; husband likes me with "meat on my bones"; never really thought of losing weight to help blood pressure; willing to try.

O 5'4", 174#, BMI 30, 34# overweight, hypertension

A Referred by physician for diet interventions for weight loss secondary to hypertension. Assessment forms indicate that she has various lifestyle concerns that are influencing her health. She is motivated by her need to remain healthy to meet her family responsibilities as well as the desire to prevent long-term complications. Empowering her to set realistic and achievable goals for realistic lifestyle changes should further motivate her.

P Rx: Client education
1. Diet and hypertension educational materials
2. DASH Food Plan and role of balanced diet
3. Seeking a support system—work and home
4. Food journaling for increased awareness of intake

Will follow up with telephone call in 1 week

Goals: 1. Fruit and vegetable intake of 3 servings per day for 1 week.
Eat a banana or orange daily at work for a snack.
Use pre-packaged cut-up veggies as snack at work.
One serving of vegetables at dinner.
Homemade soup with vegetables and a 4-ounce glass of grape juice at 3 P.M.
2. Increase awareness of food intake and aim for dietary variety.

ADIME Format

The ADIME format facilitates the American Dietetic Association Nutrition Care Process (NCP) Model as shown in Figure 5.7. Practitioners using the NCP Model are not required to use this format because components of the model can be integrated with established charting requirements. A review of the NCP follows and the ADIME documentation format will be integrated into the following discussion.

NUTRITION CARE PROCESS

The Nutrition Care Process (NCP) was developed by the American Dietetic Association in 2002 to provide a standardized framework using **evidence-based guidelines** for nutrition interventions.

The model is intended to illustrate the dynamic nature of the dietitian's relationship with the factors that influence how clients receive nutrition services, unique attributes of the food and nutrition professional, the four steps of the NCP, and a collaborative partnership with clients. Emphasis is placed on providing a reliable and systematic process to deliver an intervention to consistently produce positive outcomes. Dietitians are expected to follow the framework and use critical thinking skills to provide individualized care, not necessarily the same care, for a client or target group.[31] The NCP has been evolving to meet the needs of Registered Dietitians (RDs). Between 2002 and 2010 there have been three updates of the manual so dietitians need to continuously keep abreast of

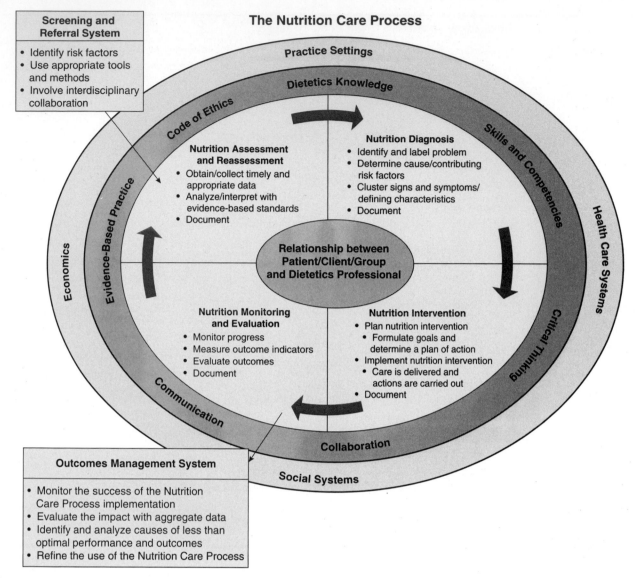

Figure 5.7 Nutrition Care Process and Model Diagram
Source: Reprinted from the *Journal of the American Dietetic Association* 108:1113–17. Writing Group of the Nutrition Care Process/Standardized Language Committee. Nutrition care process and model part I: the 2008 update. © 2008, with permission from Elsevier.

developments of the NCP. The following defines the objectives of the NCP:[5]

- To reduce variation in documentation which will improve consistency and quality of individualized care
- To provide consistent structure and framework for delivery of nutrition care
- To provide common language for practice, legislation, and reimbursement
- To generate reliable outcomes
- To focus on evidence-based practice
- To partner with electronic medical records

The NCP begins with a referral and assumes a problem has been previously suspected or identified. Because a number of healthcare practitioners may conduct a screening, the model does not assume that a dietitian conducted the screening process, so screening is outside the circle in Figure 5.7. The NCP consists of four interrelated steps with the acronym ADIME:[5] **A**ssessment, **D**iagnosis, **I**ntervention, and **M**onitoring and Evaluation. The NCP is relatively new and has been increasingly used in clinical facilities. However, the framework was designed to meet the needs of all dietitians providing

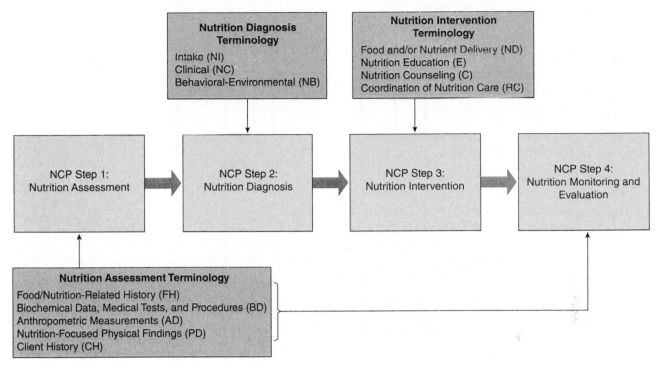

Figure 5.8 Overview of the Nutrition Care Process Domains and Standardized Language
Source: From NELMS/SUCHER/LACEY/ROTH. *Nutrition Therapy and Pathophysiology,* 2E. © 2011 Brooks/Cole, a part of Cengage Learning, Inc. Reproduced by permission. www.cengage.com/permissions.

direct services to clients or groups, regardless of their work setting. The relationship among the domains is illustrated in Figure 5.8.

Standardized Language

An important component of the NCP is standardized nutrition terminology. Many health professionals, including physicians, nurses, and physical therapists use standardized terminology, but before the development of the NCP, none existed for nutrition care. This hampered communication with other health care professionals and even among nutrition professionals. For example, dietitians may have charted "poor nutritional status," "low energy intake," "nutritional imbalance," or "provide nutrition education." These are vague terms that have multiple meanings. The American Dietetic Association reference manual, *International Dietetic and Nutrition Terminology,* has three sets of standardized terminology addressing the following: Nutrition Diagnosis, Nutrition Intervention, and Terms for Nutrition Assessment and Monitoring and Evaluation are combined. See Figure 5.8.

Each term has a unique definition that clearly indicates when the term should be used.

Step 1: Nutrition Assessment

The nutrition assessment, as described by the NCP, follows the process previously explained in this chapter. See Table 5.3 for a description of the five domains of the NCP assessment. There are two major components of the NCP assessment (1) collection of timely and appropriate data and (2) analysis and interpretation of the collected data, using relevant norms and standards.[5] Assessments need to be done at the beginning of an intervention, throughout the nutrition care intervention to check advancement of goals, and at the end to evaluate outcomes. See Exhibit 5.4 for ADIME guidelines regarding assessment documentation.

Critical Thinking During the Nutrition Assessment Process:[5]

- Determine suitable data to collect
- Differentiate relevant from irrelevant data

- Cluster data to help make decisions about the cause of the problem
- Choose valid and reliable assessment tools and procedures
- Compare data to reliable norms and standards
- Validate the data

Step 2: Nutrition Diagnosis

The purpose of the nutrition diagnosis is to "identify and describe a specific problem that can be resolved or improved through treatment and nutrition intervention by a dietetics practitioner."[5]

EXHIBIT 5.4   ADIME Documentation –Assessment (A) Guidelines

- Review the five nutrition assessment domains to guide the assessment documentation.
- Record date and time.
- Select data pertinent to clinical decision making and provide comparisons to standards, such as estimated nutrient intake and Dietary Reference Intakes. Standards could be from national, institutional or regulatory agencies. Additional charting topics may include client perceptions, motivation to change, level of understanding, food-related behaviors, cultural needs, and lab values.
- Reason for discharge or discontinuation, if appropriate. Note that a dietitian may have conducted an assessment based on a referral and determined that there was not a nutrition problem that required an intervention.

For example, the nutrition diagnosis (excessive sodium intake) differs from a medical diagnosis (hypertension).

Procedure for Defining a Diagnosis By using the data collected during the assessment procedure, you are ready to define a nutrition diagnosis using standardized terminology found in the American Dietetic Association reference manual, *International Dietetics and Nutrition Terminology (IDNT)*.[32] Frequently more than one problem exists and the dietitian needs to critically analyze which one or two are most likely to improve with a nutrition intervention.

Nutrition Diagnosis Domains (Categories) There are over 100 nutrition diagnoses available for dietitians to choose from, and they are grouped into three domains: intake, clinical and behavioral-environmental. The intake terminology often includes the definers "inadequate", "excessive", or "inappropriate" to describe an altered intake of a particular nutrient or substance. The clinical terminology relates to physical or medical conditions such as swallowing, chewing, digestion, absorption, or appropriate weight. The behavioral-environmental categories include various cognition designations and environmental factors. If you are deciding whether the intake domain or one of the other categories is the problem, the intake domain should be chosen because that domain is more closely aligned with nutrition professionals' expertise. See Table 5.10 for domain descriptions and examples of diagnosis terminology.

CASE STUDY   Nancy: ADIME Documentation–Assessment

A (Assessment): **Food Intake** DASH serving sizes 8 grains, 1–2 vegetables, 1 fruit, 0 dairy, high fat meats 10, 0 nut, seeds, legumes, 4 sweets each day. **Mineral and Element Intake** 3,600 mg sodium per day. **Energy intake** ~3,200 calories. **Total fat intake** ~142 grams/day of total fat, about 70 grams of saturated fat. **Carbohydrate intake** 20% of calories from sugar or other concentrated sweets. **Beliefs and attitudes** Diet readiness test indicated client is in preparation stage. She is concerned her about her family history of diabetes and high blood pressure. **Body composition** Ht. 5'4", weight 174#, BMI 30. **Recommended body mass index** healthy BMI 19–24, weight range 110–140#, Client is 34# above the upper limit for her healthy BMI range.

Note: In this example of ADIME assessment documentation food intake, mineral and element intake, energy intake, fat intake, carbohydrate intake, beliefs and attitudes, and body composition are standardized terms defined in the International Dietetic and Nutrition Terminology Reference Manual.

Table 5.10 NCP Diagnosis Domains, Classes, and Examples

Domain and Description	Selected Classes	Examples of Nutrition Diagnosis Terminology (Diagnostic Label)
Intake Problems related to intake of energy, nutrients, fluids, bioactive substances through oral diet or nutrition support.	Energy Balance Fluid Intake Bioactive Substances Nutrient Fat and Cholesterol Carbohydrate and Fiber	Excessive energy intake Inadequate fluid intake Excessive alcohol intake Malnutrition Excessive fat intake Inadequate fiber intake
Clinical Nutritional findings and problems identified relating to medical or physical conditions.	Functional Biochemical Weight	Swallowing difficulty Food-medication interaction Overweight and obesity
Behavioral and Environmental Nutritional findings and problems identified that relate to knowledge, attitudes and beliefs, physical environment, access to food, or food safety.	Knowledge and Beliefs Physical Activity and Function Food Safety and Access	Undesirable food choices Impaired ability to prepare foods and meals Limited access to food or water

Source: Adapted from The Practical Guide to the Identification, Evaluation, and Treatment of Overweight and Obesity in Adults. National Heart, Lung, and Blood Institute and North American Association for the Study of Obesity. Bethesda, MD: National Institutes of Health; 2000. NIH Publication number 00-4084, October 2000.

Writing a Nutrition Diagnosis Statement Similar to a nursing diagnosis, a structured sentence, called the PES Statement, is used to write the NCP diagnosis. The letters represent the three components of the statement. See Figure 5.9 for a representation of how Step 2, Nutrition Diagnosis, relates to the rest of the NCP.

- **P stands for problem** indicating the nutrition diagnostic label. The terminology used to describe the problem must come from one of the standardized terms used to describe a diagnosis as indicated in the examples column of Table 5.10.

- **E refers to etiology** reflecting the root cause or contributing risk factors. Those factors can include pathophysiological, psychosocial, situational, developmental, cultural, and/or environmental problems.[5] See Table 5.11 for a list of etiology categories.

Table 5.11 Etiology Categories Used for PES Statements

Category	Cause or Contributing Risk Factor of Problem
Beliefs-Attitudes	Confidence and feelings related to a nutrition-related belief or observation
Cultural	Customs of defining social groups including, but not limited to, religion, politics, ethnics, and social class
Knowledge	Understanding of nutrition information and guidelines
Physical Function	Physical ability, including cognitive, to engage in activities
Physiological-Metabolic	Medical or health issues impacting nutritional status
Psychological	Diagnosed or suspected mental health issues
Social-Personal	Social and personal factors related to food habits and nutritional status
Treatment	Medical treatments needed for health management and care
Access	Factors affecting availability of safe and healthful food and water
Behavior	Actions impacting nutritional status and goal attainment

Source: American Dietetic Association. International Dietetics & Nutrition Terminology (IDNT) Reverence Manual Standardized Language for the Nutrition Care Process. 3rd ed. Chicago, IL: American Dietetic Association; 2010.

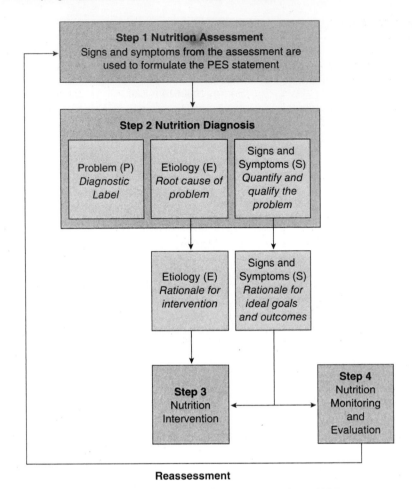

Figure 5.9 Relationship of the Nutrition Diagnosis to the Other Steps of the NCP
Source: Lacey K and Pritchett E. Nutrition Care Process and Model: ADA adopts road map to quality care and outcomes management. *J Amer Diet Assoc.* 2003;103:1061–72. ©2003 with permission from Elsevier.

- **S indicates the signs or symptoms**, defining characteristics used to determine that the client has the nutrition diagnosis specified. They can be measurable objective data, such as: decreased oral intake, consuming < 25% of meals; or obesity, BMI > 30 or subjective (but quantifiable) symptoms including observations and statements from a client or caregiver (number of bowel movements). These will be the basis for setting ideal and measurable goals in Step 3 and providing outcome measures in Step 4.
- The format for the PES Statement is: Nutrition Problem Label (using NCP nutrition diagnosis terminology) related to Etiology (root cause of the problem) as evidenced by Signs and symptoms (observable indicators that the problem exists). Simply written: "Nutrition problem label related to ____ as evidenced by ____." See

Table 5.12 for examples of nutrition diagnosis statements.

Self-Evaluation of PES Statements
You want your intervention to address the most urgent problem and your PES statement to clearly define your intent. Use the following criteria and questions to help evaluate a well-written PES statement:

Problem (P):   Can the practitioner resolve or improve the nutrition diagnosis? Consider intake as the preferred domain when in doubt about domain choice.

Etiology (E):   Is the etiology truly the "root" cause? Can the intervention eliminate the problem by addressing the etiology? If not, can the intervention alleviate the signs and symptoms?

Table 5.12   Examples of Nutrition Diagnosis (PES) Statements

Client	Problem (P) Diagnostic Label	Etiology (E) "related to"	Signs and Symptoms (S) "as evidenced by"
45-year-old male	Inappropriate intake of fats	Frequent consumption of fast-food meals	Serum cholesterol level of 250 mg/dL and 10 meals per week of fried chicken and biscuits or hamburgers
20-year-old WIC client	Breastfeeding difficulty	Poor sucking ability	Poor infant weight gain and fewer than six wet diapers in 24 hours
35- year-old female	Inadequate fiber intake	Inappropriate food preparation practices, for example, reliance on over processed foods	Estimated fiber intake of 15g/day

Source: From BOYLE/HOLBEN. *Community Nutrition in Action*, 5E. © 2010 Brooks/Cole, a part of Cengage Learning, Inc. Reproduced by permission. www.cengage.com/permissions.

Signs and Symptoms (S): Will measuring the signs and symptoms indicate if the problem is resolved? Are signs and symptoms specific enough to monitor and document?

PES Overall: Does the nutrition assessment data support the diagnosis, etiology, and signs and symptoms?

Step 3: Intervention

The purpose of an intervention is to "resolve or improve the identified nutrition problem by planning and implementing appropriate nutrition interventions that are tailored to the patient/client's needs."[5] The nutrition professional uses a specific set of activities and materials to address the problem. The selection of an intervention is based on the nutrition assessment and needs to be directed to the root cause (etiology) with the objective of relieving signs and symptoms of the diagnosis. If the intervention cannot be directed at the etiology (for example, depression) then the intervention should focus on reducing signs and symptoms. Planning and implementation are two components of the intervention process and are outlined in the sections that follow.

Planning the Nutrition Intervention

- Prioritize the nutrition diagnoses.
- Consult ADA's Evidence-Based Nutrition Practice Guidelines and other appropriate practice guides.
- Jointly determine patient-focused expected outcomes with the client and relevant care givers.
- Develop a nutrition prescription and recommend specific strategies for altering dietary intake of energy and selected foods or nutrients.
- Offer well defined, measureable, and achievable goals.
- Determine time and frequency of intervention activities.

Implementation of the Intervention

- Communicate the nutrition care plan to all relevant individuals.
- Carry out the nutrition intervention.
- Collaborate with other professionals.
- Continue data collection and monitor progress of intervention.
- Revise intervention strategies as needed.

CASE STUDY   Nancy: ADIME Documentation–Diagnosis with PES Statement

D (diagnosis): Excessive oral food and beverage intake related to knowledge deficit concerning appropriate food choices and meal planning as evidenced by BMI of 30 and inadequate intake of DASH food groups: 1–2 vegetables, 1 fruit, 0 dairy, high fat meats 10, 0 nut, seeds, legumes, 4 sweets each day.

Note in the above example of ADIME documentation of diagnosis, "excessive oral food and beverage intake" is defined, standardized terminology and "knowledge" is an etiology category in the *International Dietetic & Nutrition Terminology Reference Manual*.

Nutrition Intervention Domains There are four domains of nutrition intervention strategies and over 60 terms to choose from that address altering nutritional intake, nutrition-related knowledge or behavior, environmental conditions, or access to supportive care and services. You can choose more than one strategy. The following list contains a general description of each category. See Table 5.13 for examples of each domain.

Table 5.13 NCP Intervention Categories, Descriptions and Examples

Selected Classes (Categories) of Interventions	Description	Examples
Food and Nutrient Delivery Domain		
Meals and Snacks	Regular eating event that includes a variety of foods; food served between meals.	Specific foods, beverages or food groups for meals or snacks.
Supplements	Foods or nutrients not intended as a sole item or meal or diet, but are intended to improve nutritional status.	Commercial or prepared food supplements, vitamins, minerals, or bioactive substances (for example, sterols).
Feeding Assistance	Accommodation or assistance in eating to support adequate nutrient intake or restore eating independence and improve nutritional status.	Provide feeding assistance training for caregiver; recommend adaptive feeding equipment, feeding cues, or feeding position.
Feeding Environment	Adjustment of the factors where food is served to influence food intake.	Change in table height, meal schedule, seating arrangements, or size of utensils.
Nutrition Education		
Nutrition Education—Content	Instruction or training intended to lead to nutrition-related knowledge.	Provide fundamental purpose of nutrition education intervention; communicate relationship between nutrition and disease and health.
Nutrition Education—Application	Instruction or training leading to nutrition-related result interpretation or skills.	Teach distribution of carbohydrate intake based on blood glucose readings; explain use of glucometer.
Nutrition Counseling		
Theoretical Basis and Approach*	The theories or models used to design and implement an intervention.	Behavior change interventions using Cognitive-Behavioral Theory, Health Belief Model, Social Learning Theory, or Transtheoretical Model (Stages of Change).
Strategies	Selectively applied evidence-based methods or plans of action designed to achieve a particular goal.	Motivational interviewing, goal setting, self-monitoring, problem solving, social support, stress management, stimulus control, cognitive restructuring, relapse prevention, rewards and contingency management.
Coordination of Nutrition Care		
Discharge and Transfer of Nutrition Care to a New Setting or Provider	Discharge planning and transfer of nutrition care from one level or location to another.	Referral to appropriate agency: home delivered meals, food stamps, or food pantry.

*For more information regarding behavior change theories and models, see Chapter 2.

Source: American Dietetic Association. International Dietetics & Nutrition Terminology (IDNT) Reference Manual Standardized Language for the Nutrition Care Process. 3rd ed. Reprinted by permission.

- **Food and Nutrient Delivery.** An individualized approach for providing food and nutrients including meals, snacks, food vouchers, enteral and parenteral feeding, supplements.
- **Nutrition Education.** A formal process to instruct or train a client or group in a skill or impart knowledge to promote awareness and improve eating behavior.
- **Nutrition Counseling.** A supportive, collaborative process developed to identify priorities, set goals, and create an action plan designed to encourage self-care to treat an existing condition and promote health. If you choose nutrition counseling to implement an intervention, you must select a theoretical approach and at least one counseling strategy.
- **Coordination of Nutrition Care.** Entails consultation with, referral to, or coordination of nutrition care with other health care providers, institutions, or agencies that can assist in treating or managing nutrition-related problems. This domain includes discharge planning.

Step 4: Monitoring and Evaluation (M & E)

In this step, a practitioner evaluates the effectiveness of a nutrition intervention by engaging in monitoring, measuring and evaluating changes in nutrition care indicators. You want to determine how much progress is being made toward goals or anticipated outcomes. This process involves three activities:

- *Nutrition monitoring* refers to periodic reviews and measurement of the client's nutritional status as a result of the intervention. Possible activities for monitoring progress include checking for understanding and implementation of the nutrition prescription, recording all evidence of behavior change indicators including the signs and symptoms identified in the PES statement, and making conclusions based on evidence. Note that the monitoring does not need to be limited to the defined PES signs and symptoms but can include any positive or negative factors that may have altered due to the nutrition intervention.
- *Measuring outcomes* includes selecting appropriate nutrition care indicators to measure and using standardized indicators to increase validity and reliability of findings. The measures should interface with electronic charting and coding.
- *Nutrition evaluation* refers to a systematic comparison of current data collection and measurements against criteria, which could be the client's previous nutritional status, intervention goals, or a standard. A second component of the evaluation includes an assessment of the overall impact of the total nutrition intervention on patient outcomes.

Domains used in this step to identify changes in nutrition care indicators come from the same domains as those specified in the NCP assessment step, except for the Client History Domain. Data collected from Client History in the initial assessment would not be expected to change or be influenced by an intervention, such as gender, age, religion, or occupation. Documenting judgments

---

CASE STUDY Nancy: ADIME Documentation–Intervention

I (Intervention): Nutrition prescription DASH food group servings for 2000 calorie intake: grains, 6–8; vegetables 4–5; fruits 4–5; dairy 2–3; lean meats 6; nuts, seeds, legumes, 4–5/wk; fats and oils, 2–3; sweets ≤ 5/wk.

Nutrition Counseling based on the Transtheoretical Model and Motivational Interviewing included exploring ambivalence to change and goal setting to encourage client to move from preparation to action stage. Client described reasons for desiring changes in food intake; outlined support and barriers for change; pros and cons of current eating habits; and requested specific guidance on diet. A small achievable goal was set jointly with the client to take a fruit to work each day and a long-term goal to follow the DASH food group plan most days of the week. The client will be seen twice a month for four months. **Social Support:** Client's mother is willing to provide social support. She is scheduled to come with the client to the next counseling session.

Note: In this example of ADIME documentation, "nutrition counseling", "Transtheoretical Model", "goal setting", and "social support" are standardized terms defined in the International Dietetic & Nutrition Terminology Reference Manual.

about whether outcomes were achieved is an essential component of the NCP. This communicates progress toward achieving established goals and the effectiveness of the nutrition professional intervention to other health care practitioners.[33] Documentation of monitoring and evaluation activities could include any of the following: nutrition intervention indicator measurements and method for obtaining them, criteria used to evaluate measurements, positive or negative factors affecting outcome data; and future plans including follow-up or discharge.

NCP Documentation and Charting

NCP can be incorporated into any facility charting guidelines or the ADIME format can be followed using the standardized language for assessment, diagnosis, and intervention. You or your facility will need to obtain the reference manual, *International Dietetics & Nutrition Terminology* (IDNT), from the American Dietetic Association. The American Dietetic Association is committed to making the NCP an important standard in the profession. Standards of Education for accredited dietetic education programs require the NCP to be integrated into their curriculum, 50% of the RD exam

addresses the NCP, and the American Dietetic Association is working to incorporate the NCP Model in Joint Commission evaluations and NCP terminology in electronic medical records.[34] In order to become better acquainted with NCP, use the numerous resources on the American Dietetic Association website. Also, local professional groups and the American Dietetic Association regularly schedule hands-on seminars to practice using the process.

EXERCISE 5.12   Interview a Dietitian Utilizing the NCP

Interview a dietitian using the NCP. Address the following in your journal:

1. Record the name and date of the interview.
2. How long has the interviewee been using the NCP?
3. What resources did the interviewee use to learn the NCP?
4. What are the most beneficial components of the NCP for the interviewee?
5. Are there components of the NCP, which are difficult for the interviewee to follow?
6. What advice does the interviewee have about using the NCP?

CASE STUDY   Nancy: ADIME Documentation—Monitoring and Evaluation

M & E:

Beliefs and attitudes Criteria: Diet readiness to increase to the action stage. **Food Intake** DASH food group servings for 2000 calorie intake: vegetables 4–5; fruits 4–5 in past month. **Social Support** Client's mother will attend a counseling session with the client within 30 days. Will monitor readiness to change with an assessment graphic and food group intake with a 3-day diet record at next encounter.

Note: In this example of ADIME documentation "beliefs and attitudes", "food intake", and "social support" are defined, standardized terms in the International Dietetic & Nutrition Terminology Reference Manual.

REVIEW QUESTIONS

1. While engaging in the goal setting process, what are three messages a counselor should convey to a client?
2. Define SMART goals.
3. Name the three steps involved in completing a dietary evaluation.
4. Describe client assessment questionnaires, food diaries, 24-hour recalls, diet history interviews, food frequencies, and usual diet analysis. Explain the advantages and disadvantages of each.
5. Explain the SOAP documentation method.
6. Identify and explain the four components of the NCP.
7. Fill in the blank. "Related to" and "as evidenced by" are connectors used in a _____ Statement.

ASSIGNMENT—NUTRITIONAL ASSESSMENT

In this assignment, you will complete a **nutritional assessment**, give feedback regarding dietary evaluation, discuss broad general goals (if your volunteer wishes), and document the intervention using both the SOAP and ADIME method. Because the objective is for you to gain experience in performing common nutritional assessment procedures, you will be completing more tasks than would normally be done in one intervention. There should be no intention on your part to resolve difficulties. Volunteer clients may find some benefit in clarifying their problems through discussions and feedback; however, the participants should not be led to believe that there will be a nutrition intervention. If the person wishes to explore additional nutrition counseling, then a referral to an appropriate health care professional can be made, or the volunteer can be directed to the American Dietetic Association website search service (http://www.eatright.org/) for help in finding a nutrition counselor.

Only the involving phase, exploration-education phase, and the closing phase of the nutrition counseling motivational algorithm found in Chapter 4 will be addressed in this assignment. Ask a colleague, friend, or relative who is willing to have an assessment to be your volunteer client. Complete the following assessment forms and activities:

PART I–Nutritional Assessment of a Volunteer

Use the following interview guide/checklist to conduct the interview/assessment with your volunteer. Examples of possible counselor questions, statements, and responses are given in italics. Because the focus of this assignment is nutritional assessment, most relationship-building responses have been omitted from the checklist. You are encouraged to use these responses when appropriate. See Chapter 3 for a discussion of relationship-building responses.

Preparation

- ❏ Review the following procedures and guidelines:
  - The motivational nutrition counseling algorithm (Figure 4.2).
  - Protocol for obtaining a consent in preparation for session 1 in Chapter 14 (Exhibit 14.1).
  - Protocol for a diet history interview (Exhibit 5.1, Chapter 5).
  - Waist measurement protocols (Table 5.8, Chapter 5).
- ❏ Give copies of the Client Assessment Questionnaire and Food Frequency Questionnaire (Lifestyle Management Forms 5.1 and 5.4 found in Appendix D) to your volunteer to complete before your meeting.
- ❏ Bring copies of the following forms found in Appendix D:
  - Lifestyle Management 4.1, Assessment Graphic
  - Lifestyle Management Forms 5.1, in case your client does not bring
  - Lifestyle Management Form 5.3, 24-Hour Recall and Usual Diet Form
  - Lifestyle Management Forms 5.4, in case your client does not bring
  - Lifestyle Management Form 5.5, Anthropometric Feedback Form
  - Lifestyle Management Forms 5.7, Student Nutrition Interview Agreement, duplicate copies
- ❏ Bring Body Mass Index Chart, Appendix B.
- ❏ Bring a tape measure and a calculator.
- ❏ Minimize distractions.
- ❏ Bring visuals to estimate portion size.

Involving Phase

- ❏ Greeting
- ❏ Thank volunteer—*Thank you for participating in this interview.*
- ❏ Set agenda. Explain purpose of the interview—*This is a project I am required to do for my nutrition counseling class. The purpose of this interview is for me to work on my counseling skills, complete a nutritional assessment, give feedback to you regarding your diet, and explore your interest in making dietary changes. I will be taking some physical measurements, and I can give you feedback regarding the standards for your age. These physical*

standards are a guide for desirable weight; however, other factors such as susceptibility to kidney stones or osteoporosis should be considered before embarking on a change in dietary behavior.

❑ Review the consent form with your volunteer, follow the procedure for obtaining consent in Exhibit 14.1, Chapter 14. You and your volunteer should sign both a client copy and a clinic copy of the form. Give the client copy to your volunteer. The clinic copy should be handed in with this report.

Transition to Exploration Phase

❑ Transition statement. *Do you have any questions before we begin the assessment procedure?*

Exploration-Education Phase

❑ Ask your volunteer to describe himself or herself (age, cultural group, occupation, interests).

❑ Ask your client whether he or she has any nutritional concerns.

❑ Review the completed Client Assessment Questionnaire, Lifestyle Management Form 5.1—*Thank you for completing the Client Assessment Questionnaire. I am wondering, what came to your mind as you were filling out this form? What topics covered in this form do you think have particular importance for your food issues?*

Look Over the Form and Ask for Clarification Where Appropriate

– Health history: Inquire whether the client had any nutrition concerns related to health history responses—*I see you stated that you have a family history of heart disease and high cholesterol. Has this influenced your food selections in any way?*

– Drug history: If your volunteer is taking a medication, ask the purpose for taking it and if she or he is aware of any nutritional implications of the drug.

– Socioeconomic history: Comment on highlights of the responses of this section—*I see you frequently eat at fast-food restaurants. Is this just a habit, or is it something you really enjoy doing?*

– Diet history: Ask for clarification of any significant reporting—*You wrote that you don't eat fruits. Is there a particular reason?*

❑ Use a diet history interview (Exhibit 5.1) *Can you take me through a typical day in your life so I can understand more fully what happens and tell me where eating fits into the picture? Take me through this day from the beginning to the end.* While your client is telling you his or her story, fill in the 24-Hour Recall and Usual Diet Form, Lifestyle Management Form 5.3.

❑ Summarize

❑ Review the completed Food Frequency Questionnaire, Lifestyle Management Form 5.4. Clarify portion sizes using food models, if needed—*Thank you for completing the Food Frequency Questionnaire. I am wondering, what came to your mind as you were filling out this form? Did you feel a need to clarify or expand on anything while you were completing this form?*

❑ Measure or use the data on the Client Assessment Questionnaire to determine your volunteer's height and weight.

❑ Measure your volunteer's waist circumference (Table 5.8, Chapter 5).

❑ Use the BMI chart in Appendix B to determine your volunteer's BMI.

❑ Complete the Anthropometric Feedback Form, Lifestyle Management Form 5.5.

Provide Feedback

❑ Review the 24-Hour Recall and Usual Diet Form and the Anthropometric Feedback Form, point by point, in a nonjudgmental manner with your client. Compare the standards to your volunteer's food intake or to his or her anthropometric findings—*As you can see, your usual vegetable intake is one cup a day, and the and the estimated MyPyramid desirable intake is 2 ½ cups a day. Your body mass index is 26, and the desirable numbers range is from 19 to 25. Continue in this vein until you have reviewed all the findings.*

❑ Clarify when needed. Your client may ask about the BMI or other assessment parameters. Be sure you are familiar with the standards so that you can provide educated answers. Again, your answers should not indicate judgment. Avoid the word "you" — *Body mass index numbers are based on height and weight. Authorities have found that people who have a body mass index between 19 and 25 have a lower risk of developing high blood pressure, high blood cholesterol, diabetes, and coronary heart disease.*

❑ Ask your client his or her impression of the evaluation—*What do you think about this information?* Give your opinion if requested.

❑ If your client expresses interest in making a change, use the assessment graphic, Lifestyle Management Form 4.1 in Appendix D, to evaluate willingness to make a change.

❑ Summarize.

Closing Phase

❑ Express appreciation—Thank you very much for volunteering for this project.

PART II–Report

Answer the following questions in a formal typed report or in your journal. Number and type each question and put the answers in complete sentences under the question.

1. Record the name of the person interviewed and location, time, and date of the meeting.
2. Describe the person you interviewed—age, cultural group, gender, occupation.
3. Write a narration of the experience. There should be four titled sections to the narration—preparation, opening phase, exploration-education phase, and closing phase. Summarize what occurred in each phase.
4. Complete a Client Concerns and Strengths Log, Lifestyle Management Form 5.6.
5. Chart your experience using the SOAP and ADIME format.
6. Complete an Interview Assessment Form, Lifestyle Management Form 7.5. Do not fill out portion C of the checklist.
7. What did you learn from this experience?
8. Attach completed copies of Lifestyle Management Forms 5.1, 5.3, 5.4, 5.5, 5.6, 5.7 and 7.5.

SUGGESTED READINGS, MATERIALS, AND INTERNET RESOURCES

Cochrane Database of Systematic Reviews
www.cochrane.org/
Evidence based reviews of healthcare interventions.

Evidence Analysis Library (EAL) www.eatright.org
Provides evidence summaries of major research findings on nutritional health and interventions.

National Guideline Clearinghouse (NGC)
http://www.guideline.gov/
A public resource for evidence-based clinical practice guidelines.

USDA Food and Nutrition Information Center, Dietary Guidance
http://fnic.nal.usda.gov/nal_display/index.php?info_center=4&tax_level=1&tax_subject=256
Links to interactive assessment sites and professional resources.

Indirect Calorimetry Resources

MedGem (Microlife. Dunedin, FLI www.microlife.com)

REEVUE (Korr Medical Technologies. Salt Lake City, UT: www.korr.com)

Fitmate (Cosmed. Chicago, IL; www.cosmed.it)

Thomas PR (Ed.). *Weighing the Options—Criteria for Evaluating Weight-Management Programs.* Washington, DC: National Academy Press; 1995.
A useful assessment instrument to point out potential problems with motivation and attitudes toward weight loss diets and exercise, and a psychological assessment to determine the need for a psychological referral.

The 24-Hour Food Recall. 1998. From Oklahoma Cooperative Extension Service, Oklahoma State University, 315 HES Building, Stillwater, OK 74078-6163, videotape with facilitator's manual.

REFERENCES

[1]Answers.com. Biography: Florence Chadwick. Available at: http://www.answers.com/topic/florence-chadwick Accessed September 6, 2010.

[2]Berg-Smith SM, Stevens VJ, Brown KM, et al. Dietary Intervention Study in Children (DISC) Research Group. A brief motivational intervention to improve dietary adherence in adolescents. *Health Educ Res.* 1999; 14:101–112.

[3]Botman JA. Seven steps toward behavioral compliance. *Behav Health Treat.* 1997; 2:8.

[4]van Osch L, Beenackers M, Reubsaet A, et al. Action planning as predictor of health protective and health risk behavior: An investigation of fruit and snack consumption. *Internatl J Behav Nutr Phy Activity* 2009; 6:69–80.

[5]American Dietetic Association. *International Dietetics & Nutrition Terminology (IDNT) Reverence Manual Standardized Language for the Nutrition Care Process.* 3rd ed. Chicago, IL: American Dietetic Association; 2010.

[6]Sugerman SB, Eissenstat B, Srinith U. Dietary assessment for cardiovascular disease risk determination and treatment. In: Kris-Etherton P, Burns JH, eds. *Cardiovascular Nutrition.* Chicago: American Dietetic Association; 1998:39–71.

[7]Windhauser MM, Ernst DB, Karanja NM, et al. Translating the dietary approaches to stop hypertension diet from research to practice: Dietary and behavior change techniques. *J Am Diet Assoc.* 1999; 99 (suppl):S90–S95.

[8]Thomas PR, ed. *Weighing the Options Criteria for Evaluating Weight-Management Programs.* Washington, DC: National Academy Press; 1995.

[9]Fisher KL. Assessing psychosocial variables, a tool for diabetes educators. *The Diabetes Educator.* 2006; 32:53–57.

[10]White KV, Dwyer JT, Posner BM, et al. Nutrition screening initiative: Development and implementation of the public awareness checklist and screening tools. *J Am Diet Assoc.* 1992; 92:163–167.

[11]Bauer JM, Kaiser MJ, Anthony P, et al. The Mini Nutritional Assessment®—Its history, today's practice, and future perspectives. *Nutr Clin Prac.* 2008; 23(4):388–396.

[12]Mahan L K, Escott-Stump S, eds. *Krause's Food, Nutrition, and Diet Therapy.* 12th ed. Philadelphia: Saunders; 2007.

[13]Lee DR, Nieman D. *Nutritional Assessment.* 5th ed. St. Louis, MO: Mosby; 2009.

[14]National Cancer Institute. US National Institutes of Health. Usual Dietary Intakes: NHANES Food Frequency Questionnaire (FFQ) Available at: http://riskfactor.cancer.gov/diet/usualintakes/ffq.html Accessed August 5, 2010.

[15]Rolfes SR, Pinna K, Whitney ER. *Understanding Normal and Clinical Nutrition.* 8th ed. Belmont, CA: Cengage; 2008.

[16]Jonnalagadda SS, Mitchell DC, Smiciklas-Wright H, et al. Accuracy of energy intake data estimated by a multiple-pass, 24-hour dietary recall technique. *J Am Diet Assoc.* 2000; 100:303–308, 311.

[17]Blanton CA, Moshfegh AJ, Baer DJ, et al. The USDA automated multiple-pass method accurately estimates group total energy and nutrient intake. *J Nutr.* 2006; 136:2594–2599.

[18]Conway JM, Ingwersen LA, Moshfegh AJ. Accuracy of dietary recall using the USDA 5-step multiple pass method in men: An observational validation study. *J Am Diet Assoc.* 2004; 104:595–603.

[19]Kittler PG, Sucher KP. *Food and Culture.* 5th ed. Belmont, CA: Cengage; 2007.

[20]Cassidy CM. Walk a mile in my shoes: Culturally sensitive food-habit research. *Am J Clin Nutr.* 1994; 59(suppl):S190–S197.

[21]Block G, Hartman AM, Dresser CM, et al. A data-based approach to diet questionnaire design and testing. *Am J Epidemiol.* 1986; 124:453–469.

[22]National Institutes of Health (NIH) Obesity Health Initiative. *Clinical Guidelines on the Identification, Evaluation, and Treatment of Overweight and Obesity in Adults,* NIH Publication No. 98-4083. Washington, DC: U.S. Department of Health and Human Services; 1998.

[23]Centers for Disease Control. Prevalence and Trends Data, Overweight and Obesity (BMI)–2009. Available at: http://apps.nccd.cdc.gov/brfss/list.asp?cat=OB&yr=2009&qkey=4409&state=All Accessed on August 5, 2010.

[24]National Diabetes Education Program. Department of Health and Human Services. *Take these small steps now to prevent diabetes.* NIH Publication No. 07-5526; 2007.

[25]Deurenberg P, Deurenberg-Yap M, Guricci S. Asians are different from Caucasians and from each other in their body mass index/body fat per cent relationship. *Obes Rev.* 2002; 3(3): 141–6.

[26]World Health Organization Expert Consultation. Appropriate body-mass index for Asian populations and its implications for policy and intervention strategies. *Lancet.* 2004; 363:157–63.

[27]Obesity in Asia Collaboration. Waist circumference thresholds provide an accurate and widely applicable method for the discrimination of diabetes. *Diabetes Care.* 2007; 30(12): 3116–3118.

[28]Klein S, Allison DB, Heymsfield SB, et al. Waist circumference and cardiometabolic risk: A consensus statement from Shaping America's Health, Association for Weight Management and Obesity Prevention, NAASO, the Obesity Society, American Society for Nutrition, American Diabetes Association. *Obesity.* 2007; 15:1061–1067.

[29]Wang Y, Rimm EB, Stampfer MJ, et al. Comparison of abdominal adiposity and overall obesity in predicting risk of type 2 diabetes among men. *Am J Clin Nutr.* 2005; 81(3):555–63.

[30]Klein CJ, Bosworth JB, Wiles CE. Physicians prefer goal-oriented note format more than three to one over other outcome-focused documentation. *J Am Diet Assoc.* 1997; 97:1306–1310.

[31]Lacey K, Pritchett E. Nutrition Care Process and Model: ADA adopts road map to quality care and outcomes management. *J Am Diet Assoc.* 2003; 103(8):1061–1072.

[32]Ritter-Gooder P, Lewis NM. Validation of nutrition standardized language-next steps. *J Am Diet Assoc.* 2010; 110:832–835.

[33]Hakel-Smith N, Lewis NM, Eskridge KM. Orientation to Nutrition Care Process Standards improves nutrition care documentation by nutrition practitioners. *J Am Diet Assoc.* 2005; 105:1582–1589.

[34]Writing Group of the Nutrition Care Process/Standardized Language Committee. Nutrition care process and model part I:The 2008 update. *J Am Diet Assoc.* 2008; 108:1113–1117.

Promoting Change to Facilitate Self-Management

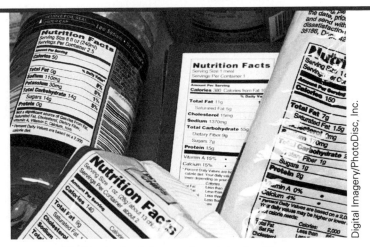

Digital Imagery/PhotoDisc, Inc.

Habit is habit, and not to be flung
Out of the window by any man,
But coaxed downstairs a step at a time.

—MARK TWAIN

Behavioral Objectives

- Use common food management tools.
- Identify various methods used to track food behavior.
- Explain the ABCs of eating behavior.
- Use common behavior change strategies.
- Describe cognitive restructuring.
- Identify three components of dysfunctional thinking.
- Demonstrate basic cognitive restructuring counseling skills.
- Identify effective ways to enhance education in a nutrition counseling session.
- Discuss factors affecting dietary adherence.

Key Terms

- **ABCs of Behavior:** antecedent, behavior, and consequence; used to describe behavior chains.
- **Barriers:** obstacles that hinder accomplishment of a goal.
- **Behavior Chains:** a sequence of events that explains recurrence of behavior.
- **Cognitions:** what and how a person thinks and perceives based on life experiences.
- **Cognitive Restructuring:** challenging destructive thoughts, beliefs, and internal self-talk and substituting self-enhancing cognitions.
- **Contract:** a formal agreement to implement a goal.
- **Countering:** substituting healthy responses for problem behaviors.
- **Cue Management (Stimulus Control):** addresses antecedents of a behavior chain; technique involves using cues to increase or decrease a particular behavior.

- **Dash Food Plan:** food group plan developed to lower high blood pressure.
- **Exchange Lists:** food management tools that organize foods by their proportions of carbohydrate, protein, and fat.
- **Journaling:** a tracking method used to analyze and modify behavior.
- **Modeling:** behaviors learned by observing and imitating others.
- **MyPyramid:** health guidance icon stressing proper balance of food groups and physical activity.
- **Problem Solving:** a systematic approach to breaking undesirable behavior chains.
- **Reinforcement or Rewards:** tangible or intangible incentives to encourage a behavior change.
- **Thought Stopping:** a technique using the word *stop* to end destructive reoccurring thoughts.

INTRODUCTION

In this chapter, we will be investigating strategies that promote behavior change. First, there is an examination of food management tools commonly used in nutrition interventions. A discussion of tracking mechanisms proceeds since clients and counselors require a method for evaluating the effectiveness of the tools. Next, factors affecting our choice of behaviors, **ABCs of behavior**, and what constitutes a behavior chain are covered. In order to break the chain of events encompassing an undesirable behavior, counselors need to know a variety of behavioral and cognitive strategies. These are reviewed as well as various professional issues to consider for producing positive intervention outcomes.

STRATEGIES TO PROMOTE CHANGE

Nutrition professionals need to become familiar with a variety of strategies for promoting behavior change to meet their clients' needs. Since counselors differ in their predisposition and ability to use certain strategies, you need to have a variety of intervention strategies available. Also, some clients may need additional options for tackling difficult situations, particularly as circumstances change.

For example, in the beginning of a dietary behavior change, avoiding exposure to foods or settings where temptations exist may be possible. However, if conditions change, clients should be ready with additional coping mechanisms. To explain the need to learn new strategies, the following script has been developed by Tsoh et al.:[1]

Counselor: *Having more than one or two strategies is very important for you. What we are trying to do is expand your toolbox, so that if one tool does not work, you will have another. In many situations, you may require a combination of various tools. (p. 25)*

FOOD MANAGEMENT TOOLS

A variety of tools are available to aid in the management of eating behavior (see Table 6.1). Both professional and commercial organizations have developed techniques to help design food plans and track eating behavior. Tools differ in their degree of structure and amount of work needed on the part of clients to use the approaches. Explore the advantages and disadvantages of the methods with your clients to select an approach that fits their interests. To encourage clients to think about this topic, the Client Assessment Questionnaire (Lifestyle Management Form 5.1 in Appendix D, has questions asking clients about the amount of structure they desire in a nutrition intervention.

Meal Replacements

I do not want to spend the whole day thinking and planning about what to eat.

Meal replacements have been used effectively in nutrition interventions for weight loss and reduction of cardiovascular risk factors and are a component of the Look AHEAD (Action For Health In Diabetes) Clinical Trial to reduce weight in individuals with type 2 diabetes.[2-4] Long-term weight and health benefits have been observed in studies using meal replacements for one or more meals.[3] Because overweight individuals have been shown to underestimate calories by 40% and average-weight individuals by 20% or more, having controlled portion sizes are an enormous advantage.[4] Meal replacements can take the form of shakes, bars, or portion-controlled

Table 6.1 Comparison of Food Management Tools

Method	Advantages	Disadvantages
Meal Replacements	• Simplify food choices • Reduces exposure to temptation • Portion size is clear	• Some may not find taste acceptable • May interfere with social plans
Detailed Menus and Meal Plans	• Clearly defined • Useful for someone who expresses a need for structure • Useful for someone who has complex dietary requirements who is not ready or not capable of following a food group plan	• Does not allow for spontaneous events • Food items needed for the plan may not be available • May be difficult to design to complement a client's lifestyle
Exchange Lists	• Offers choices • Provides structure • Allows for variety • Meal pattern is individualized	• May be too complex for some individuals
MyPyramid	• Easy to understand • Flexible	• Requires computer technology
DASH Food Plan	• Easy to understand • Flexible • Rich in various nutrients believed to benefit good health	• Some of the foods may not be part of a client's usual intake (for example, nuts, beans, and seeds)
Goal Setting	• Easy to understand • Flexible • Designed to take into consideration a client's lifestyle	• Approach may move too slowly when aggressive intervention is needed

frozen meals. By using meal replacements for some meals, clients report feeling less stressed because there are fewer temptations and a reduced need to think about selecting and preparing an acceptable meal.[3] Clients and counselors can concentrate on changing behavior patterns using a narrow focus. As clients gain knowledge and skills, the number of regular meals can be increased.

Detailed Menus and Meal Plans

Just tell me exactly what to eat. I want a detailed plan.

Sometimes clients are overwhelmed with the diagnosis of a new disease or have been frustrated

One of my clients was a man who had been told by his doctor to lose weight. He said he had been on diets before and did not want a diet where he had to choose something from columns or groups. He said, "Just tell me exactly what to eat." I told him my concerns about boredom or the need to make adjustments due to events or food availability. He was adamant and his wife, who was also at our meeting, assured me that she would be sure to always have needed food accessible. During this meeting, we were only able to design three days' worth of meals. At our next appointment, which was in two weeks, he was happy to report a loss of 6 pounds, but he said that he really wanted more variety. We developed three more days of detailed eating plans that provided some flexibility for fruits and vegetables. As he became familiar with the new eating pattern and serving sizes, his program evolved into a more general, flexible plan based on food groups. Actually this ended up being just the type of food group plan he said he did not want. This client seemed to need a highly structured plan to begin a program.

with previous attempts to follow a therapeutic diet plan on their own, and they are looking for a great deal of structure. Structured weight loss programs frequently include grocery lists, menus, and recipes to enhance adherence to low calorie diets.[5] Similar to the meal replacements, structured plans reduce the stress of making choices. Software programs and several websites can assist in developing meal plans complete with recipes and a shopping list. Many of the websites can develop meal plans for common dietary restrictions, such as low sodium, low fat, and high fiber. Care should be taken to design a plan based as much as possible on your client's food preferences and lifestyle patterns.

In fact, the best way to use websites with prede-signed meal plans is to have your client sit at the computer with you to work out a week's plan while discussing upcoming life events for the next seven days. If you have a capable client, this activity can be repeated at weekly intervals until the client can take over complete responsibility.

Exchange List for Weight Management

I want a lot of structure but freedom to select foods.

The Exchange Lists for Meal Planning in Appendix C would probably work for someone who wants considerable structure but freedom to make some choices. The American Diabetes Association and the American Dietetic Association have developed two versions of this system: one is oriented to diabetes meal planning, and the other can be applied to weight management and more general food management needs. The exchange system booklet geared to weight management meal planning emphasizes healthy eating, lifestyle changes, physical activity, and goal setting with a health professional. **Exchange lists** are organized into four main groups:

One of my clients who used journaling successfully stated, "Some people say keeping a diary is too much work, too stressful, but I found it much more stressful before journaling when I was eating well and exercising and gaining weight. Now that we developed a diary method that works for me, it does not take much of my time, and I am able to eat well and have foods that I enjoy."

- Carbohydrates–includes starch (breads, cereals and grains, crackers, beans, and starchy vegetables); fruit; milk; sweets, desserts, and other carbohydrates; and nonstarchy vegetables
- Meat and meat substitutes–includes lean, medium-fat, high-fat, and plant-based proteins
- Fats–includes monounsaturated (avocado, nut butters, nuts, oil, and olives), polyunsaturated (margarine, mayonnaise; pignolia and English walnuts; salad dressing; seeds), and saturated fats (bacon, butter, chitterlings, coconut milk, cream, cream cheese, lard, tropical oils, salt pork, shortening, and sour cream)
- Alcohol

Modifications of the exchange system are often available in clinical settings to address a variety of medical needs. For example, exchanges can be altered to take into account intake of potassium, fluid, or protein.

MyPyramid or Dash Food Plan

I want some structure and freedom to select foods.

Either the **MyPyramid** Website or the **DASH Food Plan** can provide guidance for an individual who wants some structure but freedom to select foods. Lifestyle Management Form 5.3 in Appendix D has the approximate MyPyramid recommendations for a 2,000 calorie diet. However, the MyPyramid website, www.MyPyramid.gov, provides an individualized downloadable food group plan based on gender, age, and activity level. For a person who is pregnant or breast-feeding, there are additional resources. Food groups include grains, vegetables, fruits, milk, meats and beans, oils, and discretionary calories. Physical activity is emphasized.

The DASH Food Plan has an additional food group consisting of legumes, seeds, and nuts, and there are defined serving sizes for the food groups. Also, keeping sodium intake between 1,500 and 2,300 mg per day is emphasized. See Appendix A. The DASH Food Plan was developed for individuals with high blood pressure; however, it provides guidance for healthy eating for everyone.

Goal Setting

I don't want to follow a diet. I just want to eat better.

For clients who do not want any type of structured eating plan, lifestyle changes can be made solely through goal setting. This approach complements non-dieting programs for weight management, which have a significant voice among nutrition professionals.[6] See Chapter 5 for goal setting guidelines.

TRACKING

No matter what food management system has been selected, your clients should be encouraged to keep track of their progress. Without a self-monitoring method, evaluation of food behavior goals becomes difficult. Tracking methods vary in complexity. The method selected should depend on your client's ability to work with structure and details. Some people enjoy recording everything they eat and reviewing the material at the end of the day; others become quickly frustrated with the procedure. Discuss several options with your client.

Journaling

Journaling is a tracking method that has consistently been shown to be effective in altering behavior in general and food habits in particular.[2,7] In an evaluation of self-monitoring of an 18-week standard behavioral weight loss program, individuals who consistently kept food records had a mean loss of approximately 15 kg, while those who did not self-monitor gained an average of about 4 kg.[8] Recording behavior becomes a self-management tool by increasing awareness and providing a "time-out" for making a decision. Food intake should be recorded immediately before or after eating. The short time spent on reflection may result in taking an action to maintain one's lifestyle objectives.

A number of factors related to food intake can be recorded in a journal depending on the client's motivation and journaling objectives. At the very least, a food diary would include a list of foods consumed and may also contain portion size, calories, and fat grams, and physical activity. Some behavior management programs have encouraged participants to record time, place, the presence of others, mood, thoughts, concerns, degree of hunger, activities at the time of food consumption, and health parameters (blood sugar and blood pressure). As can be expected, client resistance is likely to mount with increasing requirements for journal entries. However, the need to record the psychosocial variables probably declines for most individuals after a few weeks.

Analysis of the records can help counselors and clients develop new goals. Streit et al.[9] found the following procedures successful in guiding clients who chose journaling to help manage their food intake:

- **Provide training.** Clients gain a better understanding of the process when hands-on activities are included as well as complete instructions of how to keep a journal. Although clients should be encouraged to measure what they eat whenever possible for at least a few weeks, the training should include a review of estimation of portion size since measuring will not always be possible. An education method found to be successful in one study included a review of participants' food diary entries of a sample meal by dietitians.[9]

- **Use estimates.** Sometimes clients can become frustrated with writing down exact amounts of food consumed and then calculating caloric content or percentage of calories from fat, especially when consuming mixed dishes that the client did not prepare. Therefore, to reduce anxiety, clients should understand that approximations are acceptable.

- **Set meaningful and achievable goals.** Clients are more likely to be successful in record keeping if the first goal is modest, such as two days a week. Then gradually increase journaling activity in subsequent weeks. Streit et al.[9] found successful dietary changes with as few as two days of records per week. However, more frequent journaling is associated with even greater food behavior change.[10] The anxiety of record keeping is likely to be reduced if there is a degree of flexibility, enabling clients to feel they are "off the hook" for complicated meals or stressful days.

- **Provide a variety of record-keeping options.** Some clients prefer to use a pocket calendar with enough space to record a day's worth of food intake. Others like to use 8½ by 11-inch forms. At times, clients will devise their own record-keeping system.

- **Provide nonjudgmental feedback.** Clients should always be praised for their journaling work, and counselors should always review their clients' journals with no hint of

I am so pleased when clients keep food records because I know it is one of the best ways to change and maintain food behavior. I always review the food records with my clients, and I often put stars in the journal on entries that indicate that their goals were kept.

criticism. Client thoughts regarding the journaling activity should be sought. Analyzing the record keeping may give greater insight regarding patterns and triggers that encourage poor eating choices. In a study reported by Streit et al.,[9] the most effective technique for motivating participants to keep food records was to have dietitians review the journals and return them with brief comments. Clients were also encouraged to write questions in their diaries.

Journaling Alternatives

There is more than one approach to journaling. Counselors and clients should think creatively to find ways to make it work. Here are some ideas:

- **Checking off.** For clients following food group plans or the exchange system, checking off boxes or crossing out slashes throughout the day can be used. Names of exchanges or food groups could be written on a form, or the names could be abbreviated and written each day in a pocket calendar. This methodology may appeal to some who resist writing; however, there would be a reduction of information collected for evaluation.

- **Messaging.** Hand-held, voice-activated recorders or leaving messages on a confidential voice mail system has been reported to work successfully.[11]

- **Using art.** Drawing pictures, scribbling, or choosing colors has been used to assist clients in getting in touch with their feelings or moods while consuming food.[11]

- **Empty bowl.** Put desired food objectives in a visible spot such as a bowl in the kitchen at the beginning of the day or week. Goals are assessed according to the amount in the bowl at the end of the day or week.

- **Electronic note pads.** Tracking in electronic note pads can be used for clients comfortable with the technology. There are various

applications for cell phones that some clients may prefer to use. Clients can be encouraged to take pictures with a camera application on a mobile phone of confusing or difficult food encounters to be reviewed with their counselor.

BEHAVIOR CHANGE STRATEGIES

A number of behavior change strategies can be incorporated into a counseling program. The selection of a particular strategy to be used with clients depends on their motivational stage; resources; lifestyle, educational, and emotional needs; as well as the expertise of the counselor. The Social Cognitive Theory sets a framework for a discussion of behavior change strategies by focusing on points in the sequence of behaviors.[12] Behavior change strategies can be addressed in terms of the ABCs of behavior:

A: Antecedent (stimulus, cue, trigger)

B: Behavior (response, eating)

C: Consequence (punishment, reward)

Let's take a closer look at each of these components:

- **Antecedents.** Encountering antecedents to eating occurs normally throughout the day. For example, time of day or the feeling of hunger in the stomach stimulates the desire to eat so we eat. Usually in nutrition counseling, we are particularly interested in cues that trigger unconscious eating or consumption of large quantities of certain types of food. Behavior change strategies addressing antecedents often concentrate on physical availability of food (cookie jar), social (parties), emotional (stress), or psychological (motivation; destructive thought patterns). Behavior change strategies dealing with antecedents can focus on avoiding the cue (remove cookies from the house) or altering the cue (cover a piece of cake with pepper).

> I made cards for frequently consumed individual foods and meals. The cards contain a food or components of a meal and calories. I color coded the cards according to food groups. Often in the morning, I select the cards that I will need for the day. If there is no card for a particular food I plan to eat, I make a new card. As I eat a food, I move the corresponding card into an envelope. At midday or near the end of the day, I review what I have eaten and plan the rest of the day accordingly. This process only takes me a couple of minutes. At the end of the day I record my calorie intake and exercise in a small calendar. Sometimes I skip a day of food journaling if it is just too difficult to figure out calories. Using the cards has worked much better for me than writing down all the foods I eat in a journal.

- **Behavior.** Strategies dealing with the behavioral response to an antecedent may address the actual act of eating (speed), physical (eat in one place), emotional (do not clean your plate), awareness (pay attention to eating; no TV), or attractiveness (sparkling water in a wine glass with a slice of lemon). Behavior change strategies may focus on providing a substitute for eating (take a walk).
- **Consequences.** Consequences can be positive reinforcers or punishment; such as a reward or losing a privilege.

Behavior Chain

The sequence of events from antecedent to consequence is referred to as a behavior chain; see Exhibit 6.1 for some examples. Behavioral strategies can address all aspects of a behavior chain or can zero in on one aspect. Caban et al.[13] have determined that tailoring counseling approaches to client needs improves outcomes. Specifically, they found that clients who are highly motivated and reported few external stressors were likely to be receptive to nutritional and physical activity interventions. On the other hand, those who reported more external stressors in the initial assessment benefited from relaxation training, stress management, and problem-solving approaches in the beginning of treatment.

Asking clients to reflect on their behavior can give clues as to what behaviors are in the greatest need of change. One excellent method for evaluation, mentioned earlier, is a complete diary of food behavior. Intense journaling activity can be demanding, but clients may be willing to do an extensive recording for a few days or one week. Some clients may enjoy the intense analytical activity and want to continue this form of journaling for an extended period. See Lifestyle Management Form 6.1 (in Appendix D) for an example of a food diary form for intensive journaling.

Cue Management (Stimulus Control)

Cue management (stimulus control) deals with the antecedent component of a behavior chain by prearranging cues to increase a desired response or

---

EXHIBIT 6.1   Behavior Chain Examples

Negative Consequence

Juanita comes home from work hungry. She walks into the kitchen to fix dinner. There is a bag of her favorite potato chips on the counter (antecedent). Juanita tells herself she will eat just a few and opens the bag, leaving it on the counter as she prepares dinner. She ends up eating the whole bag of chips (behavior). When dinner is ready, Juanita picks at her food because she is not hungry. She is frustrated because she knows a bag of potato chips is not nutritious and she is not getting enough vegetables and fiber in her diet (consequence).

Positive Consequence

Juanita comes home from work hungry. There is an orange and walking clothes and shoes prominently displayed on a chair in the kitchen (antecedent). Juanita eats the orange, changes her clothes, and goes out for a half-hour walk (behavior). When she returns, Juanita is in good spirits, and she prepares dinner. She eats a healthy meal with whole grains and vegetables; she is content (consequence).

EXERCISE 6.1   Behavior Chains

Record in your journal one of your frequent **behavior chains.** Identify the cue (trigger), specify the behavior, and describe the consequence.

❑ Is this a behavior that you would like to continue? If so, is there a way to encourage the occurrence of the cue? If not, is there a way to reduce the incidence of the trigger? Explain.

---

to suppress a detrimental one. Counselors and clients work together to identify and modify social or environmental cues that trigger undesirable eating.[14] An effective method of introducing the concept of cue management to a client is by using an analogy (see Exhibit 6.2).

The first task for using this method will be to identify cues. Sometimes triggers that stimulate a particular food response are obvious (such as a stash of candy bars in an office desk), but other times the cues are inconspicuous. When this is the case, a few days or weeks of intensive journaling by using a form such

---

EXHIBIT 6.2	Example of Using an Analogy to Introduce Cue Management

COUNSELOR: Imagine that one day you are driving your [use the name of client's dream automobile] and hit a big pothole. Luckily, you have a sturdy car and it is not damaged. What would you do if you were to go to the same destination again?

CLIENT: Avoid the pothole. Drive slowly. Take an alternative route.

COUNSELOR: Very good! You simply drive differently. We are trying to apply the same idea here, which is to learn to identify and anticipate potential "potholes" in your road to meeting your food goal so that you can be prepared to "drive" differently.

Source: Janice Y. Tsoh, Jennifer B. McClure, Karyn L. Skaar, et al. Smoking Cessation 2: Components of Effective Intervention. Behavioral Medicine 1997; 23:15–28. Adapted with permission of Taylor & Francis Group, http://www.informaworld.com.

---

as Lifestyle Management Form 6.1 in Appendix D could help identify the stimuli. In the case of decreasing the occurrence of a detrimental food stimulus, the objective in cue management is to alter the triggering stimulus or reduce exposure to it.[12] Table 6.2 lists some strategies for controlling common physical environment and social cues. A discussion of regulating stress can be found in Chapter 7, and a review of cognitive factors influencing behavior is found later in this chapter.

After a strategy is developed to counteract a cue, it is generally a good idea to talk through or visualize with your client how the strategy will be implemented. The scheme should include reminders to perform the new activity, such as the following:

- Sticky notes left in strategic places*-computer, dashboard, refrigerator
- Notes on a calendar
- Cartoons or jingles posted in a book or briefcase used everyday
- Entries on a daily to-do list

---

*Note that the location of the notes should be periodically changed, or else they become part of the scenery and lose their influencing effect.

Remember when using this technique to look for cues that produce beneficial behaviors as well as cues that encourage detrimental behaviors. If the focus can be on increasing the occurrence of positive behavior producing cues, the transition to a healthier lifestyle is likely to go more smoothly with seemingly less effort. As desirable behaviors make a greater impact on one's lifestyle, the time available to engage in the undesirable is automatically reduced. Sometimes counselors place so much emphasis on solving problems and getting rid of undesirable food behaviors that a search for desirable cues is neglected.

Countering

Countering, a technique of exchanging healthy responses for problem behaviors, addresses the B (behavior) portion of the behavior chain.[15] If an individual simply stops a pattern, there is likely to be a void unless a new behavior is substituted. For example, a client may consume two cups of ice cream each night after dinner in front of the television. The transition to changing that behavior will go more smoothly if an alternative behavior is planned, such as riding an exercise bike, stretching, or mending. Ordinarily, countering is a usable behavior intervention strategy when the objective is to find an activity to substitute for eating or when an acceptable food alternative is available. General categories of substitutions include the following:

- Foods that are acceptable or a healthier alternative, such as baked chicken in lieu of fried chicken.
- Active diversions, such as knitting, writing, playing an instrument, working on a puzzle, or cleaning.
- Physical activities, such as walking, stretching, or weight lifting.
- Relaxation activities, such as deep breathing, yoga, prayer, or progressive muscle relaxation. (Relaxation is covered in Chapter 7 under stress management.)

Reinforcement: Rewards

For some people, tracking and cue management interventions are adequate to develop and sustain

Table 6.2 Strategies to Control Environmental Cues

Stimuli	Solution
Physical Environment	
Location: Some people eat in many places of their homes, resulting in numerous cues for eating–bed, TV room, kitchen sink, or in front of the refrigerator.	• Designate one place for eating all meals and snacks. Besides reducing location cues, the need to travel to a particular spot provides extra time to think about whether the food actually needs to be eaten.
Activities: Engaging in activities such as reading or driving a car while eating results in producing additional cues and encourages unconscious eating.	• While eating, do nothing else. With fewer distractions, focus can be placed on the pleasures of eating and on the degree of fullness.
Restaurant Eating: When individuals are hungry and make selections in fast food restaurants or in a sit down restaurant, they are susceptible to impulse eating.	• Bring lunch from home. • Order from a menu rather than eating from a buffet. • Decide on a selection from a menu on the Internet before going to the restaurant. • Review calories of fast-food choices on the Internet if not labelled at point of purchase.
Shopping: Degree of hunger can influence food purchases and make consumers more receptive to store cues, which encourage impulse buying.	• Do not shop when hungry. • Use a shopping list. • Bring only enough money to purchase foods on the shopping list.
Serving: Attractiveness of food can be a cue to eating and encourage food consumption. This is particularly important if foods that need to be consumed are not well liked.	• Serve foods that are being encouraged in an attractive manner; for example, put a slice of lemon in a water glass or use parsley for a pleasing effect. • Serve foods in desired portion size and put away the rest. For example, count out the desired number of nuts into an attractive bowl and replace the storage bag in the refrigerator. Take the bowl to the designated place to eat. • Learn to say "No thank you" to offers of food.
Reminders: Physical cues can be used as reminders to perform a behavioral action.	• Use sticky notes, cards, or other reminders to eat certain foods.
Storage of food: Usually food stored in sight, such as counters and tables, provides cues to eat.	• Store foods that should be avoided out of sight in inconvenient places. • Foods that should be encouraged should be highly visible. • If possible, do not bring foods into the home that should be discouraged. • Do not store undesirable foods in an office desk.
Social Environment	
Social events: Many people identify social gatherings as a particularly difficult time. There is often an array of tempting foods, and if alcohol is consumed, defenses often decline.	• Plan and rehearse how to deal with temptations at an upcoming event. • Do not go to social events hungry. Have a snack ahead of time, such as an apple or orange. • Drink water and plan to have one alcoholic drink, if any. • Volunteer to bring a vegetable platter or other acceptable food for your needs. • Do not stand near food. Move food to a distant location if it is placed near you.

(continued)

Table 6.2 Strategies to Control Environmental Cues *(continued)*

Stimuli	Solution
Social support: Family and friends can help or hinder with progress in meeting dietary objectives.	See Chapter 7 for a discussion of ways to enhance social support for lifestyle changes.

Eating Behavior

Stimuli	Solution
Food in mouth: Eating quickly may be a habit that developed encouraging mindless eating of large quantities of food.	• Put utensils down between mouthfuls. • Chew fully before swallowing. • Take small bites. • Take a break during the meal. • Swallow each bite before adding any more food.
Unconscious cues: A person may not be aware of the triggers that encourage the desire to eat certain foods.	• Move to a new place at the table or a new room. • Practice mindful eating. See Chapter 7.
Food on plate: There may be a need to clean the plate.	• Set a goal to leave food on the plate.

EXERCISE 6.2   Identifying Cues and Exploring Countering

List three common cues to unconscious eating or eating undesirable foods. Next to that list, write a countering behavior that would be acceptable to you if the cue was applicable to you. For example:

Cue	Behavior	Alternative
Movies	Large popcorn with butter	Small popcorn without butter

---

a desired behavior; however, others need an added incentive to regulate and strengthen behavior.[16] **Reinforcement** behavior change strategies provide incentives by addressing the end of the behavior chain–consequences. **Rewards** provide positive consequences, commonly thought to be more effective than negative consequences, such as punishment.[15] Generally, there is an inherent reward of feeling good following the accomplishment of achieving a goal, resisting temptation, or substituting alternative behaviors for undesirable behaviors. In fact, clients should be encouraged to get in touch with those feelings and compliment themselves with positive self-talk, such as "I am doing great!" or "I can make this work." Although self-compliments are beneficial, more tangible rewards can provide additional motivation during challenging times–for instance, in the initial attempts of making a lifestyle change.

Rewards can take many forms, and a client may need some time to think over potential options before selecting a workable reward. Usually rewards are luxury items (purchase of clothing) or pleasurable experiences (relaxing bath or reading a book). A technique for accumulating small rewards could be considered, such as placing pennies in a jar each time a particular behavior is accomplished. When the desired amount is collected, the money could then be used to buy something special. Rewards can be self-administered or given to a client by a significant other. Rewards should be contingent on the behavior. For example, if the reward was a manicure, the client would not get the manicure if the behavior was not accomplished. If you believe your clients would benefit from rewards, you should discuss the behavior change strategy with them and evaluate the response. If there is hesitation on the client's part, it may be best to allow time to think about the concept until your next meeting.

The following are some major factors to consider when establishing rewards:

- Rewards should be individualized.
- Rewards should be well defined—what and how much.
- Rewards should be timed to come after the behavior, not before.
- Rewards should be given as soon as possible after the behavior is accomplished.

Rewards can also be used as a component of a nutrition intervention program. For example, rewards can be used for attendance, completion of food or blood glucose records, or attainment of a waist circumference. Studies using financial incentives for weight loss did not reveal any benefit.[2] Since adults are likely to be intrinsically motivated, rewards may have a greater influence on children, who are more likely to be extrinsically motivated.

I generally encourage my clients to identify positive reinforcements for their behavioral contracts. However, one time I had a client insist that her contract stipulate that she would give her ex-husband a hundred dollars if she did not accomplish her task. Needless to say, she followed through on her goals.

Contracting

Rewards are often used in conjunction with contracting. A **contract** documents an agreement between a counselor and a client to implement a particular goal. The contract can cover short- or long-term goals. An example of a behavioral contract can be found in Exhibit 6.3 and Lifestyle Management Form 6.2 in Appendix D.

Contracts should be used when clients want structure and accountability,[11] and they should be recorded because written contracts are more powerful than oral ones.[15] The following is a list of factors to consider when using a contract:

- Clients should define their intended behavior change for the contract. The counselor should never impose a goal.
- Behavioral goals should be clearly defined. The same conditions apply as those described in Chapter 5 for goal setting. The goal statement should answer what will happen, how often the behavior will occur, and when it will take place.
- Time limits for reaching the goals should be delineated.
- Reinforcers should be stated. They can be rewards or punishments. Generally rewards are considered to have a greater impact and should be immediately available after the intended behavior is performed. If the contract covers a long-term goal, a reward that requires greater effort, such as a trip or shopping, can be considered. Otherwise, rewards that do not put an extra

---

EXHIBIT 6.3 **Sample Behavioral Contract**

Counseling Agreement

Name _____ Date _____

My plan is to do the following:

_____

_____

This activity will be accomplished on _____

_____

My reward will be _____

_____   _____

Client signature   Date

_____   _____

Counselor signature   Date

Behavior Change Strategies

151

EXERCISE 6.3   Make a Contract

The purpose of this activity is to give you direct experience with contracting. Think about a behavior that you would like to change or a task you would like to accomplish within the next week. Use Lifestyle Management Form 6.2 in Appendix D to complete the contract. Have a colleague discuss the contract with you and sign it. Because the goal you will be setting is not part of a total counseling program and you may not be ready to take action on a major behavior change you are contemplating, the goal you establish should be modest. Some examples of modest goals could be brushing teeth after breakfast, giving the dog a bath, or writing a letter to a friend. The goal should be something that you believe you need a little push to accomplish and should be possible to complete within the next week.

❑ At the end of one week, reflect on the experience and document any problems or successes you encountered in your journal. Explain how useful you believe this tool would be in a nutrition behavior change intervention.

EXERCISE 6.4   Exploring Encouragement

Describe in your journal a time when words of encouragement were particularly useful to you to accomplish a task. Explain anything you can bring from the experience that could help you in your nutrition counseling endeavors.

cards, notes, voice mail, personalized signs, notes, or recorded tapes. One counselor has clients recording their own relaxation tapes with their own words of encouragement to listen to before going to bed.

Goal Setting

Achievement of goals provides a pathway to actually performing the new behavior. Breaking down desirable behavior patterns into small achievable steps provides a series of successes and an improvement in self-efficacy. As the saying goes, "Nothing breeds success like success." Each success raises mastery expectations. We covered the process of goal setting in Chapter 5.

Modeling

Many of our behaviors are learned by observing and imitating others. By observing others accomplish a goal similar to their own, clients' beliefs in their ability to imitate the behavior increase. This process is referred to as modeling. Videotapes, lay counselors, written testimonials, success stories, counseling buddies, and role playing are possible models that can be used in a counseling environment to increase self-efficacy. Outside counseling sessions, models who have prestige, status, or expertise are more likely to influence behavior than those who do not have those characteristics. For example, when Oprah Winfrey used a diet drink to lose weight, sales of the product soared. However, most often people with whom we identify closely have the greatest influence on our behavior. We are more likely to imitate an individual who is similar to our age, gender, and culture.

burden on a client should be sought. For example, a contract for an individual who enjoys coffee in the morning can be as simple as, "I do not drink coffee for breakfast until I have eaten a piece of fruit—no fruit, no coffee."

• Signing and dating the contract can reinforce a client's commitment. Usually the contract is signed by only the counselor and the client; however, there are times when consideration should be given to having the document signed by a support person, such as a spouse or friend.[17]

Encouragement

Encouragement is generally well received; however, a client's past experience trying to accomplish a desired behavior change can influence the impact of encouraging remarks. Also, the effect will vary with the credibility, trustworthiness, and prestige of the person giving the words of encouragement. Some counselors have provided encouragement creatively using e-mails,

> I had a religious client who consumed a lot of low-nutrient-dense foods in her bedroom. I made a small sign for her nightstand reminding her of her goals and telling her I would be thinking about her and praying for her. When my client went on vacation, she was worried about how she would handle her food goals. I gave her an audiotape I recorded for her trip with some personal words of encouragement, a few supportive sayings I took from a book on motivation, and a prayer. This client did well, and I believe the encouragement I provided was helpful.

Modeling can have a greater impact if a client can practice the observed behavior under supervision and then receive immediate feedback.[12] For example, clients could select acceptable items from a restaurant menu after watching a video with people making desirable choices.

Problem Solving

Problem solving is a process that involves a counselor and client working together to identify a behavior chain, detecting **barriers** to change, brainstorming possible options, and weighing the pros and cons of the alternatives. The objective is to design an action plan by selecting as many breaks in the behavior chain as possible, especially focusing on the antecedents of the chain. The plan should include a reward. After the plan has been implemented, the counselor and client evaluate the outcomes of the action plan and make any needed adjustments. The client should receive the planned award if any changes were made.[2] See Exhibit 6.4 for a simplified description of this process.

Barriers are obstacles or roadblocks to achieving a desired lifestyle change. Common barriers and possible strategies to minimize or eliminate their impact are listed in Table 6.3. Perceived barriers come in many forms; for example, taste preferences, difficult food preparation, complexity of the diet, lack of social support, inadequate financial resources, job or family pressures, and time constraints. An investigation to identify barriers of middle-aged patients who experienced myocardial infarction found social and work situations, the price of food, and situations in which large amounts of food are available to be the most frequently reported challenging conditions.[18] Consumers indicate confusion if messages are negative or focus on a particular nutrient.[19] Sometimes demographic variables have been found to have an impact on barriers, too. For example, low income, low education, and being male have been found to increase barriers to consumption of fruits and vegetables.[20]

The American Dietetic Association advocates for a total diet approach to provide guidance for healthy eating to reduce barriers to behavior change. In their discussion of reducing barriers to achieve goals, Danish and Laquatra[21] identify four major obstacles: lack of knowledge, lack of skill, the inability to take risks, and lack of adequate social support. For examples of the first three, see Table 6.3. Some of the most frequently named barriers can be addressed by an increase in knowledge. Nutrition counselors can often aid clients in finding ways to make needed dietary changes palatable, assessable, and convenient. However, counseling time should not be devoted to giving unneeded information if your client already has an adequate

> One of my clients relayed a story of going to a pancake house for breakfast with friends. She had not been to this type of restaurant in years and assumed that she would order waffles or pancakes and sausages for her meal. However, her friend ordered a broccoli omelet made with an egg substitute. This observation influenced my client, and she decided that she would imitate her friend and order the same thing.

EXERCISE 6.5   Identifying Models

Divide into groups and discuss the potential impact of a counselor who is not at "ideal body weight." Do you believe a counselor who does not appear to be what is generally considered fit can be an effective nutrition counselor? Record your thoughts in your journal.

Table 6.3 Overcoming Barriers to Change

Classification	Barrier	Strategy
Lack of knowledge	The client does not know what foods can lower blood pressure.	Provide pamphlets, videos, Internet sites, and grocery store tour.
Lack of skill	Limited cooking ability.	Shared cooking; demonstrations.
Lack of risk taking	Afraid of hurting mother's feelings. The client's mother prepares a large dinner every Sunday. There are seldom foods available that meet dietary objectives.	Explore ambivalence to requesting that the mother make acceptable alternatives. Use imagery or role playing to practice making a request.

EXERCISE 6.6   Explore Barriers

Interview someone who is attempting to make a dietary change. Record in your journal the difficulties the person is encountering.

❏ Describe the desired behavior change.

❏ Identify where the barrier or barriers fit into the behavior chain–antecedent, behavior, or outcome.

❏ How will this investigation assist you with nutrition counseling in the future?

EXHIBIT 6.4   Problem Solving

Glasgow et al.[22] describe a specific technique called STOP for systematically analyzing a problem and developing a solution. This problem solving method involves the following:

S—Specify the problem
T—Think of options
O—Opt for the best solution
P—Put the solution into action

---

knowledge base. Mental and physical skills can be lacking including assertiveness, decision making, positive self-talk, time management, label reading, or estimation of portion sizes. Clients who are unable to take risks are often afraid of the negative consequences of taking action. For example, they could be afraid of hurting a friend's feelings by refusing an offer of food or afraid of dealing with the consequences of disruption for friends and family. Requesting help from family and friends or joining a support group can address social support concerns. See Chapter 7 for elaboration of social support.

COGNITIVE RESTRUCTURING

The mind is its own place and in itself can make a heaven of hell, or a hell of heaven.

— **John Milton**

Another component of maintaining a behavior change is how the troubles encountered when attempting to continue with the new behaviors are perceived. Around 55 A.D., a Roman Stoic philosopher, Epictetus, maintained that difficulties related to problems are rooted in how problems are perceived rather than the actual troubles caused by the problems.[23] Today cognitive therapists embrace this concept of **cognitive restructuring**, focusing on identifying irrational thoughts and modifying them. The premise is that because **cognitions** (what and how a person thinks and perceives based on life experiences) are learned thinking behaviors, they can be relearned.[15] The objective is to change behavior patterns by changing destructive thinking patterns. Cognitive coping strategies, such as using positive self-talk, have been shown to effectively change lifestyle behaviors,[24] and in one study on smoking cessation, these strategies outperformed behavioral methods.[25]

Thinking patterns have been categorized as *opportunity thinking* and *obstacle thinking*. Each mindset can "influence our perceptions, the way we process information, and the choices we make in an almost automatic way."[26] A pattern of opportunity thinking allows finding constructive ways to deal with difficult situations. On the other hand, engaging in obstacle thinking leads to self-destructive behavior–making a difficult situation worse or giving up and retreating from problems. For example, an opportunity thinker diagnosed with high blood cholesterol is more likely to feel inspired by the challenge and may focus on the resources available to learn about new foods, cooking techniques, and support groups. In contrast, obstacle thinkers are likely to engage in self-pity and be bogged down in the difficulties of obtaining or preparing appropriate foods or adjusting their lives to take part in a support group. Authorities have identified common cognitive distortions leading to obstacle thinking patterns that adversely affect attempts to change behavior. They are listed in Table 6.4 and Exhibit 6.5 with examples of how they could be expressed in attempts to change lifestyle behaviors. Although the categories are presented as distinct entities, it is not unusual for several of them to manifest at one time. See Exhibit 6.5 for an example.

The process of changing dysfunctional thinking addresses three factors:[25]

• **Internal dialogue.** All of us engage in an ever-constant dialogue that influences our feelings, self-esteem, behavior, and stress level.[25] By influencing this dialogue to provide self-enhancing messages, clients can better cope with difficult situations and are more likely to find the resources to take positive actions.

- **Mental images.** Athletes have used mental imagery to help produce a desired performance. By visualizing the accomplishment of an intended task, clients are more likely to attain an intended goal.

- **Beliefs and assumptions.** Core beliefs are deeply ingrained, leading to assumptions that trigger automatic thoughts. See Exercise 6.8 for an exercise to evaluate your core beliefs. For example, a core belief could be "If I do everything right, I will not have any health problems." The assumption is that everything must be done correctly, which could influence the development of some of the distorted cognitions listed in Exhibit 6.5 and Table 6.4.

Changing patterns of thinking that have been part of a person's makeup for many years can take a great deal of effort. The process of changing cognitions is a complex process, and there are psychotherapists who specialize in this type of therapy. Nutrition counselors, however, could incorporate some cognitive interventions into their sessions. The steps of this process are as follows:[26]

1. **Education.** Many individuals are not aware that thoughts are controllable. Your first step in cognitive restructuring is to educate your clients about this concept, reminding them that just because thoughts pop into their heads does not mean that those thoughts must persist. In particular, we do not want to allow self-destructive thoughts and irrational messages to remain. They will influence our actions, and our behavior is likely to be counter to our lifestyle objectives. A leading psychologist is reported to have written, "One of the most significant findings in psychology in the last twenty years is that individuals can choose the way they think."[26]

   Counselor: *Possibly your thoughts may be hampering your progress to make behavior changes. Some people are surprised to learn that we have control*

over what messages our brains deliver. It has been found that by directing our self-talk, we can improve the outcomes of our behavior change attempts.

2. **Identify dysfunctional thinking.** Analyze existing beliefs and assumptions, self-talk messages, and mental imagery patterns.

   - Sometimes clients are well aware of their negative thought patterns, but may have never thought they could be changed.
   - Show Table 6.4 and ask your client whether he or she can identify with any of the common cognitive distortions.
   - Review and analyze a situation your client identified as difficult. The following are some questions to help explore cognitions:

   Counselor: *What were your feelings before, during and after the event?*

   What were you saying to yourself?

   - Keep a journal. Recording thoughts and feelings before, during, and after the behavior change attempt can help identify obstacles or self-enhancing thinking patterns. The Eating Behavior Journal, Lifestyle Management Form 6.1 (see Appendix D), can be used for this purpose.

3. **Explore validity of self-destructive statements.** Counselors can help their clients challenge their obstacle thinking patterns by exploring the validity of the internal messages.

   - Ask self-evaluating questions. By using probing questions rather than evaluations, counselors can aid clients in discovering inconsistencies.[12] This exploration provides a template for clients to challenge their irrational beliefs on their own. See Exhibit 6.6 for a list of probing questions.
   - Use humor. Corey[27] reports that humor is one of the most popular techniques that rational emotive behavior therapy (REBT) practitioners use to illustrate the absurdity of certain self-destructive ideas and to help clients not to take themselves so seriously.

   Counselor: *Heavens! You should be boiled in oil. You ate the whole bag of chips!*

4. **Stop destructive thoughts. Thought stopping** is a technique that was developed to put an end to recurrent self-destructive thoughts and self-dialogue. It involves mentally saying the word *stop*, pushing away destructive automatic thoughts, and substituting *constructive thoughts*.[28] To enhance the forcefulness of the word *stop*, a big red stop sign can be imagined or, if an individual is alone, a book can be slammed on a table or the back of a hand can be slapped. A constructive, affirming thought is then substituted.

5. **Prepare constructive responses to substitute automatic dysfunctional cognitions.** After clients have recognized that they have obstacle-thinking patterns, a counselor should explore their openness to preparing more effective thinking patterns. The following are some intervention techniques:

   • Identify and develop constructive thoughts to substitute for dysfunctional ones. This can be done through using challenging self-evaluation questions, such as these:

   Client: *Is a hot dog the only reason for going to a ball game?*

   • Clients can also substitute opportunity thinking, such as this:

   Client: *Rain makes walking outside inconvenient; it doesn't mean I need to stop my walking program. This shows I need to prepare for a rainy day. I will buy a good raincoat.*

   • Use imagery. In this technique, an intense mental rehearsal is used to set new patterns of thinking. Ellis and Harper[29] describe this as an effective method involving clients imagining their feelings and self-talk in a worst-case scenario using previously established self-destructive thinking patterns, and allowing negative feelings to emerge. Then a plan is made for a better response, and the imagined scenario is replayed using opportunity-thinking patterns. By repeating this exercise a number of times before encountering the activating event, clients will be better equipped to respond with nondestructive thinking when the event does occur. Events a nutrition counselor may visually imagine with a client could be ordering food at a fast-food restaurant, handling desserts at a holiday meal, or reducing cups of coffee or soft drinks consumed in a day.

6. **Substitute constructive thoughts for destructive ones.** Replace destructive thoughts with previously prepared constructive thoughts—for example, "I learned that I shouldn't buy potato chips."

Table 6.4   Negative Thinking

Common Negative Thoughts		Examples
Good or Bad (no in-between)	• Divide the world into good or bad foods • See yourself as a success or failure • Being on or off a diet	*"I had potato chips. This isn't working. I give up."* *"I am a jerk for eating that candy. I am worthless."*
Excuses	• Blame something or someone else • "Can't help it"	*"I will never be able to change. I just don't have any will power. It's just no use."* *"I don't have anyone to walk with."*
Should Must Have to	• Expect perfection, no middle ground • Irrational standard • A set-up for disappointment • I must have the approval of others • Others must treat me fairly • I must get what I want	*"I really must eat fish and oatmeal every day."* *"Because I ate one potato chip, the harm has already been done. I might as well eat the whole bag."*
Give Up	• Follows other distorted thinking	*"There was no skim milk at the store. I can't take this. Forget this food plan business."*

Source: Adapted from Diabetes Prevention Program. Session 11: Talk Back to Negative Thoughts. Available from: http://www.bsc.gwu.edu/dpp/manuals.htmlvdoc. Accessed August 25, 2010.

EXHIBIT 6.5   Countering Negative Thinking

People who are attempting to make lifestyle changes need to guard against destructive negative thinking. For example, people who say such statements to themselves as the following need to find substitutions:

- *"A physical activity program is out for me. A woman at the gym said I should be ashamed of myself. She is right, not the people in my support group who say I should accept and love myself. Also I tried walking once, but I got a blister. That just goes to show that I wasn't made for exercise! Probably if I walked every day my blood pressure wouldn't come down anyway."* The talk exhibited here will certainly lead to defeat. This woman is focusing on negative feedback, generalizing that a single blister means "give up", and assuming the worst would happen. This talk could be transformed into "I am still searching for a way to make physical activity work for me."

- *"I did have a piece of fruit for a snack but that didn't mean much because I ate potato chips. First I ate just one chip and then I figured this is absolutely awful so the diet is over and I might as well eat the whole bag of chips. I guess I just don't have willpower. I am such a jerk!"* This person should have given herself more credit for eating the fruit and focused on how and why the success happened. Identifying an episode as awful is not helpful because such a label can lead to a feeling that the situation is so bad that a solution can never be found. Using the word "absolutely" compounded the negativity of the phrase. The idea that after she started eating chips, there was no use stopping often comes up when certain foods are considered off-limits. Also, blaming an indulgence on lack of willpower is always counterproductive because the characteristic is considered a personal failing so a change in lifestyle could not possibly occur. Instead this woman should ask herself what she learned from the situation and tell herself what she will do differently next time. In addition, clients should be discouraged from using derogatory terms, such as "jerk", to describe themselves. After being denigrated, a person is not likely to expect success in lifestyle change attempts in the future. Instead, people should remind themselves that they are learning so that better choices can be made in the future.

- *"I really do not want to go on the hayride because I can't have the hot chocolate afterwards. The other people in my group should be more considerate of the fact that I have diabetes. I can't stand this!"* Sometimes people focus on one small difficulty and distort the total picture. Instead of searching for acceptable options, this obstacle thinker is caught-up criticizing himself or herself or others using words such as *should, ought, must,* and *have to.* This creates an impossible standard, resulting in negative feelings that may lead to a relapse.

EXHIBIT 6.6   Self-Evaluating Questions

- Is the idea accurate?
- What evidence exists that this idea is not correct?
- Why is it so terrible that you ate that food?
- Do people learn new behaviors by performing perfectly all the time?
- Where is it written that you cannot stand a situation?
- Is there any factual evidence that supports this idea?
- What are the worst things that could happen if what you must, should, or ought to do doesn't happen?
- Are there good things that would occur if what should happen did not happen or what should not happen did happen?
- What good does it do to focus on negative thoughts?

EDUCATION DURING COUNSELING

The primary step to changing dietary behavior and maintaining dietary objectives is education.[30] Clients must understand why dietary change is important and be informed of pertinent nutrition information to be capable of making informed decisions to change their behaviors. In a study designed to measure the key determinants of satisfaction with diet counseling, patients identified knowledge along with facilitative skills as the most important components of their counseling experience.[31] Patient education was an integral part of the successful Diabetes Control and Complication Trial.[32] In the American Dietetic Association Nutrition Care Process, the nutrition education domain is divided into two categories — content and application. The content

Mario is fifty years old and has a high blood cholesterol level. He attended a holiday party after a stressful day at the office. At the party, Mario told himself it would be OK to eat some higher-fat foods because he hadn't eaten lunch. However, when he stared eating and drinking alcohol, Mario ate some fruit and what he considered too many avoid foods, including several types of high-fat cheeses and cold cuts, various pastries, fatty snacks, and vegetables with dip. After leaving the party, Mario was annoyed and went home and ate some of his daughter's Halloween chocolate. He said to himself, "I deserve to have a heart attack the way I eat. This is awful. I am a terrible person. I should have prepared a lunch. I should have eaten something before I went to the party. I should have eaten more vegetables. I'm surprised I ate any fruits at all. How did that happen? I might as well continue to blow the diet and eat the chocolate. I'll never be able to eat right."

❑ Identify cognitive distortions in the monologue.

❑ How are the distortions affecting Mario's ability to make a lifestyle change?

category is defined as, "instruction or training intended to lead to nutrition-related knowledge." The application category addresses assistance in skill development and interpreting medical results related to a nutrition prescription stating "instruction or training intended to lead to nutrition-related result interpretation and/or skills."[14] In this discussion of nutrition education, we will use these definitions to review the components of the education process during a counseling intervention. Chapters 11 and 12 will use a more robust definition of nutrition education for a discussion of nutrition education interventions.

Effective Education Strategies

The education component of nutrition counseling must be incorporated into an intervention in a manner that facilitates behavior change. In recognition of this concept, the American Diabetes Association uses the term self-management education in the National Standards for Diabetes Self-Management Education.[33]

Educational targets have been linked to specific educational interventions in Table 6.5. Funnell and Anderson[34] advocate an integrated approach to education so clients can make informed decisions about their behaviors. This means addressing psychosocial concerns and initiating behavior change strategies before pouring educational content into an "empty bucket." This methodology interfaces well with the counseling objectives covered in Chapter 4 and includes the following:

• Review role of client as self-manager and role of counselor as a source of expertise, support, and inspiration.
• Elicit client concerns and questions.
• Discuss clients' experiences and understanding of their condition.
• Identify what the client wants from the counselor. Ascertain educational topics client would like addressed.
• Explore behaviors the client wishes to alter.
• Present information to address concerns and questions.
• Discuss strategies to address the behavioral aspects of the concerns.

> *I hear and I forget,*
> *I see and I remember,*
> *I do and I understand.*
> **—Confucius**

To enhance the educational impact and to support different learning styles, varied approaches and active learning experiences can be incorporated into counseling sessions. See Exhibit 6.7 for a list of interactive activities, and Table 6.5 and Table 6.6 for a list of approaches for conveying the message that have been used successfully in nutrition counseling sessions.

Effective Education Language

Consider your use of language while giving educational instructions, particularly the use of imperatives. For example, "You should eat more fiber." "You have to increase your intake of vegetables." "You ought to reduce your intake of soda." "You need to start exercising." Do these statements feel like commands? Imperatives are likely to elicit

---

EXERCISE 6.8   Core Belief Activity

Read the following statements and put a check next to the ones that apply to you.

❑ I need to have love and approval from peers, family, and friends to be worthwhile.

❑ I must not fail or make a mistake. I must be a success.

❑ Life should be easy, and I should not be frustrated. I can achieve happiness through passivity and inaction.

❑ I should always be in control of my emotions. I should be able to control negative feelings, never showing unhappiness or depression.

❑ I should never argue with someone I love.

❑ If I am alone, I will be miserable and not feel worthwhile.

❑ It is horrible when things or people are not as I expect them to be.

❑ All evil and wicked people should be punished.

❑ If someone criticizes me, something is wrong with me.

❑ I must live up to other people's expectations.

❑ I am ugly unless I have a perfect outward appearance.

❑ My worth depends on my achievements, intelligence, status, or attractiveness.

❑ If I do everything right, I will be successful.

Answers

If you checked six or more statements, you seem to view the world as all good or all bad. You are likely to be hard on yourself and engage in obstacle thinking when attempting to make lifestyle changes.

If you checked three to six statements, you are being too hard on yourself. Your tendency to be rigid may leave you feeling bad when you make mistakes or when things fall below your expectations.

If you checked fewer than three statements, you have positive views about life. You are more likely to set realistic goals and not to be discouraged when things do not go as you had planned.

❑ In your journal describe your reaction to this activity. What did you learn about yourself? Is this an activity you would like to do with a client? Why or why not?

Source: Copyright 2007 American Diabetes Association. From The Complete Weight Loss Workbook. Modified with permission from The American Diabetes Association. To order this book, please call 1-800-232-6733 or order online at http://store.diabetes.org

---

Table 6.5   Linkage of Educational Targets and Interventions

Target	Intervention
Knowledge, beliefs	*Didactic education:* increasing awareness of risks and benefits; helping clients know how to make appropriate self-care decisions
Skills	*Demonstration and feedback:* showing how to execute skills; observing performance, correcting errors
Intentions	*Goal setting:* establishing specific and appropriate goals that are ambitious but realistic; behavioral contracting to increase commitment
Barriers	*Problem solving:* helping clients find ways to overcome barriers to implementing intentions
Self-efficacy, burnout	*Support and counseling:* helping clients maintain positive emotional well-being

Source: Adapted from Peyrot M. Behavior change in diabetes education. Diabetes Educator. 1999;25(suppl 6):62–73. Adapted with permission from SAGE Publications.

Table 6.6   Nutrition Education Approaches During Counseling

Category	Explanation
Avoid technical jargon	Tailor your use of technical terms to the background of your client. Generally, technical jargon, such as *hypertension* for *high blood pressure*, should be avoided. Use low-literacy materials when appropriate.
Simplify directions	Concise, straightforward instructions with information about actual choices (such as items or brands) are more likely to be followed than complex regimens.[35]
Incorporate self-help materials	To support the educational process, nutrition counseling programs have successfully incorporated self-help materials, such as workbook activities, for highly motivated clients who report few external stressors.[13]
Repeat important points several times	Explain important points in several ways and vary learning experiences.
Limit the number of learning objectives per session.	Too many learning objectives produce information overload, dilute important messages, and cause confusion.
Organize material in a logical manner.	Generally, the first third of an information-giving session is remembered best.[17] Use organizing terminology, such as "We will go over three ways to reduce cholesterol. First, . . ."
Check for understanding	When giving factual data, be sure that the client understands what you are saying, especially before starting to cover a new topic. Watch for verbal and nonverbal cues or ask a question, such as "Do you understand?" or "Would you like me to repeat any of this?"
Incorporate significant others	When dietary instructions are involved, supportive family members, friends, or caregivers should be included. Ask your client if there is someone who should be included in the discussions.
Utilize visuals	Discussions of important concepts can be supplemented with anatomical models, videos, displays, pictures, diagrams, charts, and so forth.
Provide meaningful support materials	To aid memory and encourage the processing of information after leaving a counseling session, supportive reading material can be beneficial. This is particularly important when the client is feeling stressed. Studies of patient education have shown that clients typically forget half the information presented to them within five minutes. However, clients should not be confounded with an abundance of fact sheets, brochures, recipes, and coupons. Feeling overwhelmed may lead to an inability to take action. Handouts should be geared to a particular educational objective and tailored to your client's needs. The lifestyle behavior change program at Health Partners, Inc., in Minneapolis leaves large blank spaces on their handouts for dietitians to write personalized messages for their clients.[36]
Disperse information over a period of time	A planned educational experience can be designated for part of a counseling session (for instance, portion size activities); however, education can be introduced or reinforced throughout a session when the need or opportunity arises. King[11] refers to these responses as *sound bites*. Used appropriately, they can affirm a client's behavior. For example, a client who enjoys eating chocolate could be told that chocolate contains caffeine and a chemical that enhances the feeling of well-being. However, a counselor should be sure of the educational value of the sound bite because overuse of this method could interfere with the progress of counseling.
Use stories, examples, personal accounts, and comparisons.	These are aids to enhance an educational experience or to understand complex material. Exhibit 6.8 provides an example of using a comparison. The learning value of educational aids will be improved when they integrate with a client's cultural orientation.

EXHIBIT 6.7 **Interactive Educational Experiences**

Here are some hands-on experiences that counselors can share with clients:

- Grocery store tour
- Cooperative cooking
- Cafeteria meal
- Fitness trail walk
- Trip to a gym
- Practice selecting items from a menu—circle high-fat foods on a menu
- Simulations

- Interpret food labels—compare the labels of two similar products
- Jointly modify recipes (have client bring recipes)
- Create menus
- Measure and weigh portion sizes
- Analyze blood glucose records of previous clients
- Role playing

EXHIBIT 6.8 **Example of Using a Comparison**

NANCY: I don't even think anything is wrong. I think they might have made a mistake. I feel good.

COUNSELOR: Actually, with high blood pressure you may not feel a thing. But that doesn't mean there isn't anything going on that can't eventually make you feel bad. When you were growing up and in school, did you ever get a callus on your finger from writing all the time?

NANCY: Yeah. As a matter of fact, I have a callus there now. I write orders all night long, and I have this ugly hard spot from my pen.

COUNSELOR: That hard spot is from the pressure of your pen pressing against your finger while you are writing. The pressure causes scarring or hard tissue to form. A similar type of thing happens to your blood vessels when you have high blood pressure for years. They harden—it is called arteriosclerosis.

---

resistance. People like to believe they have a choice and control over their lives. You want your language to support a desire to make the changes necessary for your client to obtain positive outcomes–not a rebellion.

Supportive language usually means staying away from the word "you" and providing information in a neutral manner. Often starting sentences with the word "I" and asking permission can create a helpful counselor-client interaction. Kellogg[37] provides the following example of using supportive language, "I hear you. You are frustrated with how erratic your blood sugars have been. I have a few suggestions that I believe will bring you better numbers. Would you like to hear them and then you can choose which ones you will try out?" See Table 6.7 for additional examples of supportive counselor language.

Positive or Negative Approach

The issue as to whether a health risk message regarding dietary behavior should be cast in a positive or a negative light has not been resolved. Consumers have indicated that positive messages focusing on healthy food choices were more motivating than avoidance messages.[19,38] For example, the focus on increasing fruit and vegetable intake could be an emphasis on decreasing the risk of cancer or on looking and feeling better. Brownell and Cohen[30] point out that a moderate amount of fear appears useful for motivating a client to make a change. There would be no reason to change if there is too little fear; however, too much fear provokes denial and encourages attention to be directed elsewhere. Snetselaar[17] notes that clients should not be protected from negative information and that they have a right to all relevant facts. As a counselor, you will have to evaluate the situation as to how much emphasis should be placed on the negative aspects of your client's condition.

Table 6.7   Language Shift Ideas

Instead of:	Experiment with Neutral Language:
You *should* eat less saturated fat.	Those who decrease their saturated fat intake reduce their risk of heart disease.
You *need to* eat meals at more consistent times.	Eating meals at about the same time every day contributes to more even blood sugars.
You *have to* limit your carbs at dinner.	My successful clients include about one ounce or 1/2 cup of carbohydrates at each meal.
You *have to* start exercising more often.	Some exercise at least 5 days a week helps blood pressure stay normal.
You *ought to* plan menus before going food shopping.	People who shop with a list come home with healthier foods and find they need to go to the market less often.
You *need to* test your blood sugar at least four times a day.	Those who test at least four times a day find it easier to keep their blood sugar normal.

Source: Kellogg M. Tip #39 Imperative Language. Counseling Tips for Nutrition Therapists: Practice Workbook, Vol. 2. Available at: http://www .mollykellogg.com/tipslist.html

After your client has started to implement a lifestyle change, there should be a strong emphasis on the positive. Berry and Krummel[35] advocate a positive approach with clients by emphasizing foods to include rather than those that should be avoided. Also, they advocate positive reinforcement by acknowledging, praising, and encouraging clients when they make desirable changes, no matter how small.

SUPPORTING SELF-MANAGEMENT

The following sections provide a discussion of various professional issues to consider for providing effective nutrition counseling interventions.

Terminology

Following diet orders has been a challenge since Adam and Eve struggled with the forbidden fruit restriction. Traditionally, adherence or compliance has been defined as following advice, recommendations, diet orders, or a prescribed regimen. Clients were labeled as non-compliant if they did make the recommended changes. A better definition, reflecting today's cooperative approach toward nutrition counseling, would be to describe the degree to which an individual's dietary behaviors coincide with the dietary objectives as set by clients in collaboration with their health practitioners. This definition takes into consideration a more positive and accurate change process.

Although the words *adherence* and *compliance* are often used interchangeably, the term *dietary adherence* is generally preferable. For many, the word *compliance* conjures up an image of an authoritarian counselor dictating dietary orders and expecting obedience. The American Dietetic Association uses the term adherence in the Nutrition Care Process terminology.[14]

Individualization of Therapy

Nutrition counseling should be tailored to meet client needs, goals, and living arrangements to enhance adherence.[13,31] Clients want to eat foods that taste good and develop a dietary pattern that can realistically fit into their lifestyle. King and Gibney[39] found that dietary advice to lower fat intake was more successful when existing eating frequency patterns were taken into consideration.

EXERCISE 6.9   Enhancing Learning

Write in your journal what enhances your learning experiences. Identify two things you do to aid your memory.

Dietitians can generally find ways to include favorite foods into a diet pattern. Positive emphasis can be placed on which foods to add or substitute, rather than which foods should be avoided. Evaluation of dietary satisfaction in the Modification of Diet in Renal Disease Study found that study participants who enjoyed eating their diets with the level of protein they were allowed were more likely to adhere to the regimen.[40]

Generally, gradual stepwise modifications are recommended for changing dietary patterns that will endure.[41] Slower changes allow for the nutrition counselor to help tailor adjustments to the taste preferences of the client. However, some highly motivated clients may be capable of making substantial changes. Barnard et al.[42] reviewed thirty published research trials designed to reduce risk of heart disease and found that studies setting relatively strict limits of fat intake achieved a greater degree of dietary change than those with modest dietary goals. Although food habits are highly resistant to change, it is encouraging that after a new habit has been fully adopted, our taste preferences can change.[43] Frequently people who follow a sodium-restricted diet come to prefer low-salt foods to higher-salt foods.

Length and Frequency of Counseling Sessions

Length of visits and number of sessions needed to produce favorable outcomes vary with complexity of the clients' problems. One study of patients with type 2 diabetes found that any contact with a dietitian produced better medical outcomes than no contact.[44] Nutrition practice guidelines for medical nutrition therapy for type 1 and type 2 individuals include at least three sessions lasting from 45 to 90 minutes after diagnosis for the duration of three to six months.[45] Guidelines for individuals who have disorders of lipid metabolism include an initial visit lasting from 45 to 90 minutes and at least two to six follow-up visits lasting from 30 to 60 minutes each.[45] King[11] notes that in her private practice she may see clients over a period of one to five years to help them fully achieve their goals.

Perception of Quality of Care

Perception of quality of care is highly related to adherence.[46] A warm, caring environment created by the counselor, staff, and physical setting creates an environment conducive to counseling. Attempts should be made for the same counselor see a client at each visit.

The exact physical surroundings may vary because nutrition counseling sessions are conducted in a variety of settings, including clinics, private offices, fitness centers, hospital rooms, and work site locations. Attempt to arrange a meeting place that is attractive, comfortable, quiet, well ventilated, adequately lighted, and private. Be sure sturdy seating is available for large individuals. Do not allow big obstacles, such as a desk, to be a barrier between a counselor and a client. If a desk is a necessity, then have the client sit alongside it, not behind it. If a table is needed to view materials, a round table is preferable because it avoids the head-of-the-table position. The essential consideration is to provide an environment as free of distractions as possible. If the meeting place is not ideal, search for innovative ways to rearrange the environment. In a clinic, creative placement of a bookcase or a plant could help define space and give the illusion of privacy. In a hospital room, this may mean pulling a privacy curtain, asking the client to turn off the television, and pulling a chair near the patient's bed. Every effort should be made not to stand while a client reclines in a bed. If the patient is ambulatory, a conference or counseling room may be available for meetings.

Arrangements should be made so phone calls and staff members do not disturb sessions. If a phone call or a colleague does interrupt a session, make every effort to discontinue the intrusion and then make an apology. Clients should not be made to wait for long periods of time. The health care team can contribute to the perception of care through an interdisciplinary approach. The team should present a single, unified treatment plan and meet on a regular basis to maintain good communication with each other. Sometimes clients will question several health care members about their treatment plan because they want assurance that what they are being asked to do is really necessary.

Nonadherence Counselor Issues

Working with a client who is successfully making healthful lifestyle changes is a joy; however, sometimes clients will not change, leaving the counselor feeling frustrated or even angry. It is irrational thinking on the part of a counselor to assume personal responsibility for a client's inaction. Counselors are there to assist their clients, but clients have the responsibility to make changes.[47] King[11] emphasizes, "We are not here to fix our clients but to inform, guide, encourage, understand and support them".

Changing dietary behavior is a complicated process requiring numerous lifestyle adjustments that often interfere with pleasurable activities and thereby compromise a client's motivational level. A counselor may view the required changes as extremely important, but the client may not have the same priorities. Counselors need to be willing to accept the motivational level of their clients and work with them accordingly.

Another factor to consider is that sometimes the benefits of an interaction with a nutrition counselor will not be immediately observable. Sigman-Grant[48] emphasizes that change involves a series of processes requiring time as an important dimension. The interaction with a nutrition counselor may be one of several important events that eventually lead to a significant change.

EXERCISE 6.10   Intervention Strategies for Mary

Complete the following questions after reading the case study below.
1. Complete a Client Concerns and Strengths Log, Lifestyle Management Form 5.6 in Appendix D, for Mary.
2. What are the most crucial factors (barriers) influencing Mary's potential adherences to any plan?
3. What type of food management tool guide (menus and meal plans, exchange system, MyPyramid, DASH food plan, or goal setting) do you think would benefit Mary? Why?
4. What system of tracking do you believe she would select?
5. An assessment would identify educational issues Mary would like to address. What educational materials and interactive experiences would you like to incorporate in counseling sessions with Mary?
6. Would positive or negative educational messages be more appropriate to use with Mary? Explain.
7. Review the following behavior change strategies, indicate for each whether you would consider using the strategy with Mary, and explain why or why not.

Countering	Reinforcement (rewards)	Goal setting
Contracting	Encouragement	Modeling
		Problem Solving

CASE STUDY   Mary: Busy Overweight College Student and Mother

Mary is a thirty-four-year-old university student presently carrying twelve credits. She is on campus four days a week from 9:00 A.M. to 2:00 P.M. She is committed to her schoolwork and currently has a 3.4 cumulative average.

Mary is married and has three children ages six, eight-and-a-half, and ten. She wants to be a good role model for her children, setting an example as to the importance of education. Her husband is a pharmaceutical sales representative, and due to the nature of his job he is away from home frequently. This means that the bulk of child care responsibility is hers.

Mary has come to the Lifestyle Management Program seeking assistance for weight loss and because she really does not feel as good as she thinks she should. She is 5 feet, 6 inches tall and weighs 155 pounds. Mary completed a client assessment questionnaire and a food frequency form in advance, and you review them when she arrives. The exploration phase of your interview allows Mary to elaborate on her lifestyle that has created and perpetuated her diet concerns.

(continued)

> **CASE STUDY   Mary: Busy Overweight College Student and Mother *(continued)***
>
> Mary is short on time in the morning because she must make her children's lunches, get them dressed, drive them to school, and make a 9:00 A.M. class. She does not eat breakfast until 10:30, when she grabs something in the Student Center. Lunch consists of a vending machine snack that she grabs on her way to the parking lot, and she eats it on the forty-minute drive home. She drives directly to the elementary school to pick up her children, because they do not get bused.
>
> Because she has such small children, Mary feels the need to keep a supply of snack foods on hand. Upon returning home, it is time to oversee her children's homework and give them a snack. By the time Mary's starts to make dinner, she is starving and finds herself picking at whatever is around simply to keep her sanity while she is helping with homework and separating some occasional boxing matches. Because her husband is away so frequently, she must clean up the dishes and get the children showered and into bed herself. By the time everyone is settled down and bedtime stories are finished, Mary is exhausted. It is about 9:00 P.M. before she is free to do her own homework and study.
>
> Typically, she first puts in a load of wash and folds the clothes from the dryer, puts on her robe, prepares herself a snack, and settles down to do schoolwork. It is at this time that Mary really feels the loneliness and starts to feel sorry for herself. She expresses dismay that in a few hours she will have to get up and do it all over again. After about an hour of snacking and studying, she puts in some more laundry and heads to bed.

REVIEW QUESTIONS

1. Six food management tools were reviewed in this chapter. Describe the tools and indicate the advantages and disadvantages of each.

2. What are the benefits of tracking food behavior goals?

3. Identify five topics a nutrition counselor can review with a client to provide guidance for record keeping.

4. Explain the ABCs of eating behavior and behavior chains.

5. Identify four factors to consider when guiding clients in the use of rewards.

6. Explain the following behavior change strategies: cue management, countering, rewards, modeling, problem solving, contracting, and encouragement.

7. How does obstacle thinking hamper behavior change?

8. What are cognitive distortions?

9. Explain the integrated approach to learning advocated by Funnell and Anderson.

Answers Exercise 6.7 *Cognitive Distortions included the following: I might as well continue to blow the diet and eat the chocolate. I'll never be able to eat right. This is awful. I am a terrible person. I deserve to have* another heart attack. I'm surprised I ate any fruits at all. How did that happen? I should have prepared a lunch. I should have eaten something before I went to the party. I should have eaten more vegetables.

ASSIGNMENT—FOOD MANAGEMENT TOOL USAGE

The objective of this assignment is to investigate, develop, and utilize four food management tool options and four tracking methods. You will develop a separate food plan using each of the following food management tools: MyPyramid, Exchange System for Meal Planning, DASH Food Plan, and Personal Goal Setting. You will follow each plan for three days and use a coordinated tracking method. This assignment will take 12 days to complete. The point of this assignment is not to evaluate your usual diet but to use a tool to design a food plan and then follow that plan. You want to simulate what a client would experience when given instructions to follow a prescribed diet plan. Use the following directions:

Method 1

Food Management Tool: MyPyramid
Tracking Method: Record Food Intake

1. Obtain a MyPyramid Plan and Meal Tracking Worksheet.

- Go to http://www.mypyramid.gov.

- Input your age, gender, weight, height, and level of physical activity to receive a customized food guide–MyPyramid Plan. Print a PDF version.

- Print at least three copies of the MyPyramid Meal Tracking Worksheet.

2. Follow Customized MyPyramid Plan.

- Follow your customized food guide, MyPyramid Plan, for three days.

- Record your food intake by completing (all sections) of the MyPyramid Worksheets.

3. Assess MyPyramid Worksheets.

- Go to MyPyramid home page **http://www .mypyramid.gov/** and click on MyPyramid Tracker.

- Complete the MyPyramid Tracker Tutorial.

- After completing the tutorial, use the MyPyramid Tracker to assess your food selections recorded on the MyPyramid Worksheets three times, one for each day you followed the MyPyramid Food Plan.

- Request the following assessments: Dietary Guidelines (DG) Comparison, MyPyramid Recommendations (Stats), Nutrient Intakes from Foods. Print out each page.

Method 2

Food Management Tool: Exchange System for Meal Planning
Tracking Method: Cross Out Slashes

1. Design a Food Plan Using the Exchange Lists

- Select an appropriate kilocalorie level for yourself and the percentage of fat, carbohydrate, protein, and alcohol that you would like to consume for three days.

- Use the Exchange Lists for Meal Planning to calculate a food plan. A sample calculation and template for your calculations can be found in Appendix C in this text.

- Complete a sample meal plan. The form for this can also be found in Appendix C. Note this form is an example of how the exchanges can be distributed throughout the day based on your preferences. This is not a form for writing down everything you eat for three days. This means you are not using this form for tracking. Think of how you would use this form if you were counseling an individual. You would use this form to illustrate how the exchanges could be distributed throughout the day. Therefore, you will hand in only one copy of this form to your instructor.

2. Follow Your Exchange System Food Plan

- Follow the exchange system food plan you designed for yourself for three days.

- Track your intake of exchanges (food groups) by a slash method. Note, this is not writing down everything you are eating, but instead keeping track of the food groups (exchanges) as they are consumed. See the cross out tracking method instructions below.

3. Cross Out Slash Method For Tracking

- On an index card make slashes that indicate the number of servings you expect to eat from each food group.

- Carry the card with you throughout the day. As you consume a serving (exchange), cross out one of the slashes. Note that this is your only tracking technique for this food management tool. If you consume one slice of bread (one starch exchange), you will put a line through one of the slashes you have on the index card you are carrying with you. You will not be writing down everything you eat. The point of this assignment is to become familiar with alternative tracking procedures. Use the following example as a guide:

Exchange Food Plan	Cross out slashes as exchange is consumed
Starch |||||||	Starch XXXX||||
Fruit |||	Fruit XX|
Milk (Reduced-fat) ||	Milk (Reduced-fat) X|
Other carbohydrates |||	Other carbohydrates X||
Vegetable |||||	Vegetable XX|||
Meat ||||	Meat XX||
Fat |||||	Fat XX|||

Method 3

Food Management Tool: DASH Food Plan
Tracking Method: DASH Diet Tracking Form or Appendix A Form, What's On Your Plate?

1. Obtain: Your Guide to Lowering Your Blood Pressure With Dash. Go to the National Heart Lung and Blood Institute (NHLBLI) website to download this DASH booklet, http://www.nhlbi.nih.gov/health/public/heart/hbp/dash/.

2. Follow a Dash Food Plan. Follow the DASH Food Plan for three days. The DASH Food Plan in Appendix A was designed for a 2000 kilocalorie food plan. If you prefer to use a different calorie level for implementation of the plan, refer to the NHLBI booklet for guidance and adjust the number of servings per food group accordingly.

3. Track Your Intake. Use the form in Appendix A to track your intake of DASH food groups and sodium intake. Adjust the form if you are following a calorie intake different than 2000 kilocalorie.

Method 4

Food Management Tool: Goal Setting
Tracking Method: Personal Creation

1. Design a Goal. Review the goal-setting process described in Chapter 5, and determine an appropriate food goal for yourself.

2. Create a Tracking Method. Create your own tracking method such as, marks in a calendar, tear off stubs of a sticky note, or empty fruit bowl. Hand in a copy of your method or describe the tracking method in your final report.

3. Follow Your Plan for Three Days. Describe your progress in meeting your goal for each of the three days in your final report.

Write an Evaluation

Write an evaluation of your experiences by answering the following questions:

a. Describe the advantages and disadvantages of each eating plan guide.

b. Describe the advantages and disadvantages of each tracking method.

c. What barriers did you encounter in trying to achieve your dietary objectives?

d. What did you learn from this experience?

SUGGESTED READINGS, MATERIALS, AND INTERNET RESOURCES

Weight Loss Program

Lifestyle Balance Manuals/Diabetes Prevention Program
http://www.bsc.gwu.edu/dpp/manuals.htmlvdoc 16-week weight loss behavior management program; treatment manuals, and participant handouts.

Meal Plans and Recipes

Shape Up America. A personalized meal plan can be developed in the Cyberkitchen. http://www.shapeup.org/index.html

USDA's Nutrition and Weight Management website. Provides information about shopping tips, recipes, and menu makeovers. www.nutrition.gov

Oldways Preservation and Exchange Trust. Provides menus and information about ethnic diets. www.oldwayspt.org

Food Guides

Exchange Lists for Meal Planning. This booklet can be ordered from the American Dietetic Association from the Web site.
www.eatright.org

Your Guide to Lowering Your Blood Pressure with DASH. National Heart Lung and Blood Institute (NHLBLI) website.
www.nhlbi.nih.gov/health/public/heart/hbp/dash/

MyPyramid. Website provides a host of nutrition information as well as a personalized food guide.
www.MyPyramid.gov

Self-Help Resource

Katz D. *The Way to Eat: A Six Step Path to Lifelong Weight Control. A* comprehensive guide to a lifetime of eating well in support of three goals: overall good health, weight control and enjoyment of food. Order from: American Dietetic Association, www.eatright.org

REFERENCES

[1]Tsoh JY, McClure JB, Skaar KL, et al. Components of effective intervention. *Behav Med.* 1997;23:15–28.

[2]Spahn JM, Reeves RS, Keim KS, et al. State of the evidence regarding behavior change theories and strategies in nutrition counseling to facilitate health and food behavior change. *J Am Diet Assoc.* 2010;110:879–891.

[3]Heymsfield SB, van Mierlo CA, van der Knapp HC, et al. Weight management using a meal replacement strategy: Meta and pooling analysis from six studies. *Intl J Obes.* 2003;27:537–549.

[4]National Institute of Diabetes & Digestive & Kidney Diseases. Look AHEAD Counselor Manual. Available at: www.lookaheadtrial.org Accessed August 22, 2010.

[5]Fabricatore AN. Behavior therapy and cognitive-behavioral therapy of obesity: is there a difference? *J Am Diet Assoc.* 2007;107:92–99.

[6]Herrin M, Parham E, Ikeda J, et al. Alternative viewpoint on National Institutes of Health clinical guidelines. *J Nutr Educ.* 1999;31:116–118.

[7]Foreyt JP, Goodrick GK. Attributes of successful approaches to weight loss and control. *Appl Prev Psychol.* 1994;3:209–215.

[8]Baker RC, Kirschenbaum DS. Self-monitoring may be necessary for successful weight control. *Behav Ther.* 1993;24:377–394.

[9]Streit KJ, Stevens NH, Stevens VJ, et al. Food records: A predictor and modifier of weight change in a long-term weight loss program. *J Am Diet Assoc.* 1991;91:213–216.

[10]Stevens VJ, Rossner J, Hyg MS, et al. Freedom from fat: A contemporary multi-component weight loss program for the general population of obese adults. *J Am Diet Assoc.* 1989;89:1254–1258.

[11]King NL. *Counseling for Health & Fitness.* Eureka, CA: Nutrition Dimension; 1999.

[12]Glanz K, Greene G, Shield JE. Understanding behavior. In: American Dietetic Association. *Project Lean Resource Kit.* Chicago: American Dietetic Association; 1995:142–189.

[13]Caban A, Johnson P, Marseille D, et al. Tailoring a lifestyle change approach and resources to the patient. *Diabetes Spectrum.* 1999;12:33–38.

[14]American Dietetic Association. International Dietetics & Nutrition Terminology (IDNT) Reverence Manual Standardized Language for the Nutrition Care Process. 3rd ed. Chicago, IL: American Dietetic Association; 2010.

[15]Prochaska JO, Norcross JC, DiClemente CC. *Changing for Good.* New York: Avon; 1994.

[16]Cormier S, Cormier B. *Interviewing Strategies for Helpers: Fundamental Skills and Cognitive Behavioral Interventions.* 4th ed. Pacific Grove, CA: Brooks/Cole; 1998.

[17]Snetselaar LG. *Nutrition Counseling Skills for Medical Nutrition Therapy.* Gaithersburg, MD: Aspen; 1997.

[18]Lappalainen R, Koikkalainen M, Julkunen J, et al. Association of sociodemographic factors with barriers reported by patients receiving nutrition counseling as part of cardiac rehabilitation. *J Am Diet Assoc.* 1998;98:1026–1029.

[19]American Dietetic Association. Position of the American Dietetic Association: Total diet approach to communication food and nutrition information. *J Am Diet Assoc.* 2007;107:1224–1232.

[20]Iszler J, Crockett S, Lytle L, et al. Formative evaluation for planning a nutrition intervention: Results from focus groups. *J Nutr Ed.* 1995;27:127–132.

[21]Danish SJ, Laquatra I. *Working with Challenging Clients.* Chicago, IL: American Dietetic Association; 2004.

[22]Glasgow RE, Tooberi DJ, Mitchell DL, et al. Nutrition education and social learning interventions for type II diabetes. *Diabetes Care.* 1989;12:1105–1110.

[23]Kiy AM. Cognitive-behavioral and psychoeducational counseling and therapy. In: Helm KK, Klawitter B, eds. *Nutrition Therapy Advanced Counseling.* Lake Dallas, TX: Helm Seminars; 1995:135–154.

[24]Tsoh JY, McClure JB, Skaar KL, et al. Components of effective intervention. *Behav Med.* 1997;23:15–28.

[25]Shiffman S, Paty JA, Gnys M, et al. First lapses to smoking: Within-subjects analysis of real-time reports. *J Consult Clin Psychol.* 1996;64:366–379.

[26]Neck CP, Barnard WH. Managing your mind: What are you telling yourself? *Educational Leadership.* 1996;53:24–27.

[27]Corey G. *Theory and Practice of Counseling and Psychotherapy.* 5th ed. Pacific Grove, CA: Brooks/Cole; 1996.

[28]Glanz K, Greene B, Shield J. Understanding behavior. In: American Dietetic Association. *Project Lean Resource Kit.* Chicago: American Dietetic Association; 1995:142–189.

[29]Ellis A, Harper RA. *A Guide to Rational Living.* Hollywood, CA: Melvin Powers Wilshire; 1997.

[30]Brownell KD, Cohen R. Adherence to dietary regimens: Components of effective intervention. *Behav Med.* 1995;20:155–165.

[31]Trudeau E, Dube L. Moderators and determinants of satisfaction with diet counseling for patients consuming a therapeutic diet. *J Am Diet Assoc.* 1995;95:34–39.

[32]Pohl SL. Facilitating lifestyle change in people with diabetes mellitus: Perspective from a private practice. *Diabetes Spectrum.* 1999;12:28–33.

[33]Funnell MM, Brown TL, Childs BP, et al. National Standards for Diabetes Self-Management Education. *Diabetes Care.* 2010;33:S89–S96.

[34]Funnell MM, Anderson RM. Putting Humpty Dumpty back together again: Reintegrating the clinical and behavioral components in diabetes care and education. *Diabetes Spectrum.* 1999;12:19–23.

[35]Berry M, Krummel D. Promoting dietary adherence. In: Kris-Etherton P, Burns JH, eds. *Cardiovascular Nutrition: Strategies and Tools for Disease Management and Prevention.* Chicago: American Dietetic Association; 1998:203–215.

[36]Gehling E. *Model Program: Promoting Lifestyle Behavior Changes—A Counseling Approach.* University of Minnesota Educational Videos from the 1999 National Maternal Nutrition Intensive Course. Minneapolis: University of Minnesota School of Public Health; 1999.

[37]Kellogg M. Tip #39 Imperative Language. Counseling Tips for Nutrition Therapists: Practice Workbook, Vol. 2. Available at: http://www.mollykellogg.com/tipslist.html

[38]International Food Information Council. How Consumers Feel about Food and Nutrition Messages. February 2002. Available at: http://www.foodinsight. org/Resources/Detail.aspx?topic=How_Consumers_Feel_ about_Food_and_Nutrition_Messages. Accessed August 24, 2010.

[39]King S, Gibney M. Dietary advice to reduce fat intake is more successful when it does not restrict habitual eating patterns. *J Am Diet Assoc.* 1999;99:685–689.

[40]Coyne T, Olson M, Bradham K, et al. Dietary satisfaction correlated with adherence in the Modification of Diet in Renal Disease Study. *J Am Diet Assoc.* 1995;95:1301–1307.

[41]Schiller MR, Miller M, Moore C, et al. Patients report positive nutrition counseling outcomes. *J Am Diet Assoc.* 1998;98:977–982.

[42]Barnard ND, Akhtar A, Nicholson A. Factors that influence compliance with low-fat diets. *Arch Fam Med.* 1995;4:153–158.

[43]Mattes, Richard D. (1993). Fat preference and adherence to a reduced-fat diet. *Am J Clin Nutr.* 1993;57:373–381.

[44]Franz MJ, Splett PL, Monk A, et al. Cost-effectiveness of medical nutrition therapy provided by dietitians for persons with non-insulin-dependent diabetes mellitus. *J Am Diet Assoc.* 1995;95:1018–1024.

[45]American Dietetic Association. ADA Evidence Library. Available at: www.eatright.org Accessed: August 24, 2010.

[46]Caggiula AW, Watson, JE. Characteristics associated with compliance to cholesterol lowering eating patterns. *Patient Educ Counsel.* 1992;19:33–41.

[47]Siminerio LM. Defining the role of the health education specialist in the United States. *Diabetes Spectrum.* 1999;12:152–157.

[48]Sigman-Grant M. Change strategies for dietary behaviors in pregnancy and lactation. *University of Minnesota Educational Videos from the 1999 National Maternal Nutrition Intensive Course.* Minneapolis: University of Minnesota School of Public Health; 1999.

Making Behavior Change Last

Digital Imagery/PhotoDisc, Inc.

Look to this day!
For it is life, the very life of life.
In its brief course
Lie all the verities and realities of your existence:
The bliss of growth;
The glory of action;
The splendor of achievement;
For yesterday is but a dream,
And tomorrow is only a vision;
But today, well lived, makes every yesterday
A dream of happiness,
And every tomorrow a vision of hope.
Look well, therefore, to this day!

—KALIDASA

Behavioral Objectives

- Explain significance of social support.
- Suggest ways significant others can support clients.
- Explain usefulness of social disclosure.
- Develop mental imagery and role-playing skills.
- Describe possible responses to stress.
- Identify common sources of stress.
- Distinguish three general categories of stress management strategies.

- Describe the basics of stress management counseling.
- Describe the basic components of relapse prevention counseling.
- List immediate determinants and covert antecedents of relapse.
- Describe procedures for providing a smooth transition to ending a counseling relationship.
- Identify ways to evaluate counseling effectiveness.

Key Terms

- **Mental Imagery:** a mental rehearsal of an anticipated experience.
- **Mindfulness:** being attentive and aware of the present.
- **Relapse Prevention:** systematic approach to maintaining a behavior change involving the identification of and preparation for high-risk situations.
- **Social Disclosure:** sharing information about oneself in order to enhance lifestyle change objectives.

INTRODUCTION

This chapter places emphasis on strategies to support lifestyle changes. Because humans are social beings by nature, our social environment can help or hinder our behavior change efforts. This chapter explores ways to encourage social support and minimize social hindrances to making lifestyle changes. Stress is inevitable, but the way an individual handles stress can impact behavior change efforts as well as a possible relapse. We cover how components of stress management counseling can be incorporated into a nutrition counseling intervention. Mindfulness as an approach to enhance quality of life has been taught for centuries. However, only recently the approach has been applied to changing food behaviors. We explore the basic tenets of mindful eating. Then, we address factors to consider to bring closure when ending a counseling relationship. As the American Dietetic Association Nutrition Care Process mandates, monitoring and evaluation of nutrition interventions are extremely important. Methods of monitoring counseling progress and evaluating outcomes are addressed.

> One of my clients was an elderly female who had kidney disease and had a number of factors in her diet that had to be monitored. It was my responsibility to design a discharge diet plan for her. She was a very nice person, but she told me she would never be able to understand the diet because she had never been good with numbers. We scheduled to have her son meet with us for another time and all together we designed a detailed food plan.

SOCIAL NETWORK

Because the act of eating is often a social activity, either as an element of daily life or an integral part of special events, a client's social environment can have a significant impact on attempts to change eating behavior. In addition to the actual eating experience, social factors are likely to play a role concerning availability, procurement, selection, and preparation of food.[1] Social context is a component of several behavior change theories. Recognizing the impact of the social atmosphere, the American Dietetic Association provides a definition of social network in the assessment and monitoring and evaluation component of the Nutrition Care Process: "Ability to build and utilize a network of family, friends, colleagues, health professionals, and community resources for encouragement, emotional support and to enhance one's environment to support behavior change."[2] The people closest to a person making dietary lifestyle changes will be affected by the new behavior patterns and in return, can exert a powerful influence on your client. The changes often put a stress on the dynamics of a household. For example, a Friday night activity may have been to eat fried fish at a certain restaurant that only serves fried food. If a client needs to avoid this type of food, family members may resent a change in the family pattern. In conditions where a client has many lifestyle change requirements, the feelings of resentment are likely to mount. One way of helping dispel negative feelings is by involving people closest to your client in your counseling sessions.

Social Support

Encouraging family and significant others to provide a supportive environment has been shown to have a beneficial effect on dietary change objectives.[3] In one study in an acute inpatient hospital setting, a family or friend acted as a "care partner," resulting in increased patient and family satisfaction and adherence to the patient's medical regimen.[4] Also successful was an award-winning program designed to reduce hypertension through exercise.[5] In this program, health care providers enlisted the support of a family or friend to be a designated "helper." Perceived family support for adolescents' diabetes care has been shown to increase adherence to health care protocols.[6] One explanation for the positive effects of social support on health is that clients perceive a sense of support from others leading to a feeling of a more generalized sense of control.[5] A greater sense of control often results in an increase in self-efficacy.

When meeting with family members or significant others, nutrition counselors can provide information, explore potential stress that the new food pattern could put on family interactions, and suggest ways in which family members can be supportive. Possible topics for discussion are presented in Exhibit 7.1.

Direct involvement with a client's social network is not always possible. In any case, clients should be encouraged to be vocal and request support for cooperation and assistance from their family and friends. Clients could approach potential supporters and cover any of the topics reviewed in Exhibit 7.1. To keep a social environment conducive to change,

counselors should remind clients to thank others for their involvement.

If your clients do not have readily available support in their immediate environment, possible alternatives can be explored. The following list contains some suggestions:

- **Locate a distant support buddy.** Possibly a relative, associate, or friend in a distant location could be involved through telephone conversations, e-mail, or instant messaging.
- **Join clubs or organizations.** Clients may obtain direct or indirect support by taking part in organizations compatible with their lifestyle goals. This may include joining a walking club, gym, dance troop, or vegetarian society.

EXHIBIT 7.1   Social Support Discussion Topics

- Inquire about any concerns or questions family members have about your client's dietary needs.
- If reasonable, invite family members to participate in the dietary changes and explore new tastes.
- Ask whether there are times when the new diet pattern is likely to cause unusual stress.
- Explore willingness of family members to show support for your client.
- Suggest ways in which family members can support your client:
  - Help keep undesirable foods out of sight to avoid tempting cues.
  - Purchase foods to be avoided in varieties client does not like (for example, an ice cream flavor client is unlikely to eat).
  - Do not give foods to be avoided as gifts.
  - Think of creative ways of celebrating a holiday or a birthday. For example, fresh fruit, gelatin, and fat-free sherbet can be quickly and attractively arranged on a plate for fat-free celebrations. Candles can even be placed in the composition.
  - Show support:

    ❏ Offer praise when desirable behavior is observed.
    ❏ Express appreciation for the accomplishment of difficult tasks (such as taking blood sugar for the first time).
    ❏ Provide preplanned or surprise rewards, such as, hugs, gifts, or back rubs.
    ❏ Brag to others about positive behavior changes.

  - Show patience for extra time needed to calculate or prepare foods.
  - Offer help to plan ahead when visiting or going on trips. Avoid teasing or tempting with foods that need to be avoided.
  - Avoid scolding, nagging, preaching, and embarrassing. Although such behavior may be well intentioned, the overall effect is destructive.
  - Do not give criticism unless there have been three compliments.[5]
  - Use positive statements about the new food pattern; avoid using the words *strange* or *different* to describe foods on the pattern. A motivated family member may be willing to keep a record of positive and negative statements regarding the new dietary pattern.

Source: Raab C, Tillotson JL, eds., Heart to Heart. DHHS (PHS) publication 83-1528. Washington, DC: U.S. Department of Health and Human Services; 1983; Prochaska JO, Norcross JC, Diclemente CC, Changing for Good. New York: Avon; 1994.

- **Locate a social support group.** Social support groups have been found to be effective components of behavior change interventions.[7] Available social support can be explored in local medical centers as well as the Internet, where chat rooms may be available. Perceived social support has been found to increase with Internet social support interventions.[8] Encourage your clients to become active in support groups by taking on responsibilities such as organizing a meeting, writing a newsletter, or volunteering for a committee. Active involvement increases commitment and expands the likelihood of making connections to others.

- **Take classes.** Classes related to your client's condition may be available in community education programs, supermarkets, or health centers. By taking part in these programs, clients enhance their skills and make contacts that can support their behavior change endeavors.

> One of my clients was a pleasant, overweight teenage girl. She was anxious to lose weight, but her mother seemed to sabotage her daughter's attempts by baking her favorite cakes, keeping the kitchen stocked with high-fat foods, and encouraging her daughter to eat. When her mother came to one of our sessions, she admitted that this was a problem. Although her mother loved her daughter, she voiced her fear that her daughter would become thin, be attractive to men, get married, and leave her. Eventually her mother agreed to go to a psychotherapist.

> One of my clients was a busy professional who could not easily see how she could join a support group without putting additional stress in her life. However, she did feel that social disclosure would benefit her attempts to change her eating and exercise behavior. Her solution was to post her eating and exercise records each day in her office. Periodically throughout the day someone would inquire about the records, and this would precipitate supportive short conversations.

Social Disclosure

Closely related to social support is the concept of **social disclosure.** Disclosure of behavior records and progress in meeting goals to peers or professionals has been shown to exert a powerful influence in changing behavior.[9] Even announcing the intent to engage in a new behavior can have a significant influence. Regular disclosure can be done formally in a weekly meeting with support buddies or informally during walks or while eating lunch with friends.

Social Pressures

A client's immediate social environment can sometimes exert a negative influence on eating behavior. If so, seek ways to reduce the impact without causing undue social stress. Sometimes this involves assertive behavior, such as suggesting an alternative restaurant, calling a host ahead of time to discuss potential problems, or offering to bring a vegetable platter to a social function. Scenarios for dealing with difficult issues can be developed in counseling sessions through role playing, microanalysis of the scenario, or **mental imagery**.

Role Playing   This strategy can effectively prepare clients for behavior change. For role playing to be an effective tool, a secure relationship should have been established with your client. Therefore, it is usually not advisable to use this technique during the first session. See Table 7.1 for role playing guidelines.

Microanalysis of the Scenario   This method is used to talk through an anticipated experience, identifying as many contingencies as possible and deciding on the best response. The following are examples of questions that could help lead the way in anticipation of a telephone call as illustrated in Exhibit 7.2:

- Where will you be when you make the phone call? What do the surroundings look like? What will be going through your head?
- What will you say to approach the topic of dinner on Sunday night?
- What will you say about cholesterol levels?
- What do you believe will be your mother-in-law's response?
- What do you want your mother-in-law to feel?

EXERCISE 7.1   Social Support Survey

Find two people who have made or attempted to make a lifestyle change who believe their social environment had an impact on their efforts. Ask your interviewees to describe the impact of others on their behavior change efforts.

- Did your interviewees describe the actions of significant others as social supporters or non-supporters?
- Specify what the significant others did to help or hinder the situation. Record these behaviors in your journal.

Table 7.1   Role Playing Guidelines

Prepare for the Role Play	• Analyze the concern, discuss possible scenarios to handling a situation, and decide on the best course of action.
	• Explain the goals and objectives of role playing. Usually that means preparing for a difficult encounter.
	• Assign roles.
	• Set time limits. Generally you do not want role playing to go more than five minutes; in fact, an effective role play can be as short as two minutes. You want to leave time for processing the experience.
Enact the Role Play	• Arrange chairs for appropriate interaction.
	• As a counselor, you take on the role of one or more of the characters, and your client plays him- or herself. If you believe there is a benefit to modeling a certain behavior, the role play of a scenario could be done twice—once with you taking on the role of the client and a second time with the client playing him- or herself. See Exhibit 7.2 for a role play example.
	• Stop the role play in five minutes or less.
Process the Experience	• Ask your client these questions: What went well? What could have been done differently? How did you feel about the interaction? What may happen differently in the actual situation? Do you feel confident about how you will handle this situation when you encounter this problem?
	• Provide feedback. Your comments should always be supportive and positive. For example, "It was really good that you . . ." If you have suggestions, they should be prefaced with tentative remarks, such as "You might try . . ." or "You could consider . . ."

EXHIBIT 7.2   Role Play Example

Problematic Scenario

- **Mother-in-law:** You are coming for our beefsteak dinner on Sunday, right?
- **Client:** I can't eat beefsteak—didn't John tell you?
- **Mother-in-law:** Yes, but surely you can cheat once in a while. There is nothing like beefsteak dipped in melted butter.
- **Client:** No, I can't! If you find it to be a problem that I can't eat beefsteak, then I will just stay home.
- **Mother-in-law:** I don't understand why you can't just let this be a pleasant family get-together. It isn't like you are going to die if you eat it.

Effective Scenario

Setup: Client needs to establish control. Statements or arguments to be used must make sense to the mother-in-law. When dealing with difficult people, *blending* is an initial effective behavior. This involves agreeing on common ground.

- **Mother-in-law:** You are coming for our beefsteak dinner on Sunday, right?
- **Client:** That is wonderful of you to invite us, but I can't eat beefsteak. I have to try to get my cholesterol down.
- **Mother-in-law:** Yes, but surely you can cheat once in a while. There is nothing like beefsteak dipped in melted butter.
- **Client:** Well, beefsteak sure sounds delicious, and I guess I could cheat, but I really don't want to. I have three months to try to get my cholesterol down with my diet. I don't want to end up like Aunt Joan with all her heart problems. Or become a burden to John and the kids. Maybe I can bring a platter of grilled vegetables, and then I can come and enjoy?
- **Mother-in-law:** Well, that sounds good, too. I must say you have more willpower than I do. You bring your grilled vegetables, and I can throw in some chicken, too.

EXERCISE 7.2 Practice Using Microanalysis and Mental Imagery

Practice microanalysis and mental imagery with a colleague. Each should select an anticipated encounter, preferably an uncomfortable one, and take turns assuming the role of counselor.

❏ Record your experiences and impression of the technique in your journal.

Mental Imagery A common technique for developing new behavior patterns, mental imagery involves a mental rehearsal of an anticipated experience. Clients imagine themselves thinking, feeling, and behaving in precisely the way they would like in the actual situation.[6] Counselors can help their clients reconstruct a past or potential scene in their minds and play out the scenario with a desirable ending. After microanalyzing the scenario, clients can be asked to close their eyes and play out the scene in their minds. After they have completed the exercise, ask whether any new concerns came to mind. Clients should be encouraged to do this activity several times before the actual encounter, thereby allowing them to practice their responses several times.

> *Rule Number 1: Don't sweat the small stuff.*
> *Rule Number 2: It's all small stuff.*
> —**Robert Eliot, M.D.**

STRESS MANAGEMENT

Although stress is a normal part of life and can serve as a positive force to stimulate performance, too much stress can harm health and impair attempts to make lifestyle changes. Because food is often used to provide nurturing and stress reduction, especially for women, finding alternative methods of coping with stress is important.[10] Also, stress has been found to be a major predictor of relapse, overeating, and dysfunctional eating patterns,[11,12] and is linked to six leading causes of death—heart disease, cancer, lung ailments, accidents, cirrhosis of the liver, and suicide.[13]

Dr. Hans Selye, the scientist accredited with identifying the link between health and stress, defined *stress* as a nonspecific response of the body to threats or requirements for action or change.[14] The response is considered nonspecific because various physical, mental, and emotional factors could be affected. The physiological response stems from the stimulation of the hypothalamus from an imagined or real threat.[15] The hypothalamus activates the sympathetic nervous system to increase heart rate and blood pressure in order to deliver extra nutrients to muscles and the brain. Also, blood sugar and lipids rise to meet the anticipated increase in needs for energy, while breathing accelerates to supply extra oxygen for energy metabolism. Blood supply is diverted from the skin to large muscle groups. The release of stress hormones from the adrenal glands prepares the body to be on a heightened alert to make a quick response by shutting down tissue repair, digestion, reproduction, growth, and immune and inflammatory responses. The body prepares for fight or flight. The observable symptoms can include sweaty palms, rapid breathing, dilated pupils, dry mouth, nervous or shaky speech, crying, and a feeling of butterflies in the stomach, or heart in the throat.[14] If the stress response is continually being triggered, there is a negative effect on the mind, body, and quality of life, increasing the likelihood of developing one or more of the eight physical indicators of stress (see Exhibit 7.3). However, our bodies have a countering mechanism available that turns off the stress response and allows all systems to return to their normal state. Herbert Benson[16] refers to this natural restorative process as "the relaxation response."

Major life changes, whether positive or negative, are stressful events because they often require a series of new adjustments. Life-changing events include change in marital status, job status,

---

EXHIBIT 7.3 Indicators of Stress

1. Increases in blood pressure
2. Suppressed immunity
3. Increased fat around the abdomen
4. Bone loss
5. Increases in blood sugar
6. Increased levels of cortisol
7. Weaker muscles in the abdomen
8. Increases in blood cholesterol levels

Source: McEwen BS, Protective and damaging effects of stress mediators. New England Journal of Medicine. 1998;338:171–179.

EXHIBIT 7.4   Ten Common Sources of Stress

1. Overscheduled daily calendars
2. Job stress and demands
3. Lack of play and downtime
4. Lack of time with family, friends, and significant other
5. Inequity in home responsibilities
6. Lack of time to explore own interests
7. Guilt (about everything)
8. In families: children's behavior and how to discipline
9. Lack of time
10. Lack of money

Source. Bradlye AC, Under pressure: Identifying and coping with stress. American Fitness, 1997;15:26–33. © 2000 Aerobics and Fitness Association of America.

and family issues as their top stressors, while older adults are likely to select social isolation.[20] Even within the same age category, differences exist as to appraisal of a situation as stressful. Stressed individuals are more likely to see difficulties as dangerous, difficult, or painful and are not likely to have the resources to cope with a problem.[15]

Stress management techniques have been successfully employed to enhance dietary lifestyle changes and to improve health.[21,22] Intervention strategies addressing these objectives can be divided into three general focus areas: environmental, physiological, and cognitive. See Table 7.2.

"A crust eaten in peace is better than a banquet partaken in anxiety."
—Aesop's Fables

financial status, birth or adoption of a child, death of a loved one, new residence, caring for a loved one with a debilitating illness, and diagnosis of a serious illness.[17] In addition to major life events, the National Institute of Mental Health has identified ten common sources of stress (see Exhibit 7.4). Recent research regarding increased cortisol levels among individuals following low calorie diets (1,200 kcal) and increases in perceived stress for individuals monitoring calorie intake has created concerns and discussions among health professionals.[18,19] However, what is considered stressful for one person may not be for another because perceptions of events and conditions differ among individuals. This point is illustrated by stage-of-life data. American adults usually cite work, finances,

Stress Management Counseling

As a nutrition counselor, you cannot possibly be an authority on all stress reduction methods. However, you can educate clients on the impact of stress on behavior change objectives, explore stress as an issue in your clients' lives, assist them in identifying stressors, provide information about possible stress reduction techniques, help clients locate stress reduction resources, and aid in developing stress-reducing behavior change goals. You may consider obtaining training in some of the stress management techniques[23] or acquiring expertise in stress inoculation, a comprehensive approach to stress management.[24] Table 7.3 provides general guidelines for addressing stress management in a nutrition intervention.

Table 7.2   Strategies to Reduce Stress

Category	Description	Strategies
Environmental Focus	Remove or reduce exposure to specific stressors Strategies are problem-focused	Planning ahead, cue management, time management, skills for convenient food preparation, social support, guidance for healthy eating on the run, assertiveness training, conflict management, communication skills, and engaging in distracting behaviors, such as doing puzzles.
Physiological Focus	Strategies address the physiological response	Meditation, the relaxation response, visual imagery, soothing music, prayer, humor, emotion-focused coping, breathing exercises, exercise, and biofeedback.
Cognitive Focus	Strategies deal with cognitive coping skills	Cognitive restructuring and self-acceptance

Table 7.3 Stress Management Counseling Guidelines for Nutrition Interventions

Category	Description	Example	
• *Furnish information about the impact of stress on behavior change objectives.*			
Explain reaction to stress.	Review two major components of stress: physiological arousal and internal monologue or thoughts that provoke anxiety, hostility, or pain.	**Counselor:**	*When you feel stressed, two things are going on simultaneously. Physically your heart pounds, hands sweat, breathing rate increases, and you are likely to feel tightness in your muscles. Mentally your thoughts and self-talk are either intensifying the physical arousal with self-destructive statements, such as "I'm washed up," or they are providing tension reducing counsel, such as "I have learned a lot from this situation."*
Explain impact of stress on food behavior.	Provide information on the desire to use food to reduce stress.	**Counselor:**	*Stress is an issue in many people's lives. It can severely affect health and impair attempts to change food behavior. We may be consciously or unconsciously looking for ways to calm down. Many of us naturally turn to food because we learned to associate it with nurturing.*
• *Investigate clients' stress issues.*			
Review symptoms of stress.	Review Lifestyle Management Form 7.1 (Appendix D), Symptoms of Stress.	**Counselor:**	*Stress and our reactions to stress can become so common in our lives, that we may not realize that they present us with problems. By looking over this list, you may be able to evaluate if stress is adversely affecting you.*
Investigate what is causing stress.	Sometimes people are aware of feeling stressed, but are not aware of the causes of their stress. Use Stress Awareness Journal, Lifestyle Management 7.2 (Appendix D).	**Counselor:**	*You indicated that you believe you are eating in reaction to stress, but you are not aware of the triggers. How do you feel about using this form to record your stress throughout the day for a few days?*
• *Explore possibilities for reducing stress.*			
Discuss methods and resources.	Discuss possibilities of reducing, minimizing, mastering, or tolerating stress. See Table 7.2. If appropriate, explore options by reviewing Tips to Reduce Stress, Lifestyle Management Form 7.3 (Appendix D).	**Counselor:**	*It appears that stress is hampering your ability to make changes. Would you like to explore ways to reduce the amount of stress you are experiencing?*
Provide your client with community resource options.	You need to have available a list of community resources and local referrals.	**Counselor:**	*It appears that stress is hampering your ability to make changes. Are you interested in exploring community resource options?*
Recommend books or Internet sites.	Several excellent self-help books and Internet sites are available for dealing with stress reduction. See the end of the chapter for a list of resources.	**Counselor:**	*It appears that stress is hampering your ability to make changes. Would you be interested in looking at some self-help books or Internet sites?*
• *Set behavior change goals that address stressors. (See Chapter 5.)*			

There is no failure except in no longer trying.
—Elbert Hubbard

RELAPSE PREVENTION

There are many obstacles to initiating a lifestyle behavior change, but maintaining that change is a major challenge. Mark Twain once quipped, "It's not difficult to stop smoking–I've done it dozens of times." Relapse rates for dietary regimens are disconcerting, ranging from 50 to 100 percent. However, these numbers could be misleading because they do not reflect the cumulative effect of multiple attempts to change dietary habits over time.[25] Described as a normal part of the change process in the Transtheoretical Model (see Chapter 2), relapse can in fact be part of a positive spiral that leads to enduring change. Also, relapse numbers do not account for self-changers, people who may have acquired skills in programs during previous attempts to change behavior but were not ready to follow through on their objectives at that time.

In response to this issue, Marlatt[26] developed a **relapse prevention** model that has been successfully employed for weight loss programs and addiction treatment interventions.[27,28] The premise of this model (see Figure 7.1) is to ascertain which factors are threats for relapsing and then to develop cognitive and behavioral strategies to prevent or limit relapse episodes.[29] There are two major categories of factors: *immediate determinants* and *covert antecedents*.

Immediate Determinants

Threats to relapsing that are categorized as immediate determinants include the following: high-risk situations, a person's lack of coping skills, overly positive outcome expectancies, and a negative reaction to a lapse.

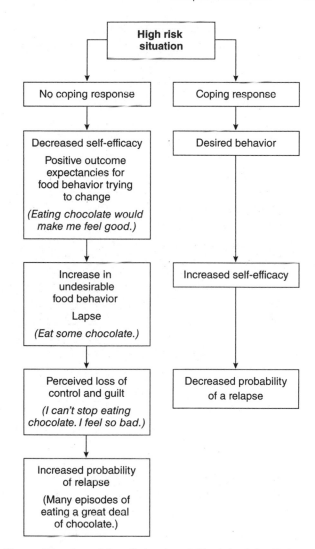

Figure 7.1   Cognitive-Behavioral Model of the Relapse Process

Source: Adapted from Marlatt BA, Gordon JR, eds. *Relapse Prevention: Maintenance Strategies in the Treatment of Addictive Behaviors.* New York, NY: Guilford Press; 1985, p. 38. Reprinted by permission of Guilford Press.

High-Risk Situation   Certain situations or events provide an alluring environment to revert to previously established behavior patterns. These high-risk conditions threaten a person's sense of control and frequently precipitate a relapse. Exhibit 7.5 lists common high risk situations and the following provides a discussion of selected ones:[28]

• **Negative emotional states.** The feelings of anger, anxiety, depression, or frustration that one may feel about an impending divorce or credit problem can precede a relapse. Also, feeling bored or lonely often leads to undesirable food intake.

- **Interpersonal conflict.** Conflicts with others, such as an argument with a friend, often trigger a relapse.
- **Social pressures.** Being around others who are eating foods that are to be avoided can result in direct and indirect pressure to relapse.
- **Positive emotional states.** Celebrations or events frequently serve as cues to eat certain foods and can provide a vehicle for a relapse.

Coping Skills Whether or not a high risk situation results in a relapse will depend on the ability and determination of the individual to cope with the difficulty.

Positive Outcome Expectancies During a stressful or high-risk situation, previous pleasurable experiences associated with particular foods and the significance of those perceptions can add to the desire to lapse.

Reaction to a Lapse A *lapse* is a single act, a slip, and a momentary return to previous behavior; and a *relapse* is a series of lapses, loss of control, a return to previous behavior. Lapses are a momentary indulgence that increases the likelihood of a full-blown relapse, but the progression is not inevitable.[28] Because no one is perfect, there should be preparation for handling a lapse. The way in which the slip is viewed is an important predictor of a relapse. If the lapse is considered a personal failure ("I am a failure. My wife is going to be so disappointed.") or due to a global attribute ("I have no willpower. I will always eat the wrong foods."), the probability of a relapse increases. On the other hand, if the slip is viewed as a learning experience providing the groundwork for developing more effective strategies for the future, relapse is less likely to occur.

Covert Antecedents

Some factors that precede a relapse are not obvious and are referred to as covert antecedents. These include apparently irrelevant decisions, lifestyle imbalances, and urges and cravings.[28]

Apparently Irrelevant Decisions (AIDs) A series of seemingly harmless decisions can provide a conduit for a relapse. For example, buying a bag of potato chips for the "children" or a bag of cookies "in case guests stop by" creates conditions that can bring an individual to the brink of a relapse.

Stress Level A person experiencing a high degree of stress is automatically generating negative emotional states, thereby creating high-risk situations (see Exhibit 7.5). In addition, there is an increased desire to relapse and connect to the satisfying emotional states with previous unhealthy eating pleasures.

Cognitions Cognitive factors such as rationalization and denial set the stage for a relapse. For example, "I deserve a whole batch of brownies after this rejection." Here rationalization that the indulgence is justified adds to the creation of a relapsing environment.

Urges and Cravings The desire for immediate gratification can take the forms of *urges,* sudden impulses to indulge, or *cravings,* subjective desires to experience the effects of an indulgence.

Relapse Prevention Counseling

Relapse prevention programs were originally designed as a follow-up treatment to maintain gains made during an intervention. Now therapists report integrating relapse prevention strategies throughout the course of interventions.[29] As one authority remarked, "Life is a chronic relapsing condition."[30] Many of the behavior change strategies previously

---

EXHIBIT 7.5   Examples of Common High-Risk Situations

- Negative emotional states: anger, anxiety, depression, frustration, boredom, loneliness
- Positive emotional states: celebrations, events
- Conflict with others, arguments
- Social gatherings, parties
- Holidays
- Traveling and vacationing
- Eating out
- Snacking
- Lack of coping skills
- Negative self-talk
- Stress
- Hunger, urges, food cravings
- Fatigue
- Lack of social support

covered can be integrated in a relapse prevention program. Most versions of such a program incorporate three main categories of strategies: skills training, cognitive restructuring, and lifestyle balancing. Let's look at some of the specific components of these strategies.

Description and Introduction To begin, you may want to describe the relapse prevention model and introduce the concept of high-risk situations. Larimer et al.[28] suggest using the metaphor of a road trip to introduce the concept of anticipating high-risk situations and preparing a "toolbox" for dealing with them.

Counselor: *Behavior change can be compared to a highway journey that has easy and difficult stretches. The difficult portions can be handled effectively by being prepared (having a road map, a spare tire, a cell phone, and so forth), paying attention to road signs (warning signals), and using driving skills learned previously for handling troublesome conditions. In your journey to change your food behavior, there are high-risk situations, too, and during our sessions together we will work on identifying them and the preceding warning signals. Also, we will focus on skills that you already have and incorporate them into coping strategies.*

Identification of High-Risk Situations To anticipate and prepare for high-risk situations, the conditions that precede a lapse must be identified. Attention should be paid to warning signs such as stress or apparently irrelevant decisions. The following list outlines some methods of identifying situations where coping difficulties could arise:

- Investigate past lapse and relapse episodes.
- Journal activities, cognitions and eating behavior. See Lifestyle Management Form 6.1 or 7.2 in Appendix D.
- Review a list of common high-risk conditions. See Exhibit 7.5.

Behavioral and Cognitive Strategies to Deal with High-Risk Situations The metaphor of a toolbox can be used to describe the availability of a variety of coping strategies. Because all strategies are not applicable to every temptation, clients need a repertoire from which to choose. Sometimes several strategies can be used at the same time. For example, behavioral

strategies for dealing with temptations at a cocktail party could include eating a small snack before the event, bringing a vegetable platter or other dish with acceptable food choices, and making a contract that a glass of water will be drunk before tasting any foods an individual is trying to avoid at a party. In addition, positive self-talk strategies could be employed. A number of the strategies previously covered can also be used to deal with high-risk situations: tracking, cue management, countering, eating behavior interventions, reinforcement, contracting, goal setting, modeling, stress management, relaxation techniques, social disclosure, and cognitive restructuring.

Strategies to Minimize the Occurrence of High-Risk Situations By minimizing exposure to stressors, the risk of encountering high-risk situations declines. This could include such activities as removing foods that one is attempting to avoid from the home, sitting in a different chair at the dining table, or taking a new route to work to avoid passing a particular bakery. These techniques include cue management and countering, previously discussed in Chapter 6.

Enhancement of Self-Efficacy One of the major objectives of the relapse prevention model is the enhancement of self-efficacy. Several ways of addressing this objective are possible:

- **Collaborative counseling style.** A collaborative counseling style is employed to encourage clients to take an objective view as well as ownership of their behaviors and goals. As a result, clients are more likely to feel a sense of accomplishment when goals are obtained, resulting in an increase in self-efficacy.
- **Set clear, modest, and achievable goals.** One of the objectives of setting goals is to encourage a behavior change to take place. Another objective is for clients to feel a sense of mastery over their problems when the goals are achieved. This increases the belief that additional or more complex goals can be achieved. The overriding theme is that behavior change will occur as new skills are acquired rather than by employing will power.
- **Providing feedback.** Providing positive feedback about the accomplishment of new tasks, even if they were not related to the dietary objectives, can increase self-efficacy.

Lapse Management The objective of lapse management is to have a plan in place for handling a slip so that it does not escalate into a full-blown relapse.[26] Sometimes a written motivational set of instructions can be useful for a client to refer to in the case of a lapse. The following example contains two possible ways to introduce the topic:

Counselor: *Because no one is perfect, I thought we should talk about lapses and relapses. Lapses are momentary slips, such as eating a bag of potato chips. A relapse is a total abandonment of the food program objectives. Research has shown that if you handle the lapse effectively, there doesn't need to be a relapse. We could compare this to going on a trip. Let's assume you were driving to from New Jersey to Florida to go to Disney World. If you got lost in Baltimore, you wouldn't simply give up and go home. The same is true for a lapse. It should be treated as a stumbling block, not a reason to give up.*

It is a good thing that you had this lapse while we were still seeing each other. It gives us an opportunity to understand what happened and work out a plan to deal with the lapse. Actually lapses are a normal part of the change process. Let me show you a diagram of the process of change.

A good way to introduce the concept of relapse to your clients is to show them a visual representation of Prochasca's States of Change model and discuss the process of change (see Lifestyle Management 7.4 in Appendix D).

Cognitive Restructuring Cognitive restructuring was covered in Chapter 6; however, self-talk related to lapses is particularly important in the relapse prevention model. Emphasis should be placed on lapses as a learning experience and the need to use or learn new skills. Clients should be warned not to fall into the trap of blaming themselves as a failure or having moral weakness and not to proclaim a prophecy that the slip means a total relapse.

One of my clients stated that she did not believe she could make a food behavior change if there was no chocolate in her life. To deal with this craving and to keep her indulgence under control, she contracted to eat twelve chocolate chips every day. The chips were kept in her freezer. Each day she counted them out, put the chips into a small bowl, and immediately returned the package to the freezer. This plan worked well to satisfy her cravings, and the rest of her behavior change goals were also accomplished.

Urge Management Techniques Clients need to realize that they are likely to experience an urge to return to previous eating styles. Ways of handling the desires should be addressed, such as the following:

- Contract to consume a certain amount of the food causing urges and cravings.
- Plan for a response when the urge arises. For example, clients can tell themselves that they can have the piece of cake after drinking a glass of water or eating an apple. Often the urge to eat the undesirable food has passed by the time the water or apple is consumed.
- Plan for a nonfood countering activity such as a relaxation response, knitting, jumping, and so on.
- Use an image technique called *urge surfing*. In this method, the client visualizes the urge or craving as a wave that crests and then washes onto a beach. Clients are told to imagine riding the wave on a surfboard rather than struggling against it or giving into the want.[28]

Let mindfulness of your body set the rhythm for your upcoming meal.

—Donald Altman

MINDFUL EATING

Mindful eating or a similar approach, intuitive eating, has become increasingly popular as a method encouraging individuals to use their inner wisdom to find joyfulness in the preparation and consumption of food.[31] **Mindfulness** has roots in Buddhist and other contemplative traditions and is defined as, "the state of being attentive to and aware of what is taking place in the present."[32] Without awareness, individuals are more likely to behave compulsively or automatically and may not be behaving in their best interests. Factors that can pull us away from awareness of the present include absorption in the past, fantasies and anxieties about the future, preoccupation with multiple tasks; or negative feelings, such as anger or jealousy.

Advocates of mindful eating encourage replacing mindless eating with conscious awareness allowing individuals to recognize the pleasures of the eating experience. Supporters of this method believe mindfulness permits attentiveness to feelings of hunger and fullness and helps bring a balanced approach to eating. See Table 7.4 for a list of components generally included in a mindful eating approach. Although results of studies have not always been consistent, benefits of this methodology with individuals seeking weight loss or who have eating disorders have been noted.[33] An evaluation of eating slowly has been shown to help maximize satiation and reduce kilocalorie intake within meals.[34] The tenets of this approach harmonize with Slow Food, an international, grassroots movement that "links the pleasure of food with a commitment to community and the environment" and with the size acceptance and intuitive eating approach to improved eating behavior.[35,36] See the resources at the end of the chapter for links to established programs helping individuals restructure their eating behavior and relationship with food.

> *Nothing so difficult as a beginning*
> *In poesy, unless perhaps the end.*
> **—Lord Byron, Don Juan**

ENDING THE COUNSELING RELATIONSHIP

Life is a journey of change. Nutrition counseling, either a brief intervention or an intensive program, is part of that journey and at some point needs to come to an end. For brief encounters, the transition to terminating the counseling experience generally would not pose any special issues. However, for clients with whom there has been long-term involvement, several considerations should be taken into account.

Reasons for termination vary. Sometimes insurance coverage or program protocols impose a time limit. In other cases, the counselor, the client, or both will feel that counseling goals have been obtained or at least reasonable progress in attaining them has been made. On occasion, a counselor may believe that a client should be referred to a psychotherapist before resuming work on nutritional concerns. Referrals would also be in order when problems emerge, such as bulimia, that could be best handled by a specialist. At times a counseling relationship needs to end because the counselor or the client is preparing to move away or take a new job. The following sections offer suggestions for ensuring a smooth transition to the end of your counseling support.

Table 7.4 Components of Mindful Eating

Component	Procedure and Description
Eat slowly	Periodically take breaks during eating to breathe and assess fullness.
Focus on eating	Remove distractions; do not eat in the car or while watching TV or working on the computer.
Recognize inner cues	Use feelings of hunger and fullness to guide eating rather than a defined diet plan.
Eat nonjudgmentally	Acknowledge likes, dislikes, and neutral feelings about food without judgment.
Be aware of senses	Use all your senses to explore, savor, and taste food.
Be in the present	Focus on the direct experiences associated with food and eating, not distant outcomes.
Reflect on mindless eating	Be aware of and reflect on the effects caused by unmindful eating (eating out of boredom or frustration, eating to the point of fullness).
Recognize interconnectedness	Recognize an interconnection of the earth, living beings, and cultural practices and the impact food choices on those systems.
Practice meditation	Make meditation practice a part of life.

Source: Adapted from Mathieu J. What should you know about mindful and intuitive eating? *J Am Diet Assoc.* 2009;1091982–1987; The Center for Mindful Eating. Available at: http://tcme.org/principles.htm. Accessed August 27, 2010.

Preparation for a Conclusion

First, provide transition time. Final meeting dates should be agreed on before the terminating session. For time-limited programs, counselors can remind clients of the upcoming final meeting. Care should be taken, if possible, not to spring a decision to end a counseling relationship at the last session. Counselors and clients may be surprised at the intensity of the emotional response to the end of the interaction.

Discuss reasons for ending the counseling relationship. Share your perception of your clients' progress, and request that your clients voice their views.

Final Session

Consider taking these steps for a smooth wrap-up session with your clients:

- **Review beginnings.** Discuss the issues that brought your client to you in the first place. You might review initial assessments.
- **Discuss progress.** Identify goals and progress in meeting them. Having your progress notes handy could be useful.
- **Emphasize success.** A review of accomplishments can be a source of much pleasure as clients remember how certain tasks were anticipated with dread (for example, taking glucose readings) or how stuck they felt in the midst

of indecision. Miller and Jackson[31] encourage a discussion of the skills a client used to bring about change. Research indicates that clients who feel a major responsibility for a transformation will continue to experience success rather than those who attribute fairing well to an extrinsic factor, such as a therapist. A counselor could ask, "What do you believe you did to bring about this change?"

- **Summarize current status.** Highlight current biochemical and physical parameters, coping skills, social support, challenges, and environmental issues.
- **Explore the future.** How will the changes that have been made be maintained? How will old challenges be addressed and what new challenges are likely to be encountered? How will those difficulties be handled? Are there additional changes to be made? Who will offer support in the future? This is also the time to discuss a referral if one is being made.
- **Discuss future involvement.** Follow-up meetings are advantageous to both clients and counselors. Such meetings provide opportunities for reassessment of nutritional status and evaluation of goals and a time to reinforce previously set behavior changes. A periodic check-in arrangement could be negotiated that involves personal meetings, phone calls, or e-mail. Sometimes recommendations are made for clients to initiate contact if changes in clinical parameters (such as a five-pound weight gain) or a significant life event occurs (divorce, marriage, or death of a close family member). These interactions help counselors document progress to implement long-term evaluation of counseling effectiveness.
- **Provide and elicit feedback concerning the significance of the relationship.** Allow time to express what the meaning of the relationship has meant to you and to your client. This generally means expressing appreciation for each other. Often clients express appreciation by saying counselors did so much for them. In this case, thank your client for the compliment, adding the reminder, "The reason you

did so well was because of your hard work." Possible lead-ins to relationship discussions include the following:

Counselor: *It has been such a pleasure working with you. I want you to know that I will miss our weekly meetings.*

I want you to know that I really appreciate what you have taught me about Cuban culture.

- **Consider holding a ceremony and exchanging symbols.** You may think about a special location or activity for the last meeting. Some possibilities include a walk in the park, a lunch, or a different meeting room. Sharing a particular food, especially if it had a significant connection to your client's diet, can make a profound impression. Exchange of gifts will be based on program or facility policy; however, it is generally considered appropriate to give and accept an inexpensive gift as a symbol of completion. Mementos symbolic of work done together can be particularly meaningful, such as a wooden apple for a client who made a major effort to increase intake of apples. One counselor regularly writes an individualized letter of support and encouragement to be given to clients on the final meeting day or to be mailed after the last meeting.
- **Final good-bye.** Generally, you would expect to acknowledge the end of the counseling experience, shake hands, and walk with your client to the door, waiting room, or usual exit. Murphy and Dillion[37] suggest a final parting by telling your clients you will be picturing them doing something they have longed for or intend to accomplish, such as running a five-mile race, taking a cooking class, or completing a degree. Such statements reinforce that your client's future welfare is important to you and that you will always be rooting in his or her corner.

Handling Abrupt Endings

Unfortunately, all closures are not tidy. Sometimes clients simply do not show up for a session, or they

EXHIBIT 7.6 Example of a Termination Letter

Dear Mary,

Because I haven't heard from you after our meeting on December 2, I wanted to touch base with you. I hope you are doing well. On two occasions I called and left a message but did not receive a reply. I am assuming that you wish to stop working together at this time.

If you wish to resume lifestyle counseling at a future date, I would be happy to work with you again. You made commendable changes in your exercise pattern as well as an increase in your fruit, vegetable, and fiber intake.

It was a pleasure to work with you. I hope you have continued success in your goals to lower your blood pressure.

Sincerely,
Sally Frank

cancel future appointments. If there is an abrupt ending to an involvement, consideration should be given to sending a termination letter or e-mail as an attempt to have a closure experience. In your communication, you may wish to reinforce achievements and leave the door open to a future association. See Exhibit 7.6 for an example of such correspondence.

COUNSELING EVALUATION

Evaluation is an important component of counseling for your growth as a counselor. After an intervention, counselors should take time to assess the quality of their skills and contemplate what could have been handled differently. In addition, there is increasing pressure from the managed care industry for health professionals to produce outcome measurements. To meet new mandates, nutrition counselors need to incorporate brief, efficient, and inexpensive assessment procedures into their counseling programs. Large facilities are likely to have a tracking system in place for evaluation. Licensing regulations differ around the country, but they often include requirements on what and how data need to be reported. For small facilities and for counselors in private practice, nutrition counselors will likely find the need to develop their own procedures for producing outcome data.

Evaluation of Client Progress

Several of the factors necessary for an effective evaluation of client progress have already been covered throughout this book. Parameters used for evaluation can include the following types of data: behavioral (food diaries, exercise records), physical (body weights, blood pressure), biochemical (cholesterol and glucose levels), and functional (length of hospital stay). The initial assessment establishes a baseline of client behaviors and problems that will be used for future comparisons. Standards, such as Recommended Dietary Intakes, established by national organizations provide a yardstick to determine normalcy. In addition, the American Dietetic Association Nutrition Care Process has monitoring and evaluation as a major component of the process.[2] See Chapter 5. The key to a client monitoring process and evaluation of outcomes is having well-defined, measurable goals and a charting process that tracks implementation of strategies and goal attainment. If done properly, counselors can continually assess whether goals were achieved and what strategies were successful.

Goal Attainment Scale

Cormier and Cormier[38] describe a goal attainment scale (GAS) rating system that a counselor and client can work together collaboratively to establish. A range of values are assigned to possible results, from a score of +2 for a most favorable result to a −2 for a least favorable outcome. A score of 0 represents the anticipated level of performance. This type of rating system is particularly useful for providing outcome results to funding agencies or for supplying useable numerical scores to determine levels of change for statistical analysis. The graduated level of desired outcomes has the added advantage of allowing success at several levels of performance. This is useful when there is a tendency to set goals that are too ambitious. See Table 7.5 for an example.

Final Client Evaluation

At the end of an intervention, a final assessment should be conducted to determine the degree of attainment of final goals. If a follow-up interaction was arranged at the termination, a post-treatment evaluation can be conducted to determine whether the benefits of counseling have been maintained. Possible implementations of these evaluations include an in-person follow-up interview, a questionnaire sent via e-mail or postal delivery, or

---

EXERCISE 7.5   Design a Goal Attainment Scale

Choose a new goal or one which you have been working on. Design a goal attainment scale representing what you would like to accomplish in the next week. Use Exhibit 7.5 as a guide.

- Write your goal attainment scale in your journal.
- Describe your experience using the goal attainment scale. Would you want to use this scale in a counseling session with a client? Explain why or why not.

Table 7.5   Goal Attainment Scale for Exercise for One Week

Value	Description of Value	Behavior
−2	Most unfavorable outcome thought likely	Do not walk at all.
−1	Less than expected success with performance	Walk less than 60 minutes during the week.
0	Anticipated level of performance	Walk for 30 minutes two times.
+1	More than expected success with performance	Walk for at least 30 minutes, three times.
+2	Best expected level of performance	Walk for at least 30 minutes, more than three times.

a telephone conversation. The follow-up has the added advantage of indicating to clients that the counselor continues to be interested in their welfare.

EVALUATION OF COUNSELING EFFECTIVENESS AND SKILLS

Counselors also need to evaluate their effectiveness. A number of methodologies can be used for this purpose.

Client Evaluation of Counselor

The CARE Measure was developed to measure a doctor's relational empathy during consultations with patients. This instrument has been modified to reflect client interactions with other health care providers.[39] The form can be given to a client to complete and hand in to office staff or mailed back to the office. See Lifestyle Management Form 7.7 in Appendix D for a copy of this validated instrument.

Assessing Client's Nonverbal Behavior

The non-verbal behavior of a client represents a key to his or her emotional state and can indicate how the counseling session is going. Any discrepancies between a client's non-verbal behavior and verbal messages should be noted.

Checking

Checking, periodic summaries, is a technique covered in Chapter 3. This method allows counselors to evaluate whether they are on target during a counseling session.

Counseling Checklists (Interview Guides)

Counseling checklists, such as those provided in Chapter 14, can serve as a rudimentary assessment tool, even though their primary function is to help organize a counseling session. At the end of a session, a counselor can review the form to assess whether all planned counseling interventions were addressed.

Charting

Charting can be a valuable tool to evaluate counseling effectiveness as well as client progress. The assumption can be made that if counseling goals have been met, then the counselor demonstrated effective skills. However, additional sources of evaluation are required because client ability or inability to meet counseling objectives does not always reflect back on the counselor. The following questions should be considered if general counseling goals were not met:

- Was the assessment adequate?
- Were major problem areas clearly identified?
- Were goals realistic and clearly defined?
- If specific goals were not achieved, was there an adequate assessment of why not?
- What behavior change strategies were attempted? How effective were they?
- What could have been done differently?

Videotape, Audiotape, or Observation Evaluations

Counselors can conduct self-evaluations of their skills by using a video- or audiotape. An alternative would be to have a colleague or mentor conduct an assessment. Generally, it is a good idea to use an assessment instrument to guide the evaluation. The Interview Assessment Form, Lifestyle Management Form 7.5 in Appendix D, addresses general counseling effectiveness including an evaluation of the flow and organization of the interview, application of interpersonal skills, and quality of client responses. This form was originally developed at Brown University School of Medicine to assess medical student interviewing skills and was modified to meet the needs of a nutrition counselor. Another assessment instrument, Counseling Responses Competency Assessment, Lifestyle Management Form 7.6 also available in Appendix D, was developed to increase understanding and awareness of basic counseling responses. Either form can be used as a self–rating instrument, without frequency tabulation, following a session to upgrade counseling skills.

CASE STUDY   Amanda: The Busy Sales Representative

Amanda, age twenty-seven, is a sales representative for a large pharmaceutical company. She lives in a western sub-urb of Chicago, but her present territory covers the north side of Chicago. Her day is quite full. Typically, she leaves her townhouse by 6:30 A.M. to make the drive into the city for an 8:00 A.M. call at a physician's office or a hospital. She loves her job, but the long hours and stressful lifestyle have taken a toll on her personal life as well as her health.

Despite efforts to eat well, Amanda cannot seem to control her weight. She is about forty pounds heavier than when she first started with the company despite her best efforts to diet. She believes that the nature of her job makes eating well impossible and has all but given up on any hope of losing weight permanently. She is out of town at least ten days a month and spends a good deal of time in airports or in her car. A shake and a burger do the trick, as she never quite knows when her next meal might be. Her many business meetings and associated social engagements have food as a central focus. When she finally gets home, she finds her cupboards are bare, including her refrigerator. She is never home long enough to use up any fresh produce or dairy.

Amanda's closest friend is Christine, who lives in her townhouse complex. She has known Christine for many years, and they have shared the efforts and woes of dieting. Christine is seventy-five pounds overweight and experienc-ing some depression at this time, mainly because of her weight. They socialize almost daily when Amanda is home and vacation together one to two times a year. Despite this great friendship, Amanda feels influenced by Christine's negative feelings toward life and is ambivalent about what she needs to do about the relationship to improve her own personal well-being.

EXERCISE 7.6   Case Study Review

What are possible sources of social support for Amanda? Suggest two changes in lifestyle that seem appropriate for her. Identify dysfunctional thinking patterns that could interfere with her attempts to change her behaviors. Identify a role-playing scenario that could be useful in a nutrition counseling session. How would you approach the topic of stress management with Amanda?

REVIEW QUESTIONS

1. Explain why dietary changes can put a stress on relationships.

2. What explanation has been given to explain the benefits of having social support for mak-ing lifestyle changes?

3. Explain the physiological response to an anticipated stressor.

4. Identify and explain the main components of mindful eating.

5. What are four immediate determinants of a relapse?

6. Name and explain four covert antecedents of a relapse.

7. Describe nine topics that could be addressed in a final session with a client.

8. Explain ways in which nutrition counselors can assess their own effectiveness.

9. Describe a goal attainment scale.

ASSIGNMENT—IDENTIFYING STRESS

The objective of this assignment is for you to become more aware of the stresses in your life. In Exercise 7.3 you identified your symptoms of stress. Use Lifestyle Management Form 7.2 (Appendix D) to record stressful activities, symp-toms of stress, and internal self-talk at the time of the stressful event and at the time of the occurrence of the symptom. Note that the stressful event and the symptom may not occur simultaneously. In your journal answer the following questions:

1. Were the symptoms you identified in Exercise 7.3 congruent with your journal findings? Explain.

2. What were the main stressors in your life over the three days?

3. Review your internal dialogue entries. Was there evidence of cognitive distortions? Explain.

SUGGESTED READINGS, MATERIALS, AND INTERNET RESOURCES

Mindful Eating

Center for Mindful Eating, http://tcme.org/about.htm

Mindfulness Based-Eating Awareness Training, www.mindfuleatingforlife.com

Center for Health and Meditation, Mindful Eating Program, www.upstate.edu

Discover Mindful Eating A Resource of Handouts for Health Professionals, Frederick Burggraf & Megrette Hammond, Charlotte Hall, MD: Day One Publishing; 2005.

Stress Management Resources

Stress Free Now. 360-5 Cleveland Clinic Wellness Site guided program. www.360-5.com

Davis M, Eshelman ER, McKay M, Fanning P. *The Relaxation & Stress Reduction Workbook*. 6th ed. Oakland, CA: New Harbinger; 2008.

Creative Arts Resources for Stress Reduction

- American Music Therapy Association, www.musictherapy.org
- American Dance Therapy Association, www.adta.org
- American Art Therapy Association, Inc., www.arttherapy.org
- National Association for Poetry Therapy, www.poetrytherapy.org
- Be More Creative, http://creativequotations.com/

Relapse Prevention

- Mirror-mirror, eating disorders, http://www.mirror-mirror.org/recovery.htm
- Relapse Prevention Therapy (RPT), http://www.nationalpsychologist.com/articles/art_v9n5_3.htm

Investigating Assessment Forms

Thomas PR (Ed.) *Weighing the Options Criteria for Evaluating Weight-Management Programs*. Washington, DC: National Academy Press; 1995. Helpful for identifying and evaluating various assessment instruments related to self-esteem, body image, eating disorders, self-efficacy, dieting readiness, stress, social support, physical activity, and diet.

REFERENCES

[1] Paisley J, Beanlands H, Goldman J, et al. Dietary change: what are the responses and roles of significant others? *J Nutr Edu Behav.* 2008;40:80–88.

[2] American Dietetic Association. International Dietetics & Nutrition Terminology (IDNT) Reverence Manual Standardized Language for the Nutrition Care Process. 2nd ed. Chicago, IL: American Dietetic Association; 2009.

[3] Kelsey K, Earp JL, Kirkley BG. Is social support beneficial for dietary change? A review of the literature. *Fam Community Health.* 1997:20:70–82.

[4] Grieco AJ, Garnett SA, Glassman KS, et al. New York University Medical Center's Cooperative Care Unit: Patient education and family participation during hospitalization—the first ten years. *Patient Educ Counsel.* 1990;15:3–15.

[5] Fishman T. The 90-second intervention: A patient compliance mediated technique to improve and control hypertension. *Public Health Rep.* 1995;110:173–179.

[6] La Greca AM, Bearman KJ. The diabetes social support questionnaire-family version: evaluating adolescents'

diabetes-specific support from family members. *J Pediatr Psychol.* 2002;27:665–676.

[7]Barrera M, Toobert DJ, Angell KL, Social support and social-ecological resources as mediators of lifestyle intervention effects for type 2 diabetes. *J Health Psychol.* 2006;11:483–495.

[8]Barrera M, Glasgow RE, McKay HG, et al. Do internet-based support interventions change perceptions of social support?: An experimental trial of approaches for supporting diabetes self-management. *Am J Community Psychol.* 2002;30:637–654.

[9]Stevens VJ, Rossner J, Hyg MS, et al. Freedom from fat: A contemporary multi-component weight loss program for the general population of obese adults. *J Am Diet Assoc.* 1989;89:1254–1258.

[10]Thayer RE, Newman JR, McClain TM. Self-regulation of mood: strategies for changing a bad mood, raising energy, and reducing tension. *J Pers Soc Psychol.* 1994;67:910–925.

[11]Foreyt JP, Goodrick GK. Attributes of successful approaches to weight loss and control. *Appl Prev Psychol.* 1994;3:209–215.

[12]Kayman S, Bruvold W, Stern JS. Maintenance and relapse after weight loss in women: Behavioral aspects. *Am J Clin Nutr.* 1990;52:800–807.

[13]Wylie-Rosett J, Segal-Isaacson CJ. The Complete Weight Loss Workbook: Proven Techniques for Controlling Weight-Related Health Problems, Leader's Guide. Alexandria, VA: American Diabetes Association; 1999.

[14]*The Vital Years—How to Grow Better and Older.* Nutley, NJ: Hoffman–La Roche, Department of Community Affairs; 1985.

[15]Davis M, Eshelman ER, McKay M. *The Relaxation & Stress Reduction Workbook.* Oakland, CA; 1995.

[16]Benson H. *The Relaxation Response.* New York: Harper; 2000.

[17]D'Arrigo T. Stress & diabetes. *Diabetes Forecast.* 2000; 53:56–61.

[18]Tomiyama AJ, Mann T, Vinas D, et al. Low Calorie Dieting Increases Cortisol. *Psychosom Med.* 2010;72:357–364.

[19]Remer T, Shi L. Low-calorie dieting and dieters' cortisol levels: don't forget cortisone. *Psychosom Med.* 2010; 72:598–599.

[20]American Psychological Association. Psychology at work. Available at: http://helping.apa.org/work/index.html. Accessed May 1, 2000.

[21]Spence JD, Barnett PA, Linden W, Ramsden V, Taenzer P. Recommendations on stress management. *Canadian Med Assoc J.* 1999;160 (suppl): S46–S51.

[22]National Institutes of Health (NIH), National Heart, Lung, and Blood Institute. Clinical guidelines on the identification, evaluation, and treatment of overweight and obesity in adults—The evidence report. *Obes Res.* 1998;6 (suppl 2):S51.

[23]Warpeha A, Harris J. Combining traditional and non-traditional approaches to nutrition counseling. *J Am Diet Assoc.* 1993;93:797–800.

[24]Meichenbaum D. Stress inoculation training: A 20-year update. In: Lehrer PM & Woolfolk RL, eds. *Principles and Practice of Stress Management.* 2d ed. New York: Guilford; 1993:373–406.

[25]Shattuck DK. Mindfulness and metaphor in relapse prevention: An interview with G. Alan Marlatt. *J Am Diet Assoc.* 1994;94:846–848.

[26]Marlatt GA. Relapse prevention: Theoretical rationale and overview of the model. In: Marlatt GA, Gordon JR, eds. *Relapse Prevention.* New York: Guilford; 1985.

[27]Foreyt JP, Poston WSC. The role of the behavioral counselor in obesity treatment. *J Am Diet Assoc.* 1998;98 (suppl):S27–S31.

[28]Larimer ME, Palmer RS, Marlatt GA. Relapse prevention: An overview of Marlatt's cognitive-behavioral model. *Alcohol Res Health.* 1999;23:151–160.

[29]Laws DR. Relapse prevention: The state of the art. *J Int Violence.* 1999;14:285–302.

[30]Miller WR, Jackson KA. *Practical Psychology for Pastors.* 2d ed. Englewood Cliffs, NJ: Prentice Hall; 1995.

[31]The Center for Mindful Eating. Available at: http://tcme.org/principles.htm. Accessed August 27, 2010.

[32]Brown KW, Ryan RM. The benefits of being present: Mindfulness and its role in psychological well-being. *J Pers Soc Psych.* 2003;84(4): 822–848.

[33]Mathieu J. What should you know about mindful and intuitive eating? *J Am Diet Assoc.* 2009;1091982–1987.

[34]Andrade AM, Greene GW. Melanson KJ. Eating slowly led to decreases in energy intake within meals in health women. *J Am Diet Assoc.* 2008;108:1186–1191.

[35]Slow Food. Available at: www.slowfoodusa.org Accessed August 27, 2010.

[36]Bacon L, Stern J, Van Loan MD, et al. Size acceptance and intuitive eating improve health for obese, female chronic dieters. *J Am Diet Assoc.* 2005;105:929–936.

[37]Murphy BC, Dillon C. *Interviewing in Action: In a Multicultural World. 3rd ed.* Pacific Grove, CA: Brooks/Cole; 2007.

[38]Cormier S, Nurius P, Osborn CJ. Interviewing Strategies for Helpers: Fundamental Skills and Cognitive Behavioral Interventions. 6th ed. Pacific Grove, CA: Brooks/Cole; 2008.

[39]Mercer SW, Maxwell M, Heaney D, et al. The consultation and relational empathy (CARE) measure: development and preliminary validation and reliability of an empathy-based consultation process measure. *Family Practice.* 2004;21(6):699–705.

Physical Activity

Digital Imagery/PhotoDisc, Inc.

An early morning walk is a blessing for the whole day.
—HENRY DAVID THOREAU

Behavioral Objectives

- List benefits of regular physical activity.
- Explain risks associated with exercise.
- Describe 2008 Physical Activity Guidelines for Americans.
- Differentiate between moderate and vigorous physical activity.
- Explain barriers to becoming physically active.
- Clarify the role of a nutrition counselor in physical activity counseling.
- Evaluate physical activity readiness using standard assessment tools.
- Demonstrate physical activity counseling approaches.
- Identify issues of concern for physical activity goal setting and action planning.
- Describe the basics of an introductory walking program.

Key Terms

- **Aerobic Activity:** physical activity requiring oxygen; usually sustained longer than three minutes; required to develop cardiorespiratory fitness.
- **Anaerobic Activity:** physical activity not requiring oxygen; utilized during high-intensity activities and at the beginning of sustained aerobic activities.
- **Cardiorespiratory Fitness:** the ability of the circulatory and respiratory systems to supply oxygen during physical activity; usually reported as maximum oxygen uptake (VO$_2$ max).
- **Ergogenic Aids:** substances, including medications and dietary supplements, or techniques and devices, that are intended to improve physical performance.
- **Flexibility:** full range of joint motion without discomfort.

- **Maximum Heart Rate:** roughly 220 beats per minute minus age.
- **Maximum Oxygen Uptake:** (VO_2 max): the body's capacity to transport and use oxygen during a maximal exertion involving dynamic contraction of large muscle groups. It is also known as aerobic power and cardiorespiratory endurance capacity.
- **Moderate Physical Activity:** use of large muscle groups that are at least equivalent to brisk walking. On a scale relative to an individual's personal capacity, moderate physical activity is usually a 5 or 6 on a scale of 0 to 10.
- **Muscular Endurance:** repetitive muscle contractions over a prolonged period.
- **Muscular Strength:** ability to generate appropriate force.
- **Physical Activity:** any movement produced by the contraction of skeletal muscle that increases energy expenditure above a basal level.
- **Physical Fitness:** set of attributes relating to the ability to perform physical activity; often viewed as cardiorespiratory fitness; components include flexibility, suitable body composition, muscular strength, and muscular endurance.
- **Vigorous Physical Activity:** repetitive activities using large muscle groups at 70 percent or more of maximum heart rate. On a scale relative to an individual's personal capacity, vigorous physical activity is usually a 7 or 8 on a scale of 0 to 10.

INTRODUCTION

Improving the health of Americans through physical activity and good nutrition must become a national priority.
—Martha N. Hill, R.N., Ph.D., past president, American Heart Association[1]

Although media coverage of the need for exercise gives the general appearance of widespread interest, the actual number of North Americans actively involved in regular **physical activity** remains rather low. Authorities in the United States and Canada have voiced concern about a growing lack of exercise,[2,3] and some have referred to the problem as an "epidemic of inactivity."[4,5] The 2009 National Health Interview Survey provided estimates that 55% of the population never participates in any "periods of vigorous leisure-time physical activity."[6] In general, people become less physically active as they age, and women are less likely to be active than men. Physical activity levels decline dramatically during adolescence, particularly among females, with the decline more apparent among African American and Hispanic than White students.[5,6] In this chapter, we review the basics of physical activity and wellness and discuss the role of a nutrition counselor in providing physical activity guidance.

PHYSICAL ACTIVITY INITIATIVES

Several initiatives have been instituted to address the inactivity problem. The National Physical Activity Plan, developed by leading government, professional, research, and community organizations, is a comprehensive set of policies, programs, and initiatives that aim to increase physical activity in all segments of the American population. For example, The President's Challenge Program, a project of the President's Council on Fitness, Sports and Nutrition seeks to engage, educate. and empower all Americans to adopt a healthy lifestyle that includes regular physical activity and good nutrition. This program partners with a variety of organizations to offer an on-line physical activity tracking log leading to a Presidential Active Lifestyle Award.[7]

Additional national, regional, state and local programs, and initiatives that aim to get Americans active and healthy include: Let's Move!, a national campaign launched by First Lady Michelle Obama to address the issue of childhood obesity;[8] America's Great Outdoors, an interagency partnership to advance better use and conservation of our outdoor environments across the country;[9] MOVE!, a national weight management program designed for veterans,[10] and Healthy People 2020, a Department of Health and Human Services initiative which identifies physical activity as a component of a healthy lifestyle.[11] In addition, the National Association for Health and Fitness, a nonprofit organization supporting a network of state and governor's councils, promotes worksite fitness programming.[12]

ROLE OF NUTRITION COUNSELOR IN PHYSICAL ACTIVITY GUIDANCE

Physical activity is an integral part of overall health. Because the combination of nutrition and physical activity is a primary strategy for reducing the risk of coronary heart disease, hypertension, diabetes, and osteoporosis, nutrition counselors need to be knowledgeable about the principles of exercise and utilize these principles with clients.[12,13] To incorporate physical activity planning into nutrition counseling effectively, a counselor needs the following:

1. **A basic knowledge about the relationship between physical activity and health.** Such information can be obtained from course work, readings, and attendance at professional meetings. Most of the physical activity professional organizations, government health programs, and nonprofit health associations have educational websites and materials (See the end of the chapter; Internet Resources). Nutrition counselors could consider enhancing their physical activity counseling skills by obtaining additional training and credentials in the area of physical activity. Possibilities include the American College of Sports Medicine (ACSM) certification, the American Council on Exercise (ACE) certification, the Commission on Dietetic Registration board-certified specialist in sports dietetics.
2. **Collaboration with physical activity professionals.** Many nutrition professionals use the services of physical activity professionals for obtaining information and guidance. Opportunities for interaction are often available in fitness facilities and at physical activity professional meetings.
3. **Referral resources.** Nutrition counselors should keep a list of physical activity professionals, diabetes educators, professional organizations, fitness facilities, and fitness-knowledgeable physicians.
4. **Educational resources for clients.** Nutrition counselors should have educational materials on hand for their clients. Fact sheets and pamphlets are available from many government and professional organizations, including the Cooperative Extension Service, also known as the Extension Service of the USDA, which is a non-formal educational program designed to help people use research-based knowledge to improve their lives. See resources at the end of the chapter.
5. **Medical approval.** Clients should receive a medical evaluation before beginning or increasing physical activity.

A survey of registered dietitians regarding the extent of their involvement in physical activity counseling revealed the following:

- Most (64 percent) provide physical activity advice and guidance regarding the role of physical activity in a person's life.
- Most (65 percent) felt they know enough about physical activity to get their client started on a program of physical activity.

PHYSICAL ACTIVITY AND FITNESS

Physical activity is defined as any bodily movement provided by the contraction of skeletal muscle that increases energy expenditure above a basal level.[14] **Physical fitness** relates to a set of performance- and health-related attributes connected to the ability to perform activity.[15] Performance-related attributes include agility, balance, coordination, power, and speed.[11] Health-related attributes include body composition, cardiorespiratory function, **flexibility**, **muscular strength**, and **muscular endurance**.[11]

Sometimes physical fitness is viewed as specifically **cardiorespiratory fitness** and is expressed as physical activities **maximum oxygen uptake** (VO_2 max).[16] **Aerobic** activities involve large muscles over a sustained period, generally thirty minutes or more, and require oxygen to provide energy. **Anaerobic** activities do not need oxygen for energy as would occur during high-intensity exercise or at the beginning of sustained aerobic activities.

Benefits of Regular Physical Activity

Physical activity has a favorable effect on musculoskeletal, cardiovascular, respiratory, and endocrine systems.[15] Poor diet and lack of exercise have been identified as the second leading actual cause of death in the United States[17](see Figure 8.1). Regular physical activity increases quality of life and decreases risk of developing or progression of many of the leading causes of illness and death in the United States. Regular physical activity can benefit health in the following ways:

- **Lower risk of early death.** Strong scientific evidence shows that physical activity reduces the risk of premature death[6] (dying earlier than the average age of death for a specific population group). Even moderately active individuals have a lower mortality than sedentary individuals. One study found that a daily two-mile walk (approximately 4,000-5,000 steps per day) could add years to a person's life.[18] It is also estimated that people who are physically active for approximately 7 hours a week have a 40 percent lower risk of dying early than those who are active for less than 30 minutes a week.[14]

- **Improve cardiovascular health.** Physically active people have a substantially lower overall risk for coronary heart disease. This favorable effect is thought to be due to larger coronary arteries and increased heart size and pumping capacity.[19] As the heart becomes more efficient, heart rates decline, resulting in resting rates of about sixty beats per minute in trained athletes.[20] Also, regular physical activity decreases serum triglyceride and increases levels of high-density lipoprotein (HDL), a cholesterol-carrying blood protein associated with a lower risk of cardiovascular disease.[21,22]

- **Positive effect on blood pressure.** Recurrent physical activity has been shown to prevent

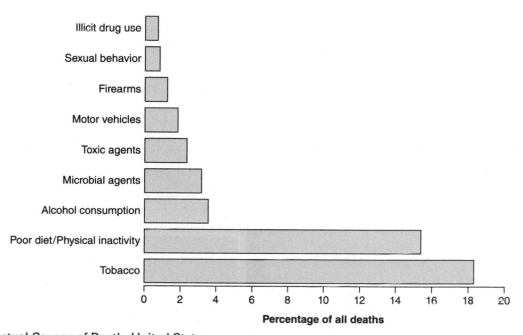

Figure 8.1 Actual Causes of Death, United States
Source: Mokdad AH, Marks JS, Stroup DF, Gerberding JL. Actual causes of death in the United States, 2000. *J Am Med Assoc.* 2004;291:1238–1245.

or delay the onset of high blood pressure, and regular exercise reduces blood pressure in people with hypertension.[15] This positive effect is thought to be due in part to a reduction in circulating low-density lipoprotein (LDL) cholesterol and fats, thereby increasing flexibility of blood vessel.[20]

- **Reduced risk of cancer.** In general, regular exercise is associated with lower rates of cancer.[20] There is a reduced risk of colon cancer, in particular, and possibly cancers of the prostate and endometrium[15] (the lining of the uterus). There was evidence for a dose-response effect found in the majority of studies that examined the role of physical activity in breast cancer prevention. The strongest associations were found for recreational and household activities and for activity that was of at least moderate intensity and sustained over a lifetime.[23]

- **Lower risk of developing type 2 diabetes.** Physically active muscles enhance insulin sensitivity, readily accept glucose, and lower blood glucose levels. As a result, regular physical activity lowers the risk of developing type 2 (non-insulin-dependent) diabetes and helps control blood sugar levels for those who have the condition.[15] A prospective twin pair study established that even small amounts of physical activity compared with sedentariness play a significant role in reducing or postponing the occurrence of type 2 diabetes.[24]

- **Weight control.** Because energy intake has not increased over the past few decades, several experts believe the expanding prevalence of obesity in North America is attributed to a progressive decrease in the amount of calories expended for work, transportation, and personal chores.[25] Regular physical activity is inversely related to rate of weight gain with age and plays a key role in long-term weight control and/or maintenance of weight loss.[25] Another advantage of regular physical activity for adults and children is the favorable effect on body fat distribution away from the abdominal area, which is associated with the development of several chronic disease states.[26]

- **Promotion of optimal bone density.** Physical activity, especially weight-bearing exercise, produces a force on the bones that contributes to building bone and helps protect against osteoporosis.[27] Examples of bone-strengthening activity include jumping jacks, running, brisk walking, and weight-lifting exercises.[14]

- **Enhanced psychological well-being.** Physical activity counters anxiety and depression and improves mood and the ability to cope with stress.[19] These benefits may be due to chemical alterations in concentration and activity of dopamine, norepinephrine, and serotonin. Also, the release of a morphine-like substance may contribute to the relief of pain and a feeling of euphoria.

- **Enhanced immune system. Moderate physical activity** is associated with resistance to colds and other infections; however, strenuous activity can have an adverse effect on the immune system.[19]

- **Reduced risk of falls.** Falls among older individuals can be debilitating. Increasing evidence indicates that balance, strength, and flexibility training can reduce the risk of falling and allow maintenance of an independent living status.[15] Research studies on physical activity to prevent hip fracture show that participating in 120 to 300 minutes a week of moderate levels of activity, including walking, is associated with substantially lower risk of hip fracture in postmenopausal women[28] and older adults.[29]

EXERCISE 8.2 Survey of Perceptions of Benefits of Physical Activity

Interview three people who do not exercise regularly. Ask each his or her beliefs about the benefits of exercise. Review the Benefits of Regular *Moderate* Physical Activity fact sheet, Lifestyle Management Form 8.1 (Appendix D), with the interviewee. Ask the individuals for their reactions. Record your observations in your journal.

Injury Risks Associated With Exercise

Although there are numerous health benefits to a physically active lifestyle, some injury risks and other adverse events (overheating and dehydration) are associated with exercise.

- Sudden cardiac deaths, though extremely rare, are a serious concern. These occurrences are most often associated with primarily sedentary individuals who have preexisting coronary heart disease and engage in vigorous activity.[16] The risk of myocardial infarction or sudden death also slightly increases for well-trained athletes; however, the net effect of regular exercise is a lower risk than the risk of cardiac death associated with a sedentary lifestyle.[15] One study found the overall risk of cardiac arrest of habitually active men was only 40 percent of that of sedentary counterparts.[30]
- Musculoskeletal injuries are the most common harmful side effect of physical activity. Such injuries can generally be prevented by gradually working up to the desired intensity level and avoiding excessive physical activity.[15]

Note that moderate physical activity is not associated with a significant risk of sudden cardiac death or musculoskeletal injury.[16] Scientific evidence shows that physical activity is safe for almost everyone. The health benefits of physical activity far outweigh the risks.[14]

Exercise Myths

Clients may use several myths about exercise to rationalize their inactivity:

- **Exercise causes arthritis.** Moderate physical activity is not associated with joint damage and in fact is recommended for individuals with arthritis during nonacute phases.[15]
- **Working out with weights is only for men.** Strength training tones muscles without making women appear overly muscular or "bulked up" like a professional male bodybuilder. There are numerous benefits for both men and women, including greater bone density, muscle strength, and balance.

- **It is dangerous for older people to start exercising.** All people can benefit from regular physical activity. Even people in a nursing home as old as ninety-eight have been able to improve their walking speed and ability to climb steps.[31]

PHYSICAL ACTIVITY GOALS

In October 2008, the Department of Health and Human Services (DHHS) published the Physical Activity Guidelines for Americans.[14] For the first time one document provided guidelines for people in a wide range of ages (age 6 on up), and those with special needs. This comprehensive document gives information and guidance on the types and amounts of physical activity that afford substantial health benefits based on sound scientific evidence. Significant conclusions are summarized in Exhibit 8.1

EXHIBIT 8.1   Major Conclusions of the Physical Activity Guidelines for Americans

- Regular physical activity reduces the risk of many adverse health outcomes.
- Some physical activity is better than none.
- For most health outcomes, additional benefits occur as the amount of physical activity increases through higher intensity, greater frequency, and longer duration.
- Most health benefits occur with at least 2 hours and 30 minutes (150 minutes) a week of moderate-intensity physical activity, such as brisk walking. Additional benefits occur with more physical activity.
- Episodes of activity that are at least 10 minutes long count toward meeting the guidelines.
- Both aerobic (endurance) and muscle-strengthening (resistance) physical activity are beneficial.
- Health benefits of physical activity occur for children and adolescents, young and middle-aged adults, older adults, and those in every studied racial and ethnic group.
- Health benefits of physical activity are attainable for people with disabilities.
- The benefits of physical activity outweigh the risks of injury and heart attack.

Source: U.S. Department of Health and Human Services, Physical Activity Guidelines for Americans, Available at: www.health.gov/paguidelines.

and a summary of the guidelines can be found in the Table 8.1.

Moderate Physical Activity

The minimum caloric expenditure able to produce health benefits is approximately 750 kilocalories per week. To achieve this goal, the general recommendation is to engage in at least thirty minutes of moderate physical activity five days a week. However, the nature of the activity alters the time requirement. The time needed for less intense activities needs to be longer to expend 150 kilocalories, and the time required to utilize this amount of energy during strenuous activities is reduced (see Exhibit 8.2). Although sustained physical activity for at least thirty minutes is preferable, intermittent episodes of ten minutes of exercise that add up to thirty minutes can be beneficial.

Vigorous Physical Activity

Although moderate activity is beneficial, there are cardiovascular advantages to engaging in vigorous physical activities involving large muscle groups. **Virgorous physical activity** has been defined as that resulting in the individual reaching 70 percent of **maximum heart rate**; however, authorities caution sedentary, unfit individuals to aim for a lower level of 50 percent[32] or 55 to 64 percent.[33]

Table 8.1 A Summary of the Physical Activity Guidelines for Americans

Category	Key Guidelines

Children and Adolescents	• Physical activities should be age appropriate, enjoyable and offer variety.
	• Physical activities should be 60 minutes or more daily and meet the following qualifications:
	  - Most of the physical activity should be either moderate or vigorous intensity aerobic activity.
	  - Vigorous intensity physical activity should occur at least 3 days a week.
	  - Muscle and bone strengthening activity should take place on at least 3 days of the week.
	• Children and adolescents with disabilities should attempt to meet the physical activity guidelines for their age group as their abilities and conditions allow. Health care providers should be consulted to identify appropriate types and amounts of physical activity.

Adults	• Inactivity should be avoided. Some physical activity is better than none. • Moderate aerobic physical activity should be performed at least 150 minutes a week or 75 minutes of vigorous aerobic physical activity, or an equivalent combination of the two. • For additional health benefits, moderate aerobic physical activity should be performed 300 minutes a week, or 150 minutes a week of vigorous aerobic activity, or an equivalent combination of the two. • Muscle strengthening activities that are moderate or high intensity and involve all major muscle groups should occur on 2 or more days a week. • Adults with disabilities should attempt to meet the adult physical activity guidelines as their abilities and conditions allow. Health care providers should be consulted to identify appropriate types and amounts of physical activity. • Older adults should do exercises that maintain or improve balance, if they are at risk of falling. • Older adults should determine their level of effort for physical activity relative to their level of fitness.
Adults with Chronic Conditions	• Adults with chronic conditions obtain important health benefits from regular physical activity. • Physical activity is safe when performed according to their abilities. • Adults with chronic conditions should be under the care of a health-care provider and be consulted regarding appropriate types and amounts of physical activity.
Women During Pregnancy and the Postpartum Period	• Healthy women who are not already highly active or doing vigorous intensity activity should get at least 150 minutes of moderate intensity aerobic activity a week during pregnancy and the postpartum period. • Pregnant women who are highly active can continue physical activity during pregnancy and the postpartum period. They should discuss with their health care provider how and when activity should be adjusted over time.

Source: Adapted from 2008 Physical Activity Guidelines for Americans, U.S. Department of Health and Human Services, Physical Activity Guidelines for Americans. Available at: www.health.gov/paguidelines.

EXHIBIT 8.2   Examples of 150 Kilocalorie Activities

A moderate amount of physical activity is roughly equivalent to physical activity that uses approximately 150 kilocalories of energy. Some activities can be performed at various intensities; the suggested durations correspond to expected intensity of effort.

Less Vigorous, More Time (in descending order)

Washing and waxing a car for 45–60 minutes

Washing windows or floors for 45–60 minutes

Playing volleyball for 45 minutes

Playing touch football for 30–45 minutes

Gardening for 30–45 minutes

Wheeling self in wheelchair for 30–40 minutes

Walking $1^3/_4$ miles in 35 minutes (20 minutes/mile)

Basketball (shooting baskets) for 30 minutes

Bicycling 5 miles in 30 minutes

Dancing fast (social) for 30 minutes

Pushing a stroller $1^1/_2$ miles in 30 minutes

Raking leaves for 30 minutes

Walking 2 miles in 30 minutes (15 minutes/mile)

Water aerobics for 30 minutes

Swimming laps for 20 minutes

Wheelchair basketball for 20 minutes

Basketball (playing a game) for 15–20 minutes

Bicycling 4 miles in 15 minutes

Jumping rope for 15 minutes

Running $1^1/_2$ miles in 15 minutes (10 minutes/mile)

Shoveling snow for 15 minutes

Stair walking for 15 minutes

More Vigorous, Less Time (in ascending order)

Source: U.S. Department of Health and Human Services (USDHHS), Physical Activity and Health: A Report of the Surgeon General. Atlanta, GA: Department of Health and Human Services, Centers for Disease Control and Prevention, National Center for Chronic Disease Prevention and Health Promotion, 1996 (http://www.cdc.gov/nccdphp/sgr/sgr.htm).

These activities generally include bicycling (10 miles per hour or faster), cross-country skiing, jump roping, race-walking, running, stair walking, swimming, and vigorous dancing.

Methods to Determine Level of Exertion

Perceived Exertion A method used to determine if a vigorous level of exertion has been attained is the perception of an activity feeling "somewhat hard" to "hard" as indicated by breathlessness and a fatigue level.[34] Generally a physical effort should allow conversation, not singing, but if one is out of breath quickly or the activity must be stopped to catch breath, then the exertion is too much.

Heart Rate Another common method used to determine level of exertion is to measure heart rate during or immediately after exercise and compare that rate to a target heart rate or a target zone. Your *target heart rate* is your recommended heart rate for exercising. Some authorities prefer to recommend a range of heart rates referred to as a target zone. If you are exercising above the target heart rate or zone, you are exercising too vigorously; if you are exercising below that level, you are not exercising strenuously enough.

Although various monitoring devices are available to measure heart rate, the usual procedure is to simply take a ten-second pulse at the base of the neck or the wrist. The following steps explain how to determine an appropriate target heart rate or target zone:

1. First ascertain the maximum heart rate (MHR). You can determine the MHR from a chart (see Table 8.2) or by direct calculations. The following is an easy, conservative formula for estimating maximum heart rate:

   - 220 - your age = MHR

   - Note that some medications, such as beta-blockers, lower the MHR. If you have a client taking a beta-blocker, a physician should be contacted to determine whether the calculations need to be adjusted. To avoid the calculation difficulties for such people, perceived exertion is often the preferred method of monitoring intensity of physical activity.

2. Next, select a desired level of exertion. Less fit individuals should generally work at an intensity level of 50 to 70 percent of maximum heart rate, whereas physically fit individuals can aim for higher bouts of intensity of 70 to 85 percent of maximum heart rate.[16]

3. A heart rate that corresponds to a particular percentage of MHR can be calculated or read off a standardized chart to determine

the target heart rate or the target zone. See Table 8.2 for a chart and Exhibit 8.3 for calculation instructions.

Muscular Strength

The American College of Sports Medicine recommends performing activities that maintain or increase muscular strength for a minimum of two days each week.[33] Recommendations include at least eight to ten different exercise sets with a resistance (weight) on two non-consecutive days with eight to twelve repetitions each (arms, shoulders, chest, trunk, back, hips, and legs).[33]

All age groups can benefit from strength training; however, the advantages to older individuals seem particularly significant. These include maintaining independence in performing activities of daily life, preserving bone, and reducing the risk of falling.[15]

Table 8.2   Maximum Heart Rate (MHR) and Target Heart Rate Zone

AGE	MHR	Target Zone 50%–70% MHR		Target Zone 70%–85% MHR	
		BPM*	10 S	BPM	10 S
20	200	100–140	17–23	140–170	23–28
30	190	95–133	16–22	133–162	22–27
40	180	90–126	15–21	126–153	21–26
50	170	85–119	14–20	119–145	20–24
60	160	80–112	13–19	112–136	19–23
70	150	75–105	13–18	105–128	18–21

*BPM = beats per minute; S = seconds.

For weight training, healthy adults should be referred to an exercise specialist to receive proper guidance. Adults with disabilities and chronic conditions should be under the care of a health care provider. They should consult their health care provider about the types and amounts of activity appropriate for them.[14]

Flexibility

Stretching exercises improve flexibility (range of motion), preventing the development of rigid joints by improving the elasticity of muscles, tendons, and ligaments.[35] Flexibility affects many aspects of life including walking, stooping, sitting, avoiding falls, and driving a vehicle.[35] Older adults particularly benefit from continued physical function contributing to independent living.

Static stretching is often incorporated into warm-up and cool-down periods of aerobic activity. Popular longer programs that can greatly impact flexibility include yoga and T'ai Chi Chuan. Such stretching exercises are recommended at least two times a week.[33]

EXHIBIT 8.3   Calculation of Target Heart Rate and Target Zone

Target Zone

Example: Calculation of the target zone of a less fit individual who is fifty-two years old.

First, maximum heart rate (MHR) must be determined: 220 − 52 = 168 MHR. Less fit individuals should be working at 50 to 70 percent of MHR.

50%MHR = 168 × 0.05 = 84 beats per minute
(10 seconds = 84 ÷ 6 = 14 beats)
70%MHR = 168 × 0.70 = 118 beats per minute
(10 seconds = 118 ÷ 6 = 20 beats)

Target Zone = 84 to 118 beats per minute or 14 to 20 beats in 10 seconds

Target Heart Rate

A specific heart rate could be selected to aim for during exercise. The fifty-two-year-old may have been rather sedentary and should aim for a lower level. If 52 percent of MHR was selected, then the target heart rate can be calculated as follows:

52%MHR = 168 × 0.52 = 87 beats per minute
(10 seconds = 87 ÷ 6 = 15 beats)

Target Heart Rate = 87 beats per minute or 15 beats in 10 seconds

---

EXERCISE 8.3   Calculate a Target Heart Rate and Target Zone

Select a desired intensity level of exertion for yourself. Calculate the corresponding target heart rate and a target zone.

❏ Record the calculations in your journal.

(Appendix D), Physical Activity Options, provides some suggestions for overcoming barriers.

EXERCISE 8.4   Record Your Physical Activity Patterns

Record your physical activities for seven days in the Physical Activity Log, Lifestyle Management Form 8.2 (see Appendix D). Compare your level of activity to the examples of moderate amounts of activity listed in Exhibit 8.2.

❏ Record your reaction to the activity and the evaluation in your journal. What did you learn from this assignment? Do you meet the 2008 Physical Activity Guidelines for your age group? (see Table 8.1) Explain. Do you have any interest in making changes in your activity pattern? Explain.

BARRIERS TO BECOMING PHYSICALLY ACTIVE

A major reason for the decline in physical activity in the past few decades is our high-tech society.[36] A great deal of time is spent behind a computer monitor or in front of a television instead of toiling on a farm or doing physical chores. The most common barriers to becoming physically active identified are lack of time, access to convenient exercise facilities, and safe environments in which to be active.[35] Other common barriers are: lack of energy, motivation, and skills, fear of injury, family obligations. and weather conditions.[37]

By discovering what matters most to your clients, you will be able to provide messages and guide your clients towards physical activity programs and messages that your specific audience wants and needs. Lifestyle Management Form 8.3

EXERCISE 8.5   Survey of Physical Activity Barriers and Benefits

Survey five people who do not engage in regular physical activity. Ask them why they are not physically active. Survey another five people who have a regular exercise program. Ask them how they have overcome barriers and what they believe motivates them personally.

❏ Record responses to your survey in your journal. How did your findings compare to the most common physical activity barriers identified in this chapter?

PHYSICAL ACTIVITY COUNSELING PROTOCOLS

Counseling approaches for changing physical activity behaviors need to be tailored to a client's motivational level. A protocol developed for physicians[16] has been adapted here for nutrition counselors, and selected motivational interviewing strategies have been incorporated.[38,39] See Figure 8.2.

In Chapter 4, the nutrition counseling algorithm branched into three preaction counseling approaches addressing the needs of the majority of individuals who seek guidance for a nutritional lifestyle change. Although many clients are also in a preaction readiness stage for physical activity, the physical activity algorithm presented in Figure 8.2, has a fourth branch to accommodate clients who are already physically active and may even be participating in vigorous sports. The counseling branches are presented as distinct approaches, but counselors need to be flexible to accommodate fluctuation in readiness to change that may occur during a counseling intervention.[38] Such cases will require a cross-over of counseling strategies among the four approaches.

ASSESSMENTS OF PHYSICAL ACTIVITY

One of the first tasks for physical activity counseling will be to assess clients' physical activity status, medical readiness, and their motivation to change.

Physical Activity Status

Some clients come to counseling who are already physically active and may even be participating in vigorous sports such as marathon running or dancing. If this is the case, the nutrition counselor's responsibility will be to address any special nutritional needs related to their physical activity. However, an assessment of their physical activity status may still be appropriate because they may not be meeting all the standards. Lifestyle Management Form 8.6 in Appendix D can be used as a quick assessment of

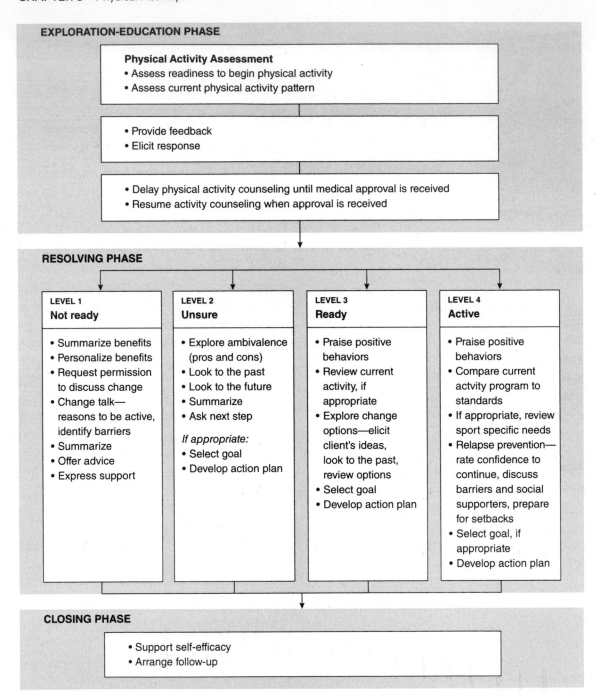

Figure 8.2 Physical Activity Motivational Counseling Algorithm
Source: Adapted from Long B, Woolen W, Patrick K, Calfas K, Sharpe D, Sallis J, Project PACE Phsyician Manual. Atlanta, GA: Centers for Disease Control; 1992.

physical activity status. Motivated clients who are not sure of their activity level may be willing to log their physical behaviors for a week. (See Lifestyle Management Form 8.2 in Appendix D.)

Considering the low activity level of many people, nutrition counselors more often deal with an individual who is contemplating an increase in

activity level. For these individuals, you have two readiness assessment responsibilities—medical and motivation. The medical assessment will guide you as to whether a referral to a physical activity professional is essential, and the assessment of motivation allows you to tailor a counseling intervention approach.

Medical Assessment

Depending on motivation level, your role may be to help your client become aware of the importance of physical activity, or it may be to provide guidance towards making physical activity lifestyle changes. If the latter is the case, you will need to assess whether a medical release or a referral is in order. The following provides some guidelines for handling this matter:

- **Physical Activity Readiness Assessment Questions.** A form frequently used to assess readiness is the Physical Activity Readiness Questionnaire, PAR-Q. It was developed by the British Columbia Ministry of Health to guide individuals in deciding whether a medical assessment is necessary.[40] See Lifestyle Management Form 8.4 in Appendix D. If a client answers yes to any of the questions, then medical approval is required. However for your professional safety, a medical release is always advisable. The advantage of using these questions in your practice, even if a medical evaluation was a prerequisite, is to be sure that the medical professional is advised of any yes answers to provide a better evaluation of the situation.
- **Medical release form.** If you are using a medical release form for all clients to whom you are supplying nutrition counseling services, you may include a question asking whether there is any reason why an increase in moderate physical activity would be prohibited. See Lifestyle Management Form 8.5 in Appendix D for an example.
- **Delay physical activity.** Physical activity should be delayed in the following conditions:[40] Women in the third trimester of pregnancy or experiencing a high-risk pregnancy or clients experiencing acute symptoms (such as fever) during an illness or infectious disease.
- **Referral to physical activity professional.** Referral to an exercise specialist is always useful, but it is essential in the following cases after receiving medical approval:
  - Clients with any symptoms of cardiovascular or metabolic disease

  - Clients recovering from coronary heart disease
  - Clients with severe bone or joint problems
  - Clients who begin an exercise program and experience dizziness, chest pain, undue shortness of breath, difficulty breathing, or unusual discomfort

- **Referral to a diabetes educator.** Coordinating food intake, medication, and the demands of physical exertion can be complex. The integration can usually be done successfully, but the guidance of a diabetes educator is advisable.

Motivational Level

As with counseling about food issues, physical activity counseling techniques need to be adjusted to a client's motivational level. See the physical activity counseling algorithm in Figure 8.2. A motivation assessment as described in Chapter 4 can aid in determining motivation for engaging in physical activity and, if needed, evaluating desire to increase activities that would address endurance, flexibility, strength, or balance.

Assessment Feedback

The procedure for giving feedback about physical activity status is similar to the guidelines given in Chapter 4 for explaining dietary status to a client. A physical activity evaluation form to aid in that feedback is provided in the Physical Activity Assessment and Feedback Form, Lifestyle Management Form 8.6 (see Appendix D). The American College of Sports Medicine position,[33] and 2008 Physical Activity Guidelines[14] provide the national standards for assessing personal physical activity habits in this form.

While giving feedback, the objective is to create an environment of self-discovery in which clients can judge whether their present physical activity habits are congruent with where they would like to be. If an individual perceives a discrepancy, his or her motivation to change is likely to increase.[38] The procedure for giving feedback follows:

- **Give clear, concise nonjudgmental feedback.** The completed assessment form should not be simply handed to a client, but gone over point by point with no hint of criticism.

- **If appropriate, elaborate on the assessment.** If your client shows concern or interest, certain components of the assessment may require elaboration. For example, clients may wish to have a description of the organizations that set the national standards or an explanation of the stages of change model. At this time it may be helpful to have on hand Lifestyle Management Form 8.1 (Appendix D), Benefits of Regular Moderate Physical Activity, and Lifestyle Management Form 7.4 (Appendix D), Prochaska and DiClemente's Spiral of Change.

Counselor: *The national standards used for this form were taken from the 2009 position statement of the American College of Sports Medicine, the premier international organization of fitness experts and sports medicine scientists, and from the 2008 Physical Activity Guidelines for Americans, a document of the Department of Health and Human Services that provides science-based recommendations that physical activity offers substantial health benefits.*

- **Elicit client response.** This is an opportunity for clients to make self-motivational statements. Counselors should not inform clients as to how they ought to feel, but rather should encourage clients to express their thoughts and be allowed to form their own conclusions.

Counselor: *What do you think about this evaluation?*

Does this information surprise you?

- **Summarize.** At the end of an assessment feedback, Miller and Rollnick[38] suggest that a counselor summarize what transpired during this period. The summary should include (1) risks or problems that were identified; (2) client reactions, including self-motivational statements; and (3) a request for the client to add or correct your summary.

Counselor: *Daniel, the medical release form signed by your doctor did not indicate a need to avoid an increase in moderate physical activity, nor did your* answers on the Physical Activity ParQ Form. As for most people in North America, your present level of physical activity fell short of national standards for endurance, flexibility, and strength. Your motivation level for increasing physical activity was found to be unsure, which means you are thinking about a change but have not made a firm commitment to increase physical activity. You also said, "I guess I have a problem here." Does this seem to be where we are? Is there anything you would like to add or correct?*

RESOLVING PHASE PROTOCOLS

After completing an assessment of your client's physical activity status, medical readiness to begin or increase physical activity, and motivation to make an activity behavior change, you are ready to begin the resolving phase. Your counseling approach for this phase will be determined by a motivational readiness assessment. Table 8.3 presents a summary of resolving phase approaches for physical activity counseling for four levels of motivation.

Level 1—Not Ready to Change

(Numbers 1–3 on a 10 point motivational assessment)

The major objective of working with clients in this stage is to furnish a warm, nonjudgmental atmosphere while providing information and encouraging consideration of beginning an activity program.

Summarize Benefits of Physical Activity David Satcher,[7] former director of the Centers for Disease Control and Prevention, has stated, "Many Americans may be surprised at the extent and strength of the evidence linking physical activity to numerous health improvements." Lifestyle Management Form 8.1 (Appendix D), a fact sheet addressing this concept, could be reviewed during your session to increase awareness.

Counselor: *Even though exercise seems to be a media buzzword, many people are not aware of the many ways in which even a moderate amount of physical activity can benefit health.*

Table 8.3   Resolving Phase: Physical Activity Counseling Approaches for Four Levels of Readiness

Readiness to Change		Counseling Approach
Not ready Level 1	**Goal**	**Raise doubt about present level of physical activity.**
	Major task:	**Inform and facilitate contemplation of change.**
	Approach:	<ul><li>Summarize benefits of physical activity.</li><li>Personalize benefits to health status.</li><li>Request permission to discuss change.</li><li>Ask key open-ended questions to promote change talk.</li><li>Elicit personal reasons to be active.</li><li>Elicit identification of barriers to physical activity.</li><li>Summarize.</li><li>Offer professional advice, if appropriate.</li><li>Express support.</li></ul>
Unsure Level 2	**Goal:**	**Build motivation and confidence.**
	Major task:	**Explore ambivalence.**
	Approach:	<ul><li>Ask key open-ended questions to explore ambivalence.</li><li>Client identifies advantages of not changing.</li><li>Client explores consequences of inactivity.</li><li>Client identifies hoped-for benefits.</li><li>Look to the past.</li><li>Look to the future.</li><li>Summarize ambivalence.</li><li>Ask about next step.</li></ul>
Ready Level 3	**Goal:**	**Negotiate a specific plan.**
	Major task:	**Resolve ambivalence and elicit a firm commitment.**
	Approach:	<ul><li>Praise positive behaviors.</li><li>Review current activity program, if appropriate.</li><li>Explore change options.</li><li>Elicit client's ideas for change.</li><li>Look to the past.</li><li>Review options that have worked for others, if needed.</li><li>Client selects an appropriate activity goal.</li><li>Develop an action plan.</li></ul>
Active Level 4	**Goal:**	**Continue the activity program.**
	Major task:	**Prevent relapse.**
	Approach:	<ul><li>Praise positive behaviors.</li><li>Review current activity program.</li><li>Review sport-specific nutrient needs.</li><li>Prevent relapse.</li><li>Explain relapse prevention.</li><li>Client rates confidence in ability to continue.</li><li>Identify potential barriers.</li><li>Explore solutions to barriers.</li><li>Prepare for setbacks.</li><li>Identify social supporters.</li><li>Set goal and develop an action plan, if appropriate.</li></ul>

Source: Adapted from Long B, Woolen W, Patrick K, Calfas K, Sharpe D, Sallis J, Project PACE Phsyician Manual. Atlanta, GA: Centers for Disease Control; 1992.

Personalize Benefits to Health Status Clients probably know that being physically active is a good idea, but they may feel with so many important things to do in life, it is hard to make each one a priority. Your clients may not have thought how increasing their activity level could be a benefit for them personally.

Counselor: *Mrs. Bernstein, physical activity could be particularly beneficial to you to help lower your blood pressure and aid in your effort to lose weight. Brisk walking just three times a week could produce results. You said you had some trouble sleeping at night. Increasing your activity could also help you to fall asleep more quickly and to sleep well.*

Ask Key Open-Ended Questions to Promote Change Talk Helping clients think and talk about change can aid in the development of motivation to change.[39] However, because the client has already indicated no desire to change, it is generally best to begin this discussion with a tentative approach, requesting permission to discuss the issue.

Counselor: *Would you be willing to continue our discussion and talk about the possibility of a change in your physical activity level?*

- **Elicit personal reasons to be active.** People are often more likely to be persuaded to change if they are presenting the arguments for change.[35]

Counselor: *We've looked at a list of reasons to have at least a moderate amount of physical activity in your life. If you were to start an exercise program, which of the benefits do you think would most likely apply to you?*

On a scale of 1 to 10, with number 1 being no increase in physical activity and 10 being "yes, I will start," what number would you select? That is interesting that you chose the number 2. How come not number 1?

- **Elicit identification of barriers to physical activity.** People at this motivational level may have given little serious thought to why they do not become more involved in physical activity. Having clients identify reasons could

lead to discussions of overcoming physical activity barriers. Physical Activity Options, Lifestyle Management Form 8.3 (see Appendix D), could be useful to review during this discussion. King suggests assessing a client's psychological relationship to physical activity as a possible barrier.[41] Negative feelings about exercise may have arisen from past embarrassments, punishments, or weight loss failures. Sometimes a negative physical experience, such as an injury or violation of the body, can hamper exercise attempts. If this is the case, then the issue should be explored with a psychotherapist. King also encourages the use of the term *joyful movement* for exercise to enhance a pleasurable perspective.[41]

Counselor: *You chose number 2 on the scale. How come you didn't choose 4? What would have to change for you to select 4? Is there something I could do to help you get to 4?*

What would have to change for you to consider increasing the amount of joyful movement in your life?

Summarize Summaries help reinforce what has been said, tying together various aspects of a discussion. They indicate to clients that you have been closely listening to them and prepare them to move on to a new topic. They encourage clients to rethink their position.

Give a summary of reasons not to change first and then give a summary of reasons to change. End your summary with any self-motivational statements your client may have made during the discussion and a "what's next" question.

Counselor: *Juanita, I'd like to review what we have talked about regarding physical activity. We looked at some of the benefits of being physically active, and you said you were surprised that health benefits could be attained with much less effort than you thought. You indicated that your biggest obstacles to beginning a physical activity program are finding time in your already busy schedule and that you have never found exercise to be enjoyable. I gave you some reading material to help you think about how to fit physical activity into your life, and you said you would look it over. You* *also said that increasing physical activity is something you should probably give some thought to. Was that a fair summary? Did I leave anything out? Where does all this leave us now?*

Be prepared to fully accept any answer to the "Where does all this leave us now?" question. Do not jump on a half-hearted ready-to-take action statement. In our eagerness to guide clients toward making a change, we may shift gears and move to the action stage too quickly. Instead, probe the issue further with more open-ended questions.

Counselor: *When you came in today, you indicated that you were not ready for a change, but now it appears that you are having a change of heart. Is that correct? So we don't start going down the wrong path, I want to be sure that is what you really want to do. What can you say to convince me?*

Offer Professional Advice, If Appropriate Well-timed and compassionate advice can aid in motivating behavior change.[38] In the ideal situation, a client asks for advice, but if that is not the case, then the counselor can ask permission to give advice.

Counselor: *Obviously you know my opinion that starting a physical activity program is what you should do, but of course only you really know what can work for you. Physical activity is important for everyone's health, and specifically for you I believe this would help you with your desires to lose weight and lower your blood pressure. A common procedure is to start walking ten minutes, three times a week, and to increase walking time gradually. What do you think?*

Express Support Several authorities have documented that counselor expectations for a client's ability to change can produce a favorable outcome.[38]

Counselor: *Only you know what will work for you and when you will actually be ready to start an exercise program. When the time is right, I know you will be able to do it. I want you to know that I respect your decision and anytime you are ready, I'm available to help. If it is OK with you, I would like to bring up the topic periodically during our future sessions.*

Level 2—Unsure About Changing

(Numbers 4–6 on a 10 point motivational assessment)

At this stage, clients have not made a firm commitment to change. The task at hand for clients and counselors is to build motivation and confidence to change.[38] The following provides a number of strategies to explore ambivalence for your review. How many you use in a counseling session is likely to vary depending on the desire of your client to explore the issues and what fits best with your personal counseling style.

Ask Key Open-Ended Questions to Explore Ambivalence Exploring ambivalence is very important to help clients tip the scales in favor of making a change.[38] This involves exploring the pros and cons of changing physical activity patterns. Barriers to increasing physical activity need to be discovered in order to explore ways to effectively handle them. The fact sheet on physical activity options, (Lifestyle Management Form 8.3 in Appendix D), could be useful to review during this discussion.

- Client identifies disadvantages of changing.

Counselor: *What do you believe are the reasons you haven't increased your level of physical activity?*

What problems would occur if you increased your level of physical activity?

What are some reasons why you would like things to stay just like they are?

- Client explores consequences of inactivity.

Counselor: *What concerns do you have about not increasing your level of physical activity?*

- Client identifies hoped-for benefits.

Counselor: *What are some reasons for increasing physical activity?*

What do you hope would be better for you if you increased your level of physical activity?

What are some good things that would happen if you were more physically active?

- **Look to the Past.** There may be skills and resources that have worked for your client in the past that can be used in the present to build a more physically active lifestyle.

Counselor: *Can you think of physical activities you have done in the past that have been enjoyable?*

- **Look to the Future.** By imagining the future, clients may be able to work out some ways physical activity could fit into their world and everyday life.

Counselor: *I can see why you have some concerns about increasing your level of physical activity. Could we just stand back and imagine that you have an optimum amount of physical activity in your life? What would your life be like?*

Summarize Ambivalence and Reiterate Self-Motivational Statements Periodic summaries throughout a counseling session can be useful for exploring ambivalence and for selectively emphasizing issues or self-motivational statements that could tip the balance in favor of making a change.

Counselor: *It sounds as if you have opposing feelings about exercise. On the one hand, you have difficulty seeing how you can find time to exercise, and when you do have time you don't have the energy or the motivation to begin. On the other hand, your weight and blood pressure have been creeping up over the last ten years, and you know that some exercise would help both problems. Before you had children, you used to enjoy bike rides with a cycling group on the weekends. You said, "With all the ways that exercise could help me, I should be able to find a way to make an exercise program work." Did I miss anything? What are you thinking about this?*

Ask About Next Step Ideally a client will be ready to select a goal and develop an action plan for increasing physical activity. However, you need to be ready to accept whatever direction your client wants to go about planning a change. For your client to experience intrinsic motivation to change, he or she must perceive the freedom to choose a course of action.[38] The fact is that few people like to be told what they must do. Feeling coerced into taking a certain course of action is likely to produce resistance,

counterproductive to your counseling goals. If your client indicates that he or she would like to make a behavior change, then setting a goal and developing an action plan would be appropriate. See Chapter 5 for an explanation of the process.

Counselor: *What do you think your next step should be?*

Level 3—Ready to Change

(Numbers 7–9 on a 10 point motivational assessment)
Individuals at this level of readiness have indicated that they are ready to make a change. When working with these clients, you should recognize that the objectives are to resolve any ambivalence, elicit a firm commitment to change, and develop specific goals and action plans.

Praise Positive Behaviors Some clients who are classified in this category have begun limited amounts of physical activity, and those attempts should be praised.

Counselor: *It is really great that you have been doing some physical activities and are ready to increase your level of exercise.*

Review Current Activity Program, If Appropriate Some people in the category have indicated that they are physically active on occasion. These activities may be expanded to develop a comprehensive program.

Counselor: *I see from your assessment form that you are physically active sometimes. What kinds of things do you do?*

Explore Change Options While exploring viable options for increasing physical activity, Berg-Smith et al. emphasize the need for a counselor to remain neutral while conveying the following messages to a client:[39]

1. There are a number of physical activity options to choose among.
2. You are the best judge of what will work.
3. We will work together to review options.

- **Elicit client's ideas for change.**

Counselor: *I have a list of possible options for people who are initiating a physical activity program that we could go over, but you may already have definite ideas of what would work for you. After all, you are the best judge of what will work for you.*

- **Look to the past.**

Counselor: *When have you enjoyed being physically active in the past?*

- **Review options that have worked for others, if needed.** Reviewing the Physical Activities Options Fact Sheet, Lifestyle Management Form 8.3 (Appendix D), could be useful.

Counselor: *Would you like to review the types of physical activities that have worked for others?*

Client Selects an Appropriate Physical Activity Goal The goal-setting process covered in Chapter 5 can be applied to physical activity. Emphasis should be placed on activities that your client finds convenient and enjoyable.

Develop an Action Plan The process of developing an action plan was covered in Chapter 5 and should be applied to physical activity.

Level 4—Pysically Active

(Number 10 on a 10 point motivational assessment)

Some clients come to counseling sessions who are already physically active and may even be participating in vigorous sports. When working with these clients, your major goal will be to have them continue with their program, and the major task will be to prevent relapse.

Praise Positive Behaviors As discussed in the previous stage, positive behaviors should be praised to encourage continuation of a physically active lifestyle.

Counselor: *It is wonderful that you have such a physically active lifestyle.*

Review Current Activity Program A physically active person may not be meeting all the guidelines

for frequency, duration, intensity, flexibility, and strength training as set by national standards. Any problem areas should be identified and addressed. If Lifestyle Management Form 8.6 (Appendix D) was not used for an assessment during the exploration-education phase of the counseling session, then you may wish to use it at this point to ensure that all aspects of fitness are covered.

Counselor: *You are doing so well with aerobic and flexibility activities; however, your strength training does not meet the national standards. What do you think about this?*

Review Sport-Specific Nutrient Needs Clients may be engaging in a particular sport or at a level of intensity that gives rise to special nutritional needs.[42, 43] These should be investigated and addressed with the client. Clients may also have questions about possible **ergogenic aids** (substances, including medications and dietary supplements, or techniques and devices, that are intended to improve physical performance)[44] they have heard about. As in many areas of nutrition, there is a continual influx of new claims and supplements in the sports arena. One person cannot possibly keep up with every one of them. When these issues arise, you should tell your client that you are not familiar with the specific claim but you will use your professional resources to investigate the claim. Many professional organizations involved in physical activity have websites with educational reviews of new claims (see the end-of-chapter resources for some possible websites to visit). Professional organization listservs can also be helpful for posting a question.

Counselor: *Do you have any questions about foods or supplements related to your physical activity program?*

Prevent Relapse A major task of counseling people at this level of motivation is relapse prevention. This can be addressed by helping your clients understand that perfection in a physical activity program should not be expected and setbacks occur, but they do not need to cause an abandonment of a physical activity program. Identifying potential barriers, preparing possible solutions for anticipated problems, and enlisting social support can also aid in relapse prevention.

- **Explain relapse prevention.** Often roadblocks that could interfere with continuation of a physical activity program can be anticipated, and preparations can be made to prevent the problem from resulting in a relapse.

Counselor: *You are doing really well in your physical activity program. The major task we have now in counseling is to work on making sure that your behavior continues. First, let's look at how confident you are that you can keep up this activity pattern.*

- **Client rates confidence in being able to continue.** Confidence in ability to be physically active is an important component of a person's ability to maintain an active lifestyle.[35] This confidence can be assessed with the aid of the confidence assessment graphic, Lifestyle Management Form 4.1 (Appendix D).

Counselor: *If we use the numbers on this picture to represent how confident you are that you can maintain your present level of physical activity for the next three months, what number would you select? Number "1" indicates not at all, and number "10" denotes very confident.*

- **Identify potential barriers.** An important component to preventing relapse is identification of potential barriers to continuing a physical activity program. By planning for an effective response to an anticipated problem, the difficulty will not lead to a complete abandonment of the program. Key questions about the client's number selection on the continuum can identify what concerns may be interfering with his or her self-confidence.

Counselor: *How come you chose 8 instead of 10?*

What would have to change for you to select 10 instead of 8?

What has prevented you from keeping with your physical activity program in the past?

- **Explore solutions to barriers.** When potential problems have been identified, your client should be encouraged to explore possible solutions.

Counselor: *You said vacations are a difficult time to maintain an exercise program. Do you have any ideas of how to overcome difficulties while traveling?*

- **Prepare for setbacks.** One important concept to explore with your client is that setbacks are common, are to be expected, and do not mean a total program should be abandoned.

Counselor: *In anyone's physical activity program, setbacks are to be expected. This could be because of illness, family responsibilities, work demands, house guests, or travel. Sometimes you can anticipate the difficulties and prepare for them, but other times continuing your physical activity program is simply is not feasible. If this is the case, what is important is to just start up again as soon as possible.*

- **Identify Social Supporters.** Receiving social support has a key influence on physical activity levels.[35] Asking clients to identify social supporters and suggesting to clients to seek out their assistance is part of the successful PACE program.[16] Social support could come from loved ones giving words of encouragement or having someone to jog with on occasion. Social support can also come from organized groups, such as a canoe club. Counselors need to use their judgment about exploring this issue for someone who is heavily involved in a team sport such as soccer or one such as dance in which social support is intrinsic to the organized activity. Social support should be encouraged, but the point should be made that a physical activity program cannot rely on another person.

Counselor: *You are doing so well with your physical activity program, and I hope you continue to do well. Getting support from friends or relatives has been found to be helpful in maintaining a program. Can you think of someone who could be supportive? It would really be great if there was someone who was also interested in working on increasing or maintaining his or her level of exercise with whom you could discuss your progress.*

Set Goals, if Appropriate Since some clients at this motivation level are heavily involved in a sport, goal planning may not be indicated. Follow Chapter 5 guidelines for goal setting, if necessary.

Develop an Action Plan, if Appropriate Development of an action plan would only be appropriate if a goal was selected. Follow Chapter 5 guidelines for developing an action plan.

ISSUES PERTINENT TO PHYSICAL ACTIVITY GOAL SETTING AND ACTION PLAN DEVELOPMENT

The following are some factors to take into consideration that specifically apply to physical activity planning:

- **Initial goals should be modest.** To avoid soreness and injury and to maintain motivation, a sedentary individual contemplating an increase in physical activity should start with short sessions (five to ten minutes). The activity could be as simple as getting off a bus at an earlier stop or parking a car at a distant part of the parking lot and walking the extra distance.[12]
- **Increase gradually.** Injuries can be prevented by gradually building up to the desired amount of physical activity.
- **Take into consideration sustainable factors.** The physical activities chosen should meet the following criteria:
  Enjoyable—Clients should be encouraged to think creatively. Enjoyable activities could include folk dancing, bike touring, gardening, stair climbing, kickboxing, or family fitness activities (examples: camping, dancing, flying a kite).

  Safe—Safe areas can include community parks, gyms, pools, malls, and health clubs. If a safe location cannot be identified, then areas of the home should be evaluated for providing space for exercise equipment, such as a stationary bike or treadmill.

  Convenient—For some people, increasing physical activity works best by including short activities throughout the day, such as using steps instead of an elevator or taking a short walk after lunch. If this procedure is chosen, then goal selection and monitoring may require special consideration. Some people have used a pedometer attached to a shoe or belt. Depending on the quality of the device, clients can set goals and monitor their progress in terms of miles covered in a day or number of steps taken. Meta-analyses focused on pedometer-based programs conclude that the use of a pedometer is associated with significant increases in physical activity and significant decreases in body mass index and blood pressure.[45]

Affordable—Usually a variety of community programs and facilities are located in schools, community colleges, and universities. In the United States the Young Men's and Women's Christian Association provide an array of exercise programs for a modest cost. Also, many malls are available to walkers early in the morning before stores open.

Walking Basics

Walking is frequently suggested as an introductory exercise for sedentary individuals.[46] This activity provides excellent cardiovascular and endurance advantages;[47] however, walking does not improve muscle power or reduce age-related muscle loss.[48] For most people, walking is inexpensive, accessible, safe, and enjoyable. The only expense is a pair of good walking shoes. Several commercial and noncommercial organizations have developed guidelines for individuals embarking on a walking program. (See suggested resources at the end of the chapter.) The American College of Sports Medicine (ACSM) has three protocols for individuals beginning a walking program based on their fitness level; these are presented in Table 8.4.

Table 8.4 ACSM Walking Program

	Daily Walking Times*		
Week	**Level 1 (Not Walked Aerobically In Years)**	**Level 2 (Slightly More Fit Than Level 1)**	**Level 3 (Participated In Some Aerobic Walking)**
1	10	20	30
2	12	20	30
3	15	25	35
4	15	25	35
5	20	30	40
6	20	30	40

*Walking sessions should be preceded by two-minute walking in place warm-ups plus basic stretches. See *ACSM Fitness Book,* 3d ed. American College of Sports Medicine, 2002.

Source: Adapted, with permission, from American College of Sports Medicine, 2003, *ACSM Fitness Book,* 3rd ed. (Champaign, IL: Human Kinetics), 128–133.

EXERCISE 8.5 Incorporating Physical Activity

Read Case Study – Officer Bill and explain how physical activity would benefit Bill. If you were Bill's dietitian, what role would you take in counseling him about physical activity? Describe a specific physical activity goal that would be appropriate for Bill to begin making a change in his lifestyle.

- ❏ **Plan for variety.** A physical activity program should include a variety of activities to maintain motivation and decrease the risk of injury due to overuse of any particular muscle group.
- ❏ **Consider the daily routine.** A plan should have physical activity as part of a daily routine. Clients should be encouraged to use more physical activity when feasible, such as using stairs rather than an elevator or walking to a store rather than driving.
- ❏ **Plan for sustained activity.** To maintain the benefits of exercise from both endurance and resistance training, the activities must be continuous. Health benefits will decrease within two weeks if physical activity is considerably reduced, and if the inactivity is sustained, the gains will entirely disappear within two to eight months.[15]

CASE STUDY Officer Bill

Bill Melia has been on the Los Angeles police force for twenty-six years. He has a good family life–married for thirty years with three grown children. He enjoys his job and is happy with his career, although at times the stress can be a challenge. Overall, Bill copes by planning great vacations with his wife Lola; they often go to Mexico to visit their parents. Bill's recent physical exam revealed metabolic syndrome. The physician indicated that this is an early stage and if left unattended would evolve into type 2 diabetes. His doctor warned Bill that he needed to make significant lifestyle changes, including weight loss, to control the situation. An appointment was made with the office dietitian to discuss lifestyle modifications that would be helpful in preventing progression of the disease.

This was Bill's first experience with the possibility of a major health problem, and the news was difficult for Bill to handle. He started feeling depressed about the possibility of being on a strict diet and taking insulin injections. Bill remembered his grandmother in Mexico taking shots every day, and the thought of this being part of his life greatly worried him.

As Bill drove to meet the dietitian, he thought of all the changes he would have to make. Working balanced meals into his schedule as a police officer would not be easy. Because he is never quite sure when he will have time to eat, Bill has large meals when the opportunity arises. Any given call over his radio could result in hours without being able to eat, so the larger the meal, the better. However, his interest in returning to working out would be something positive to share with the dietitian. He used to love to lift weights for increased muscle strength, and surely this would be a positive activity toward controlling his diabetes. Yes, that would do it, he thought.

REVIEW QUESTIONS

1. Describe physical activity and physical fitness.
2. Identify the two most common risks associated with exercise and explain how to best prevent the harmful effects.
3. Explain the major conclusions of the 2008 Physical Activity Guidelines (see Table 8.1).
4. Explain two ways to monitor the level of physical exertion.
5. Explain how to determine maximum heart rate, target heart rate, and target zone.
6. Why do older adults particularly benefit from strength and flexibility training?
7. Explain when a client should obtain a physician's approval before engaging in physical activity counseling.
8. When should physical activity be delayed?
9. Explain the physical activity counseling algorithm.

10. Identify physical activity counseling strategies for each of the four motivational levels discussed in the section on physical activity counseling protocols.

ASSIGNMENT—PHYSICAL ACTIVITY ASSESSMENT AND COUNSELING

The objectives of this assignment are to gain experience using the physical activity readiness assessment forms and the physical activity protocols. Pair off with a colleague and take turns counseling each other.

PART I. Use the following interview guide checklist to conduct the counseling session with your colleague. Possible counselor questions, statements, and responses are given in italics. Additional examples can be found in the chapter.

Preparation
❑ Review the physical activity counseling algorithm in Figure 8.2 and protocols in this chapter, as well as goal setting and action plan development (see Chapter 5).

❑ You and your partner should each complete copies of physical activity assessment forms found in Appendix D:

  ○ Client Assessment Questionnaire, Lifestyle Management Form 5.1.
  ○ Physical Activity ParQ Form, Lifestyle Management Form 8.4.
  ○ Physical Activity Assessment and Feedback Form, Lifestyle Management Form 8.6.

❑ Exchange completed assessment forms.

❑ Bring a blank card and copies of physical activity fact sheets found in Appendix D:

  ○ Benefits of Regular Moderate Physical Activity, Lifestyle Management Form 8.1.
  ○ Physical Activity Options, Lifestyle Management Form 8.3.

Interview
Because the focus of this assignment is practicing physical activity counseling skills, not all phases of a counseling session will be addressed. One of you should take on the role of counselor and the other to play the role of client. After completing the counseling experience, reverse the roles.

Feedback After reviewing your colleagues physical activity assessment forms, provide feedback by using the following guidelines:
❑ Point-by-point, clear, concise, nonjudgmental.

❑ Elicit response.

❑ Give summary.

  ○ Identify problems.
  ○ Reiterate any self-motivational statements—*I never thought much about weight training.*
  ○ Ask for additions or corrections.

❑ Elicit response—*What do you think about this?*

Evaluate need for physician approval This activity is a simulation. The physical activity counseling practice session should continue. However, if you or your partner requires an evaluation from a physician, it should be understood that any goal setting or action plan discussed would not be implemented until receiving proper medical clearance.

Customize the counseling approach Tailor your counseling approach to the motivational level of your volunteer client. Use the motivational assessment graphic, Lifestyle Management Form 4.1 in Appendix D to determine motivational level: Level 1(not ready) = 1–3 on graphic, Level 2 (unsure) = 4–6 on graphic, Level 3 (ready) = 7–9 on graphic, and Level 4 (active) = 10 on graphic.

Not Ready (Level 4)
❑ Summarize benefits of physical activity—see Benefits of Regular Moderate Physical Activity fact sheet, Life Management Form 8.1 in Appendix D.

❑ Personalize benefits—use the completed Client Assessment Questionnaire, Life Management Form 5.1 in Appendix D, as an aid.

❑ Request permission to discuss the possibility of a change.

❑ Ask key open-ended questions—reasons to be active, barriers.
❑ Summarize.
❑ Offer advice if requested, or request permission to offer advice.
❑ Express support.
❑ Support self-efficacy.

Unsure (Level 2)
❑ Explore ambivalence—ask key open-ended questions.

○ Advantages of not changing.
○ Consequences of not changing.
○ Hoped-for benefits.

❑ Look to the past—*Have you ever been physically active?*
❑ Look to the future—*What would your life be like?*
❑ Summarize ambivalence and reiterate self-motivational statements.
❑ Ask about next step.
❑ Set goal and develop an action plan, if appropriate. See Ready action plan checklist in the section that follows.
❑ Support self-efficacy.

Ready (Level 3)
❑ Praise positive behaviors.
❑ Review current activity program, if appropriate.
❑ Explore change options to develop a broadly stated goal.

○ Elicit client's thoughts.
○ Look to the past.
○ Go over list of possibilities if requested—see Physical Activity Options fact sheet, Life Management Form 8.3 (Appendix D).

❑ Client selects an appropriate activity goal.
❑ Develop an action plan.

○ Investigate physical environment—*Do you have everything you need? Do you have walking shoes?*
○ Examine social support—*Is there anyone who could help you achieve your goal?*
○ Review cognitive environment—*What will you be saying to yourself if you miss a day that you planned to walk?*

○ Explain positive coping talk, if necessary.
○ Select tracking technique—chart, journal, and so forth.
○ Ask your client to verbalize goal.
○ Write down goal on a card and give it to your client.
○ Support self-efficacy.

Active (Level 4)
❑ Praise positive behaviors.
❑ Review current activity program; compare to standards.
❑ Review sport-specific nutrient needs, if necessary.
❑ Prevent relapse.

○ Explain need to discuss relapse.
○ Use Assessment Graphic, Life Management Form 4.1 in Appendix D, to rate confidence to continue.
○ Identify potential barriers.
○ Explore solutions to barriers.
○ Explain that setbacks are common.
○ Identify social supporters.

❑ Set goal and develop an action plan, if appropriate. (See previous motivational level.)
❑ Support self-efficacy.

PART II. Answer the following questions in a formal typed report or in your journal. For formal reports, number and type each question, and put the answers in complete sentences under the question. For journal entries, number each answer.

1. Write a narration of the experience when you had the role of counselor. There should be three titled sections in the narration: preparation, feedback, and counseling approach.

2. What counseling strategies had the greatest impact?

3. Explain the impact of the Lifestyle Management Forms and the picture of the assessment graphic, if used.

4. Describe the experience of being counseled as compared to being a counselor.

5. What did you learn from the experience?

SUGGESTED READINGS, MATERIALS, AND INTERNET RESOURCES

Guiding Clients in Designing a Fitness Program

ACSM Fitness Book. 3rd ed. American College of Sports Medicine. Champaign, IL: Human Kinetics; 2003; a basic skills approach providing a fitness test and guidance for developing a personal fitness program.

American Heart Association. *Fitting in Fitness: Hundreds of Simple Ways to Put More Physical Activity into Your Life*. New York: Time-Life Books; 1997; provides an abundance of suggestions for putting physical activity into daily life.

Kratina K, King NL, Hayes D. *Moving Away from Diets—New Ways to Heal Eating Problems & Exercise Resistance*. Lake Dallas, TX: Helm Seminars. Expands on the concepts of exercise resistance and joyful movement.

Miller A. *Action Plan for Arthritis*. Champaign, IL: Human Kinetics; 2003; an exercise guide for individuals with arthritis.

U.S. Department of Health and Human Services. *Promoting Physical Activity A Guide for Community Action;* Champaign, IL: Human Kinetics; 1999; a guide for promoting physical activity in the community.

Rosenbloom, CA, ed. Sports Nutrition: Client Education Handouts. Chicago: American Dietetic Association; 2006. This CD-ROM contains client education handouts for nutrition professionals who work with athletes at all levels and physically active individuals.

American Medical Association. Roadmaps for Clinical Practice, Number 5, Physical Activity Management; 2003; www.ama.org

Walking and Fitness

American Heart Association. *The Healthy Heart Walking CD: Walking Workouts for a Lifetime of Fitness 2004;* beginner and intermediate walking workouts. Available at www.amazon.com

American Heart Association. *Fitting in Fitness: Hundreds of Simple Ways to Put More Physical Activity into Your Life*. Clarkson Potter; 1997; pocket guide tells how to fit exercise into people's daily schedule by giving helpful hints.

Selected Internet Resources

American Heart Association
http://startwalkingnow.org/home.jsp

America on the Move
https://aom3.americaonthemove.org/default.aspx

Healthier Worksite Initiative
http://www.cdc.gov/nccdphp/dnpao/hwi/index.htm

Exercise is Medicine™
http://www.exerciseismedicine.org

Gatorade Sports Science Institute
http://www.gssiweb.com

HealthierUS.gov
http://www.healthierus.gov/exercise.html

National Institutes of Health/We Can!
http://www.nhlbi.nih.gov/health/public/heart/obesity/wecan/

President's Council on Physical Fitness and Sports
http://www.presidentschallenge.org

Shape Up America!
http://www.shapeup.org/index.html

US Department of Health and Human Services, Smallsteps, Adults and Teens, Kids
http://www.smallstep.gov

eXtension, Families Food and Fitness
www.extension.org
Search families, food and fitness

Pep Up Your Life: A Fitness Book for Mid-Life and Older Persons
http://www.fitness.gov/pepup.htm

REFERENCES

[1]U.S. Department of Health and Human Services (USDHHS). *Physical Activity and Good Nutrition: Essential Elements for Good Health.* Atlanta, GA: Department of Health and Human Services, Centers for Disease Control and Prevention, National Center for Chronic Disease Prevention and Health Promotion; 1999.

[2]Bryan SN, Katzmarzyk PT. Are Canadians meeting the guidelines for moderate and vigorous leisure-time physical activity? *Appl Physiol Nutr Metab.* 2009;34:707–715.

[3]Pleis JR, Lethbridge-Cejku M. Summary health statistics for U.S. adults: National Health Interview Survey, 2006. *Vital Health Stat 10.* 2007;235:1–153.

[4]Rippe JM. The role of physical activity in the prevention and management of obesity. *J Am Diet Assoc.* 1998;98 (suppl 2):S31–S38.

[5]Guthold R, Ono T, Strong KL, Chatterji S, Morabia A. Worldwide variability in physical inactivity: A 51-country Survey. *Am J Prev Med.* 2008;34:486–494.

[6]U.S. Department of Health and Human Services. (2010). Vital and Health Statistics - Summary Health Statistics for U.S. Adults: National Health Interview Survey, 2009. DHHS Publication No. (PHS) 2011-1577. Available at: http://www.cdc.gov/nchs/data/series/sr_10/sr10_249.pdf

[7]President's Challenge is a program of the President's Council on Fitness, Sports, and Nutrition. The Presidents Challenge Program. Available at: http://www.presidentschallenge.org Accessed September 17, 2010.

[8]The Partnership for a Healthier America, Let's Move. Available at: http://www.letsmove.gov/ Accessed September 17, 2010.

[9]America's Great Outdoors. Available at: http://www.doi.gov/americasgreatoutdoors, Accessed September 17, 2010.

[10]United States Department of Veteran Affairs, MOVE! http://www.move.va.gov/ Accessed September 19, 2010.

[11]U.S. Department of Health and Human Services (USDHHS). *Healthy People 2020.* Available at: http://www.healthypeople.gov/ Accessed September 17, 2010.

[12]American College of Sports Medicine, American Dietetic Association, International Food Information Council. For a healthful lifestyle: Promoting cooperation among nutrition professionals and physical activity professionals. *J Am Diet Assoc.* 1999;99:994–997.

[13]Goodrich DE, Larkin AR, Lowery JC, Holleman RG, Richardson CR. Adverse events among high-risk participants in a home-based walking study: A descriptive study. *Int J Behav Nutr Phys Act.* 2007: 4: 20.

[14]U.S. Department of Health and Human Services (USDHHS). *2008 Physical Activity Guidelines for Americans.* ODPHP Publication No. U0036, October, 2008. Available at: http://www.health.gov/paguidelines/ Accessed September 17, 2010.

[15]U.S. Department of Health and Human Services (USDHHS). *Physical Activity and Health: A Report of the Surgeon General.* Atlanta, GA: Department of Health and Human Services, Centers for Disease Control and Prevention, National Center for Chronic Disease Prevention and Health Promotion; 1996 (http://www.cdc.gov/nccdphp/sgr/sgr.htm).

[16]Long B, Woolen W, Patrick K, Calfas K, Sharpe D, Sallis J. *Project PACE Physician Manual.* Atlanta, GA: Centers for Disease Control; 1992.

[17]Mokdad AH, Marks JS, Stroup DF, Gerberding JL. Actual causes of death in the United States, 2000. *J Am Med Assoc.* 2004; 291:1238-1245. [Correction: Actual causes of death in the United States, 2000. *J Am Med Assoc.* 2005; 293:293-294.]

[18]Hakim AA, Petrovitch H, Burchfiel CM, Ross GW, Rodriguez BL, White LR, Yano K, Curb JD, Abbott RD. Effects of walking on mortality among non-smoking retired men. *N Engl J Med.* 1998;338:94–99.

[19]Davis JM. What can a physically active lifestyle promise? In: Wardlaw GM. *Perspectives in Nutrition.* 4th ed. Boston: WCB McGraw-Hill; 1999:314–315.

[20]American Medical Association Fitness Basics. In: *AMA Health Insight.* http://www.ama-assn.org/insight/gen_hlth/fitness/fitnes2.htm; 1995–1999.

[21]Fletcher GF, Balady G, Blair SN, Blumenthal J, Caspersen C, Chaitman B, Epstein S, Froelicher DSS, Froelicher VF, Pina IL, Pollock ML. Statement on exercise: Benefits and recommendations for physical activity programs for all Americans. *Circulation.* 1996;94:857–862.

[22]Goldstein LB. Physical activity and the risk of stroke. *Expert Rev Neurother.* 2010;10:1263–1265.

[23]Friedenreich CM. The role of physical activity in breast cancer etiology. *Semin Oncol.* 2010: 37: 297–302.

[24]Waller K, Kaprio J, Lehtovirta M, Silventoinen K, Koskenvuo M, Kujala UM. Leisure-time physical activity and type 2 diabetes during a 28 year follow-up in twins. *Diabetologia.* 2010 [Epub ahead of print] DOI: 10.1007/s00125-010-1875-9.

[25]National Institutes of Health, National Heart, Lung, and Blood Institute. Clinical guidelines on the identification, evaluation, and treatment of overweight

and obesity in adults—the evidence report. *Obes Res.* 1998;6(suppl 2):S51.

[26]Saelens BE, Grow HM, Stark LJ, Seeley RJ, Roehrig H. Efficacy of increasing physical activity to reduce children's visceral fat: A pilot randomized controlled trial. *Int J Pediatr Obes.* 2010: [Epub ahead of print] DOI:10.31 09/17477166.2010.482157.

[27]American College of Sports Medicine. *ACSM's Guidelines for Exercise Testing and Prescription.* 5th ed. Baltimore, MD: Williams & Wilkins; 1995.

[28]Feskanich D, Willett W, Colditz G. Walking and leisure-time activity and risk of hip fracture in post-menopausal women. *JAMA,* 2002: 288(18): 2300–2306.

[29]Karlsson MK, Nordqvist A, Karlsson C. Physical activity, muscle function, falls and fractures. *Food Nutr Res.* 2008; 52. doi: 10.3402/fnr.v52i0.1920.

[30]Siscovick DS, Weiss NS, Fletcher RH, Lasky T. The incidence of primary cardiac arrest during vigorous exercise. *N Engl J Med.* 1984;311:874–877.

[31]Fiatarone MA, O'Neill EF, Ryan ND, Clements KM, Solares GR, Nelson ME, Roberts SB, Kehayias JJ, Lipsitz LA, Evans WJ. Exercise training and nutritional supplementation for physical frailty in very elderly people. *N Engl J Med.* 1994;330:1769–1775.

[32]American Heart Association. *Physical Activity and Cardiovascular Health: Fact Sheet.* www.justmove.org/; 1999.

[33]American College of Sports Medicine (ACSM). *ACSM's Guidelines for Exercise Testing and Prescription.* 8th ed. Philadelphia (PA): Lippincott, Williams & Wilkins: 2009.

[34]Borg GA. Psychophysical bases of perceived exertion. *Med Sci Sports Exerc.* 1982;14:377–381.

[35]U.S. Department of Health and Human Services (USDHHS). *Healthy People 2010.* Atlanta, GA: Department of Health and Human Services, Office of Disease Prevention and Promotion; 2000.

[36]American Dietetic Association. *Project Lean Resource Kit.* Chicago: American Dietetic Association; 1995.

[37]U.S. Department of Health and Human Services (USDHHS). *Promoting Physical Activity A Guide for Community.* Human Kinetics, Champaign, IL; 1999.

[38]Miller WR, Rollnick S. *Motivational Interviewing Preparing People to Change Addictive Behavior.* New York: Guilford; 1991.

[39]Berg-Smith SM, Stevens VJ, Brown KM, Van Horn L, Gernhofer N, Peters E, Greenberg R, Snetselaar L, Ahrens L, Smith K. A brief motivational intervention to improve dietary adherence in adolescents. *Health Education Research.* 1999;14:399–410.

[40]Thomas S, Reading J, Shepard RJ. Revision of the Physical Activity Readiness Questionnaire (PAR-Q). *Can Sport Sci.* 1992;17:338–345.

[41]King NL. *Counseling for Health and Fitness.* Eureka, CA: Nutrition Dimension; 1999.

[42]Berning JR, Steen SN. *Nutrition for Sport and Exercise.* 2d ed. Gaithersburg, MD: Aspen; 1998.

[43]Dunford M, Doyle JA. *Nutrition for Sport and Exercise.* Florence, KY: Cengage Learning, Inc.; 2007.

[44]Brooks GA, Fahey, TD, Baldwin K. *Exercise Physiology: Human Bioenergetics and Its Application.* 4th ed. New York, NY: McGraw-Hill Humanities Social; 2004.

[45]Bravata DM, Smith-Spangler C, Sundaram V, Gienger AL, Lin N, Lewis R, Stave CD, Olkin I, Sirard JR. Using pedometers to increase physical activity and improve health: A systematic review. *JAMA.* 2007: 298: 2296-304.

[46]Fletcher GF. How to implement physical activity in primary and secondary prevention. *Circulation.* 1997;96:355–357.

[47]Clark KL. Promoting a healthful lifestyle through exercise. In: Kris-Etherton P, Burns JH. *Cardiovascular Nutrition: Strategies and Tools for Disease Management and Prevention.* Chicago: American Dietetic Association; 1998:127–134.

[48]Evans WJ, Spokas D. *Fitness from 50 forward: A Manual Describing How to Begin and Continue an Exercise Program.* Chicago: American Dietetic Association; 1998.

9

Communication with Diverse Population Groups

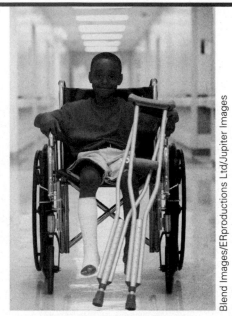

Blend Images/ERproductions Ltd/Jupiter Images

Father, Mother, and Me
Sister and Auntie say
All the people like us are We,
And every one else is They
And They live over the sea,
While We live over the way,
But – would you believe it?
– They look upon We
As only a sort of They!
We eat pork and beef
With cow-horn-handled knives
They who gobble Their rice off a leaf,
Are horrified out of Their lives;
And They who live up a tree,
Feast on grubs and clay,
(Isn't it scandalous?) look upon We
As a simply disgusting They!

—RUDYARD KIPLING, WE AND THEY

Behavioral Objectives

- Define cultural competence.
- Describe the importance of cultural competence for health providers.
- Describe demographic trends in North America.
- Explain four cultural competence models.
- Explain the advantages of using trained medical interpreters.
- Identify useful nutrition intervention strategies for various stages of the life cycle.
- Describe weight bias issues.
- Identify communication essentials for people who have disabilities.

Key Terms

- **Aphasia:** an impairment of any language modality.
- **Cultural Competence:** a set of knowledge and interpersonal skills that allows individuals to increase their understanding and appreciation of cultural differences and similarities.
- **Cultural Groups:** nonexclusive groups that have a set of values in common; an individual may be part of several cultural groups at the same time.
- **Culture:** learned patterns of thinking, feeling, and behaving that are shared by a group of people.
- **Disability:** an umbrella term, covering impairments, activity limitations, and participation restrictions.
- **Discrimination:** behavior that treats people unequally because of their group memberships.
- **Eating Disorders:** a group of conditions characterized by abnormal eating habits.
- **Ethnocentric:** believing a particular cultural view is best.
- **Health Disparities:** population-specific differences in the presence of disease, health outcomes, or access to health care.
- **Interpreter:** a person who transfers the meaning of one spoken language to another one.
- **Prejudice:** a biased opinion, preconception, or attitude about a group or its individual members.
- **Stereotype:** an exaggerated belief, image, or distorted truth about a person or group.
- **Weight Bias:** discriminatory actions and attitudes towards individuals who are overweight.

INTRODUCTION

This chapter is devoted to working on gaining **cultural competence** and reviewing communication essentials for working with selected population groups. Good communication skills are necessary tools for nutrition counselors and educators to provide effective interventions for all population groups. They are the heart of developing relationships with clients and guiding behavior change.[1] Chapter 1 included a review of the impact of **culture** on how we perceive the world, and a culturally sensitive approach to nutrition counseling was presented in Chapter 4. In this chapter, we will examine the meaning of cultural competence and review four cultural competence models. In addition, we will examine special counseling, education, and communication issues related to selected population groups including individuals in various stages of the lifespan and people with disabilities.

GAINING CULTURAL COMPETENCE

Gaining cultural competence in health care means developing attitudes, skills, and levels of awareness enabling the development of culturally appropriate, respectful, and relevant interventions. We will first review some of the reasons for focusing on cultural competence and then examine cultural competence models with emphasis on methodology for increasing cultural sensitivity.

Reasons to Focus on Cultural Competence

- **Demographic Trends.** Population changes in North America require that health care professionals acquire skills in communicating across cultures. Since the 1970s, the United States has been moving toward a cultural plurality, where no single ethnic group is a majority. By 2050 non-Hispanic whites will become less than 50 percent of the total population of the United States.[2] As of 2006, minorities made up more than one third of the population with Hispanics as the largest minority group accounting for 15 percent of the U.S. population.[2] A review of Canadian immigrant patterns also shows substantial demographic changes. In 1980, the leading country of origin for immigrants was Great Britain, whereas by 2006, the five top countries supplying Canadian immigrants included China, India, the Philippines, Pakistan, and the United States.[3]

> A classmate in my nutrition class was pregnant and an immigrant from Africa. One day she brought in clay pellets that had been sent to her from home. She said they tasted good and that all women in her country eat them when they are pregnant to be sure their children are born healthy. This surprised me. I had read about the practice, but I didn't think that an educated woman who was majoring in nutrition would eat clay. I guess I was being ethnocentric.

- **Health Disparities.** Inequalities exist in regard to access to healthcare, delivery of quality healthcare, as well as health outcomes. Substantial health inequalities exist based on age, gender, race, ethnicity, education, income, **disability**, residence, or sexual orientation. African Americans, Hispanic-Americans, American Indians, Asian-Americans, Alaska Natives, and Pacific Islanders, typically experience higher incidences of chronic disease, disability, and mortality as compared to non-Hispanic whites.

- **Quality Care.** An understanding of the components and importance of culture provides a foundation for good health care practices. We need to understand and appreciate the health practices of the cultures of our clients in order to implement meaningful client-centered nutrition interventions. By working towards gaining cultural competence, we can increase client satisfaction and improve health outcomes.[4–6]

- **Legislative and Accreditation Requirements.** Government and professional organizations mandate culturally appropriate services.[7] The United States Department of Health and

Human Services (DHHS) has national standards for culturally and linguistically appropriate services in health care.[8] Also, Title VI of the Civil Rights Act of 1964 reads, in part, "No person in the United States shall, on ground of race, color or national origin, be excluded from participation in, be denied the benefits of, or be subjected to **discrimination** under any program or activity receiving federal financial assistance." In addition, suppliers of health care need to include "reasonable steps to provide services and information in appropriate languages other than English to ensure that persons with limited English proficiency are effectively informed and can effectively benefit."[9]

CULTURAL COMPETENCE MODELS

Cultural competence is a "developmental learning process that requires time, effort, active awareness, practice, and introspection."[10] A number of cultural competence models have been developed to guide the process.[1] Chapter 1 used an Explanatory Model approach to guide a cross-cultural interview. Here, we will review models that cover a total intervention. We will explore four models: Cultural

---

EXERCISE 9.1   Exploring Health Disparities

Go to your state website and do a search on **health disparities**. Record three findings in your journal. Then go to Food Environment Atlas of the Department of Agriculture, Economic Research Service, http://ers.usda.gov/foodatlas/. Find your state and county and record the following in your journal: pounds per capita package sweet snacks, gallons per capita soft drinks, adult diabetes rate, adult obesity rate, low income preschool obesity rate, percent adults meeting activity guidelines, poverty rate, and child poverty rate. Evaluate your findings of the two websites (similarities and differences) and record three comparisons in your journal.

Competence Continuum, The LEARN Model, The ETHNIC Model, and the Campinha-Bacote Model of Cultural Competence. The last one will be covered in more detail to allow greater exploration of components of cultural competence. As you continue in your development of cultural competency skills, you may find components of certain models more useful for visualizing the progress of an intervention and certain models may have more relevance for your client or target audience.

> One of my clients was a forty-two-year-old overweight woman who lived in a group home. She had a lovely personality. Having no articulation ability, she could only communicate by facial expressions and body language. This client's diet instructions were hand-drawn pictures on a page kept on the refrigerator. As she consumed appropriate foods, she would put an x through the picture. A new diet page was posted every day. This client did beautifully with the program, but her overweight sister was not happy. The sister took my client home on weekends and encouraged her to overeat. It wasn't until I included her sister in our counseling sessions that my client was truly successful.

Cultural Competence Continuum

This model provides a series of stages to assess the act of gaining cultural competence starting with a low level of competence, cultural destructiveness, and ending with a high level of cultural proficiency. Individuals may be at different levels of awareness, knowledge, and skills along the cultural competence continuum. See Table 9.1. Uniform movement through stages may not occur for all cultural groups. For example, a person may be at a high level of proficiency when working with obese individuals but at a lower stage for working with people who are deaf.

The LEARN Model

The LEARN Model was developed by Elois Ann Berling and William C. Fowlkes[11] for health care providers to elicit cultural, social, and personal information relevant to a given illness episode. This model helps to reduce communication barriers and entails five steps to guide an intervention. See Table 9.2.

The ETHNIC Model

The ETHNIC Model was developed by Stevin J. Levin, Robert C. Like, and Jan E. Gottleib at the Center for Healthy Families and Cultural Diversity Department of Family Medicine UMDNJ-Robert Wood Johnson Medical School. This model incorporates the client's explanation and beliefs and guides an intervention to a culturally acceptable plan of action. See Table 9.3.

Table 9.1 Cultural Competence Continuum

Stage	Description
Cultural Destructiveness	Attitudes, practices, and policies destructive to other cultures.
Cultural Incapacity	Paternalistic attitude towards the "unfortunates." No capacity to help.
Cultural Blindness	Belief that culture makes no difference. Treat everyone the same. Approaches of the dominant culture are applicable for everyone.
Cultural Precompetence	Weaknesses in serving culturally diverse populations are realized, and there are some attempts to make accommodations.
Cultural Competence	Differences are accepted and respected, self-evaluations are continuous, cultural skills are acquired, and a variety of adaptations are made to better serve culturally diverse populations.
Cultural Proficiency	Engages in activities that add to the knowledge base, conducts research, develops new approaches, publishes, encourages organizational cultural competence, and works in society to improve cultural relations.

Source: Cross T., Bazron R, Dennis K & Isaac M (1989) *Toward a Culturally Competent System of Care.* Volume I. Washington, DC: Georgetown University Child Development Center.

Table 9.2   The LEARN Model

Sequence and Description of Interactions	Example
Listen Active listening is an extremely important skill for a counselor to develop. Some factors related to listening merit emphasis for successful counseling across cultures. You should listen carefully to a client without assumptions or bias and recognize the client as the expert when it comes to information about his or her experience.[12] Not only are you learning, but you are demonstrating to your clients that what they have to say is important to you. Make sure you come to a common understanding of their issues and problems. All of this information will be important when designing intervention strategies.	Request clarification when necessary by saying, "I didn't quite understand that."[13] Listen carefully to how food decisions are made. Probe to find out who does the food preparation and shopping and determine whether an additional person should be included in the next counseling session.
Explain To clarify that your understanding of the issues is accurate, you should explain back to the client your perception of what has been related. The explanation creates an opportunity to clarify any misunderstandings.	"You feel that diarrhea is a hot ailment, and your baby should not be given a hot food like infant formula, but should drink barley water, a cool food. Did I understand you correctly?" Balancing intake of hot and cold foods is believed to help heal among several Caribbean and Asian cultures.
Acknowledge The nutrition counselor should acknowledge the similarities and differences regarding the causes, symptoms, and treatment of the problem.	"Both you and your doctor feel that what your baby drinks will help her feel better. You feel your baby needs a cool food like barley water, and the health care providers at this clinic feel that your baby needs a drink with minerals like Pedialyte to get better."
Recommend The client should be given several options that are culturally sensitive.	An Indian woman who is a vegetarian who wishes to lose weight might be given the following options: "You could start a walking program, reduce the amount of oil or butter used to make lentil dishes, use skim milk for making yogurt, or eat fruits instead of fried snacks."
Negotiate After reviewing the options, the counselor and client should develop a culturally sensitive plan of action. By understanding the powerful influence of the client's culture as well as the equally powerful culture of biomedicine, then the need for compromise and mediation becomes obvious. When the condition is life threatening or the cultural differences are enormous, Kleinman[15] recommends a cultural anthropologist or a respected member of the client's community to aid in the negotiation. The health practitioner should decide what is critical and be willing to compromise on everything else.	Look to your client to select a starting point: "Which of these options do you think would be a good place to start?" After selecting an option, discuss how that option will be implemented.

Campinha-Bacote Cultural Competency Model for Healthcare Professionals

Campinha-Bacote's model provides a bridge between cultures in order to achieve mutual understanding and meet unique needs.[16] This model views cultural competence as a process rather than an end result–"the process in which the healthcare provider continuously strives to achieve the ability to effectively work within the cultural context of a client (individual, family, or community)". Five

Table 9. 3   The ETHNIC Model

Order	Interventions and Possible Questions
Explanation	How do you explain your illness? Why do you believe you have these symptoms? Do other people have your illness? Have you heard or read about this problem in the media?
Treatment	What treatments have you tried? Are there foods or beverages you consume or avoid to stay healthy? Are there foods or beverages you consume or avoid to treat the problem? What treatments do you desire from the practitioner?
Healers	Have you sought any advice from healers or alternative medical sources? How has the advice worked out?
Negotiate	Agree on acceptable options. Do not contradict but incorporate client's beliefs.
Intervention	Choose a viable option. How do you feel about the plan? Are there parts of the plan that you are concerned about?
Collaboration	Collaborate with key supporters, family members, healers, and client.

interdependent constructs include cultural awareness, cultural knowledge, cultural skill, cultural encounters, and cultural desire. Although health care professionals can work on any one of the constructs to improve balance of the others, the pivotal key construct is cultural encounters having the greatest influence over all others. See Table 9.4 and Figure 9.1.

- **Development of Cultural Self-Awareness**
For successful intercultural interventions, you need to develop an awareness of the cultural beliefs and values that influence your conscious and unconscious thoughts and understand that these attributes create a bias for your view of acceptable behavior. Without cultural self-awareness, there is a tendency to be ethnocentric,

Table 9.4   Constructs of the Campinha-Bacote Model of Cultural Competence

Cultural Construct*	Description
Awareness	Health care providers become appreciative of the influence of culture on the development of values, beliefs, values, practices, and problem solving strategies. A basic requirement for cultural awareness is an in-depth exploration of one's own cultural background, including biases and prejudices toward other cultural groups.
Skill	Health care providers learn to perform culturally sensitive assessments and interventions.
Knowledge	Health care professionals develop a sound educational foundation concerning various worldviews in order to understand behaviors including food practices, health customs, and attitudes toward seeking help from health care providers. They also acquire knowledge of physical needs such as common health problems and nutrition issues of different cultures.
Encounters	Providers seek and engage in cross-cultural encounters.
Desire	In order to appear genuine and to be an effective cross-cultural health care provider, there must be a true inner feeling of wanting to engage in the process of becoming culturally competent.

*The mnemonic ASKED can assist nutrition professionals in assessing their level of cultural competence.

Source: Campinha-Bacote J. *The Process of Cultural Competence in the Delivery of Healthcare Services.* 5th ed. Cincinnati, OH: Transcultural C.A.R.E. Associates; 2007.

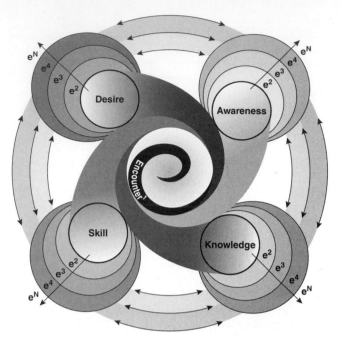

Figure 9.1 The Process of Cultural Competence in the Delivery of Healthcare Services
Source: Campinha-Bacote J. The Process of Cultural Competence in the Delivery of Healthcare Services. http://www.transculturalcare.net; 2010. Reprinted with permission from Transcultural C.A.R.E. Associates.

devaluing alternative cultural practices, blindly imposing your own cultural procedures. In addition, you may miss perceiving opportunities for successful interventions. A nutrition professional needs to approach cross-cultural interactions with a nonjudgmental attitude and a willingness to explore and understand different values, beliefs, and behaviors. One of the best ways to become aware of differences is to immerse yourself in the perceptual world or culture of others as can be done by traveling or working in other countries.

- **Development of Cultural Knowledge**
  Gaining knowledge about our clients' cultures helps us to avoid **stereotypes** and biases as we learn to appreciate the positive characteristics of various cultures. As we grow to understand more about the cultural group of interest, there is a greater likelihood that we will develop culturally effective and relevant programs. Without cultural understanding, there is a risk that the intervention you develop could conflict with common beliefs, values, and customs of the group. For some examples, see Table 9.5.

- **Development of Cultural Skills**
  Working on developing cultural skills takes time and requires technique flexibility. See Table 9.6 for some practical considerations for cross-cultural communication. Skills include ability to use respondent-driven interview questions to effectively conduct a culturally sensitive assessment. The counselor asks simple, open-ended questions to initiate conversation, prompts clients for a better understanding when necessary, but for the most part exerts little control over responses. By showing an unbiased and sincere desire to understand and accept traditional views and practices, it is hoped that clients will not fear criticism or ridicule. Exhibit 9.1 contains examples of open-ended questions to acquire information to direct

---

EXERCISE 9.2   Test Your Hidden Bias

Project Implicit is a collaborative reasearch project between researchers at Harvard University, the University of Virginia, and University of Washington. Their "Implicit Association Tests" (IATs) can be used as a tool to help you begin thinking about your hidden biases, which may translate into behavior under stress, distraction, relaxation, or competition.[17] Go to the Project Implicit website, https://implicit.harvard.edu/implicit/, and take one of the tests. In your journal, record your reaction to the test. Did you agree or disagree with the way the test was conducted? Explain how you feel the experience may help you in your cross-cultural counseling experiences.

Table 9.5 Conflicting Values in Health Program Interventions

- Messages that emphasize eating certain foods to prevent specific diseases may not have much of an effect if ill health is viewed as "God's will."
- Prevention may be viewed as a useless attempt to control fate. Doing good deeds and requesting forgiveness from a spiritual leader, may appear to be the best courses of action for those who believe that illness is a curse for sins.
- Food programs that require that only particular family members eat donated foods may not be well received in cultures where the welfare of the group is placed before the individual.

Source: Schilling B & Brannon E. (1990). *Cross-Cultural Counseling A Guide for Nutrition and Health Counselors.* Washington, D.C: U.S. Government Printing Office.

Table 9.6   Practical Guidelines for Cross-Cultural Communication

Interpersonal Considerations

- Smile, show warmth, and be friendly.
- Attempt to learn and use key words, especially greetings and titles of respect, in languages spoken by populations serviced by your organization.
- Thank clients for trying to communicate in English.
- Suggest that clients choose their own seat (to make comfortable personal space and eye contact possible).
- Articulate clearly; speak in a normal volume. Often people mistakenly raise the volume of their voice when they feel others are having difficulty understanding them.
- Paying attention to children appeals to women of most cultures; however, some believe that accepting a compliment about a child is not appropriate, especially in front of the child.
- When interacting with individuals who have limited English proficiency, always keep in mind that limitations in English proficiency do not reflect their level of intellectual functioning. Limited ability to speak the language of the dominant culture has no bearing on ability to communicate effectively in their language of origin. Clientele may or may not be literate in their language of origin or in English.
- Explain to clients that you have some questions to ask and that there is no intention to offend. Request that they let you know if they prefer not to answer any of the questions.
- If you are not sure how to interpret a particular behavior, Magnus[13] suggests that you should ask for clarification. For example, you could ask, "I notice that you are mostly looking down. Would you tell me what that means for you?"
- Follow your intuition if you believe something you are doing is causing a problem. Magnus[13] suggests that you ask, "There seems to be a problem. Is something I am doing offending you?" When informed of a difficulty, immediately apologize and admit, "I am sorry. I didn't mean to offend you."
- Ask clients to identify their ethnicity. The specific terms used to identify an individual's ethnicity can be a touchy issue. For example, Asian, Oriental, Chinese, and Chinese-American have been used to describe individuals of a similar background, but not all of these terms are acceptable to all individuals. Magnus[13] suggests that to avoid alienation, a counselor should directly inquire about heritage with phrases such as "How do you describe your ethnicity?"

Communicating Information

- Consider using a less direct approach than is common among Americans. Gardenswartz and Rowe[18] suggest some communication approaches that may lower the risk of misunderstanding and hurt feelings:
  - Make observations rather than judgments about behaviors. For example, do not say, "Your dairy intake is low." Instead say, "You eat one dairy food a day and the authorities tell us to eat three."
  - Refrain from using "you." For example, say "People who have a low intake of calcium are at an increased risk for osteoporosis" rather than "You are at an increased risk for osteoporosis."
  - Be positive, saying what you want rather than what you do not want. For example, say, "Use a pencil to fill out the form." rather than "Don't use a pen to complete the form."
- Use visual aids, food models, gestures, and physical prompts during interactions with those who have limited English proficiency.
- If answers are unclear, ask the same question a different way.
- Consider using alternatives to written communications, because word of mouth may be a preferred method of receiving information.
- Write numbers down, just as they would appear in recipes, because spoken numbers are easily confused by those with limited skills in a language.

Source: Bauer, Kathleen. *Gaining Cultural Competence in Community Nutrition*. In: Boyle, Marie and Holben, David. Eds. *Community Nutrition in Action An Entrepreneurial Approach*. Belmont, CA: Wadsworth/Thomson Learning; 2010, pp.543–544.

the flow of conversation for cross-cultural counseling. These questions aid in understanding an illness from a client's perspective. However, each question is not appropriate for every cross-cultural encounter, so counselors must use their judgment to select suitable ones.

You should also be able to work collaboratively with clients to institute culturally acceptable and effective interventions. Speaking your clients' language is especially conducive to successful approaches. However, this is not always possible and you need to know how to work effectively with an **interpreter** to improve outcomes. Conventional helping theories often have gender and cultural limitations and are likely to disregard family issues. However, to

---

EXHIBIT 9.1 Culturally Sensitive Open-Ended Questions to Encourage a Response-Driven Interview

Questions to Understand the View and Treatment of Health Problems

1. What name do you call your problem? What name do you give it?
2. What do you feel may be causing your problem?
3. Why did it start when it did?
4. What does your sickness do to your body?
5. Will you get better soon, or will it take a long time?
6. What do you fear about your sickness?
7. What problems has your sickness caused for you personally? for your family? at work?
8. What kind of treatment will work for your sickness? What results do you expect from treatment?
9. What home remedies are common for this sickness? Have you used them?
10. Are there benefits to having this illness?

Questions to Aid Understanding about Traditional Healers

11. How would a healer treat your illness? Are you using that treatment?

Questions to aid understanding of food habits and to assist in completing a nutritional assessment

12. Can what you eat help cure your sickness? Or make it worse?
13. Do you eat certain foods to keep you healthy? To make you strong?
14. Do you avoid certain foods to prevent sickness?
15. Do you balance eating some foods with other foods?
16. Are there foods you won't eat? Why?
17. How often do you eat your ethnic foods?
18. What kinds of foods have you been eating?
19. Is there anyone else in your family who I should talk to?

Table 9.7 Values and Behaviors of Various Cultural Groups

In order to behave in a culturally competent manner, your attitudes need to convey an understanding and acceptance of diverse values and behaviors as those listed in this table.

- Family is defined differently by different cultures (for example, extended family members, fictive kin, godparents).
- Individuals from culturally diverse backgrounds may desire varying degrees of acculturation into the dominant culture.
- Male-female roles in families may vary significantly among different cultures (for example, family decision making, play activities, and social interactions expected of male and female children).
- Age and lifecycle factors must be considered in interactions with individuals and families (for example, high value placed on the decisions of elders or the role of the eldest male or female in families).
- Meaning or value of medical treatment, health education, and wellness may vary greatly among cultures.
- Religion and other beliefs may influence how individuals and families respond to illnesses, disease, and death.
- Folk and religious beliefs may influence a family's reaction and approach to a child born with a disability or later diagnosed with a disability or special health care needs.
- Customs and beliefs about food, its value, preparation, and use are different from culture to culture.

Source: Adapted from material developed by T D. Goode, National Center for Cultural Competence, Georgetown University Child Development Center.

- **Cultural Encounters**

  By exploring cultures different than your own, you learn about new ways of interpreting reality and develop alternative lenses on which to base your interactions with those who appear different than you. Engaging in cross cultural encounters helps you create a vehicle for developing attitudes congruent with cultural competency, such as appreciation, respect, and understanding of people who have cultural beliefs and behaviors different than your own. However, during your investigations, it would be natural to experience some discomfort as you learn about values and beliefs that conflict with yours.

  When you think about diversity and culture, interesting activities are likely to come to mind. For example, learning about various forms of clothing, architecture, language, special foods, meal patterns, and cooking equipment may be engaging activities. However, cross-cultural

---

provide a culturally sensitive approach, you must explore who should be involved in the intervention process. Spouses, family elders, or extended family members may be key decision makers and have a major impact on the success of the intervention. See Table 9.7 for a list of values and behaviors that may need to be explored in your interventions. Development of cultural skills can also include a much broader perspective by focusing on integrating target groups in planning, implementing, and evaluating interventions. Also, advocating to incorporate cultural sensitivity in all components of organizational structure is necessary for individuals to work effectively.

communication produces special challenges because cultures influence how we view the world and interact with others. The more alike we are to individuals we work with, the more easily communication will flow because we have learned similar communication styles and taboos. Cultural encounters help us to learn the communication styles of other groups and increase the likelihood that communication will flow more smoothly.

- **Cultural Desire**

  Valuing diversity and having the ability to view the world through multiple cultural lenses are the heart of cultural competence. Engaging in the other constructs of the Campinha-Bacote Model of Cultural Competence is likely to increase your cultural desire.

CROSS-CULTURAL NUTRITION COUNSELING ALGORITHM

To understand how a culturally sensitive approach to nutrition counseling interfaces with the motivational nutrition counseling algorithm presented in Figure 4.2, a cross-cultural nutrition counseling algorithm is presented in Figure 9.2. This algorithm incorporates the response-driven interview questions in Exhibit 9.1 and components of the cultural competency models reviewed in this chapter.

> I once counseled a Hindu couple; the wife was about three months' pregnant and was fluent in English. Throughout the interview, the husband responded to every question I asked—no matter how detailed. The wife remained quiet during the whole session. At first I found myself getting angry; however, the wife did not seem to be bothered by this arrangement. After I realized there were obvious cultural behavior patterns that I should respect, the interview went more smoothly for me.

WORKING WITH INTERPRETERS

Due to radical shifts in the U.S. demographic profile, the need for interpreters (spoken) and translators (written) has grown rapidly. According to the 2010 report by the U.S. Census Bureau, about 20% of the population speak a language other than English at home, and fewer than half of this group is fully proficient in English.[19] Medical interpreters, also known as healthcare interpreters, provide language services to patients with limited English proficiency. Such individuals have a good understanding of medical and colloquial terminology in both languages, as well as cultural sensitivity to relay concepts and ideas. See Table 9.8 for guidelines for working with an interpreter. Sign-language interpreters foster communication between people who are deaf and those who can hear.[19]

Too often health care providers resort to using non-professional interpreters, such as friends or relatives of clients or housekeeping staff.[14] This approach has been shown to present numerous problems. Sometimes patients are reluctant or embarrassed to discuss certain problems in front of close relations, or the non-professional interpreter may decide that certain information is irrelevant or unnecessary and does not provide complete information. Such an interpreter may be unfamiliar with medical terminology and unknowingly make mistakes. One study of untrained interpreters showed that 23 to 52 percent of phrases were not communicated correctly; for example, *laxative* was used to describe diarrhea.[14] If a professional medical interpreter is not available on site, there are reliable telephonic interpreter services that most medical centers use. The difficulties of communication across cultures is illustrated in the story of a Hmong child, Lia, with epilepsy:[20]

Lia developed an infection and severely seized ("like something out of The Exorcist") continuously for nearly two hours. Doctors in the local community hospital had a difficult time stopping her seizures, and when they finally did, Lia was unconscious. Because her problems were complex, arrangements were made to transport her to a children's hospital with an intensive care unit. With the help of an interpreter, the situation was explained to Lia's non-English speaking parents. The attending physician charted, "Parents spoken to and understood critical condition." Later investigation revealed that the parents thought their child had to go to another hospital because the doctors at the community hospital were going on vacation.

LIFE SPAN COMMUNICATION AND INTERVENTION ESSENTIALS

Since developmental tasks and learning needs vary according to stage of life, nutrition intervention approaches should be tailored to specific segments of the population. Professional and governmental

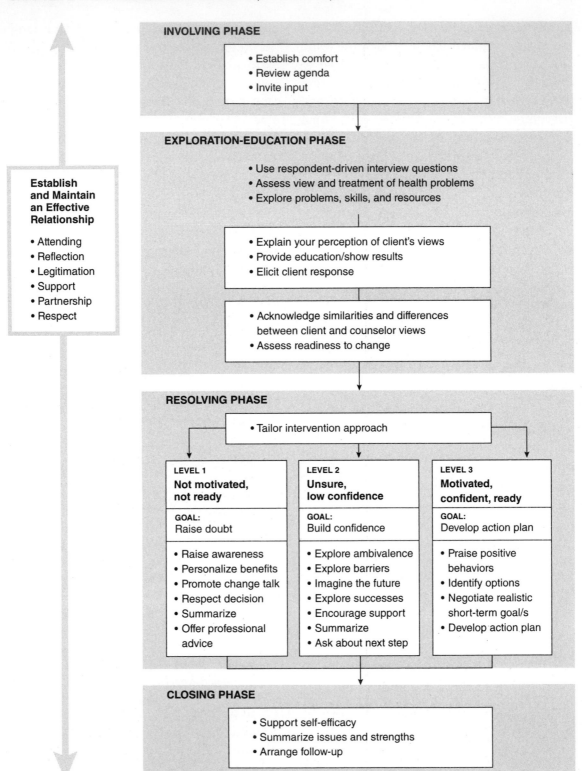

Figure 9.2   Cross-Cultural Nutrition Counseling Algorithm
Source: Adapted from Snetselaar L, Counseling for Change. In: Mahan LK, Escott-Stump S, eds. Krause's Food, Nutrition, & Diet Therapy, 10th ed.

Table 9.8   Guidelines for Working with an Interpreter

- Request an interpreter of the same gender and similar age. (Be sensitive and flexible in your selections since interpreters who are considerably older than a client may receive greater respect.)
- Decide before the meeting what questions will be asked.
- If possible, go over the questions with the interpreter before the meeting. A professional interpreter should be able to assist you in formulating new questions if certain ones are deemed offensive.
- Try to learn a few phrases of the client's language to use at the beginning or the end of the interview.
- Remember that sessions will take extra time. Schedule adequate time.
- Look at and speak directly to the client, not the interpreter.
- Speak clearly in short units of speech. Do not ask more than one question at a time.
- Avoid using slang, similes, metaphors, and idiomatic expressions. For example, do not say, "Do you have your ups and downs?"
- Listen carefully and watch body language for any changes in expression.
- Do not just follow prepared questions, but ask clients to expand upon new issues.
- To avoid misunderstandings, begin some of your sentences with "Did I understand you correctly that . . ." or "Tell me about . . ."
- To check on the client's understanding and the accuracy of the interpretation, ask the client to back-translate important dietary instructions or guidelines. This technique may also open the conversation to questions by the client.
- Be aware that interpreters come to sessions with their own cultural biases and may not completely convey everything that has been said.

Source: Adapted from: Munoz C, Luckmann, J. *Transcultural Communication in Health Care.* Belmont, CA: Delmar Cengage Learning, 2004.

organizations have developed a number of age appropriate education media and materials. Resources can be found at the end of the chapter.

Preschool-Aged Children (2 to 5 Years)

- **Determinants of Food Behavior.** Family, culture, and media have a major impact on young children's eating habits and nutritional health.[21] They learn to enjoy foods that their family and cultural environment provide. Media messages have been shown to be a significant variable influencing nutritional status.[22] Young children watch approximately 20 hours of television per week and are exposed to numerous food commercials advertising unhealthy food selections.[22–24] Interestingly, preschoolers are able to recognize a McDonalds' logo even before they learn to read.[25]

To the consternation of parents, many preschoolers are defined as picky eaters, having food jags (only want to eat certain foods), and elicit a reluctance to taste new or unfamiliar foods. As an infant, a child may have been good eater, but the desire to test independence, a decrease in the rate of growth, and a taste preference for sweet and familiar foods contribute to young children's food issues.[26] Parents should be encouraged to keep introducing new foods

because a child may need to be exposed to a food fifteen or more times before acceptance occurs.[27] Research of Mexican children's preference for the taste of chili peppers indicates that introduction to new foods should occur in a positive atmosphere; for example, threats should be minimized and encouragement provided when a new food is tasted.[28] Children are able to physiologically recognize fullness and should not be forced to eat or clean their plate.[29]

- **Developmental Factors.** Preschoolers enjoy a creative and fanciful cognitive world. They are beginning to think symbolically allowing them to make simplified drawings to depict people, houses, and other familiar objects. Although they have the ability to understand cause and effect, their thinking abilities do not permit them to discern between food advertising and regular television programming.[30]

In this developmental stage, learning is accomplished by exploring the environment rather than passive listening. Preschoolers need to be given opportunities to touch, feel, manipulate, question, compare, and identify objects.[30] They are capable of classifying foods based on color, shape, and function rather than by nutrient content.[31] By observing, modeling, and role playing behaviors of parents, teachers, and other children, they accumulate and process information. Parents need to know the importance of being good role models regarding their own food behavior as well as their attitudes toward food.

- **Nutritional Risks.** The USDA's Healthy Eating Index (HEI) report provides an overall synopsis of food choice quality for various age groups, and the report for 2001 indicates at least 64% of young children need to improve their diets.[32] A HEI evaluation of 2003-04 diets showed that only milk and fruit groups were adequate for young children.[32] Common nutrient deficits in early childhood are iron, zinc, and calcium. Increased intake of soft drinks and juice, displacing milk over the last two decades, has contributed to calcium deficiencies in young children.[33]

Intervention Strategies

- **Involve family and caregivers in interventions.** Because family and caregivers are major role models and control home food environment and food choices, engaging significant others in working to change food behaviors of young children is essential.
- **Provide action-oriented behavior change activities.** Because children learn best through hands-on activities, interventions should be creative and fun, such as, food tastings and food parties. Young children can help design meals, prepare foods, and shop for food. A counselor can also role play these activities with props.
- **Creative food records.** If keeping a diary of food is essential, consider having the child draw pictures of food consumed or putting a line through pictures of foods after they are eaten. Parental food diaries can also be used when more detailed information is needed.
- **Parental advice.** Parents can use a number of methods to improve their children's eating behavior. For example, have children choose between two good choices. Encourage a one bite rule: "You need to take just one bite." Introduce new foods at the beginning of a meal when children are most hungry. Presentation of food should be attractive and colorful.

Minimize using food as a reward. Maintain a positive atmosphere during mealtime. Uphold regular mealtimes for a consistent pattern.

Middle Childhood (Age 6 to 11)

- **Determinants of Food Behavior.** Family, school, and screen media time (about 4.5 hours per day) are major factors related to food behavior for this age group.[34] As children increase in age, their use of various forms of screen viewing (such as television, computer games, DVDs, and the Internet) also increases. Screen time is associated with high calorie snacking and poor food choices; Furthermore, television viewing is associated with a reduction in rate of metabolism that is even lower than sleeping metabolism.[24,35] Popular children's websites are prime marketing sites for candy, cereal, fast-food restaurants, and snacks.[36] In addition, parents' eating behaviors as well as the quality of food available in the home environment are important determinants of children's dietary intake.[37] Finally, the dietary habits of many children are formed in school-based settings. Policy makers looking for solutions to the growing obesity problem have increasingly paid closer attention to school food environments. The USDA Food and Nutrition Service guidelines and many state regulatory agencies have instituted healthier food standards for schools.[38] However, the 10 to 15 hours per academic year spent on nutrition education instruction is not considered adequate by authorities.[39]
- **Developmental Factors.** Middle childhood is marked by major cognitive, social, and physical development. Children in this age period are eager to understand the world in which they live. They acquire knowledge and begin to think causally, and are capable of theorizing why things happen. They are able to understand the function of food and how it influences health and growth. They are likely to accept adult viewpoints about food choices. Middle childhood is also characterized by the desire for autonomy allowing for self-regulatory intervention techniques.

Figure 9.3 Healthy Eating Report Card for Children Aged 2 to 9

Source: From BOYLE/HOLBEN. *Community Nutrition in Action*, 5E. © 2010 Brooks/Cole, a part of Cengage Learning, Inc. Reproduced by permission. www.cengage.com/permissions.

- **Nutritional Risks.** As indicated in Figure 9.3 and according to the USDA's 2005 HEI Report, the diets of children 6 to 11 years of age are significantly worse than children 2 to 5 years old. In general, children in this age group need to increase consumption of whole fruit, whole grains, dark green and orange vegetables, and legumes. They also need to decrease intake of saturated fat, sodium, and extra calories from solid fats and added sugars.[32] As for all developmental age groups, overweight and obesity are major issues.

Intervention Strategies

- **Behavioral interventions.** Several reviews have indicated successful results using behavioral strategies for moderate (26–75 hours), or high intensity (>75 hours) interventions.[40–41] Useful behavioral management techniques included: self-monitoring of diet and physical activity, cue elimination, stimulus control, goal-setting, action-planning, modeling, limit setting, contingency management, positive reinforcement, and cognitive modifications. Another technique, differential attention, has been found helpful. This involves giving positive reinforcement (praise) for desired behavior, such as choosing a nutritious snack, and ignoring undesirable behavior, such as a complaint about television restriction.

- **Multicomponent, family-based interventions.** Family support is necessary as many of the behavioral interventions need to be supported by significant individuals in a child's life. In addition, parenting can influence attitudes, preferences and values regarding food behavior in children.[42] Family-based interventions including diet, physical activity, behavior modification, and family counseling for reducing weight of overweight and obese school-age children have been successful.[39]

- **Limit media screen time.** The American Academy of Pediatrics recommends limiting recreational screen time to no more than two hours a day.[43]

- **Use activity-oriented nutrition interventions.** There are numerous colorful books and games available for promoting nutrition education. See the end of the chapter for resources. For older children in this age group who are technology savvy, handheld devices, Internet programs, and add-ons for mobile phones may be considered.

Adolescence (12 to 19 Years)

- **Determinants of Food Behavior.** Multiple interconnecting factors impact eating patterns and food choices for adolescents. As teens strive for autonomy and gain independence, family influence on food choice tends to decline, and there is greater reliance on school food, snacking, vending machines, fast food, and convenience stores. Almost fifty percent of eighth and tenth-grade students eat three or more snacks a day, and most of these snacks are high in fat, sugar, or sodium.[44] Fast food establishments frequently employ teenagers allowing them to have easy access to reduced price or free tasty low nutrient dense foods. In addition, going out to eat with friends to fast food restaurants is a common social and recreational activity. School food supplies 35 to 40% of high school students' total energy intake.[45] It is hoped that recent changes in federal and state school food guidelines will make a positive impact on adolescents' nutritional status.

Adolescents, having access to large amounts of discretionary income, are the targets of aggressive marketing campaigns.[46] Marketers are aware of teenagers' purchasing power, especially for desserts, snacks, and beverages. As a result, adolescents are surrounded with messages to buy numerous types of unhealthy foods.[47] Additionally, high schools advertise soft drinks, candy, snack foods, and fast food on school buses, free book covers, yearbooks, sport score boards, and even in daily in-class television news programs. Entertainment media screen time accounts for 6.5 hours of a teen's total day leaving little time for physical activity.[34] However, adolescents report they are stressed for time, and desire to sleep longer, so they frequently skip breakfast and believe they are too busy to worry about eating well.[44]

- **Developmental Factors.** During adolescence, dramatic physical, cognitive, and psychosocial changes occur. This is a period in which family values and standards are scrutinized and may be rejected. Adolescents go through three stages of development, which have implications for guiding interventions. Early adolescence is "characterized by respect for adult authority, discomfort with the physical changes of puberty, lack of future time perspective, and concrete, or "black and white," reasoning skills."[48] The best approaches for individuals in this stage include family involvement and educational materials with clear messages. The next stage is "characterized by recurrent challenges to family or parental authority and belief systems, reliance on peers for standards in appearance and behavior, increasing capacity for abstract reasoning, and experimentation in dating and sexual behavior."[48] Problem identification, role-playing and using their abstract thinking capability to evaluate "what if" possibilities are useful interventions. During the last stage of adolescence, there is greater confidence in their own internalized values and "fewer challenges to adult authority; less reliance on peer standards; . . . and increased capacity to solve complex life problems."[49] Interventions addressing the complexity of their health issues and

focusing on the pros and cons of current choices can be helpful approaches during late stage adolescence.

- **Nutritional Risks.** Poor quality diets of adolescents are putting them at risk for cardiovascular disease, cancer, and osteoporosis.[48] An analysis of the HEI report indicates high intakes of saturated fat, total fat, sodium, calories, and soft drinks and inadequate consumption of fruits, vegetables, fiber, and calcium.[32] Overweight, obesity, smoking, disordered body image, low levels of physical activity, and disordered eating are major issues for this age group.[49] The National Eating Disorders Association estimates that approximately 80% of young adolescents are fearful of being fat, and many are struggling with **eating disorders** including anorexia, bulimia, and binge eating.[47]

Intervention Strategies

- **Motivational interviewing (MI).** Although there has been limited use of MI with adolescents, the technique appears promising.[42] This strategy encourages critical thinking skills thought to be important for adolescent interventions.[50] See Chapter 2 for a discussion of this counseling method.
- **Use of behavioral strategies.** Useful behavioral strategies include development of decision-making skills, self-regulation and self-evaluation, personal action plans, and goal setting.[51]
- **Multicomponent, school-based interventions.** School-based interventions incorporating multiple strategies have shown to be successful for this age group.[39] These may include a comprehensive school policy; healthy vending machine, school store, and cafeteria options; training for teachers, administrators and staff; referrals for nutritional problems; integration of food service with classroom education; development of a comprehensive school health curriculum; involvement of community and parents; and scheduled periodic evaluations of the intervention.[44]
- **Consider multivariable outcomes measures.** Using behavioral, psychosocial, and medical endpoints as targets in addition to weight or

as an alternative to weight measures may be advantageous. These could include dietary intake, nutritional status, physical activity levels, self-esteem or body image assessments, blood pressure, blood lipids, or blood glucose concentrations.[39]

- **Possible intervention activities and topics.** Potential strategies include analysis of media and peer-based messages and restaurant menus. Consider role playing scenarios for problem behaviors. Create food demonstrations and taste tests. Adolescents may be responsive to messages addressing increased energy, athletic performance, and physical appearance.[51] Many teens are technology savvy and may respond to interventions using Facebook, instant messaging, taking pictures of their foods with a cell phone or making minivideos of food experiences to be reviewed in a counseling or group session. In addition, adolescents can be directed to helpful teen friendly educational websites.

- **Use a collaborative approach.** For treatment of eating disorders, nutritionists should collaborate with psychologists and medical specialists.

Older Adults

Issues concerning older adults vary with age, and the U.S. Census divided older adults into three categories: ages 65 to 74 (young-old), ages 75 to 84 (old), and ages 85 and older (oldest-old). A dramatic increase in numbers of all three categories is expected in the near future with the fastest growth in the oldest-old category.[52] After WWII, between 1946 and 1964, a large number of "baby boomers" were born and they began to reach the young-old category in 2011. See Figure 9.4 for projected increases over the next few decades. In 2006, the older generation comprised approximately 10 percent of the U.S. population, but by 2030 that number will climb to 20 percent.[53] This segment of the population will use many of the nation's health care resources. There will be an increasing need for nutritionists who have expertise in working with older adults.

- **Determinants of Food Behavior.** A variety of physical, social, economic, cultural, and psychological factors affect food behavior of older Americans.[54] Approximately 36% of older adults are considered to be individuals who are low income or living below poverty.[52]

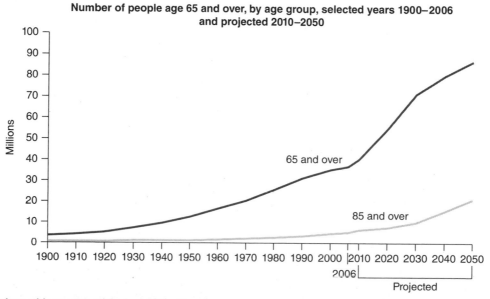

Number of people age 65 and over, by age group, selected years 1900–2006 and projected 2010–2050

Figure 9.4 Projected Increase of Aging Baby Boomers
Source: US Census Bureau, Decennial Census Data and Population Projections. Available online at www.agingstats.gov/ Accessed June 23, 2010.

This group faces the challenge of providing adequate quantity and quality of food. One third of older adults live alone.[55] Loss of spouse and other close relationships may lead to loneliness, depression, and social isolation, thus, affecting appetite and consumption of nutrient-dense foods.[56] Additionally, individuals may shy away from social occasions due to hearing loss. A loss of taste buds, reducing the ability to taste, decline in smell, dentures or other oral difficulties may also reduce desire to eat.[21] Individuals with chronic diseases often take multiple medications which may affect appetite and restrict meal times. Older adults are more likely to suffer from disabilities. Arthritis is the most frequent cause of disability among all older adult categories.[57] Arthritis can reduce mobility and severely impact buying, preparing, and consuming food.

- **Developmental Factors.** Even though life expectancy in the United States is 78 years, the average healthy life span is only 70 years.[58] In all likelihood, nutritional factors play a role in the development of disease and disabilities that plague the final years. Eighty-seven percent live with diabetes, hypertension, abnormal blood lipid profiles, or a combination of these chronic disorders.[59] After age 60, the incidence of obesity declines, but obesity is a greater problem for present day older adults than past generations.[60] For the old and the oldest-old groups, extreme thinness becomes a new concern.

A profile of older adults, those born before 1945, indicates that they are generally concerned about improving their diets and attribute health to a moderate diet, daily activity, and not smoking. They shop frugally, know how to prepare and store food, but still enjoy eating out frequently. Older adults are generally not concerned with body image. They are not likely to act on a chronic health problem until a definite diagnosis is made. Before retirement, their approach to work was "live to work", and they have had to learn multiple new technologies. As a group, they are better educated than the past generation of older adults.[61] Although most are retired, belief in the importance of work and the need for additional income have encouraged some to seek part-time employment.

The aging process reduces physiological processes on a number of levels. Changes in gastrointestinal function can reduce nutrient absorption and utilization. For example, lower stomach acidity reduces vitamin B12 and calcium absorption. Vitamin D is also a concern because of decreased absorption and a reduced ability of the kidneys to convert it into the active form. Medications and specific diseases, such as kidney disease, can exacerbate these effects. Older adults especially need to choose nutrient-dense foods. Because resting energy expenditure declines and there is often a reduction in activity level, older adults need fewer calories. However, the need for nutrients remains the same or even increases.[62] In particular, the need for vitamin D and calcium increases with age and older adults are encouraged to take a synthetic form of vitamin B12, which can be found in fortified foods and supplements.

- **Nutritional Risks.** According to the HEI report, 83% of older adults do not have a good quality diet.[63] Older Americans need to increase their intake of whole grains, dark green and orange vegetables, legumes, and dairy foods. In addition, they need to consume fewer foods high in saturated fats, sugars, and sodium.[64] As a result, their intake of zinc, iron, folate, and antioxidants are often inadequate.[64] Another concern, especially for the oldest-old is dehydration due to reduced kidney function, inadequate thirst regulation, and possible side effects from medications.[62]

Intervention Strategies
- **If food assistance is needed, make a referral.** Various food assistance programs for older adults have been shown to benefit the nutritional health and interest in healthy foods among participants.[65–66] See Table 9.9.
- **Encourage social interactions.** Explore your clients' social life and encourage daily interactions. This increases morale, health status, and

Table 9.9 National Programs Promoting Better Nutrition among Older Adults

Older Americans Act (also known as Elderly Nutrition Program)	Congregate meals served at least five days a week in churches, schools, senior centers, or other facilities. Transportation, shopping assistance, and home meals available.
Supplemental Nutrition Assistance Program (SNAP)	Provides coupons or electronic benefits transfers (EBT) to low-income individuals including the elderly to buy food.
SNAP Nutrition Education	Provides information about making healthful food choices.
Meals on Wheels Association of America	Home delivered meals are available to elderly individuals.
Senior Farmer's Market Nutrition Program (SFMNP)	This USDA program provides coupons for low-income seniors to buy eligible foods at farmers' markets, roadside stands, and community-supported agricultural programs.
Adult Day Care Food Program	Meals and snacks are provided to participating day care programs.

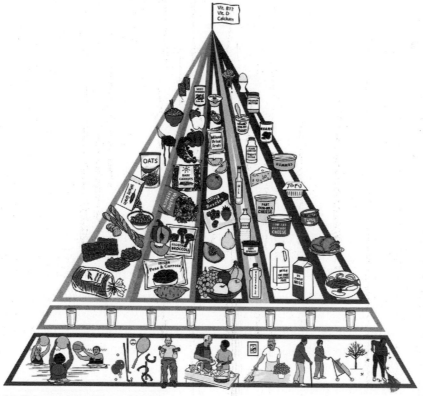

Figure 9.5 Modified MyPyramid for Older Adults
Source: From BOYLE/HOLBEN. *Community Nutrition in Action*, 5E. © 2010 Brooks/Cole, a part of Cengage Learning, Inc. Reproduced by permission. www.cengage.com/permissions.

food intake. The congregate food program addresses the need for social support.

- **Use Modified MyPyramid for Older Adults.** As shown in Figure 9.5, this MyPyramid shows pictures of good food choices for each food group category for older adults. The foundation of the pyramid is physical activity

and water. The flag on the top emphasizes the need to be sure to receive adequate calcium, vitamin D, and vitamin B12 either through diet or supplements. The official MyPyramid website, www.MyPyramid.gov, must be used to determine the amounts of food needed from each category for an individual older adult.

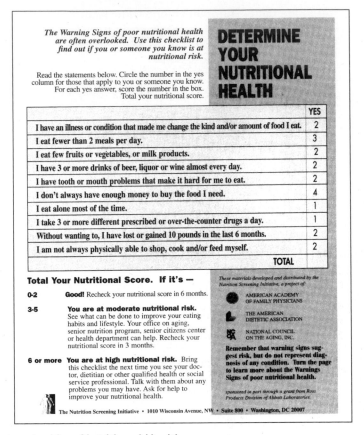

Figure 9.6 Checklist to Determine Your Nutritional Health
Source: Reprinted from the Nutrition Screening Initiative, a Project of the American Academy of Family Physicians, The American Dietetic Association, and the National Council on Aging, Inc., and funded in part by a grant from Ross Laboratories, a division of Abbott Laboratories.

- **Use Nutrition Screening Initiative checklist.** The American Dietetic Association, the American Academy of Family Physicians, and the National Council on Aging collaborated to create a 10-question self-assessment checklist to identify non-institutionalized older persons at risk for low nutrient intake and health problems. See Figure 9.6.
- **Use Mini Nutritional Assessment MNA®.** This MNA is a simplified validated assessment tool which can be administered by a nutritionist relatively quickly.[67] See Figure 9.7.
- **Suggestions for effective interventions.** Nutrition education messages should address physical, emotional, and social needs of older adults. Keep message content clear and direct with practical dietary applications, such as, "eat green leafy vegetables every day." Use visuals and nutrition education materials which

are clearly visible and easy to read. Engage audiences in discussion, respecting older adults' knowledge and prior experiences. Be a good listener as well as an effective facilitator of discussion. Use action-oriented activities to stimulate older adults' attention and motivation for learning.

EXERCISE 9.3 Analyze Marketing for Each Age Group

Search for a newspaper or magazine advertisement promoting a food product or targeting a body image for each of the four life span categories: preschool, middle childhood, adolescence, and older adults. Analyze what stereotypes or developmental issues are being addressed in the advertisements. In your journal, write your impressions of the impact of each advertisement on the targeted age group.

Mini Nutritional Assessment
MNA®

Last name:	First name:	Sex:	Date:

Age:	Weight, kg:	Height, cm:	I.D. Number:

Complete the screen by filling in the boxes with the appropriate numbers.
Add the numbers for the screen. If the score is 11 or less, continue with the assessment to gain a Malnutrition Indicator Score.

Screening

A Has food intake declined over the past 3 months due to loss of appetite,
digestive problems, chewing or swallowing difficulties?
0 = severe loss of appetite
1 = moderate loss of appetite
2 = no loss of appetite ☐

B Weight loss during the last 3 months
0 = weight loss greater than 3 kg (6.6 lbs)
1 = does not know
2 = weight loss between 1 and 3 kg (2.2 and 6.6 lbs)
3 = no weight loss ☐

C Mobility
0 = bed or chair bound
1 = able to get out of bed/chair but does not go out
2 = goes out ☐

D Has suffered psychological stress or acute disease
in the past 3 months
0 = yes   2 = no ☐

E Neuropsychological problems
0 = severe dementia or depression
1 = mild dementia
2 = no psychological problems ☐

F Body Mass Index (BMI) (weight in kg) / (height in m^2)
0 = BMI less than 19
1 = BMI 19 to less than 21
2 = BMI 21 to less than 23
3 = BMI 23 or greater ☐

Screening score (subtotal max. 14 points) ☐ ☐
12 points or greater   Normal – not at risk – no need to complete assessment
11 points or below   Possible malnutrition – continue assessment

Assessment

G Lives independently (not in a nursing home or hospital)
0 = no   1 = yes ☐

H Takes more than 3 prescription drugs per day
0 = yes   1 = no ☐

I Pressure sores or skin ulcers
0 = yes   1 = no ☐

Ref. Vellas B, Villars H, Abellan G, et al. Overview of the MNA® - Its History and Challenges. J Nut Health
Aging 2006;10:456-465.
Rubenstein LZ, Harker JO, Salva A, Guigoz Y, Vellas B. Screening for Undernutrition in Geriatric
Practice: Developing the Short-Form Mini Nutritional Assessment (MNA-SF). J Geront 2001;56A:
M366-377.
Guigoz Y. The Mini-Nutritional Assessment (MNA®) Review of the Literature - What does it tell us?
J Nutr Health Aging 2006; 10:466-487.

© Nestlé, 1994, Revision 2006. N67200 12/99 10M

For more information : www.mna-elderly.com

J How many full meals does the patient eat daily?
0 = 1 meal
1 = 2 meals
2 = 3 meals ☐

K Selected consumption markers for protein intake
• At least one serving of dairy products
(milk, cheese, yogurt) per day   yes ☐ no ☐
• Two or more servings of legumes
or eggs per week   yes ☐ no ☐
• Meat, fish or poultry every day   yes ☐ no ☐
0.0 = if 0 or 1 yes
0.5 = if 2 yes
1.0 = if 3 yes ☐.☐

L Consumes two or more servings
of fruits or vegetables per day?
0 = no   1 = yes ☐

M How much fluid (water, juice, coffee, tea, milk...) is consumed per day?
0.0 = less than 3 cups
0.5 = 3 to 5 cups
1.0 = more than 5 cups ☐.☐

N Mode of feeding
0 = unable to eat without assistance
1 = self-fed with some difficulty
2 = self-fed without any problem ☐

O Self view of nutritional status
0 = views self as being malnourished
1 = is uncertain of nutritional state
2 = views self as having no nutritional problem ☐

P In comparison with other people of the same age,
how does the patient consider his/her health status?
0.0 = not as good
0.5 = does not know
1.0 = as good
2.0 = better ☐.☐

Q Mid-arm circumference (MAC) in cm
0.0 = MAC less than 21
0.5 = MAC 21 to 22
1.0 = MAC 22 or greater ☐.☐

R Calf circumference (CC) in cm
0 = CC less than 31   1 = CC 31 or greater ☐

Assessment (max. 16 points) ☐ ☐.☐

Screening score ☐ ☐

Total Assessment (max. 30 points) ☐ ☐.☐

Malnutrition Indicator Score

17 to 23.5 points	At risk of malnutrition	☐
Less than 17 points	Malnourished	☐

Figure 9.7 Mini Nutritional Assessment

Source: ®Société des Produits Nestlé S.A., Vevey, Switzerland, Trademark Owners. © Nestlé, 1994, Revision 2009.
N67200 12/99 10M. Vellas B, Villars H, Abellan G, et al. Overview of the MNA®—Its History and Challenges. *J Nutr Health
Aging* 2006;10:456–465. Rubenstein LZ, Harker JO, Salva A, Guigoz Y, Vellas B. Screening for Undernutrition in Geriatric
Practice: Developing the Short-Form Mini Nutritional Assessment (MNA-SF). J Geront 2001;56A: M366–377. Guigoz Y.
The Mini-Nutritional Assessment (MNA®) Review of the Literature—What does it tell us? *J Nutr Health Aging* 2006;
10:466–487.

EXERCISE 9.4 Lifespan Interview

Interview an individual or parent of a young child who is from one of the four life span categories: preschool, middle childhood, adolescence, or older adults. Use the Typical Day Strategy found in Chapter 4 to gather information for a 24-hour recall. Record food intake information on Lifestyle Management Form 5.3 in Appendix D. Use the MyPyramid website to complete a computerized assessment of the 24-hour recall. Print out both the nutrient analysis and food group evaluation. In your journal, identify the date of the interview, the name and age of the interviewee. Summarize your findings. How did your findings relate to your readings?

EXERCISE 9.5 Counselor Interview

Interview an individual who counsels one of the four lifespan age groups reviewed in this chapter. In your journal, provide the following information:

- Identify the date of the interview, name, and address of the organization where the interviewee works.
- Job description of the interviewee.
- What special skills does a counselor need to possess in order to work effectively with this age group?
- Are there counseling techniques this counselor finds especially effective when working with this age group?
- What advice can this counselor give regarding involvement with this age group?
- Report two things you learned from this interview and explain how this will help you in your own counseling activities.

EATING DISORDERS

There are three main categories of eating disorders–anorexia nervosa, bulimia nervosa. and eating disorders not otherwise specified (EDNOS), including binge eating disorder.[47] People with anorexia nervosa strive for extreme thinness, for example, 85% below normal weight. They have a pronounced fear of becoming fat and exhibit endocrine difficulties, loss of menstruation in females, and lack of sexual potency in males. Individuals with bulimia nervosa also fear becoming overweight, but they have an overwhelming urge to consume large quantities of food. In response, they perform inappropriate compensatory behaviors including vomiting; excessive exercise; severe calorie restriction, or misuse of laxatives, enemas, or diuretics. EDNOS may be described as a partial syndrome because symptoms vary and do not entirely match those of anorexia nervosa or bulimia nervosa. This category includes binge eating disorder in which individuals binge eat but do not use compensatory behaviors such as vomiting.[68]

An interdisciplinary team is needed to treat the emotional, physical, and interpersonal issues that characterize eating disorders.[68] Nutrition professionals are an essential component of this team.[47] The American Dietetic Association advocates for registered dietitians to collaborate with other professions during assessment and treatment across the continuum of care. At the present time, there are no best dietetic practice models for treatment of eating disorders, but theory-based interventions hold the best promise for success.[47] For nutrition professionals who desire to work with this population group, specialized training is needed to understand the complexities of treatment.[69]

As nutrition educators, we should be cognizant that nutrition messages, especially from parents, emphasizing weight, rather than health, may influence the development of an eating disorder.[70] In addition, dieting may be the beginning of a struggle with eating disorders.[71] In their position statement, the American Dietetic Association supports a total diet approach to reduce risk of developing an eating disorder.[47,72]

WEIGHT BIAS

Approximately two-thirds of Americans are overweight or obese. Nutrition professionals need to learn communication essentials and treatment options for working with individuals who have weight issues. An important first step in developing effective communication skills is investigating your attitudes and possible bias towards this population group. See Table 9.10 for a list of questions to begin the process. Studies indicate that dietetic students and dietitians hold negative

Table 9.10 Self Assessment Questions to Evaluate Attitudes towards Overweight and Obese Individuals

- Do I make assumptions about a person's character, intelligence, health status, or lifestyle behaviors based only on body weight?
- Am I comfortable working with patients of all sizes?
- What kind of feedback do I give obese patients?
- Am I sensitive to the needs and concerns of obese patients?
- What are common stereotypes about obese people? Do I believe these to be true or false? What are my reasons for my beliefs?

Source: Puhl RM. *Treating Obese Patients: The Importance of Improving Provider-Patient Interaction.* 05/27/2010 Medscape Public Health & Prevention. Available at: http://www.medscape.com/viewarticle/722041. Accessed July 26, 2010.

attitudes and beliefs, similar to other health professionals including physicians, medical students, nurses, psychologists, physical education instructors, and even health professionals who specialize in obesity.[73–79] Stereotypes within Western cultures stigmatize the overweight and obese as lazy, ugly, unattractive, unintelligent, dishonest, and unsuccessful. These negative attitudes and beliefs have serious implications for nutrition counseling and education interventions. Clinical treatment and health outcomes for overweight and obese individuals treated by biased health professionals have been compromised.[80] Individuals who feel stigmatized by **weight bias** are more likely to experience psychological issues including depression, lower self-esteem, anxiety, body dissatisfaction, and suicide.[81] As a result, they are less likely to feel motivated to make healthy lifestyle behavior changes and may be more likely to eat poor quality food and shun physical activity.[82,83]

Intervention Essentials

- **Use sensitive language.** Using hurtful language to describe your client's condition can set an intervention off to a bad start. Research indicates the terms "weight", "excess weight", or "body mass index" are viewed more favorably than "large size", "weight problem", or "unhealthy body weight."[84] If you are unclear about the proper terminology, ask your clients for their preference.

- **Incorporate motivational interviewing approaches.**[85] As discussed in Chapters 2, 3, and 4, use open-ended, non-judgmental questions such as the following:

  - How ready do you feel about changing your eating patterns and lifestyle behaviors?
  - How is your current weight affecting your life at this time?
  - What have you done in the past to change your eating?
  - What strategies have worked for you previously?
  - On a scale from 1-10 (with 1 being not ready to change and 10 ready right now), what number would represent your readiness to make changes in your eating patterns?[82]

- **Provide bias free care.** Too often clients feel shame, experience **prejudice**, and believe they are stigmatized by their interactions with health care professionals. Overweight and obesity are the result of a number of behavioral, environmental, physical, genetic, social, and public policy factors. Placing blame on the individual is not fair and not conducive to a successful intervention. Your approach, body language, and expressions will reflect negative bias and will not encourage a successful outcome. No one wants to work with a disapproving health care practitioner.

- **Set specific, realistic, and measureable goals.** Goals do not always need to be a number on a scale. Weight change should be viewed as a long term outcome. As discussed in Chapter 5, short term measureable behavioral goals, such as eating breakfast foods or engaging in physical activity are more likely to lead to successful interventions. In addition, better health outcomes can be measured in addition to weight or instead of weight, such as, waist circumference, blood pressure, blood sugar control, or cholesterol readings.

- **Discuss benefits of modest weight loss.** Health benefits are likely to occur with even 5% or 10% reduction in weight.[82] This is one reason to focus on additional parameters for assessing an intervention rather than weight.

A healthier lifestyle can provide observable outcomes that can be motivational to the client.

- **Update skills.** There are a number of professional publications, continuing education programs, and website resources for professionals.

INDIVIDUALS WITH DISABILITIES

After signing the Americans for Disabilities Act in the White House Rose Garden in 1990, President George Bush handed the pen to Harold Wilke, a minister, who deftly accepted it with his toes.[86] Reverend Wilke, who is armless, offered the following words: "From ancient times to today we celebrate the breaking of the chains holding your people in bondage . . . new access to the Promised Land of work, play and service."[87] Since the passage of the Disabilities Act, policies have been shifting to an emphasis on inclusion, independence, and empowerment for people with disabilities. For example, in 2010 the American Dietetic Association issued a position statement regarding providing nutrition services for people with developmental disabilities and special health care needs.[59]

There is a high probability that you will encounter people with disabilities in your nutrition practice because, according to the U.S. Census Bureau, approximately one in five Americans has a disability.[88] Nutrition professionals have expertise that is relevant for this population group. People with disabilities may have an increased risk of certain conditions, such as weight gain or opportunistic infections, due to decreased mobility or side effects of medications. In order to provide the most accessible assistance, consideration needs to be given to arranging the physical environment to allow safe and free movement. Also, you should explore availability of resources, such as speech augmentation devices.

In order to provide engaging and effective interventions, nutrition professionals need to be familiar with communication essentials when working with individuals who have disabilities.[89] Finding appropriate terminology to describe disability conditions or people who have the disability can be a challenge if you have limited experience with the conditions. You do not want to use terminology that is demeaning, and you want your interactions to be respectful and uplifting. As Eileen Quann stated in her book, *By His Side Life and Love after Stroke,* labels should be used to describe a condition but not define the person.[90] She grew to understand this after her husband's catastrophic stroke left him with a variety of physical and communication limitations. This emphasizes the importance of cultural encounters, one of the constructs of Campinha-Bacote Cultural Competency Model. Mary J. Yerkes of the Illness-Disability-Healthcare-Caregiver Ministry Network encourages individuals not to hold back from interactions with people who have disabilities. She states that response of others to their disability is one of the biggest hindrances they face in the workplace.[92]

Realize that words have power. In general, you should use person-first language putting emphasis on the individual rather than the condition.[92] Referring to an individual as a diabetic, a paraplegic, or a disabled person defines the person by the condition or disability. This is dehumanizing. People do not want to be defined by any single characteristic including their disability. However, there are exceptions to this rule. There are those who feel deafness and autism are traits and not conditions that need a cure. For those who have this frame of mind, using the terms deaf people and autistics is acceptable. If in doubt, ask what terminology your clients prefer.[93,94] Table 9.11 provides a guide for appropriate behavior and language to use when interacting with people who

Table 9.11   Disabilities Etiquette 101

Disability	What You Need to Know	Interacting With People With Disabilities	Talking To or Writing About People With Disabilities
People who use wheelchairs or have mobility impairments	People with mobility impairments have varying abilities. Some can get out of their wheelchair, walk for short distances, and use their arms and hands. Others "look fine" but experience ambulatory difficulties when their symptoms flare or they grow fatigued.	• Don't push, touch, or lean on the person's wheelchair. • Consider physical obstacles (curbs, stairs, hills) when giving directions. • Keep halls, corridors, and aisles clear. • Ask before you help. • Offer to shake hands, usually possible for those with limited hand use or artificial limb. • Don't make assumptions. • Keep floors dry and slip free. Use rubber mats to prevent falls. • Position yourself at eye level when talking with a person in a wheelchair. • Don't grab someone's arm – even to help. Some people with mobility impairments use their arms for balance. • Position computers, telephones, and equipment within a wheelchair user's reach. • Provide assistive or adaptive devices, such as mouth sticks, head wands, oversized trackball mouse, adaptive keyboards, voice-recognition software, or eye-tracking devices.	**Outdated language:** • Handicapped • Crippled • Lame • Confined to a wheelchair • Wheelchair bound • The disabled **Current language:** • Wheelchair user • Person who uses a wheelchair • Person who walks with crutches • Person with limited mobility • Person with disabilities
People who are blind, visually impaired, or partially sighted	People with visual impairments are generally able to live independently, travel, maintain a career, read and write, have an active social life, and more. Not all people who are visually impaired use canes or guide dogs. Some use auditory or tactile cues or echolocation to navigate their environment.	• Identify yourself and others with you. • Never touch a person's cane or guide dog. • When walking alongside someone with a visual impairment, note obstacles, such as stairs, revolving doors, hanging plants, and so forth. • Describe the location of objects. (There is a desk four feet in front of you at two o'clock.) • Excuse yourself before leaving a person who is blind. Leave him near a desk, chair, or other landmark. • Offer to read written information, such as menus, instructions, or agreements. • Provide magnification devices or writing guides for computer screens. • Use accessibility guidelines when designing your website. • Consider speech-recognition software, smartphones, and low-vision, adaptive devices for people with visual impairments.	**Outdated language:** • The blind • Afflicted **Current language:** • Person who is blind • Person who is visually impaired • Person with low vision **In general:** • Understand that "visually impaired" is the generic term to refer to all types of vision loss. Avoid other generic labels. • Contact the National Federation of the Blind (www.nfb.org) for more information.
People who are deaf or hard of hearing	People who are deaf or hard of hearing have a range of communication	• If appropriate, use a qualified sign-language interpreter for complex exchanges of information, such as a job interview.	**Outdated language:** • Deaf mute • Deaf and dumb

(continued)

Table 9.11   Disabilities Etiquette 101 *(continued)*

Disability	What You Need to Know	Interacting With People With Disabilities	Talking To or Writing About People With Disabilities
	preferences and styles. Not everyone who is deaf or hard of hearing uses American Sign Language (ASL). ASL is a visual language that is completely different from English. If ASL is a deaf person's first language, lip reading can be difficult. However, people who are hard of hearing or late-deafened adults communicate in English and often use amplification or assistive devices, along with lip reading, to communicate.	• Speak directly to the person who is deaf, not the interpreter. • Look directly at the person when speaking. Use simple, easy-to-understand sentences. • Avoid smoking, chewing gum, or obscuring your mouth. • Speak clearly. Some people who are hard of hearing watch people's lips as they speak. • Use meaningful facial expressions and gestures. • Gain the person's attention before speaking. Gently wave your hand or tap the person on the shoulder or arm. • Rephrase, rather than repeat, words, phrases, or sentences the person doesn't understand. • People who are deaf or hard of hearing make and receive telephone calls with a TTY (a teletypewriter). If you don't have a TTY, dial 711 to reach the national telecommunications relay service. They can facilitate a telephone call between you and an individual who uses a TTY. • When working in a group, ask people who are deaf or hard of hearing how they prefer to communicate (sign-language interpreter, read lips, write back and forth, and so forth).	**Current language:** • Person who is deaf/ profoundly deaf (no hearing capability) • Person who is hearing impaired (some hearing capability) • Person who is prelingually (deaf at birth) deaf • Person who is post-lingually (deaf-after-birth) deaf **In general:** • Understand that "hearing impaired" and "hearing loss" are generic terms sometimes used to refer to all degrees of hearing loss. However, some people object to the terms and prefer terms such as "deaf" or "hard of hearing." • Contact the National Association of the Deaf (www.nad.org) for more information.
People with speech disabilities	A person who is hearing impaired, who has had a stroke, or has cerebral palsy may have a speech impairment or disability. Some choose to communicate in sign language or writing, while others use their voice or use assistive technology.	• Don't assume a person with a speech disability has a cognitive impairment. • Try to find a quiet environment in which to communicate. • Give the person your complete attention. Never interrupt or pretend to understand when you do not. • Be patient. Never finish a person's sentences for him. • When possible, ask questions that require short answers. • Repeat for verification if you are not sure you understand. • If repeated attempts to understand the person fail, find another method to communicate. For example, ask him to write down what he is saying. • If you have difficulty understanding someone on the telephone, use a speech-to-speech relay service. • If a person uses a communication device, make sure it is within easy reach. • If a person uses an interpreter or attendant, look directly at the person who is speaking, not the attendant.	**Outdated language:** • Deaf and dumb • Dumb • A mute **Current language:** • Person with a speech impediment • Person with a speech disability • Person who is unable to speak • Person who uses synthetic speech **In general:** • Avoid negative attitudes and connotations. • Never tease or laugh at a person with a speech impairment.

(continued)

Table 9.11   Disabilities Etiquette 101 *(continued)*

Disability	What You Need to Know	Interacting With People With Disabilities	Talking To or Writing About People With Disabilities
People with invisible (hidden) disabilities	According to the U.S. Census Bureau, 24 million people in the United States have a severe disability, yet separate research from the University of California reports that only 6.8 million people used a visible assistive device. Thus, a disability cannot be determined solely on whether a person uses visible assistive equipment. If a person makes a request or acts in a way that seems strange to you, such as standing during a meeting while others are sitting, understand that the behavior may be disability related. This person may be in pain, fatigued from a condition like rheumatoid arthritis, lupus, or multiple sclerosis, or may be feeling the effects of medication. Medications taken for conditions such as these are potent and often have undesirable side effects.	• Realize physical appearances can be deceiving. It is possible to "look good" but still have a serious illness. • Understand pain and fatigue, common symptoms with invisible disabilities, may limit a person's ability to walk, sit, or stand for long periods. • Recognize people with hidden disabilities may manage their condition through medication and self-management (limiting stress, alternating demanding activities with periods of rest, self-pacing). Good self-management may prevent disease progression. • Understand that simple tasks, such as shaking hands, pouring coffee, and walking up and down steps, may be painful for a person with an invisible disability. Be sensitive and respond positively to requests for help. • Work with the individual to modify tasks. She is an "expert" in what works and what doesn't. • Recognize physical symptoms and limitations may change based on fluctuations in the disease process. • Understand changes in medication often result in changes in health. • Understand someone with a hidden disability may be physically unable to participate in social activities and events, such as dancing, golfing, or other activities. • Know that people with invisible disabilities often require more rest, which makes late nights difficult. • Know that people with invisible disabilities may require special accommodations under the Americans With Disabilities Act, such as limited travel, flexible work hours, workstation modification or placement, or telecommuting. Contrary to popular opinion, most accommodations are not expensive.	**Outdated language:** • The disabled • Deformed • Victim • "Suffers with . . . " • "Overcame" his disability • Admits she has a disability **Current language:** • Person who has multiple sclerosis (or muscular dystrophy, rheumatoid arthritis, cerebral palsy, and so forth) • Person with a disability • Person with invisible chronic illness • Successful, productive • Says she has a disability **In general:** • Avoid condescending euphemisms, such as "physically inconvenienced" or "physically challenged." Instead, say "woman with rheumatoid arthritis" or "man with multiple sclerosis." • Avoid saying, "But you look so good." Although meant as a compliment, it implies, "If you had a real disability, it would show."

In General
- Avoid negative, disempowering words like "victim." Instead, use empowering, "people first" language.
- Don't use trendy terms, such as "differently abled."
- Omit stereotypes. People with disabilities are not "brave," "courageous," or "heroic" for working, using public transportation, or traveling to an event.
- Avoid pity.
- Don't be embarrassed if you happen to use a common expression, such as "See you later" or "Did you hear about this?" that seem to relate to a person's disability.
- Contact The National Organization on Disability (www.nod.org) for more information.

Source: Reprinted with permission, copyright December 2007, ASAE & The Center, Washington, DC. http://www.asaecenter.org Contributed by Mary J. Yerkes, communications manager and staff writer for NACHA—The Electronic Payments Association.

have disabilities. The following sections highlight selected factors regarding interactions with people who have disabilities.

Mobility Impairment

Often people with mobility impairment have assistance devices such as wheel chairs or artificial limbs. These apparatus should be treated as if they are part of an individual's body. Leaning on a wheel chair or resting your feet on the foot rest is not appropriate. When conversing with a person who is at a much lower level, be sure to place yourself at eye level. Having to keep looking up hinders communication and can cause neck difficulties. Any act of assistance should be given only after a person has indicated that the help would be appreciated. Not only is this respectful, but your aid may in fact hinder their balance. If shaking hands is a standard practice in the setting in which you are working, do not hesitate to shake hands with an under-developed or artificial limb.

Visual Impairment

There are various degrees of visual impairment. As people age, often their ability to see boundaries is reduced. For example, seeing a white plate on a white table cloth or the end of white step and the beginning of a white wall may be a challenge. Make sure there are clear contrasts to define boundaries when working with people who have reduced visual acuity. Be sure to identify yourself when meeting someone who is visually impaired. Sometimes when talking to a person who is blind, there is concern about using some common expressions, such as, "See you later." If you happen to use such an expression, you should not feel a need apologize. People who are blind often use such terminology themselves.[95]

Deaf or Hard of Hearing

Use a tap on the shoulder or a wave to capture the attention of a person who is deaf. For individuals who read lips, be sure to face the light source, keep hands away from the face, and speak clearly. Do not raise your voice. Shouting can distort the message. If you have the assistance of a sign language interpreter, be sure to look at your client rather than the interpreter.

Speech Disabilities

There are diverse factors that can be the reason for reduced ability to use language, such as hearing impairment, cerebral palsy, or **aphasia**. We will concentrate on aphasia because it is a complex speech condition and the communication guidelines are often transferable to individuals who have speech difficulties for other reasons.

Aphasia is a disorder that impairs the expression and understanding of language as well as reading and writing but not intelligence. Aphasia can occur suddenly, from a stroke or head injury or can develop slowly due to a brain tumor, infection, or dementia. About 23 to 40% of people who have strokes develop aphasia.[96] Hypertension is the most common cause of a stroke, and there are a number of dietary measures that can be taken to decrease blood pressure so nutrition professionals can provide meaningful interventions to decrease risk of a first or second stroke. Because of their language issues, family members often need to be included in nutrition interventions. However, the client should not be overlooked during consultations. Usually a speech-language pathologist will do an evaluation and can give guidance regarding communication possibilities, such as picture cards, hand gestures, or word cards. Your facility may have electronic or computer-assisted augmentation devices that can be used to assist communication. The following are some communication tips for interacting with individuals with aphasia:

- Make sure you have the individual's attention.
- Minimize background noise.
- Speak to the person as an adult, not a child.
- Permit a reasonable amount of time to respond.
- Simplify sentence structure and rate of speaking.
- Focus on one message at a time.
- Do not attempt to finish patient statements.
- Do not turn conversation into therapy by correcting the patient.
- Try to involve the patient in decision-making.
- Augment your speech with gestures and visual aids.
- Consider using yes or no questions or thumbs up or down.

Invisible Disabilities

Some disabilities are not readily apparent but do impair normal daily activity. Diverse conditions may fall into this category, such as mental health issues, Asperger syndrome, fibromyalgia, dexterity difficulties, or chronic conditions that cause disabling pain or fatigue. Sometimes the medication that a person needs to take may cause side effects that result in confusing behavior, such as standing or the need to walk when others are sitting. If you have clients behaving in a way you find confusing, inquire in a curious manner as to the reason for the behavior. Ask if there is something you can do to provide additional comfort for their situation.

CASE STUDY   Counseling in a WIC Program

You are working as a nutrition counselor for a WIC program. Carmen, a 16-year old high school sophomore, is three months pregnant and was referred to the WIC office during her recent prenatal visit at a local health clinic. Carmen immigrated to the United States with her family from Mexico when she was five years old. Presently, she lives in an apartment with her sister, brother, grandmother, mother, and father. After school, she works at a fast food establishment three days a week including Saturday. Carmen brought a referral form from the clinic which contains the following information:

- Height: 5 feet tall
- Prenatal weight: 100 pounds
- Present weight: 101 pounds
- Hemoglobin: 10 mg/dL (normal 12-16 mg/dL)
- Hematocrit: 33% (normal 36-48%)
- Symptoms: Vomiting in the morning, and usually nauseous through late morning
- Diagnosis: Iron deficiency anemia

A review of Carmen's usual diet indicates the following:
Breakfast: one cup of coffee, Mexican soup (noodles with a little chicken), fried tortilla or Mexican muffin
Lunch (school days): Soda, chips or French fries, pizza or cheeseburger
Dinner and Lunch (working days): Soda, fried chicken or cheeseburger, French fries
Dinner (at home): Lemonade, beans and rice or enchilada with hot sauce, bunuelos (deep fried cake)
Night time snack at home: one glass of milk and cookies
Food with friends: Fast foods, tortilla chips, soda
Carmen is looking forward to the birth of her baby. She said she would like advice to alleviate nausea.

CASE STUDY   Activities

Explore MyPyramid for Moms website. Go to www.mypyramid.gov and search specific audiences for pregnancy. Complete a MyPyramid Plan for Moms for Carmen.

1. What food groups are not adequate in Carmen's diet?

Explore WIC

2. Go to Works Resource System(WWRS), an online education and training center for health and nutrition professionals serving in the Special Supplemental Nutrition Program for Women, Infants and Children Program at: http://www.fns.usda.gov/wic/benefitsandservices/wwrs.htm and explore the site. Explain two things you learned that would be helpful for a counselor working with Carmen.

3. Go to WIC Foods for Pregnancy Fact Sheet at: http://www.state.nj.us/health/fhs/wic/documents/wic_foods%20_for_pregnancy_en.pdf. How will this food package affect Carmen's nutritional status?

Explore Iron deficiency anemia and nausea:
Go to the March of Dimes website http://www.marchofdimes.com/ and search for anemia and nausea and answer the following questions:

4. Identify two complications for pregnant women who have iron deficiency.
5. What are four common symptoms of iron deficiency?

(continued)

CASE STUDY   Activities *(continued)*

6. What are side effects of taking iron supplements?
7. Give two examples each for animal and plant sources of iron.
8. What can you tell Carmen about her deficiency that would encourage her to take the iron supplements?
9. What advice can you give Carmen about her nausea?

Counseling Approach:

10. Make up four open-ended questions reflecting the spirit of motivational interviewing that would be appropriate for a counseling session with Carmen.
11. Describe one behavioral approach and explain why it could be useful when working with Carmen.

Explore Mexican Foods:

Use the Internet or other sources to investigate Mexican foods.

12. Describe two commonly consumed foods in the Mexican culture and how they influence nutritional status.

REVIEW QUESTIONS

1. Explain four reasons for working towards gaining cultural competence for health care professionals.
2. Compare and contrast the LEARN Model and the ETHNIC Model for nutrition interventions.
3. What is the pivotal construct of the Campinha-Bacote's Model?
4. What is the highest level of cultural competence in the Cultural Competence Continuum?
5. Name two determinants of food behavior for preschool children, middle childhood, adolescents, and older adults.
6. Name two developmental factors characteristic of preschool children, middle childhood, adolescents, and older adults.
7. Identify a nutritional risk for each of the following age groups: preschool children, middle childhood, adolescents, and older adults.
8. Identify an appropriate intervention strategy for preschool children, middle childhood, adolescents, and older adults.
9. Describe three main types of eating disorders.
10. Why should health care professionals strive to use the services of a trained medical interpreter when interpretation is needed?
11. Explain using person-first language when describing a characteristic of an individual.
12. What is the most common cause of aphasia?
13. Give one communication tip for working with individuals for each of the following disability categories: use a wheel chair, are deaf, are blind, have speech difficulties, or have hidden disabilities.

CASE STUDY ANSWERS

1 = Fruits, vegetables, dairy, and possibly protein (meat/beans); 3 = supply missing food groups: 4 = preterm birth and low birth weigh; 5 = fatigue, weakness, dizziness, headache, numbness or coldness in your hands and feet, low body temperature, pale skin, rapid or irregular heartbeat, shortness of breath, chest pain, irritability, not doing well at work or in school; 6 = heartburn, constipation or nausea; 7 = poultry (dark meat), dried fruits (apricots, prunes, figs, raisins, dates), iron-fortified cereals, breads and pastas, oatmeal, whole grains, blackstrap molasses, liver and other meats, seafood, spinach, broccoli, kale and other dark green leafy vegetables, baked potato with skin, beans and peas, nuts and seeds

ASSIGNMENT—CONDUCTING AN INTERVIEW ACROSS CULTURES

Locate a volunteer who is from a culture substantially different than your own. The volunteer should be willing to talk to you about a health problem he or she is experiencing or has experienced or who can describe the health care practices of a particular individual from his or her culture who experienced an illness. Depending on the accessibility of such individuals, you may need to select a person who is different from you culturally because of age or religion. The objectives of this assignment are to work on developing counseling skills, to gather information, and to learn something about the volunteer's health care beliefs and practices. The intention is not to resolve the difficulty. The volunteer may find some benefit to him or herself by clarifying his or her problem through the discussions; however, the person should not be led to believe that there will be an intervention. Therefore, only the involving phase, part of the exploration-education phase, and the closing phase of the cross-cultural counseling algorithm will be addressed in this assignment. Consider audio- or videotaping this experience for later evaluation.

PART I. Use the following interview guide/checklist to conduct the interview with your volunteer. Examples of possible counselor questions, statements, and responses are given in italics.

Preparation
- ❏ Review Exhibit 9.1 Culturally Sensitive Open-Ended Questions to Encourage a Response-Driven Interview, Figure 9.3 Cross-Cultural Nutrition Counseling Algorithm, and the discussion of cultural competence models found in this chapter.
- ❏ Bring two copies of Lifestyle Management Form 5.7 Student Nutrition Interview Agreement found in Appendix D.
- ❏ Bring a completed Certificate of Appreciation

Involving Phase
- ❏ Greeting
  - ○ Verbal greeting—*I am happy to meet you.*
  - ○ Shake hands.
  - ○ Introduce self—*My name is Mary Smith. How should I address you?*
- ❏ Small talk, if appropriate
- ❏ Thank volunteer—*Thank you for participating in this interview.*

- ❏ Explain purpose of the interview—*This is a project I am required to do for my nutrition counseling class. The purpose of this interview is for me to work on my counseling skills, gather information about your health concern, and learn something about your culture, particularly how it relates to health care.*
- ❏ Review the consent form (Lifestyle Management Form 5.7 in Appendix D) with your volunteer, follow the procedure for obtaining a consent in preparation of Session 1 of Chapter 14 of this text, and you and your client should sign both a client copy and a clinic copy of the form. Give the client copy to your volunteer.

Transition to Exploration Phase
- ❏ Transition statement—*Do you have any questions before we go over the interview questions?*

Exploration Phase As you go through the interview questions, you will have to make a judgment regarding which ones are appropriate for your particular client. Do not ask a question if you believe it will provide repetitive information.
- ❏ Ask your client to describe him- or herself (age, cultural group, occupation, interests).

Cause of Illness
- ❏ Explain desire to learn about your volunteer's culture—*As we go through the questions, if there is anything you think I am missing about your health problem or treatment as it relates to your culture, please let me know.*
- ❏ *What do you call your problem? Is there any other name given to this condition in your culture?*
- ❏ *What do you feel may be causing your problem? Do you and your doctor agree about the cause?*
- ❏ *Why do you believe the problem started when it did?*
- ❏ Briefly summarize your client's perception of the cause of his or her illness to check for understanding—*From what you said, it appears that your elevated blood pressure is caused by too much blood and is the result of your fate in life and because of your family history. Your mother*

and father also had high blood pressure. Your family calls the problem "high blood."
- ❏ Make a reflective statement, if appropriate.

Process of Illness
- ❏ *What does the sickness do to your body?*
- ❏ *Do you have any idea when (whether) the problem will get better?*
- ❏ *What do you fear about your sickness?*
- ❏ *What problems has your sickness caused for you personally? For your family? At work?*
- ❏ Briefly summarize your client's perception of how the illness is affecting him or her—*Let me make sure I understood you correctly. You feel that the blood pressure problem will always be with you. You are feeling good now, but both your mother and father had a stroke, and you worry that might happen to you.*
- ❏ Make a reflective statement—*Although you are feeling good, it seems to me that you are feeling somewhat fearful of the future. (If confirmed, then use a legitimation statement.) Considering what happened to your mother and father, it is understandable that you would feel that way.*

Treatment of Illness
- ❏ *What kind of treatment will work for your sickness? Have you been using them? What results do you expect from the treatments?*
- ❏ *Are there home remedies for this sickness? Have you used them?*
- ❏ Briefly summarize what you understand about the volunteer's treatment—*It is interesting that some of your relatives have found help from staying away from "rich" foods and trying to eat more acidic foods. But you feel more comfortable trying to do what your doctor says is best and to take your pills.*

Healers and Future
- ❏ *Are there any benefits to having this illness?*
- ❏ *Is there anyone in your family that helps make decisions about what you should be doing to treat your sickness?*
- ❏ *Are there any healers in your culture who could treat this problem? Have you used any of their treatments?*

- ❏ Make an appropriate reflective or summarization response—*Prayer is important to you. So much is being written about prayer and healing today.*

Foods and Illness
- ❏ *Can what you eat help cure your sickness or make it worse?*
- ❏ *Do you eat certain foods to keep you healthy? To make you strong?*
- ❏ *Do you avoid certain foods to prevent sickness?*
- ❏ *Do you balance eating some foods with other foods?*
- ❏ *Are there foods you won't eat? Why?*
- ❏ *How often do you eat your ethnic foods?*
- ❏ *What kinds of foods have you been eating?*
- ❏ Make an appropriate reflective or summarization response—*You are lucky you have such a supportive wife who is trying to help you eat more fruits and vegetables.*

Explore Culture and Illness—General
- ❏ Explain that you would like to learn about your volunteer's views on illness, healing, and food—*We have been focusing on your particular illness and culture. I am wondering whether people from your culture have other views about illness, healing, or food and health for other illnesses that are different than what is generally believed in this country. Can you tell me about them? Do you know anyone who has used those methods?*

Closing Phase
- ❏ Express appreciation—*Thank you very much for letting me talk to you about your blood pressure and how you are treating the problem. I learned a great deal.*
- ❏ Use a relationship-building response (respect)—*I am very impressed with all you know about your blood pressure problem and the steps you have taken to control it.*
- ❏ Express hope for the future—*I hope you have continued success with controlling your blood pressure.*
- ❏ Shake hands.

❏ Give a certificate of appreciation—*As a show of gratitude for your willingness to participate in this project, I have a certificate of appreciation to give to you from me and the director of the project.*

PART II. Answer the following questions in a formal typed report or in your journal. For formal reports, number and type each question and put the answers in complete sentences under the question.

1. Record the name of the person interviewed and location, time, and date of the meeting.

2. Describe the person you interviewed—age, cultural group, gender, and occupation.

3. Write a narration of the experience. There should be four titled sections in the narration: preparation, opening phase, exploration-education phase, and closing phase. Summarize what occurred in each phase.

4. Write each question in sequence, and give your volunteer's response to the question. Indicate if you did not use the question.

5. Explain the use of relationship building responses and summarizations. What do you believe was the effect of using these responses? Were you comfortable using them?

6. Explain how you believe this person's cultural orientation affects his or her perception and treatment of the illness.

7. Complete an Interview Assessment Form, Lifestyle Management Form 7.5—omit Resolving Phase.

8. How useful were the respondent-driven interview questions for counseling someone from a culture different than your own? Explain your answer.

9. What did you learn from this experience?

SUGGESTED READINGS, MATERIALS, AND INTERNET RESOURCES

http://gucchd.georgetown.edu/nccc
National Center for Cultural Competence

http://www.omhrc.gov/CLAS
National Standards for Culturally and Linguistically Appropriate Services (CLAS) in Health Care Office of Minority Health Resource Center

http://www.ethnomed.org EthnoMed

http://www.diversityrx.org
Resources for Cross Cultural Health Care

http://cccm.thinkculturalhealth.org
A Family Physician's Guide to Culturally Competent Care, 8-modules curriculum free to all health care practitioners Developed by the Office of Minority Health

https://www.thinkculturalhealth.org
Office of Minority Health

http://food.oregonstate.edu/culture
Oregon State University Extension Website. Contains a host of cultural diversity materials, and links to other sites containing culturally diverse materials.

http://www.yaleruddcenter.org
On line toolkit for health professionals providing evidence-based learning regarding weight bias

Children and Adolescent Resources

www.mypyramid.gov/preschoolers/index.html
MyPyramid for Preschoolers
www.mypyramid.gov/kids/index.html
MyPyramid for Kids
MyPyramid websites provide animations, healthy eating information, suggested activities, problem eating behavior advice, and role model guidelines.

www.actionforhealthykids.org
Action for Healthy Kids is a national-state initiative to reduce obesity and improve nutrition and physical activity in schools; provides a variety of resources for school-based change.

http://kidshealth.org/teen
KidsHealth.org/Hey Teens

www.aedweb.org
Academy for Eating Disorders

Older Adult Resources

www.hospitalmedicine.org/geriresource/toolbox/determine.htm
Society of Hospital Medicine Clinical Toolbox for Geriatric Care contains downloadable Checklist to Determine Your Nutritional Health

http://www.mna-elderly.com
Official Mini Nutritional Assessment (MNA) Website Includes downloadable form, user guide and training video

http://nutrition.tufts.edu
MyPyramid for older adults

Disability Websites

http://codi.buffalo.edu
Cornucopia of Disability Information

www.nichcy.org
National Dissemination Center for Children with Disabilities

www.nia.nih.gov/alzheimers
Alzheimer's Disease Education and Referral Center

Books

Alvord LA. The Scalpel and the Silver Bear-The First Navajo Woman Surgeon Combines Western Medicine and Traditional Healing. Bantam Books; 2000.

Delgado JL, National Hispanic Women's Health Initiative. *SALUD! A Latina's Guide to Total Health-Body, Mind and Spirit*. New York, NY: Harper Collins; 1997.

Drago L, Goody CM. Diabetes Care and Education Dietetic Practice Group. *Cultural Food Practices*. Chicago, IL: American Dietetic Association; 2009.

Provides culturally appropriate counseling recommendation, practical information for 15 cultures, and client education handouts.

Dunn R, Griggs S. *Multiculturalism and Learning Style: Teaching and Counseling Adolescents*, Praeger Paperback, 1998. Provides a variety of educational interventions that accommodate diverse learning style preferences.

Fadiman A. *The Spirit Catches You and You Fall Down*. New York, NY: Farrar, Straus & Giroux; 1998.

Fairburn C. *Cognitive Behavior Therapy and Eating Disorders*, New York, NY: Guilford Press; 2008. Considered the "definitive" work on eating disorders for clinicians; contains numerous reproducible handouts.

Kittler PG, Sucher KP. *Food and Culture*. 5th ed. Belmont, CA: Thomson/Wadsworth; 2008.

REFERENCES

[1]Stein K. Moving cultural competency from abstract to act. *J Am Diet Assoc*. 2010; 110:180-187.

[2]US Census Bureau. Available at: http://www.census.gov. Accessed June 10, 2010.

[3]Statistics Canada. Available at: http://www.statcan.gc.ca. Accessed June 10, 2010.

[4]Betancourt JR. Cultural competence and medical education: Many names, many perspectives, one goal. *Acad Med*. 2006; 81:499–501.

[5]Juckett G. Cross-cultural medicine. *Am Fam Physician*. 2005; 25:2–3.

[6]Liburd LC, Namageyo-Funa A, Jack L, Gregg E. Views from within and beyond: illness narratives of African-American men with type 2 diabetes. *Diabetes Spectrum*. 2004; 17:219–224.

[7]The Joint Commission. Standards in Support of Language and Culture. Available at: http://www.jointcommission.org/PatientSafety/HLC/HLC_Joint_Commission_Standards.htm. Accessed July 24, 2010.

[8]U.S. Department of Health and Human Services, OPHS Office of Minority Health. *National Standards for Culturally and Linguistically Appropriate Services in Health Care Final Report* (Washington, D.C.: U.S. Government Printing Office, March 2001).

[9]Goode T, Sockalingam S, Brown M, Jones W. Policy Brief 2: Linguistic Competence in Primary Health Care Delivery Systems: Implications for Policy Makers, U.S. Department of Health and Human Services, OPHS Office of Minority Health. *National Standards for Culturally and Linguistically Appropriate Services in Health Care Final Report*; 2003.

[10]Robin LS. Cultural competence in diabetes education and care, in *A Core Curriculum for Diabetes Education*, 4th ed. Chicago: American Association of Diabetes Educators, 2002.

[11]Berlin EA, Fowkes WC. A teaching framework for cross-cultural health care. Application in family practice. *West J Med*. 1983; 139:934–938.

[12]Cassidy, CM. Walk a mile in my shoes: Culturally sensitive food-habit research. *Am J Clin Nutr*. 1994; 59(suppl.):190S–197S.

[13]Magnus M. What's your IQ on cross-cultural nutrition counseling? *The Diabetes Educator*. 1996; 96:57–62.

[14]Fernandez A, Schenker Y. Time to establish national standards and certification for health care interpreters. *Patient Education and Counseling*. 2010; 78:139–140.

[15]Kleinman A, Eisenberg L, Good B. Culture illness, and care: Clinical lessons from anthropologic and cross-cultural research. *Ann Intern Med*. 1978; 88:251–258.

[16]Campinha-Bacote J. *The Process of Cultural Competence in the Delivery of Healthcare Services*. 5th ed. Cincinnati, OH: Transcultural C.A.R.E. Associates; 2007.

[17]Teaching Tolerance, A Project of the Southern Poverty Law Center. Available at: http://www.tolerance.org/activity/test-yourself-hidden-bias Accessed July 31, 2010.

[18]Gardenswartz L, Rowe A. *Managing Diversity in Health Care*. San Francisco, CA: Jossey-Bass; 1998.

[19]Skinner C, Wight VR, Aratani Y, et al. *English Language Proficiency, Family Economic Security, and Child Development*. New York, NY: National Center for Children in Poverty; 2010.

[20]Fadiman A. *The Spirit Catches You and You Fall Down*. New York, NY; Noonday; 1997.

[21]Worthington-Roberts BS, Williams SR. *Nutrition Throughout the Life Cycle*. 4th ed. New York: McGraw-Hill, 1999.

[22]Mink M, Evans A, Moore CG, et al. Nutritional Imbalance Endorsed by Televised Food Advertisements, *J Am Diet Assoc*. 2010; 110: 904–910.

[23]Mendoza JA, Zimmerman FJ, Christakis DA. Television viewing, computer use, obesity, and adiposity in US preschool children. *Int J Behav Nutr Phys Act*. 2007; 4: 44.

[24]Program for the Study of Media and Health. The Role of Media in Childhood Obesity. *Issue Brief*. Menlo Park, CA: Henry J Kaiser Family Foundation; 2004.

[25]McAlister AR, Cornwell TB. Children's brand symbolism understanding: Links to theory of mind and executive functioning. *J Psych Market*. 2010; 27:203–228.

[26]Birch LL, McPhee L, Steinberg L, et al. Conditioned flavor preferences in young children. *Physiol Behav*. 1990; 47:501–505.

[27]Cooke L. The importance of exposure for healthy eating in childhood: A review. *J Human Nutr Diet*. 2007; 4: 294–301.

[28]Rozin P, Schiller D. The nature and acquisition of a preference for chili pepper by humans. *Motivation and Emotion*. 1980; 4(1).

[29]Rolls BJ, Engell D, Birch LL. Serving portion size influences 5-year-old but not 3-year-old children's food intakes. *J Am Diet Assoc*. 2000; 180:232–234.

[30]Piaget J, Inhelder B. *The Psychology of the Child*. New York: Basic Book; 1977.

[31]Matheson D, Spranger K, Saxe A. Preschool children's perceptions of food and their food experiences. *J Nutr Ed Behav*. 2002; 34:85–92.

[32]Fungwe T, Patricia M, Guenther PM, et al. The Quality of Children's Diets in 2003-04 as Measured by the Healthy Eating Index-2005 April 2009 Nutrition Insight 43. Available at: http://www.cnpp.usda.gov/Publications/NutritionInsights/Insight43.pdf. Accessed August 1, 2010.

[33]Jahns L, Siega-Riz, AM, Popkin, BM. The increasing prevalence of snacking among U.S. children from 1977 to 1996. *J Pediatrics*. 2001; 138:493–498.

[34]Rideout V, Roberts DJ, Foehr UG. Generation M: Media in the lives of 8-18 year-olds. Henry J Kaiser Family Foundation, http://www.kff.org/entmedia/entmedia030905pkg.cfm 2005. Accessed June 18, 2010.

[35]Committee on Nutrition. *Pediatric Nutrition Handbook*. 5th ed. Elk Grove Village, IL: American Academy of Pediatrics, 2003.

[36]Alvy LM, Calvert S. Food marketing on popular children's web sites: A content analysis. *J Am Diet Assoc*. 2008; 108:710–713.

[37]Neumark-Sztainer AC, Hannan DP, Van den Berg P, et al. Parental eating behaviors, home food environment and adolescent intakes of fruits, vegetables and dairy foods: Longitudinal findings from Project EAT. *Public Health Nutr*. 2008; 10:1257–1265.

[38]United States Department of Agriculture. Food and Nutrition Service. *Healthy Options USDA Foods*. 2009 http://www.fns.usda.gov Accessed June 18, 2010.

[39]Ritchie LD, Crawford PB, Hoelscher DM, et al. Position of the American Dietetic Association: Individual-, family-, school-, and community-based interventions for pediatric overweight. *J Am Diet Assoc*. 206; 106:925–945.

[40]Whitlock EP, O'Connor EA, Williams SB, et al. Effectiveness of weight management interventions in children: A targeted systematic review for the USPST. *J Pediatrics*. 2010; 125:e396–e418.

[41]Spahn JM, Reeves RS, Keim KS, et al. State of the evidence regarding behavior change theories and strategies in nutrition counseling to facilitate health and food behavior change. *J Am Diet Assoc.* 2010; 110:879–891.

[42]Rovner AJ, Tonja R. Are children with Type 1 Diabetes consuming a healthful diet? A review of the current evidence and strategies for dietary change. *Diabetes Educ.* 2009:35(1): 97–107.

[43]American Academy of Pediatrics. Committee on Public Education. American Academy of Pediatrics: children, adolescents, and television. *Pediatrics.* 2001; 423–426.

[44]Anonymous. Guidelines for school health programs to promote lifelong healthy eating. *J Sch Health.* 1997; 67:1; Story M, Newmark-Sztainer D, French SI. Individual and environmental influences on adolescent eating behaviors. *J Am Diet Assoc.* 2002; 102(S3):S40–S51.

[45]Burghardt J, Gordon A, Chapman N, et al. *The School Nutrition Dietary Assessment Study: School Food Service, Meals Offered, and Dietary Intake.* Princeton, NJ:Mathematica Policy Research, Inc.; 1993.

[46]Zollo P. *Wise Up to Teens: Insight into Marketing and Advertising to Teenagers.* 2nd ed. Ithaca, NY:New Strategist Publications, Inc.; 1999.

[47]Henry BW, Ozier AD. Position of the American Dietetic Association: Nutrition intervention in the treatment of anorexia nervosa, bulimia nervosa and other eating disorders. *J Am Diet Assoc.* 2006; 106:2073–2082.

[48]Sturdevant MS, Spear RA. Adolescent development. *J Am Diet Assoc.* 2002; 102:S30–S31.

[49]Lytle LA. Nutritional issues for adolescents. *J Am Diet Assoc.* 2002; 102:S8–S12.

[50]Sigman-Grant M. Strategies for counseling adolescents. *J Am Diet Assoc.* 2002; 102:S32–S39.

[51]Contento, IR. *Nutrition Education: Linking Research, Theory, and Practice.* Sudbury, MA: Jones and Bartlett Publishers, 2010.

[52]Profile of older Americans, 2009. US Administration on Aging Web site. http://www.aoa.gov/AoARoot/Aging_Statistics/Profile/index.aspx. Accessed June 21, 2010.

[53]Federal Interagency Forum on Aging-Related Statistics. Older Americans 2008: Key Indicators of Well-Being. Federal Interagency Forum on Aging-Related Statistics Web site. March 2008. www.agingstats.gov/agingstatsdotnet/main_site/default.aspx. Published March 2008. Accessed June 21, 2010.

[54]Niedert KC. Position of the American Dietetic Association: Liberalization of the diet prescription improves quality of life for older adults in long term care. *J Am Diet Assoc.* 2005; 105:1955–1965.

[55]Determine Your Nutritional Health Nutrition Screening Initiative. Available at: http://www.aafp.org/afp/980301ap/edits.html. Accessed August 1, 2010.

[56]Amarantos E, Martinez A, Dwyer J. Nutrition and quality of life in older adults. *J Gerontol Biol* Sci Med Sci 2001; 56A:54–64.

[57]National Center for Health Statistics, Health, United States, 2008; data from the National Health Interview Study, Available at: http://www.cdc.gov/nchs/nhis.htm Accessed August 1, 2010.

[58]World Health Organization. World Health Statistics 2010. Available at: http://www.who.int/whosis/whostat/2010/en/index.html Accessed June 20, 2010.

[59]American Dietetic Association. Position of the American Dietetic Association: Providing Nutrition Services for People with developmental Disabilities and Special Health Care Needs. *J Am Diet Assoc.* 2010:110:296–307.

[60]Flegal KM, Carroll MD, Ogden CL, et al. Prevalence and trends in obesity among US adults, 1999–2000. *J Am Med Assoc.* 2009:288:1723–1727.

[61]Jarrat J, Mahaffie JB. The profession of dietetics at a critical juncture: A report on the 2006 environmental scan for the American Dietetic Association. *J Am Diet Assoc.* 2007; 107:S39–S57.

[62]Kuczmarski MF, Weddle DO. Position of the American Dietetic Association: Nutrition, aging, and the continuum of care. *J Am Diet Assoc.* 2000; 100:580–595.

[63]Erin RB. Healthy Eating Index scores among adults, 60 years of age and older, by sociodemographic and health characteristics: United States, 1999-2002. Centers for Disease Control and Prevention Website. http://www.cdc.gov/nchs/data/ad/ad395.pdf. Published May 20, 2008. Accessed June 21, 2010.

[64]Juan WY, Guenther PM, Kott PS. Diet Quality of Older Americans in 1994-96 and 2001-02 as Measured by the Healthy Eating Index. 2005 Nutrition Insight 41 http://www.cnpp.usda.gov/Publications/NutritionInsights/Insight41.pdf Accessed June 20, 2010.

[65]Wang MC, Dixon LB. Socioeconomic influences on bone health in postmenopausal women: Findings from NHANES III, 1988-1994. *Osteoporosis Int.* 2006; 17: 91–98.

[66]Johnson DB, Beaudoin S, Smith LT, et al. Increasing fruit and vegetable intake in homebound elders: The Seattle Senior Farmers' Market Nutrition Pilot Program. *Preventing Chronic Disease.* 2004; 1(1):A03.

[67]Bauer JM, Kaiser MJ, Anthony P, et al. The Mini Nutritional Assessment®-Its history, today's practice, and future perspectives. *Nutr Clin Prac.* 2008; 23(4):388–396.

[68]Academy of Eating Disorders. Available at: http://www.aedweb.org. Accessed July 25, 2010.

[69]Cairns J, Milne RL. Eating Disorder Nutrition Counseling: Strategies and Education Needs of English-Speaking Dietitians in Canada. *J Am Diet Assoc.* 2006; 106:1087–1094.

[70]Golan M, Crow S. Parents are key players in the prevention and treatment of weight-related problems. *Nutr Rev.* 2004; 62:39–50.

[71]Spear BA. Does dieting increase the risk for obesity and eating disorders? *J Am Diet Assoc.* 2006; 106:523–525.

[72]Nitzke S, Freeland-Graves J. American Dietetic Association. Position of the American Dietetic Association: Total diet approach to communicating food and nutrition information .*J Am Diet Assoc.* 2007; 107:1224–1232.

[73]Puhl R, Wharton W, Heuer C. Weight Bias among Dietetics Students: Implications for Treatment Practices. *J Am Diet Assoc.* 2009; 109:438–444.

[74]Davis-Coelho K, Waltz J, Davis-Coelho B. Awareness and prevention of bias against fat clients in psychotherapy. *Professional Psychology: Research and Practice.* 2000; 31:682–684.

[75]Brown I. Nurses' attitudes towards adult patients who are obese: Literature review. *J AdvNurs.* 2006; 553:221–232.

[76]Hebl MR, Xu J. Weighing the care: Physicians' reactions to the size of a patient. *Int J Obes.* 2001; 25:1246–1252.

[77]Schwartz MB, Chambliss HO, Brownell KD, et al. Weight bias among health professionals specializing in obesity. *Obes Res.* 2003; 11:1033–1039.

[78]O'Brien KS, Hunter JA, Banks M. Implicit anti-fat bias in physical educators: Physical attributes, ideology, and socialization. *Int J Obes.* 2007; 31:308–314.

[79]McArthur L, Ross J. Attitudes of registered dietitians toward personal overweight and overweight clients. *J Am Diet Assoc.* 1997; 97:63–66.

[80]Amy NK, Aalborg A, Lyons P, et al. Barriers to routine gynecological cancer screening for White and African-American obese women. *Int J Obes.* 2006; 30:147–155.

[81]Trust for America's Health (TFAH). *F as in Fat Report.* Washington DC, 2010.

[82]Puhl RM. Treating Obese Patients: The Importance of Improving Provider-Patient Interaction. Medscape Public Health & Prevention. http://www.medscape.com/viewarticle/722041.Accessed July 27, 2010.

[83]Puhl RM, Heuer CA. Weight bias: a review and update. *Obesity.* 2009; 17:941–964.

[84]Wadden TA, Didie E. What's in a name? Patients' preferred terms for describing obesity. *Obes Res.* 2003; 11:1140–1146.

[85]DiLillo V, Siegfried NJ, Smith-West D. Incorporating motivational interviewing into behavioral obesity treatment. *Cogn Behav Pract.* 2003; 10:120–130.

[86]Pietsch R. Becoming the kingdom of god: Building bridges between religion, secular society, and persons with disabilities: The ministry of Harold Wilke. *Journal of Religion, Disability & Health*, 1522-9122, Volume 2, Issue 4, 1996, Pages 15–25.

[87]Wilke H, "Signs of Liberation and Access," *Any Body, Everybody, Christ's Body: A Congregational Guide for Becoming Accessible to All*, edited by Jo Claire Hartsig, Available at: www.uccdm.org/A2A/anybody.pdf. Accessed July 24, 2010.

[88]U.S. Census Bureau. Census brief: Disabilities affect one-fifth of all Americans. U.S. Census Bureau Web site http://www.census.gov/prod/3/97pubs/cenbr975.pdf Accessed July 24, 2010.

[89]Lipscomb L. Person-first practice: treating patients with disabilities. *J Am Diet Assoc.* 2009; 109:21–25.

[90]Quann E. *By His Side Life and Love after Stroke.* Hyland, MD: Fastrak Press, 2002.

[91]Yerkes MJ. Disabilities Etiquette 101. Associations Now, December 2007 Available at: http://www.asaecenter.org/Resources/ANowDetail.cfm?ItemNumber=38765 Accessed February 12, 2011.

[92]Snow K. People first language. Disability is Natural. Available at: http://www.disabilityisnatural.com. Accessed July 24, 2010.

[93]National Service Inclusion Project. Fact sheet: Person-first language. Community Inclusion Available at: http://www.communityinclusion.org/projectdocs/nsip/watch-yr-language.doc. Accessed July 24, 2010.

[94]Autistic spectrum disorder fact sheet: Community, politics & culture of autism. Available at: http://www.autism-help.org/autism-politics-culture-community.htm Accessed July 24, 2010.

[95]*The 10 Commandments of Communicating with People with Disabilities*. Challenge Publications Limited; 2001.

[96]National Aphasia Association. Communicating With People Who Have Aphasia. Available at: http://www.aphasia.org/index.html. Accessed July 24, 2010.

10

Group Facilitation and Counseling

Andy Crawford/Dorling Kindersley/Getty Images

A well run group is not a battlefield of egos.

—LAO TZU

Behavioral Objectives

- Explain characteristics of three common communication styles.
- Use questions appropriately in a group setting.
- Identify desirable characteristics and behaviors of group facilitators.
- Summarize a group facilitator's responsibilities.
- Describe selected techniques for organizing a group meeting.
- Explain advantages and disadvantages of group counseling.
- Use an emotion-based approach in a group setting.
- Implement a behavior change group counseling session.

Key Terms

- **Brainstorm:** technique to generate as many ideas as possible for consideration.
- **Emotion-Based:** feelings and emotional benefits that drive the behavior change approach.
- **Facilitator:** uses group processes to keep members focused on content and guide the flow of a meeting.
- **Ground Rules:** set of guidelines for group members.
- **Group Counseling:** using group support to find solutions for lifestyle problems.

INTRODUCTION

As a nutrition professional, you are likely to need group facilitation skills for diverse objectives. You may be the leader of a community group or working with colleagues to develop a program or implement a nutrition intervention. In another case, you could be called upon to present in-service training to a group of health care professionals with the objective of creating awareness, such as knowledge of nutrition factors related to the DASH food plan. Offering group counseling to clients has been found to work effectively and efficiently, either as a stand-alone program or in combination with personal counseling. The following discussion of group work is divided into two categories. The first, facilitating groups, focuses on leading a number of people with a particular goal in mind, such as developing a program for nutrition month at a long term care facility. The second section, group counseling, provides guidance for working with groups where the objective is behavior change. Factors related to giving effective presentations to groups are provided in Chapter 12. We will start with reviewing some factors common to a leader of any group including a review of communication styles and using questions effectively in a group setting.

COMMUNICATION STYLES

No matter what type of group you are working with, knowledge of communication styles and understanding the impact they have on others will help you to be an effective group leader. Table 10.1 describes three communication styles you are likely to encounter. Both submissive and aggressive styles provide challenges. As we review group process strategies, you will gain knowledge of possible ways to work with these styles. The most effective style for leading a group is an assertive communication style.

USING QUESTIONS IN A GROUP

Whether you are facilitating a group towards a common goal or conducting a group counseling intervention, having an arsenal of effective questions and knowing when to use them is essential. Appropriately used questions can help keep a group on task and moving toward the desired goal. Using questions during an individual counseling intervention was covered in Chapter 3. Here we will explore types of questions and their significance in group interventions.

Table 10.1 Communication Styles

Category	Submissive	Assertive	Aggressive
Characteristics	May be emotionally dishonest, indirect, self-denying, inhibited	Is appropriately: emotionally honest, direct, self-enhancing, expressive	Is inappropriately: emotionally honest, direct, self-enhancing at the expense of another, expressive
Your feelings when you engage in this behavior	Hurt, anxious at the time, and possibly angry later	Confident, capable, self-respecting at the time and later	Righteous, superior, powerful at the time, and possibly guilty later
The other person's feelings about self when you engage in this behavior	Guilty or superior	Valued, respected	Hurt, humiliated
The other person's feelings about you when you engage in this behavior	Pity, irritation	Generally respected	Angry, vengeful

Source: Adapted from: Alberti & Emmons. Stand Up, Speak Out, Talk Back! New York: Pocket Books; 1975; Katz & Lawyer. Communication and Conflict Resolution Skills. Dubuque, IA; Kendall/Hunt; 1985.

EXERCISE 10.1   Evaluate Past Group Experiences

Consider your best experience as the member of a small group. Describe the setting and the function of the group. What made the experience go well? What was your role in the group? Is there anything in that encounter that you would like to emulate in your work as a group facilitator? Explain.

Types of Questions

There are various types of questions and the ones you chose will depend upon your objective. You want your questions to meet the needs of the group, help elucidate matters under discussion, and promote participation. As previously discussed in Chapter 2 regarding motivational interviewing, getting people to talk about possibilities encourages ownership of outcomes. As a **facilitator**, you need to be cognizant of this fact and encourage participation of all members of the group.

Facilitators choose the content and the focus of a question by evaluating what is needed to move the discussion forward. You may observe that group members are stuck because they do not have adequate knowledge. In that case, the need is for questions that seek facts or clarification of concepts. If the discussion appears disjointed, a process question may be in order asking participants to choose a solution, predict what will happen, or to compare and contrast two situations. If a need to address emotions arises, affective questions can be used probing for opinions, feelings, attitudes, or beliefs. Behavior questions are useful if a plan of action is under discussion while several members have indicated they are not ready for that step. If that is the case, the focus of questions should be on the application of new knowledge, what they learned from past experiences, or how they can solve a problem.[1] Table 10.2 provides examples of categories of questions with examples of application for specific group situations. Questions that put individuals "on the spot" by asking for justification for their actions

Table 10.2   Categories of Questions for Groups

Category	Description
Factual Questions	Assist in obtaining additional details and are likely to provide answers to one of the five W's: who, what, when, where, and why.
Example: Who has worked with individuals diagnosed with aphasia?	
Explanatory Questions	Aid in the search for reasons and explanations.
Example: Does anyone know what happened?	
Justifying Questions	Help in challenging previous procedures and encourage giving consideration to new ones.
Example: How about if we scheduled the event on the weekend rather than a weekday this year?	
Leading Questions	Assist in focusing and advancing an idea. Also, they can provide a conclusion and move a discussion toward closure.
Example: How do others feel about placing the focus on fast food?	
Theoretical Questions	Help to introduce another idea or redirect the flow of the discussion.
Example: Let's suppose that the health department wants to participate?	
Alternative Questions	Assist in making a choice. They help the facilitator take control by providing only two options.
Example: What are the advantages and disadvantages of having the event on a Saturday afternoon or a Friday night?	
Exploratory Questions	Aid in exploring areas not previously addressed.
Example: Has anyone had a different experience you would like to share?	

Source: Adapted with permission from Soil and Water Conservation District Outreach: A Handbook for Program Development, Implementation and Evaluation. Ohio Department of Natural Resources, Division of Soil and Water Conservation, 2003.

or identifying blame are not likely to promote positive group dynamics. See Table 10.3 for a list of ineffective questions and possible alternatives. Exhibit 10.1 provides a list of questions generally found to be effective in a group setting.

Never doubt that a small group of thoughtful citizens can change the world. Indeed, it is the only thing that ever has.

—Margaret Mead

EXHIBIT 10.1   Questions Generally Found to be Effective in Groups

- Can you tell me more?
- How would that work here?
- What results do you want?
- What can we expect?
- What would be the advantages or benefits of this approach?
- What options do you have for getting past this obstacle?
- How can you do that even better?
- What will it ideally look like when it's complete?
- What's your reaction?
- What has worked most effectively in similar situations?
- What was particularly effective about the way that worked?
- How would you do it differently another time?
- What would be the benefit of doing it differently?

Source: Adapted with permission from Soil and Water Conservation District Outreach: A Handbook for Program Development, Implementation and Evaluation. Ohio Department of Natural Resources, Division of Soil and Water Conservation, 2003.

FACILITATING GROUPS

In order for a group (team or committee) to accomplish a task, there needs to be someone designated to be the facilitator. Groups meet for various reasons including to talk about a concern, exchange information, identify issues, complete a task, build consensus, develop plans, make a decision or solve problems. See Exhibit 10.2 for a list of desirable characteristics of a group facilitator that are useful in guiding a group to accomplish these tasks. The word facilitate derives from "facile", a French word which means "to enable, to make easy."[2] However, getting a group to work together effectively is not always easy. The role of a facilitator is to use knowledge of group processes to provide structure allowing the group to remain focused on content and work effectively to bring about results. The processes include creating an open and inclusive environment using methods that allow group members to interact productively with each other. The content addresses the issues under discussion needed to reach the ultimate goal or goals of the group.

Preparation

The first task at hand is preparation. There should to be an understanding of why there needs to be a meeting. Goals that are specific, concrete, positive, realistic, and practical must be defined. Other factors a leader needs to consider include who needs to attend, when the meeting can occur, where the meeting will take place, and what potential problems could occur that need to be addressed before

Table 10.3 Formulating Effective Questions

Less Effective Questions	Effective Questions
Who made that decision?	Where do we go from here?
Why didn't you finish?	What needs to be completed?
What's your problem?	What else?
Why did you do that?	What were your specific objectives?
Who wants to tell ___ about this problem?	What is the best way to handle this?
You don't know better than that?	What support do you need?

Source: Adapted with permission from Soil and Water Conservation District Outreach: A Handbook for Program Development, Implementation and Evaluation. Ohio Department of Natural Resources, Division of Soil and Water Conservation, 2003.

EXHIBIT 10.2   Desirable Characteristics and Behaviors of a Good Facilitator

Actively listens and observes	Asks probing questions	Uses humor
Shows respect and empathy	Thinks quickly	Knows a variety of techniques
Appears honest and fair	Assertive	Energizes the group
Accessible	Flexible	

Desired Outcomes of the Process Component of an Intervention

• To keep the process on track and moving forward with all participants engaged, making best use of time and resources.

• Balance participation with objectives of the meeting

Source: Adapted from: Lawson SL. A quick reference guide for facilitators. Ministry of Agriculture Food and Rural Affairs, 2002. Available at: http://www.omafra.gov.on.ca/english/rural/facts/95-073.htm

Burke DW, Donahoe M, Hirzel R, et al. Basic Facilitation Skills. The Human Leadership and Development Division of the American Society for Quality, The Association for Quality and Participation, The International Association of Facilitators; 2002.

the meeting.[3] Consideration should also be given to evaluating the need for equipment and supplies, soliciting input of participants, and identifying support roles (time keeper, recorder, etc.). In addition, an agenda should be developed and sent to participants before the meeting. Your plan should take into consideration the organization and flow of the meeting. There are a number of intervention possibilities depending on desired outcomes. The following reviews some useful strategies for decision making and problem solving:[4]

Pair-Share   Pair-Share works well with a large group. This process provides an opportunity for all participants to discuss their thoughts and feelings about a topic with another individual. When the topic is opened up for general discussion, the comments are likely to be more concisely and coherently formulated after sharing with small groups.

Process

1. The facilitator should supply one to three questions for discussion.
   Examples: Can you name two possible businesses to invite to the health fair?
   What food or nutrition-related activities should be at the fair?
2. Participants should be asked to work with a partner to formulate answers.

3. After an appropriate time period, ask participants to share their ideas with the group.

Corners   Corners tends to work well with a group when there are distinct tasks that need to be addressed. Participants are allowed to choose their task.

Process

1. Post the name of each task in a corner of the room.
   Example: For a health fair, the tasks may be divided into developing publicity, locating and contacting participating agencies and businesses, organizing the facility the day of the event, and coordinating volunteers.
2. Participants are asked to consider what part of the project interests them most.
3. Individuals are asked to move to the corner that has that task identified on the sign.
4. At each corner, there will be specific questions to address related to the task.
5. After corner discussions, a speaker from the corner group will be asked to report back to the whole group.

ORID   The ORID discussion method has a progression of questions that takes a group through four consecutive stages: Objective, Reflective, Interpretive, and Decisional. The facilitator asks probing questions that follow the natural sequence people generally use to contemplate an issue. This process is useful for reflecting on experiences and invites a variety of perspectives in a non confrontational manner. The questions should flow naturally from one stage to the next.

Process

Example: A group has just finished implementing a health fair.

1. *Objective Discussion.* Questions focus on getting the facts. Possible questions could include:
   How many agencies participated?
   What did you observe?
2. *Reflective Discussion.* Questions focus on emotions and feelings. Possible questions could include:
   How do you think the health fair went compared to previous fairs?
   What was the most challenging part of organizing the event?
3. *Interpretive Discussion.* Questions focus on values, meaning, purpose, and its significance to the group. Possible questions could include:
   What did you achieve by organizing this event?
   What would you say about this event to someone who was not there?
4. *Decisional Discussion.* Questions focus on making a group decision or personal response to the experience.
   Should we organize a health fair next year?
   Are health fairs something you want to be involved with in the future?

Consensus

Consensus is a method for making group decisions by encouraging members to share their thoughts, feelings, and suggestions. In order to develop a sustainable agreement, a group facilitator needs to lead the group through four stages: "gathering diverse points of view, building a shared framework of understanding, developing inclusive solutions, and reaching closure."[5]

Process

1. Explain the purpose of the discussion.
2. Review the values important to a good group discussion.

3. Explain that the goal is to reach an acceptable agreement in a defined time frame.
4. Repeat purpose of discussion.
5. Ask for someone to start the discussion. "Who would like to begin?"
6. After discussion, the group comes to a consensus agreeing on a course of action.

The above discussion provided a review of a few proven techniques, but numerous methods are available. You should choose or modify one that would best fit the needs of your group. See resources at the end of the chapter to locate additional descriptions of more techniques.

Group Management

Now, we will review factors to consider at the beginning, middle, and end of a group meeting.

Beginning a Group Meeting At the beginning of a meeting, there are some factors to consider to encourage openness and trust. You may want to use an icebreaker, especially if you are leading a group of people who do not know each other. This is a way to dispel anxiety, help participants to get to know each other, and possibly find areas of commonality. A fun and organized process for conducting introductions is a human treasure hunt. See Exhibit 10.3. Other possibilities include having participants interview each other and report findings back to the group or, depending on the age level, using a bean bag to toss to group members to determine the next speaker to introduce self. Openings could include humor (possibly a cartoon on an overhead projector), an open-ended question, or an interesting story. Depending on the purpose and the composition of the group, you may wish to set some **ground rules**. See Exhibit 10.4 for a list of common ground rules. In addition, you may wish to go over the agenda and agree on topics to be covered during the group session. You may ask participants to describe their expectations of the group process.

Starting a meeting on time is important to show respect for the individuals who made an effort to arrive on time. Inevitably there will be participants

EXHIBIT 10.3 Human Treasure Hunt Guidelines

This activity is a good icebreaker, but probably would not work well for groups with fewer than eight members. Besides helping participants become acquainted with each other, it also encourages the process of sharing experiences and coping strategies.

1. Before your meeting, find out something interesting or special about each participant. Preferably identify a fact related to the group concern. For a group of individuals who experienced a heart attack, an example of a statement could be: "Find a person who enjoys eating oatmeal for breakfast."
2. Compose a human treasure hunt sheet by writing a list of the facts on a sheet without names.
3. At the beginning of the meeting, hand out the form and ask group members to search for the member who meets that description.
4. Close the activity by reading the facts and identifying the person who corresponds to each fact.

EXHIBIT 10.4 Common Ground Rules for Meetings

- Attend all meetings and be on time.
- Start and end meetings on time.
- Be willing to share with the group.
- Listen to and show respect for the opinions of others.
- Look for value in every idea (listen openly).
- Follow the agenda—stay on track.
- Adhere to rules of confidentiality.
- Refrain from engaging in side conversations.
- Turn off cell phones and pagers.

who arrive late. If that is the case, smile and greet the late arriver warmly, give a ten-second update on the progress of the meeting, and encourage participation.[6]

Guiding the Flow of a Group Meeting In order to run an effective meeting, a facilitator will need to guide the flow of participation and keep the focus on content. See Table 10.4 for suggestions to enable the process. Also if you planned to use a specific strategy as described under preparation in this chapter, the technique should be implemented after the introduction phase of the meeting. If the flow of the meeting appears to be lagging, the following provide some stimulation ideas:

- What's making the most sense about what we've covered so far?
- Summarize group decisions and ask "What else?" rather than "Anything else?"
- Consider silence. As the facilitator of a group, a novice leader can feel pressure to fill silence. Often silence occurs early in a session when no one wants to initiate conversation. However, the silence can also indicate to the group that the facilitator does not intend to dominate. Ordinarily, if the silence is long enough, someone will take the initiative for beginning a discussion.

Closing the Meeting Factors for closing a meeting will in part depend on the purpose of the meeting. The end may occur at a preset time or after a goal has been achieved. The following contains general guidelines for closing a meeting:

1. **Summarize.** You could do the summary yourself or enlist group members to contribute to the summary. A summary should provide a synopsis of what occurred during the meeting highlighting challenges and successes. Be sure to include key points generated by individuals of the group.
2. **What's next?** Review plans for the future, such as the time and date of the next meeting.
3. **Thank you.** Thank the group for their participation and congratulate them for their accomplishments.

Follow-Up Again depending on the composition and purpose of the meeting, there may be a need to perform follow-up activities. They may include the following:[1]

- Maintain contact with members through websites, email, etc.
- Review the accomplishments and concerns of the meeting with colleagues.

Table 10.4 Strategies to Guide Participation and Flow of Content

Category	Description	Examples
Paraphrasing	This can be used to highlight an important point that may need clarification. It also can have a calming effect if a participant expressed an idea in an offensive manner See Chapter 3 for elaboration of paraphrasing.	Begin with "What I hear you saying is . . ." End with, "Does that correctly reflect what you expressed?"
Drawing People Out	If an individual has expressed a vague or unclear idea, paraphrasing is used and then there is a request for clarification.	"Can you elaborate on this idea?" "How so?"
Stacking	When there are several participants who have indicated a desire to speak, a list is verbalized to indicate speaking order.	"OK first we will hear from . . ., then from . . ."
Tracking	Various lines of thought are tracked and summarized.	"So what I am hearing are comments on the type of event that should be planned, the amount of money available for an event, and possible topics to be covered during the event. Do I have this right?"
Encouraging	Provides an opening for all to participate without calling on a particular non-participating individual.	Has this discussion raised any questions for anyone?
Balancing	By asking for additional opinions, individuals are encouraged to express their thoughts.	"So far we have heard from two people who believe . . . Are their additional viewpoints we should consider?"
Making Space	Encouraging input from a participant who has indicated by facial expression or body language a desire to speak.	"Do you have something you would like to add . . . ?" Is there something you would like to say?"
Identifying Common Ground	If group members are expressing contrasting opinions and appear to be losing sight of the common goal, this is a good time to (1) Indicate you are going to summarize differences; (2) Summarize the opinions; (3) Identify common ground; and (4) Check for accuracy.	"We appear to have very different opinions about this matter. I would like to summarize the differences . . . However, we seem to agree on . . . Is this accurate?"

Source: Adapted from: Kaner, S. Facilitator's Guide To Participatory Decision-Making. Gabriola Island, British Columbia: New Society Publishers; 1996.

- Write thank you notes, if appropriate.
- Provide minutes of the meeting to participants.
- Provide information about the meeting to people who were absent.

Group facilitation can be a rewarding experience, which requires an integration of knowledge, skills, intuition, and attitudes. Knowledge of the group process can be gained through educational experiences. Facilitation skills and intuition will evolve and improve through experience. A desire to create, explore, and work with people will lay the foundation for developing a positive attitude towards facilitating groups.[4]

Leadership is getting someone to
do what they don't want to do,
to achieve what they want to achieve.
                                    —Tom Landry

GROUP COUNSELING

Group counseling provided by nutritionists is intended to elicit behavior change related to nutrition issues. Your job is to provide a group atmosphere that encourages curious exploration and consideration of behavior modification alternatives. An American Dietetic Association review of the state of the evidence regarding facilitating health and food behavior change identified three studies showing group counseling to be more effective than individual counseling.[7] A review of twenty years of diabetes education literature found advantages of combining one-on-one and group counseling.[8] In a weight loss study that began by asking individuals a preference for individual or group counseling, group counseling produced greater weight losses even when individuals expressed a preference for individual treatment.[9]

Advantages of Group Counseling

Group counseling affords many advantages:

- **Emotional support.** Groups help clients feel as if they are not alone in dealing with their nutritional concerns. Sharing experiences with others who really know what it is like can provide a great deal of emotional support. A cohesive group helps participants feel accepted and special.
- **Group problem solving.** Participants motivate each other to change as they share coping strategies and problem-solve together. As illustrated in the DOVE activity in Chapter 1, two heads are better than one. Sharing supplies additional ideas and generates suggestions that neither person would have thought of individually (that is, a synergy effect ensues) for finding solutions for overcoming obstacles to behavior change.
- **Modeling effect.** Participants learn from each other by observing the accomplishments of others with similar problems. By observing and taking part in behavior changes of others who are experiencing similar problems, all group members are likely to feel hopeful for themselves.

- **Attitudinal and belief examples.** As participants describe their attitudes and beliefs regarding health behavior challenges and perceived failures, other group members tend to reevaluate their own belief systems.

Disadvantages of Group Counseling

Unfortunately, group counseling also presents some potential drawbacks:

- **Individual responsiveness.** Some people do not easily share in a group setting, and as a result their issues may never be addressed.
- **Group member personalities.** The dynamics of a group are heavily influenced by members' individual personalities. The ability of leaders to handle domineering, demoralizing, or needy individuals who may tend to monopolize time will impact on the counseling environment for all group members.
- **Possibility of poor role models.** Poor role models can create additional burdens for a counselor to counteract.
- **Meeting the needs of all group members.** It may be difficult to organize a group with similar issues and health concerns. Meeting the needs of participants who widely differ in age, gender, ethnic background, and specific health problems can be a challenge. If this is the case, there is limited opportunity to tailor an intervention for an individual participant.

Group Process

The first session is crucial because a group's personality evolves early and is difficult to change at a later time. Therefore, interactive and fun activities should be planned according to the participants' maturity level and interests. The principal objective is to address participants' primary concern of feeling accepted and being acknowledged as worthy.

The composition of a group could be open or closed. *Open groups* are generally considered support groups where participants are encouraged to participate, but there is no commitment to a set number of sessions. In this case, participants generally generate the topics and share their own experiences.

The leader's role is to facilitate the process. This type of group works well in Women, Infants, and Children (WIC) programs, diabetes clinics, or dialysis units. See Table 10.5 for a review of facilitator responsibilities for this type of group. However, open groups can also be theme guided. The successful Touch Hearts, Touching Minds (THTM) project provides a series of theme-based lesson plans using an emotion-based approach to guide group meetings with WIC clients. They aim to provide an engaging and memorable experience encouraging

behavior change. Exhibit 10.5 provides an example of a group lesson plan regarding family meals for WIC counselors to use.

The remaining discussion in this section pertains to running *closed groups*—that is, groups that do not accept new members after the first or second session. Generally closed groups of 8 to 12 members allow for greater bonding to take place and provide a more suitable environment for behavior change to take place. In the ideal cohesive group, each member feels a sense of belonging and

Table 10.5 Summary of Facilitator Responsibilities

Category	Descripton and Examples
Gather and identify useful resources.	Locate or make handouts or have a list of referrals. For example, if the planned topic is healthy snacks, the facilitator may give participants a list of easy to prepare and healthy snacks.
Identify and support needs of the group.	Bring together individuals who have common issues such as diabetes, breast-feeding, or gastric by-pass. Encourage participants to voice their needs.
Plan icebreakers that relate to the group.	For example, if you are leading a group of individuals who are following the DASH food plan, you may begin by asking participants to introduce themselves and to tell everyone their favorite low sodium food.
Make sure all group members feel safe.	Set rules at the beginning of the session and ask that all members agree that the discussion will be kept confidential.
Arrange a comfortable room.	Set chairs in a circle so group members can see and talk easily with each other.
Keep discussion "on track."	Make sure the discussion stays focused to the agreed upon topic.
Keep discussions moving towards change talk.	Encourage change talk as described in Chapter 2. "You each came up with a breakfast plan this week. Take a moment to think how confident you are on a scale of 1 to 10, where 1 is not confident at all and 10 is highly confident that you will do this."
Make sure all members feel their contributions are important.	Acknowledge contribution through body language or compliment.
Encourage all members to contribute.	Ask opinions of quieter members or ask open-ended questions.
Actively listen.	Encourage all participants to actively listen and have only one person talk at a time.
Correct misinformation.	Make corrections in a comfortable manner. For example, "I'm glad you brought that up. There is so much about that topic in the media. Research hasn't been able to support the claim." Or name a highly respected organization, "The American Diabetes Association recommends . . ."
Provide structure for the group.	This may include recording, guiding, and summarizing. Depending on the objective of the group, you may wish to write main points on an easel or writing board.

Source: Adapted from Module 20: Facilitated Group Discussion, Arizona WIC Training Manual. Available at http://www.azdhs.gov/azwic/documents/local_agencies/trainingmanual_pdf/module_20.pdf

---

EXHIBIT 10.5 Touching Hearts. Touching Minds Lesson Plan

Set the table (Family meals)

What is the key message?

- Family meals provide emotional, physical, intellectual and spiritual nourishment.
- Family meals connect families in a powerful way.

Who should receive this message?

- Any parent.

How can this message be used?

Open:

- Parents face challenges today that didn't exist or weren't common when your parents were raising you. What are some of the challenges you face raising children, challenges that exist because we live in changing, turbulent times?

  Sample responses:

– Unsafe communities	– War and nuclear threats	– Threats of attacks, even in
– Internet predators	– Parents working multiple jobs	schools
– Gang activity	with little quality time with	– Street crime
– Violence on TV	children	– Uncertain times

- Suppose I could take out a prescription pad and write a prescription for something that would help protect your child from the scary times in which we live. Would you be interested?
- The prescription I would write would be this: Eat meals together as a family. Family meals have enormous power. They can be the family lifeline during turbulent times. Eating together gives you and your children a sense of belonging, a connection that allows them to be strong when challenged.

Idea for a group:

- Have any of you played "Fly-on-the-Wall" before? It's a fun game. Ready to play? Imagine I was a "Fly-on-the-Wall" during your most recent mealtime with your family. A fly-on-the-wall has a way of getting around and seeing things you might not see when you're involved in a certain situation. What would I see or hear during mealtimes at your home?

Dig:

- How can busy parents find time to sit down and eat together with their families?
- What makes it difficult for your family to eat together?
- What can parents say and do at meals that give everyone—even babies—a chance to connect?
- Family meals can be pressure cookers or oases of peace in a busy day. What can parents do to make them peaceful, fun experiences for all?

Connect:

- What memories of your family meals do you hope your children will cherish?
- Are there rituals or traditions that you could start today that might be something they will share with *their* children?
- What are some things that children can learn from family meals?
- How do you feel, as a parent, after connecting with your child in a powerful way?

Act:

- Without a lot of additional effort or time, simply eating together as a family could actually change the direction of your child's and family's life.
- What's for dinner at your house this week—and who will be enjoying it with you?
- What are some things you can do this week to make eating together possible?
- What can you do to adapt your schedules to make family meals more frequent?

Source: Pam McCarthy and Associates. Emotion Based Messages. Touching Hearts, Touching Minds. 2008;THTM #16: Set the table (Family meals). Available at: http://www.touchingheartstouchingminds.com/ Accessed on October 3, 2010.

EXERCISE 10.2   Implement Set the Table Lesson Plan

First view the WIC Parent Connections Video at: http://www.touchingheartstouchingminds.com/tools_video.php

Work with a group of colleagues and follow the Women, Infants and Children Set the Table Lesson Plan in Exhibit 10.5. One person should volunteer to be the facilitator. For group individuals who do not have children, use your previous family experiences eating together and imagine the questions applying to a child of a relative or close friend.

1. Facilitator welcomes participants: "Hello everyone. Thank you all for participating in this activity. I am happy that we will be working and learning together."
2. Facilitator requests introductions: "I'd like to start by having each person give your name and say the ages of your children or of a close connection."
3. Facilitator encourages participation: "Each of your thoughts, opinions, and feelings are equally important. All of your contributions are valuable. If you have something to say, I hope you will feel free to say it."
4. Facilitator chooses questions: Follow the sequence of the lesson plan in Exhibit 10.4. Choose one or two questions from each category.
5. Summary: The facilitator should make a short summary and encourage group members to contribute to the summary. "Can anyone add to this summary?"

DEBRIEFING:

After the activity, evaluate the outcomes as a group. Discuss the following questions:

❏ What part of the process did you find the most effective?

❏ This lesson plan is one of a series of **emotion-based** materials available from the Touching Minds Touching Hearts Project of the Massachusetts WIC Program. What do you think was the impact of encouraging an emotion-based discussion?

❏ What did you learn from the activity?

❏ Are there components of this process that you can see yourself using in the future with personal or group counseling?

---

acceptance. A counselor guides a group on a journey of self-discovery and shared problem solving. To run effective groups, it is necessary to build on skills developed for conducting individual counseling sessions. The following six steps have been identified as important for the development of cohesive, well-functioning groups.[10,11]

Step 1: *Establish an open, warm environment and productive leader–participant relationships.* The same rules for establishing rapport in one-on-one counseling apply to group counseling. Facilitators need to show empathy, appear warm and genuine, and use relationship-building responses, attentive behavior, and effective body language. You should radiate positive energy indicating that you are looking forward to your time with the group members. Note this does not mean high energy which could be difficult to sustain.

Facilitators with low energy are perceived as having low self-confidence. One way of getting the flow of energy in the right direction is to stay focused on the group members and not let your mind drift to troublesome life issues such as deadlines, childcare issues, and car problems.[6] The counselor is a model of trusting behavior for all group members to emulate to promote openness and interpersonal communication. You may wish to start each meeting with an expression of intent to create an environment conducive to acceptance and open expression.

Ground rules for the group counseling program should be established during the first session. This can be an informal discussion or a written copy of guidelines that each participant signs. Leaders can ask the group to formulate ground rules with the leader making informal suggestions, or to save time the facilitator can provide a preset list of guidelines for the group to comment on and modify. For a list of guidelines generally found in ground rules, see Exhibit 10.4.

Step 2: *Balance facilitator-generated and group-generated information.* The challenge for nutrition counselors is to cover a preset curriculum and integrate client needs and experiences and allow the group to generate solutions and problem solve. Often counselors have a list of tasks identified as essential for clients to understand, but the facilitator-generated information will fall on deaf ears if group members have other concerns on their minds. For example, a person who has diabetes and is worried about amputations due to complications may have trouble focusing on other issues such as glucose monitoring if the complication issue is not addressed first. One way to handle this potential problem is to ask participants in the first session to identify their pressing concerns. Then cover the most pressing problems first.

Step 3: *Design problem-solving strategies.* Many opportunities for group problem solving should be provided rather than having the counselor tell participants what they need to do. Specific guidelines for group problem solving can be found in Exhibit 10.7. However, there are times when giving advice is appropriate and has the most impact if there is a request for advice. The guidelines for giving advice were presented in Chapter 3. However, an additional factor to consider regarding the acceptance of advice is the likeability of the advice giver. Behaviors that encourage likability are provided in Exhibit 10.6.

As covered in Chapter 7 social disclosure is a powerful force for behavior change. Group counseling provides an ideal setting for coupling group problem solving with this process. Participants design goals with feedback from the group and disclose their intent to accomplish certain tasks before the next group session. Depending on the size of the group, you may wish to break up into smaller groups for designing individual goals. In the following session each participant reports back to the group his or her progress in meeting those goals.

Many of the interactive activities identified in Exhibit 6.7 in Chapter 6 can be applied in a group setting and lend themselves to group problem-solving discussions. In fact, some of the activities and demonstrations are more practical for a group setting. For example, groups can list advantages and disadvantages of maintaining blood glucose levels, select a low sodium meal from printed restaurant menus, or role-play placing special orders at a restaurant. An effective way to use role playing is to verbalize probable or possible self-talk while making selections from a holiday buffet or after making poor selections from a smorgasbord of tempting foods.

EXHIBIT 10.6 Facilitator Likeability Behaviors

- Acknowledge and compliment.
- Let people know you like and enjoy them.
- Be enthusiastic, always allowing your joy for living to be visible.
- Listen fully, without interpreting, rather than waiting for someone to finish so you can talk.
- Show a genuine interest in participants and their lives.
- Take time to build relationships rather than being task-oriented.
- Accept each person with unconditional, positive regard.
- Smile.
- Like yourself—it's contagious.

Source: Pam McCarthy and Associates. Emotion Based Messages. Touching Hearts, Touching Minds. 2008; Available at: http://www.touchingheartstouchingminds.com/tools_tips.php.

EXERCISE 10.3 Evaluate Your Facilitator Qualities

Review the three communication styles described in Table 10.1, desirable characteristics of a good facilitator in Exhibit 10.2, and likeability behaviors in Exhibit 10.6. In your journal, answer the following questions:

1. What do you believe are the two most useful components of your communication style, facilitator characteristics, or likeability behaviors that will support your endeavors to be a good facilitator of groups? Explain.

2. What do you believe are the two weakest components of your communication style, facilitator characteristics, or likeability behaviors that could detract from your endeavors to be a good facilitator of groups? Explain.

Step 4: *Provide the opportunity for group members to practice new skills.* In step 3, the group worked as a whole to problem-solve or develop new strategies. Opportunities for each member to rehearse the skill can occur if members are divided into groups of two or three. For example, participants could each practice measuring portion sizes, analyzing blood glucose records of previous clients, or jointly modify a recipe. The new skill should be something clients can use before the next group session so members can report on their experiences using the skill in "real life."

Step 5: *Use positive role models and pacing to keep the group motivated.* Spending time reviewing and understanding the successes of group members can provide a model for other participants to make alterations in their lifestyle. Successful members inspire others to follow their example and stay in the group. However, clients who are having difficulty and appear to be monopolizing the group time with their problems can be frustrating for the rest of the group. If counterproductive behaviors emerge, the counselor needs to block them from disrupting the group process. Disruptive behaviors include scapegoating, personal attacks, aside jokes, unrelated stories, and gossiping. The focus of blocking should be on the behavior, not on the person.[12] Specific techniques for handling difficult clients can be found in Table 10.6. If necessary, the counselor should tell the client to stay after the session to receive personal attention.

Step 6: *Ask for evaluation and feedback.* Throughout the counseling process and after trying out a new activity or strategy, the facilitator should elicit feedback from the group. For example, "Since we began meeting, what did you find particularly useful?" or "Did you find analyzing glucose records of previous clients useful?"

Table 10.6 Suggestions for Dealing with Difficult Group Participants

Participant Behavior	Problem or Possible Motive	Possible Actions
Participant statement is definitely wrong.	• Making an obviously incorrect comment	Must be handled delicately. Say, "That's great that it worked for you, but others have found . . ." or "That's one way of looking at it, but there are authorities that believe . . ." or "I see your point, but let's try looking at it this way."
Searching for leader's opinion Example: *"What do you snack on?"*	• Trying to put you on the spot • Trying to have you support one view • May be simply seeking your advice	Generally, you should avoid solving problems. However, there are times you should give a direct answer. Before you do so, you may want to open up the discussion to other participants by asking, "Let us get some other opinions about this issue. How do you view this point?" (Direct your question to a particular person.)
Silent	• Bored • Indifferent • Feels superior • Timid • Feels input not wanted or valued	• Your action will depend on what you believe is motivating the participant. • Use eye contact to encourage participation. • Arouse participant's interest by asking for his or her opinion. "I do not want to miss what you have to say. What do you think about this?" • Break into small groups.
Griper Example: *"All this talk is useless."*	• Has a pet peeve • Professional griper • Has legitimate complaint	• Point out that we can't change policy here and that we must operate as best we can under the system. • Say you'll discuss problems with person privately later. • Ask group, "How do the rest of you feel?"

(continued)

Table 10.6 Suggestions for Dealing with Difficult Group Participants *(continued)*

Participant Behavior	Problem or Possible Motive	Possible Actions
Side conversations	• May be related to the subject • May be personal • Distracting to the group and you	• Friendly reminder: *"Please, one conversation at a time."* *"Let's get down to business."* • Pause until disruption stops. • Ask talker a question, or restate last opinion expressed and ask for an opinion. • Stroll over and stand casually behind or next to the participants who are talking. • If there are numerous side conversations, ask *"Do we need to take a break?"*
Inarticulate	• Lacks ability to put thought in order	Say, "Thank you—let me repeat that." Then put the idea in better language.
Attacker Example: *"That is a stupid idea."*	• Appears hostile • Seems angry • Can be abrasive • Seeks to discredit an idea	• Remind the person that we need to respect all opinions. • Remind the person that we agreed there will be no personal attacks. • Ask them about their feelings. • Ask them how their behavior helps the group. • Try humor, "Do we need body armor for this meeting?"
Expert Example: *" You should just refuse to allow any soda in the house."*	• Wants to control • Wishes to be seen as the expert	• "Sounds as if you know a lot about this. Now, let's hear what others think. Who would like to share next?" • "Grace, I'd like to stay with our discussion of self-talk and hear from others in the group before we cover new ground."
Dominator Example: *"We should be talking about the new Salmonella outbreak."*	• Just likes to talk • Wants to control • Avoiding a topic	• Generally easy to spot; talking as entering the room; if possible, have person sit next to you; slowly turn your back toward dominator and say, "Thank you Nurgis, does anyone view this differently?" • "Miguel, it appears that you want to change the topic when we talk about monitoring food intake." • "Hank, you are really telling an interesting story about your job, but let's plan to finish the story at the end of the session so we can review how everyone did with their goals this week."
Rambler	• Just likes to talk • May be related to wanting to control	Discontinue eye contact; interrupt with a new question; "I think we need to move on."
Playful Behavior *"Let's end early today."*	• Distracted	Compliment and move on. "You bring a lot of energy to this group, but we are getting off-track." Remind member of the ground rules.

Source: Adapted from: © 1995, The American Dietetic Association. "Cardiovascular Nutrition." Used with permission.

Ending

You will need to bring closure to the session. Provide a summary which may include what you want to emphasize as well as problems, solutions and some of the change talk brought up by participants.[13] Consider requesting that members of the group contribute to the summary by voicing the significance of the meeting to them. Some possible questions include the following:

• What has made the greatest impact on you from the time we have spent together today?

EXERCISE 10.4   Match Problem Statement with Response

Review the following problem statements or behaviors and match with a facilitator response that could be effective:

	Participant Statement	**Counselor Response**
1. _____	"That doesn't make any sense."	A. "How do others feel about the discussion so far?"
2. _____	"If you want to stop diarrhea, don't drink anything.	B. "That is one way of handling the problem. Does anyone else have an idea of what to do?"
3. _____	"I think our discussion isn't getting us anywhere."	C. "Do we need to take a break?"
4. _____	Numerous side conversations	D. Ask participant, "What do you think about this issue?"
5. _____	"If he doesn't eat dinner, just don't give him anything else to eat for the rest of the night."	E. "Remember we agreed to respect all opinions."
6. _____	Silent participant	F. "I am glad you brought that up. It seems so obvious, but actually that practice is dangerous."

EXERCISE 10.5   Challenging Group Scenarios

Review the following scenarios with your colleagues and discuss possible ways to deal with the situations.

Scenario 1: Dialysis Unit

Jamie is facilitating a group discussion in a kidney dialysis unit on dietary needs for individuals with diabetes who are receiving dialysis treatment. One of the participants, who looks bored and has previously attended a similar program states, "This is a waste of time. How is this supposed to help us?"

How should Jamie respond?

Scenario 2: Rehabilitation Center

LaTonya is facilitating a group of patients who have experienced a heart attack. One of the participants is monopolizing the conversation. None of the other group members appear engaged or anxious to talk.

How could LaTonya handle this situation?

Scenario 3: Sports Club

Christopher is holding a session on healthy eating at a sports club. One of the participants looks bored and is looking out the window.

What should Christopher do?

- What do you think has benefitted you the most from our meeting today?
- What strategies have you learned here that you can picture using?[13]

Special consideration should be given to the final meeting of an ongoing group in order to encourage the power of group support to have a continuing effect. Before the last meeting, tell the

---

EXHIBIT 10.7   Group Problem-Solving Guidelines

1. *Identify* a problem of one or more group members
2. *Assess* the conditions that contribute to the problem and identify factors that promote healthful practices and alleviate the problem. Evaluate the following:
   - *Physical environment*—aspects of the external environment that cue poor eating habits, as well as aspects that remind the person to eat appropriately
   - *Social environment*—social situations that support poor eating habits and identify people who could support good eating habits
   - *Cognitive environment*—thoughts and feelings that get in the way and positive ideas that can be used to promote positive habits
3. ***Brainstorm*** *solutions.* The objective of brainstorming is to generate as many ideas as possible for consideration. There is one major rule: no censorship. No idea is rejected, no matter how silly or useless it may appear initially.
4. *Select a solution.* After all ideas are listed, the person selects one and plans the details of implementation and evaluation.

Source: Raab C, Tillotson JL, Heart to Heart: A Manual on Nutrition Counseling for the Reduction of Cardiovascular Disease Risk Factors. NIH Publication No. 83-1528. Washington, DC: U.S. Department of Health and Human Services; 1983.

---

participants that you would like the group to honor and acknowledge accomplishments of the individuals. Ask members to either write or think about their journey of change and address the next chapter of the journey in their story, that is, what they plan to do to continue the new behaviors and possibly include some additional healthy changes. At the final meeting, summarize the evolution of the group and ask participants to share their stories.[13]

Practical Considerations for Successful Groups

The following is an overview of practical matters that must be handled when organizing a group:[10,14]

- **Allow adequate time for organization.** Generally six to eight weeks are needed to arrange for a meeting location, publicize, and develop curriculum.
- **Plan for adequate meeting time.** In order to receive the advantages of group counseling, sessions should be at least 60 and preferably 90 minutes in length.
- **Select a comfortable meeting room and location.** The room should be large enough to accommodate all participants to sit around a table or in a U-shaped arrangement. A circular table provides a better environment for exchange among all group members rather than a rectangular table, at which participants are more likely to limit interaction to those directly across from them. Additional space to allow participants to break up into pairs or small groups would also be an advantage. If large people are likely to attend, be sure the room has sturdy chairs of adequate size. The room should have satisfactory lighting, temperature control, and ventilation. Generally, people feel more comfortable talking in a room that has a closed door. In addition, make sure that the location is easily accessible via public transportation and car.
- **Ideal group size for closed groups is eight to twelve participants.** This appears to be a good size for the development of a group identity. If the group gets too large, there will not be a free flow of conversation, and giving individual attention will be difficult. The tendency will be to revert to a lecture format. If the size is too small, the dynamics of the group will be severely impaired if one or two people are missing.
- **Contemplate collecting fees or refundable deposits before the first meeting.** Fees encourage better attendance. If a periodic pay schedule is used, fees should be paid before attendance at sessions. Some programs use refundable deposits that are gradually returned as participants attend meetings. In some cases, the refunds are graduated so that the final payment is the largest.
- **Appraise target group needs for selecting a meeting time.** If most members are working,

then late afternoon, evening, or a weekend day should be considered. If possible, a survey could be taken of interested parties. A daytime meeting may work best for retired individuals.

- **Consider composition issues of the potential group.** Groups with common health concerns, age, and gender are likely to share similar needs and goals. However, some diversity can provide valuable perspectives. If the decision is to mix the composition, care should be taken to keep numbers somewhat balanced so no one feels left out.
- **Interview prospective group members.** Interviews could be handled over the phone, but it would be better to meet individuals in person. By assessing prospective participants, you can determine suitability for their participation and design a program that better meets their needs. Also, you can receive assurance that the people understand the specifics about the group purposes and procedures.
- **Group leaders should remain the same.** Group leaders should not be considered interchangeable. New leaders break continuity and cohesion and create disruption.
- **Be responsible.** Start meetings on time and always be well prepared. Test-run activities, such as measuring portion sizes. Make a list of equipment and supplies that will be needed. Arrive early to be sure the room is arranged properly and equipment is working. Always follow through on promises to locate information or resources or to make contact with members between sessions.
- **Plan sessions carefully.** Make an outline of topics, activities, and the amount of time expected to spend on each.
- **Consider refreshments.** Generally, sharing of food encourages bonding. The selections could provide exposure to foods promoted for the group. It is not unusual for participants to offer to bring in samples of foods. Group members could sign up to act as host or hostess to bring in refreshments and to make reminder phone calls to other group members.

EXERCISE 10.6   Interview a Group Counselor

Interview an individual, either on the telephone or in person, who runs group counseling programs, preferably nutrition programs. Ask the following questions, and record the answers in your journal.

1. What type of groups do you facilitate? How often do you run them?
2. How do you view your role in the group process?
3. What skills do you find particularly useful for facilitating groups?
4. Do you establish ground rules? If so, how are they set?
5. How do you set the agenda for the sessions?
6. How do you evaluate the effectiveness of the groups?
7. What advice do you have for a novice group counselor?
8. What did you learn from this interview?
   - ❏ Record the name of the person you interviewed and the date and time of the interview.

A restaurant meal or a potluck could be considered for the last meeting.

- **Call members who miss meetings.** If a participant misses a meeting, encourage group members to call to review what was covered and to encourage attendance at the next meeting. You may even design a buddy system. Be sure to request permission before distributing phone numbers.

EVALUATION OF GROUP INTERACTIONS

Running group meetings can be rewarding and challenging, and having a mentor to review progress can be beneficial. If possible, ask a colleague to sit in and provide feedback. Written evaluations from members can supply useful information. Table 10.7 provides a group assessment form to aid in evaluation.

Answers to Exercise 10.4: 1 = E, 2 = F, 3 = A, 4 = C, 5 = B, 6 = D

Table 10.7 Group Assessment

		Not at all	Somewhat	A Great Deal
• For each statement fill in the box under the MOST APPROPRIATE heading that best describes the group during the four sessions. • Please mark only ONE box for each statement.				
Facilitator Evaluation				
1. Warmly greeted participants.		❑	❑	❑
2. Showed genuine interest and empathy.		❑	❑	❑
3. Encouraged participation, for example, "All of your contributions are valuable."		❑	❑	❑
4. Radiated positive energy.		❑	❑	❑
5. Established a clear agenda.		❑	❑	❑
6. Made best use of time and resources.		❑	❑	❑
7. Encouraged participation from all group members.		❑	❑	❑
8. Used questions effectively.		❑	❑	❑
9. Provided a clear and concise summary at the end.		❑	❑	❑
Group Interaction Evaluation				
10. Group members appeared interested and engaged in discussions.		❑	❑	❑
11. Group members freely discussed issues of concern.		❑	❑	❑
12. Members respected and encouraged contributions of others.		❑	❑	❑
13. Members refrained from engaging in side conversations.		❑	❑	❑
14. Members of the group appeared comfortable with each other.		❑	❑	❑
15. Individual members did not try to dominate or demoralize the group.		❑	❑	❑
16. Intellectual needs of all group members were addressed.		❑	❑	❑
17. Group members offered supportive comments to each other.		❑	❑	❑

> **CASE STUDY   Group Facilitation at a Diabetes Camp for Adolescent Girls**
>
> Donna Vente, RD, works as a nutrition counselor at a summer camp for forty pre-adolescent (ages 9 to 12) girls who have diabetes. You are heading a group to plan a final farewell celebration for the end of a two-week camp for participants and their families. The planning group consists of you, the assistant camp director, the recreation counselor, the head cook, psychotherapist, two junior counselors, and four pre-adolescent camp attendees. The group needs to plan decorations; a buffet; activities and entertainment; and possible vendors or resources for sale, such as books.

> **EXERCISE 10.7   Case Study Planning Activities**
>
> Review the chapter discussion of facilitating groups. Go to the following website and read the Basic Facilitation Primer, http://www.uiowa.edu/~cqi/2002BasicFacilitationPrimer.pdf
>
> ❏ Record in your journal two things you learned in the Basic Facilitation Primer that could be useful to Donna for leading her group.
> ❏ Review the four strategies described under preparation and select one that may be useful to implement Donna's meeting.
> ❏ Besides planning a strategy for the flow of the meeting, identify two additional factors that should be part of Donna's planning process. Describe why addressing them would be useful.

REVIEW QUESTIONS

1. Describe three communication styles.

2. What guides the facilitator choices regarding content and focus of questions?

3. Identify four techniques for organizing a meeting for group decision making and problem solving.

4. What do emotion-based counseling materials emphasize?

5. Identify and explain four advantages of group counseling.

6. Identify and explain four disadvantages of group counseling.

7. Explain the six steps of the group process.

ASSIGNMENT—PRACTICE GROUP COUNSELING

Work with a small group of colleagues and plan to meet for three group sessions. During the first meeting, each person in the group should identify a healthy lifestyle behavior change. Possibilities include but are not limited to healthy snacking, eating breakfast, consuming more vegetables, late night eating, reducing sodium intake, drinking more fluids, strength training, increasing aerobic activities, or practicing yoga or relaxation exercises. Having an actual issue rather than a role-play will make the experience much more meaningful and realistic. Plan to set goals around this behavior during the group session. Start by identifying a facilitator. You may decide to rotate the facilitator position for each meeting to give opportunities for others to have experience with the facilitator's role. The following provides guidance for the facilitator to guide the flow of the first meeting:

First Session:

1. The facilitator should provide guidance on the following for the first meeting:

   • Decide on meeting time and location.

   • Arrange chairs and reduce distractions.

2. Begin with introductions and provide an icebreaker.

   For example, "I am happy we all have this experience to work together. Let's begin by introducing ourselves and answering the following, "What is your favorite healthy food?"

3. Agree on ground rules. All members should indicate agreement on the final rules. See Exhibit 10.4 for guidance. Review each and add or subtract as the group sees fit.

   For example, "Before we begin discussing our desired health behavior changes, we need to set ground rules for our meetings. Here is a list of common ground rules. What do you think?"

4. After receiving an agreement on the ground rules, the facilitator may confirm the purpose of the group and ask others for their objectives regarding participation.

   For example, "Obviously we are all here as part of an assignment so we can experience first-hand organizing and implementing a behavior change group. So learning something about the group process is a common goal. But I am wondering if there are other objectives individuals are hoping to obtain from participation?" (A possibility may be to get to know colleagues better.)

5. Express your intention for the group process.

   For example, "My hope is to create an environment conducive to acceptance and open expression."

6. Have each person identify a desired health behavior change.

   "We will begin by identifying a desired health behavior change. Because this was an assignment, our actual commitment to making a change may not be as high as the action stage, so let's also identify a readiness to make that change. On a scale of one to ten, with one indicating no change and ten meaning very ready, how motivated are you to making a change? Who would like to start?"

7. Explore experiences.

   For example, "Has anyone else attempted or accomplished any of these desired behavior changes?" "As with many people, you seem ambivalent about making a change. Can anyone offer suggestions?"

8. Experiment with some behavior change talk encouragements.

For example, for low confidence numbers, discuss simple or reduced plans. Ask the group for suggestions. "You chose a low number, so maybe the objective should be restated. Is there a smaller step that could be taken to work towards the goal?"

"Imagine the future–next fall for example–(see Chapter 4) when you are actively involved in this new behavior. What is the best part of this new behavior? What helped you get to this point?"

9. For those ready to make a behavior change, use Exhibit 10.7 to guide the formulation of goals. Be sure to give each member an opportunity to set a goal, if desired. Note that a goal could be to explore options, such as possible walking trails. Participants should not feel as if a goal would only be appropriate if they are ready to take action to implement the behavior change.

10. Ending. Bring closure to the group. Summarize the high points of the meeting and ask others for their input.

    For example (first meeting), "What was it like taking part in this group?"

    "What has made the greatest impact on you from the time we have spent together today?"

    "What do you think has benefitted you the most from our meeting here today?"

11. Evaluation. As a group review the group process.

    a. As a group, review the questions in Exhibit 10.1 and the responses in Table 10.6. Were any of these used? How effective were they?

    b. What were the high points of the group process?

    c. What were the low points of the group process?

    d. Was there a monopoly on talking? If so, what could have been done to change the dynamics of the group?

Subsequent Sessions

1. Welcome everyone back to the group meeting.

2. Provide an intention for the meeting.

For example, "My hope is to create an environment conducive to acceptance and open expression."

3. Ask individuals to explain their progress with their desired behavior change.

   Example: "Let's review the progress with the desired behavior changes. Who would like to start?"

4. Explore dilemmas. As individuals explain difficulties in handling behavior change. Invite others in the group to explore the difficulties.

   For example, "Can anyone relate to this issue?"

   "How do you feel when that happens?"

5. Repeat Steps 8 through 11 of the first session.

Written Report of the Group Counseling Experience

Using skills to adeptly lead groups takes years of experience. If you were the facilitator, do not judge yourself too harshly if the progress of the group was rocky. After completion of the group meetings, write a report answering the following questions:

1. Complete the group assessment questionnaire in Table 10.7 after each meeting. Explain what you perceived to be highlights of the assessment. Give examples.

2. Describe how the first, second, and third meeting evolved. Be sure to give examples.

3. Give two examples of questions or responses given by the group facilitator found in Exhibit 10.1 and Table 10.6. Identify the session in which each response was used and describe the circumstances of the response. What was the effect of each question or response?

4. How helpful was the group process for clarifying your desired behavior change or for making your desired behavior change? Explain.

5. If you were to repeat the group experience, give two suggestions that may have improved effectiveness.

6. Review the advantages and disadvantages of group counseling identified in this chapter. Did any of these apply to this group experience? Explain.

SUGGESTED READINGS, MATERIALS, AND INTERNET RESOURCES

http://www.managementhelp.org/
Free Management Library, many resources for working with groups.

http://www.azdhs.gov/
Arizona Department of Health Services, search for facilitated group discussion, Module 20.

http://www.brookes.ac.uk/
Oxford Brooks University, search for characteristics of groups.

http://www.touchingheartstouchingminds.com/
WIC emotion-based resources.

http://www.nova.edu/gsc/online_files.html
Center for Psychological Counseling, Southeastern University.

Kaner S, Lind L, Toldi C. *Facilitator's Guide to Participatory Decision-Making.* Jossey-Bass; 2007.

Abusbha, R, Peacock, J, Achterberg, C. How to make nutrition education more meaningful through facilitated group discussions. *J Am Diet Assoc.* 1999; 99:72–76.

Corey G. *Theory and Practice of Group Counseling.* 8th ed. Belmont, CA: Brooks/Cole Pub Co; 2011.

REFERENCES

[1]Division of Cooperative Extension of the University of Wisconsin-Extension. Facilitating Access to Resources and Best Education Practices; 2010, Available at: http://wateroutreach.uwex.edu/education/Facilitation.cfm Accessed September 10, 2010.

[2]Doyle M. Foreword. In: Kaner S, Lind L, Toldi C, et al. *Facilitator's Guide to Participatory Decision-Making.* Philadelphia, PA: New Society Publishers; 1996.

[3]Burke DW, Donahoe M, Hirzel R, et al. *Basic Facilitation Skills.* The Human Leadership and Development Division of the American Society for Quality, The Association for Quality and Participation, The International Association of Facilitators; 2002. Available at: http://www.uiowa.edu/~cqi/2002BasicFacilitationPrimer.pdf

[4]Lawson, SL. A quick reference guide for facilitators. Ministry of Agriculture, Food and Rural Affairs, 2008; Available at: http://www.omafra.gov.on.ca/english/rural/facts/95-073.htm Accessed on September 10, 2010.

[5]Kaner S, Lind L, Toldi C, et al. *Facilitator's Guide to Participatory Decision-Making.* Gabriola Island, BC; 1996.

[6]Pam McCarthy and Associates. Emotion Based Messages. Touching Hearts, Touching Minds. 2008; Available at: http://www.touchingheartstouchingminds.com/ Accessed on October 3, 2010.

[7]Spahn JM, Reeves RS, Keim KS, et al. State of the evidence regarding behavior change theories and strategies in nutrition counseling to facilitate health and food behavior change. *J Am Diet Assoc.* 2010;110:879–891.

[8]Macario E, Emmons KM, Sorensen G, et al. Factors influencing nutrition education for patients with low literacy skills. *J Am Diet Assoc.* 1998;98:559–564.

[9]Renjilian D, Perri M, Nezu A, et al. Individual versus group therapy for obesity: Effects of matching participants to their treatment preferences. *J Counsel Clin Psychol.* 2001;69:717–721.

[10]Helm KK. Group process. In: Helm KK, Klawitter B. eds. *Nutrition Therapy Advanced Counseling Skills.* Lake Dallas, TX: Helm Seminars; 1995:207–213.

[11]Raab D, Tillotson JL. *Heart to Heart: A Manual on Nutrition Counseling for the Reduction of Cardiovascular Disease Risk Factors.* NIH Publication No. 83-1528. Washington, DC: U.S. Department of Health and Human Services; 1983.

[12]Brownell K. The psychology and physiology of obesity: Implications for screen and treatment. *J Am Diet Assoc.* 1984;4:406–413.

[13]Kellogg M. Counseling Tips for Nutrition Therapists. Available at: www.mollykellogg.com Accessed on October 3, 2010.

[14]Klein L, Axelrod B. Using the workbook in a group setting. In: Wylie-Rosett J, Segal-Isaacson CJ, eds. *Leaders Guide: The Complete Weight Loss Workbook.* Alexandria, VA: American Diabetes Association; 1999:19–22.

Keys to Successful Nutrition Education Interventions

Alex Mares-Manton/Asia Images/Getty Images

You cannot teach a man anything; you can only help him to find it within himself.

—GALILEO

Behavioral Objectives

- Explain the process of conducting a comprehensive needs assessment for a target audience.
- Describe three nutrition education approaches used by researchers and health professionals.
- Understand the importance of a theory-based approach in designing nutrition education interventions.
- Use the three domains of learning in the formulation of goals and objectives.

Key Terms

- **Generalizations:** clear statements that are universally true.
- **Goals:** broadly stated learner outcomes.
- **Needs Assessment:** the collection of comprehensive data which may encompass health, educational, resource, and developmental needs of a target audience.
- **Objectives:** specifically stated learner outcomes.
- **Social Marketing:** application of commercial marketing technologies to promote the voluntary adoption of behavior that is beneficial to a target audience and/or society.

INTRODUCTION

As defined in Chapter 1, nutrition educators and counselors design interventions to influence knowledge, skills, and attitudes. This chapter explores the basic components of the nutrition education process for a target population, which could be an individual, group, or community. We will first review common locations for nutrition education interventions. Then, we will explore components (keys) of successful interventions: **needs assessment**, educational philosophy, theory-based interventions, and **goals** and **objectives**. Additional keys, learning strategies, audiovisual materials, and evaluation are covered in Chapter 12. In order to visualize the process, a model of the process is presented in Figure 11.1. We will journey through the keys by following an interactive and continuous case study. Discussion of each key is followed by a relevant component of the case study and related exercises to use and integrate core concepts. We will work together through this interactive case study to explore the development of a nutrition education intervention.

NUTRITION EDUCATION SETTINGS

Effective nutrition interventions have occurred in a variety of public, government, and commercial settings. Nutrition education programs provide

opportunities to target multiple population groups including significant people who control access to food choices in homes or community settings. The following section describes some of the common locations for interventions:

- **Consumer marketplace:** Nutrition education initiatives may take place anywhere consumers purchase food items, such as grocery stores, on-line market sites, fast food establishments, and restaurants. Surveys of consumers indicate a great deal of interest in purchasing healthy foods.[1] Using nutrition education to market food can be a win-win situation for consumers and retailers alike as long as the nutrition information is reputable. Interventions have included cooking classes, taste tests, coupons, videos, point of purchase labeling, large posters, interactive website games, and nutrition labeling on menus. Usefulness of marketplace nutrition education is illustrated by lower calorie intake of patrons of restaurants and fast food establishments displaying point of purchase calorie information.[2,3] Recognizing the benefits of marketplace education, the U.S. government has responded in at least 16 states by mandating point of purchase calorie information in chain restaurants.[4]

- **Communities:** Nutrition education interventions in community settings may be a component of organized programs, such as congregate meal programs for senior citizens or Women, Infants and Children Special Supplemental Nutrition Program for low-income pregnant mothers and young children. These community-based programs target specific population groups using media and interpersonal strategies to encourage healthy food behaviors. More recently, religious organizations have created health initiatives including nutrition seminars, weight loss groups, and fitness classes. Nutrition professionals may also become involved in the development and management of grass root programs to improve nutritional health, such as food pantries, soup kitchens, community gardens, or farm to table initiatives (programs dedicated to improving community access to nutritious, affordable, and locally grown food).

Figure 11.1   Nutrition Education Process Model
Source: Designed by author.

- **Health Care Settings:** Clinical settings can include physician's offices, health maintenance organizations, hospitals, public health clinics, nursing homes, and various assisted living facilities. In many of these sites, nutrition professionals must be registered dietitians and are likely to address treatment of disease as well as prevention. A variety of nutrition education initiatives can occur in health care sites. For example, a nutrition educator may be involved with a single client or a group of thirty. The focus of an intervention could be community outreach, outpatient services, high risk patients, or families of patients. Employees of the facility may also be the target population for in-service training or worksite wellness education.

- **Worksites:** Nutrition education is often a component of corporate wellness programs. Such programs are mutually beneficial for employees and employers.[5] Employees have a convenient location to pursue wellness activities, and healthier employees often translates into improved productivity, reduced absenteeism, and lower healthcare costs.[6] Effective worksite programs have reduced cardiac risk factors in employees.[7] Nutrition educators provide information and organize programs for weight reduction and reducing risk of chronic diseases, such as high blood pressure, metabolic syndrome, diabetes, and heart disease.

- **Schools:** Nutrition education is an important part of a comprehensive school district health intervention targeting children, families, administrators, staff and food service operations.[8,9] School curricula may include a variety of learning strategies directed at improving nutrition knowledge, attitudes, skills, and diet-related behaviors.

KEYS TO NUTRITION EDUCATION

Comprehensive reviews of nutrition education research reveal key factors contributing to effective nutrition education interventions.[10] These factors provide a guide for all nutrition educators, whether planning a corporate wellness program or designing an education course for a diabetic population in a clinical setting. The following seven keys incorporate the major factors as a guide for implementing successful nutrition education interventions:

Key #1   Know Your Audience–Conduct a Thorough Needs Assessment

Key #2   Determine Your Educational Philosophy

Key #3   Design Theory-Based Interventions

Key #4   Establish Goals and Objectives

Key #5   Provide Instruction Planning and Incorporate Learning Strategies

Key #6   Develop Appealing and Informative Mass Media Materials

Key #7   Conduct Evaluations

KEYS TO SUCCESS #1–KNOW YOUR AUDIENCE, CONDUCT A THOROUGH NEEDS ASSESSMENT

To ensure effectiveness of nutrition education interventions, a planning process is required. After a specific target audience is clearly defined, accurate and comprehensive data collection should be conducted as part of a **needs assessment**.[11] You need to know your audience! The following list provides categories and sources of information for a thorough investigation of needs. However, not all sources will be available, and time may be limited in certain situations. For example, you may be requested to provide a one-hour nutrition program for a substance abuse facility in your medical center in two days. Although time would be limited, doing what you can to explore the nutrition education needs of the group will likely improve the quality of your program. Discussion of planning for a large-scale community or national nutrition education interventions are beyond the scope of this book, which would require planning for sample size, data management, quality control and statistical analysis.

Needs Assessment Categories

- *Health Needs:* Explore disease prevalence, mortality rates, disability issues, and physical symptoms of the target audience via a review of literature, results of national and local surveys, and findings from epidemiological studies and health records. Health can broadly encompass the physical, mental, nutritional, and spiritual wellness of an individual. Epidemiological data describe how a disease or problem is distributed in a population. For example, the prevalence and incidence (new cases) of osteoporosis is particularly high among older, petite, and sedentary women. In addition, inquires of other health professionals working with the population group, distribution of a questionnaire, or direct interaction with the target group can provide valuable information about health needs.

- *Educational Needs:* An educational needs assessment includes an evaluation of knowledge, attitudes, motivational level, self-efficacy, and skills of the target population in order to plan an appropriate intervention. Of course, you also want to explore what your clients want to learn. Motivation to participate in an education intervention will be enhanced by addressing their interests. You may feel there are educational matters that are more important to address than indicated in your needs assessment; however, the needs of an educator can be incorporated into a program highlighting the concerns of the target audience.

- *Resource Needs:* Plans for nutrition education programs need to take into consideration factors that could enhance or hinder the adoption of new food behaviors. A needs assessment should include an evaluation of food availability, income, lifestyle factors, food prices, transportation resources, cooking facilities, social support, and availability of food assistance programs. The nutrition educator should also differentiate between a genuine lack of monetary need and the failure of the individual to efficiently use available resources. The evaluation should also include possible channels for promoting nutrition education, such as media outlets including newspapers, television, web-based resources, and radio stations. You may also explore possible collaborations with community or private organizations.

- *Developmental Needs:* In order to develop relevant nutrition education programs, planning needs to take into consideration the intellectual, social, emotional, and physical development of the target audience.[12] To be sure to address developmental needs and concerns of the learners in your target population, you should be able to answer the following questions: What is relevant and important to the learner at this life stage? What life events are affecting the target audience?

Developmental needs of children, adolescents and older adults were covered in Chapter 9. Here we explore the needs of adults. Characterizing the adult learning process as *andragogy* (as compared to pedagogy), researchers[13] have identified six major assumptions regarding adult learner needs:

1. Adults are *relevancy-oriented* and need to know why, what, and how. In contrast to children, who are expected to learn predetermined curriculum, adults are not likely to put in the time and effort unless they are convinced that a need exists. It is best for adults to discover gaps in learning between where they are now and where they want to be. For example, having an individual evaluate nutrient intake by comparing a dietary assessment to a standard, such as the Dietary Reference Intakes, will be more relevant than just handing an evaluation to an adult.

2. Adults are *autonomous* and *self-directed learners.* They need to feel free to direct themselves and take responsibility for their learning. Adults expect to be treated as independent, responsible individuals. Educators should solicit participants' perspectives on topics to cover.

3. Adults bring a vast amount *of experience* and *knowledge* into the learning environment. By designing programs which draw upon these factors, you increase the likelihood of providing an enriching and motivational intervention.[14] Group discussions, chat rooms, and values clarification activities are possible techniques.

4. Adults are *practical*. For adults to be motivated to acquire new skills, they need to believe that the outcome of the educational experience will lead to the development of useful skills.

5. Adults are *task-centered* and *problem-oriented* learners, especially when learning is related to real life situations. Activities such as menu selection for lower calorie entrees or learning to cook tasty foods without salt may be useful for specific adult populations.

6. Adult learners tend to be *intrinsically motivated*. Potent motivators for adults include internal pressures such as desiring an improved quality of life or increased self-esteem, as compared to extrinsic motivators such as higher salaries and promotions. Nutrition educators need to articulate to clients how the quality of their lives is likely to improve with adopting lifestyle changes.

DATA COLLECTION METHODS

Various methods can be used to gather data, but choosing which ones will work best in your situation will depend on your target population and a realistic appraisal of your resources. Choice of assessment procedures will also be contingent on practical matters such as time and money. A simple questionnaire or blood pressure screening may be all that is practical in a given situation.

- **Review of Published Data:** Carefully explore prior research, government statistical data, census data, and even statewide morbidity rates, which can reveal useful information when properly extrapolated to the target audience. Evaluation of findings will provide insight regarding behavioral, psychosocial, and environmental factors relevant for planning a nutrition education intervention for your population group. This investigation will serve as a cornerstone for the design of your nutrition education intervention.[15]

- **Use Facility Records:** Clinical records can provide data on patient knowledge, dietary behavior, biochemical profiles, types of medication, and compliance issues with medical recommendations.

- **Interviews with Target Audience:** Face-to-face interviews with members of the target population, friends, family members, and co-workers may be the most valid source of information about the educational needs of the target audience.

- **Interviews with Key Informants:** Professionals and staff personnel who consistently interact with the target population can be interviewed as they are likely to provide the educational needs of the target audience. Employers, teachers, supervisors, health professionals, and clergy may be influential in explicating the knowledge, attitudes, and behaviors of the target audience.

- **Nutrition Assessments:** A variety of methods are available to assess quality of food intake including use of food frequency questionnaires, 24-hour recalls, and photos or written records of food intake. In addition, biochemical tests can be administered to identify nutrient deficiencies, anthropometric measurements can detect moderate and severe malnutrition, and a medical history and physical examination can reveal nutritional concerns. Refer to Chapter 5 for more information about these methods.

- **Qualitative Research:** Qualitative research involves the collection and analysis of non-numerical, unstructured information from sources, such as interview transcripts, emails, notes, feedback forms, photos, and videos. Data collection can come from in-depth interviews, behavioral observations, and focus groups. Focus groups usually consist of 6 to 12 people who represent your target population and sessions last from one to three hours. Participants voice their concerns, beliefs, and experiences about an issue, a product, or service in order to obtain insight about an intended nutrition education intervention. Analysis of data reveals common themes and relationships. For example, a qualitative analysis of the cultural interface of psychosocial variables related to obesity risk among Chinese Americans revealed a pressure to over-consume calories from both the American fast-food culture and Chinese elders who viewed plumpness as a sign of status.[16]

Qualitative research can reveal real-life issues and actual thinking patterns from the target members' point of view.

- **Quantitative Research:** Quantitative research collects and analyzes numerical data that can be measured for statistical significance. Methods include analysis of standardized surveys, systematic observations, experiments, census, or epidemiological data.

- **Survey:** A survey is a systematic investigation designed to describe and quantify characteristics of a target population. Collection of data can be obtained via questions by a trained interviewer in person or over the phone. Surveys can be self-administered questionnaires on paper or on-line. Questions may request information about individuals' past, present, or future behavior and their underlying attitudes, beliefs, and intentions regarding the behavior under investigation. Surveys are relatively easy to manage and can be used to collect qualitative or quantitative data. Formal research surveys tend to involve a team of experts with knowledge in research design, statistics, epidemiology, public health, and nutrition. However, simple questionnaires developed for your population group can provide meaningful data. See Table 11.1 for a list of questions to ask yourself when designing a survey instrument. Pre-testing the survey with select members of the target group and obtaining feedback would also help to refine the usefulness of the questionnaire.

Table 11.1 Questions to Ask When Designing a Survey

Is the survey valid and reliable? Will it measure what it is intended to measure, and assuming that nothing changed in the interim, will it produce the same estimate of this measurement on separate occasions?

Are norms available? Are reference data or population standards available for comparison against the data you collect about your target group.

Are the survey questions easy to read and understand? Survey questions must be geared to the target population and its level of literacy, reading comprehension, and fluency in the primary language. Having a readable survey is especially important if it is to be self-administered.

Is the format of the questionnaire clear? If the questionnaire is not laid out carefully, respondents may become confused and inadvertently skip questions.

Are the responses clear? A variety of scales and responses may be used in designing a survey. Some questions may require filling in blanks or providing simple yes or no, or true or false answers. Others may ask respondents to rank-order their responses from "seldom to never use" to "use often to always." The trick when selecting such scales is to choose one that allows you to differentiate between responses but does not provide so many categories that respondents are overwhelmed.

Is the survey comprehensive but brief? Respondents should be able to complete a survey in a reasonable amount of time. If there are too many questions, respondents become fatigued. They may attempt to answer questions quickly and may even mark the same answer to most questions.

Does the survey ask "socially loaded" questions? Each survey question should be evaluated for how it is likely to be interpreted. Questions that imply certain value judgments or socially desirable responses should be rewritten. This is especially important when dealing with respondents from cultures other than your own.

Source: Adapted from; Fallowfield L. The Quality of Life. London: Souvenir, 1990, pp.40–45.

CASE STUDY   Nutrition Education Intervention for a Congregate Meal Program

Case Study Procedure: This is an interactive case study. You will be given information about a target audience. As we explore the development of an education intervention for this group, you will be asked to perform various activities and make observations along the way.

Description: You are responsible for implementation of a nutrition education program for a government supported congregate food program for seniors 60 years and older located in a suburban senior community center. Approximately 50% of the participants are white, 30% are African-American, and 20% are Hispanic. The program provides hot, nutritious noon-time meals each day of the week. The program services approximately 50 individuals and generally there are about 40 in attendance. Participants are often given left-over foods to take home. The program is required to offer a health

(continued)

CASE STUDY   Nutrition Education Intervention for a Congregate Meal Program *(continued)*

education or socialization activity after meals. You will provide a nutrition education intervention on a weekly basis for four weeks. Program sessions are implemented after meals are served and last approximately one hour. There is also a nurse practitioner who provides health services including blood pressure and urinary glucose screenings once a month.

CASE STUDY   Keys to Success #1—Know Your Audience, Conduct a Thorough Needs Assessment

Following the Keys to Success guidelines, you have a clearly defined target group. Your first responsibility is to conduct a needs assessment. So let's go through the data collection process.

Health Needs

Start by reviewing available literature and government statistics regarding health issues for this population group. Go to Chapter 9 of this text and review the section on Older Adults.

Discussions with the nurse practitioner indicates that approximately 60% of the participants are taking medications for high blood pressure, 50% have arthritis, 24% have diabetes, and 40% take medication to lower blood cholesterol levels. Using the Mini Nutritional Assessment tool (See Chapter 9), five participants scored in the "at risk of malnutrition" category and no one scored "malnourished."

Educational Needs

After assessing health status data, a short questionnaire was developed for congregate food participants to complete. In addition, ten participants volunteered to be a part of a focus group to explore possible education interventions. Feedback indicated that participants wanted nutritional guidance for management of high blood pressure.

Resource Needs

Feedback from the focus group indicated that the most salient issues for participants were income, chewing difficulties, and arthritis, which reduced participants' cooking abilities. The community center has audiovisual equipment, Internet access, and a kitchen that would permit food activities.

Developmental Needs

Findings from the focus group indicated a desire for hands-on activities to address their developmental challenges, such as taste tests, price comparisons of desired food items, and cooking lessons.

EXERCISE 11.1   Case Study Keys to Success #1 Related Activities

1. Go to the Center for Disease Control and Prevention site on Healthy Aging, www.cdc.gov/aging click on Data and Statistics to gather health needs statistics about older adults. In your journal, identify two national surveys cited on this website and give one finding of each for this age group. Also, under Data and Statistics, click on The State of Aging and Health in America Report and explore information about older adult health status for your state and site two of your findings in your journal.
2. Go to the Center for Disease Control and Prevention site on Healthy Aging, www.cdc.gov/aging click on Health Information for Older Adults and research arthritis and high blood pressure. Record two findings for each condition in your journal.
3. Go to http://www.mna-elderly.com/user_guide.html download a MNA tool, user guide and watch the MNA video. Work with a volunteer 60 years of age or older and complete a MNA Mini Nutritional Assessment. Record your findings in your journal.

KEYS TO SUCCESS #2—DETERMINE YOUR EDUCATIONAL PHILOSOPHY

Your educational philosophy should provide you with an overall strategic vision and general direction for all components of your nutrition education intervention. A review of basic principles of educational psychology can provide a basis for determining your philosophy. They include a positive relationship between active learning and retention, positive reinforcement, relevancy of the topic addressed, and the characteristics of the learning environment.[17] After you have conducted a needs assessment, tailor your nutrition educational philosophy to address the needs of your target population, availability of resources, and any guidelines of the organization sponsoring the intervention. As you develop your own philosophy for the target population, a review of prior guiding approaches of the nutrition education profession may be useful.

- **Focus on information dissemination:** For many decades, major government providers of nutrition education focused on increasing knowledge and relied on lecture, group discussions, and distribution of literature to increase nutrition awareness and "how-to skills", such as food preparation skills, label reading, and identification of food groups and sources of nutrients.[18] The assumption was made that simple dissemination of information leads to new knowledge, changes in attitude, and improved dietary behavior (knowledge-attitude-behavior [KAB] model). However, evaluation of this approach showed that knowledge improved but little dietary behavior changed.[19] Research documented that knowledge accounts for "4% to 8% of the variance in eating behavior, leaving 92% to 96% of the behavior to be accounted for by other influences."[20] The other influences can be cultural, psychological, or even environmental barriers inhibiting dietary change. Increasing knowledge via dissemination of information continues to be a vital component of any nutrition education program, and is essential for individuals with high levels of overall health concern and motivation to eat healthfully, but for others, it may be merely "information."

- **Focus on behavior change:** More recently, as evidence emerged that lifestyle factors including dietary, smoking, and physical activity behaviors clearly contributed to heart disease, obesity, and cancer, a major shift in the concept of nutrition education occurred. The focus of highly funded educational interventions investigating the effectiveness of strategies to reduce risk of developing common chronic diseases shifted to evaluation

of behavioral and cognitive strategies. See Chapters 2 and 6 for a review of these behavior change approaches. Research studies have demonstrated that many successful nutrition education programs focused on behavior change as their primary goal.[10] In addition, the most effective programs occurred when behavioral change strategies were derived from specific theories. These interventions specifically focused on psychosocial factors related to behavior such as personal factors and behavioral capabilities. These programs included activities such as self-assessments; identifying healthful eating behaviors; establishing goals; learning cognitive, affective, and behavioral skills; and providing incentives and reinforcements.[10]

- **Focus on the environment and public policy:** Presently, there is an increased recognition of the effect of the physical and socioeconomic environment on health behaviors.[21] No longer is the individual considered the only component in the equation for behavior change.[22,23] Analysts are increasingly viewing the obesity epidemic in terms of corrupted eating practices caused by a "toxic food environment" due to global economic development and modernity.[24] The paradigm of nutrition education is now often directed at social, political, and physical environments as well as the individual. In recent years, there have been increasing attempts by policy makers to change eating environments, such as banning soft drinks in schools, regulating location of fast food establishments, and encouraging purchase of foods from farmer's markets. Nutrition professionals are likely to focus on changing social norms, public policy, the food supply, and physical entities to support healthful behaviors. Strategies may be aimed at consciousness-raising and empowerment of individuals to encourage community activism to alter the structure of power in their localitities to change food policies and make environmental changes.[25]

CASE STUDY   Keys to Success #2–Determine Your Educational Philosophy

Your nutrition education philosophy should take into consideration the needs of the target audience, the educational setting, and the supporting institution or agency. When formulating your philosophy statements, consider the following questions:

1. What is the role of a nutrition educator?
2. What type of the learning environment do you want to provide?
3. What do you want to be the focus of your intervention–information dissemination, behavior change, environmental change, or some combination of the three?
4. What are the target audience's needs for learning?
5. What are your goals for the target audience?

EXERCISE 11.2   Case Study Keys to Success #2 Related Activity

Write a paragraph articulating your nutrition education philosophy for the case study. Refer to the nutrition education approaches described in this section.

KEYS TO SUCCESS #3– DESIGN THEORY-BASED INTERVENTIONS

Nothing is as useful as a good theory.

—**Kurt Lewin**

Nutrition educators use social psychological and behavioral theories to identify determinants of dietary behaviors, discover potential mediators of behavior change, and guide the design of an intervention.[26] Determinants of dietary behavior are the factors influencing food behavior, such as, perceptions, beliefs, attitudes, and environmental factors.[27] A theory presents a systematic explanation of events or situations. In essence, theories provide road maps for understanding problems, developing interventions, implementing programs, and evaluating their effectiveness.

Researchers have documented that a major shortcoming of early nutrition education studies was the failure to base research on theoretical models.[28,29] The studies did not have a clear theoretical base from which to elicit variables explaining the impact of nutrition education programs on knowledge, attitudes, and behavior. More recently theory-based research has been used effectively.

Three categories of theory-based interventions describe how they are often used.

- *A theory-driven intervention* refers to theory elements that are systematically used to design, implement, and evaluate the intervention.

- *A theory-informed intervention* pertains to the partial use of theory elements to design intervention components.

- *A grounded theory intervention* involves the use and application of qualitative data derived from the target audience to guide the design of an intervention. Grounded theory is a research method underscoring the generation of theory from qualitative data analysis.[30]

Nutrition educators draw from a combination of contemporary models of individual, social, and environmental change to design an intervention. A useful theory provides assumptions that are logical and consistent with daily observations. Factors to consider when selecting a particular theory or combination of theories are the characteristics of the target population, the health problem under investigation, and which theories were used successfully in prior related research.[31] Table 11.2 describes how key constructs from theories frequently used in nutrition education interventions can be applied to design programs. Please see Chapter 2 for more information about these theories and The Transtheoretical Model, which is also a commonly used approach for nutrition education interventions. Note it is not unusual to select the most salient constructs and incorporate compatible variables from several related theories into a research study or a nutrition education intervention.[21,32]

Table 11.2 Summary of Practical Applications of Theory-Based Constructs in Educational Settings

Construct	Theory-Based Strategies	Educational and Learning Strategies
Health Belief Model		
Perceived Susceptibility	Address health risks via persuasive communications to convey personal threat.	Provide personal testimonies, role-playing, and visuals of vivid images of threat. Report striking statistical trends.
Perceived Severity	Incorporate impact and consequences of disease to convey personal threat.	Present real life experiences and dramatic outcomes of disease.
Perceived Benefits	Information highlighting benefits of taking action to reduce threat of disease or condition.	Use persuasive messages. Engage in activities that depict personal gain and health benefits (for example, increased energy levels, enhanced appearance, blood cholesterol reduction).
Perceived Barriers	Reduce perception of barriers in engaging in health behavior.	Brainstorming activities and group discussion of hindrances to behavior change. Identify strategies to overcome these barriers.
Cues to Action	Use relevant and effective cues or stimuli to prompt action.	Media-generated messages, billboard advertisements, public service announcements, magazine articles, stickers, and fliers.
Self-Efficacy	Increase self-confidence to perform health behavior via social modeling of behavior and guided practice for mastery of behavioral skills.	Provide step-by-step instructions and demonstration of behavior by credible role models. Engage in direct experience such as food preparation. Provide positive reinforcements on achievements and successful performance.
Theory of Planned Behavior		
Behavioral Intention	Increase desire and resolve to engage in health behavior. Analyze pros and cons of adopting behavior.	Use hands-on worksheets analyzing advantages and disadvantages of behavior. Generate group consensus on goals for action.
Attitudes	Reflection on personal feelings and predisposition to object or behavior.	Assess attitudinal predisposition, personal affect, and provide open discussion. Use emotion-based communication strategies.
Social Norms	Identification of salient others and their social norms and expectations.	View and analyze media-based advertisements and messages from social environment.
Social Cognitive Theory		
Behavioral Capability	Provide nutrition- and food-related knowledge and behavioral skills to perform behavior.	Use PowerPoint presentations, visuals, and demonstrations to teach food- and nutrition-related knowledge and skills to enact behavior. Provide step-by-step demonstration of food preparation and cooking skills.
Reinforcements	Provide internal and external reinforcements.	Provide prizes, gift certificates, verbal praise and recognition for positive behavior.
Observational Learning	Learn to perform new behaviors via family, peer, and media models.	Incorporate positive role models to demonstrate and advocate healthful behaviors.
Social Support	Helping relationships and social network.	Foster supportive social environment. Encourage accountability partners and buddy systems among peers.

CASE STUDY   Keys to Success #3–Design Theory-Based Interventions

Now that you have completed the first two Keys to Success–a needs assessment and written a philosophy statement—you are ready to select an appropriate theory to use with the target population. After reviewing your needs assessment and educational philosophy, you determined that the Health Belief Model (HBM) may be a good fit for the group of older adults at the congregate site.

EXERCISE 11.3   Explore Health Beliefs

1. Review Chapter 2 and Table 11.2 for theories and constructs most often used in nutrition education interventions. The HBM was selected in the case study to design an intervention for the target population.

   ❏ Give two reasons why you believe this selection was made. Compare your answer with that found at the end of this chapter.

2. Explore the health beliefs of overweight and obese individuals. Go to the following website: http://www.mdpi.com/1660-4601/7/2/443/pdf

Read: Lewis, S et al. Do health beliefs and behaviors differ according to severity of obesity? A qualitative study of Australian adults. *Int J Environ Res Public Health*. 2010 Feb; 7(2):443–59; Epub 2010 Feb 3.

❏ Record three facts or observations about the health beliefs of this population group.

COMMUNITY LEVEL AND PLANNING MODELS: SOCIAL MARKETING

"Why can't you sell brotherhood like soap?"
                                    —G D Weibe

Social marketing as a discipline is generally thought to originate in the 1970s, when Philip Kotler and Gerald Zaltman coined the term and advocated using commercial marketing strategies to "sell" ideas, attitudes, and behaviors for the common good.[33] Social marketing has enormous appeal. Clearly, commercial marketing has a major impact on consumer behavior. Commercial campaigns are usually heavily funded and supported by a large infrastructure. Social marketing programs are often led by nonprofit and government organizations so that funding and resources are limited.[34] However, social marketing interventions have achieved notable outcomes. A prime example was the successful 5 A Day for Better Health program which was one of the nation's largest public and private initiatives for increasing awareness and intake of fruits and vegetables.[35] This initiative established a model for public and private partnerships and significantly increased awareness of the importance of eating fruits and vegetables. Research findings of the program indicated that self-efficacy, knowledge of the recommendation, and taste preferences were the best predictors of fruit and vegetable intake behavior. Social marketing has become a major player in behavior change methodologies. In 2004, the Centers for Disease Control and Prevention inaugurated the National Center for Health Marketing with the mission "to promote the public's health through collaborative and innovative health marketing programs, products, and services that are customer-centered, science-based, and high-impact."[36]

Definition of Social Marketing

Social marketing is a systematic and strategic process for communication planning using client-centered methods to facilitate changes in behaviors, values, and attitudes.[37] Communication models have an eminent role in nutrition education for they elucidate the process occurring between a message sender and the message receiver.[38] Social marketing is frequently described as:

"The application of commercial marketing technologies to the analysis, planning, execution, and evaluation of programs designed to influence voluntary behavior of target audiences in order to improve their personal welfare and that of society."[39]

Basic Principles of Social Marketing

- Social marketing interventions use commercial marketing strategies to influence health behaviors. In both social and commercial marketing, the primary focus is on the consumer, but in social marketing, the objective is to learn what people want and need rather than to sell a product.
- Social marketers are behaviorally-focused. The objective is to influence and change voluntary behaviors, not just increase awareness or knowledge.
- The ultimate objective of social marketing is to benefit the target audience or society and not the marketer. Social marketing is highly client-centered.
- The development of program strategies always begins with target audience members and their perceptions. Social marketers do not strive to target the general public. They specify the target audience as precisely as possible. This method of segmentation divides the audience into different subgroups which can be based on demographic, geographic, or behavioral characteristics. For instance, middle-aged male smokers residing in San Francisco, California, may be the focal point of an intervention to increase vitamin C intake via fruits and vegetables.
- To use a comprehensive strategy, social marketers incorporate the "marketing mix" into their program planning. The "four P's" of the marketing mix are Product, Price, Place, and Promotion.[34,40] Beyond the four Ps of traditional marketing, social marketing incorporates four additional Ps referred to as Publics, Partnership, Policy, and Purse strings.[40] See Table 11.3.

Table 11.3 The Marketing Mix

Four Ps of the Marketing Mix

Product	The social marketing product is not necessarily a physical offering. The product can represent a continuum of tangible products (food stamps), services (nutrition counseling), and practices (eating a plant-based diet) or more intangible products (school food policy). To market a viable product, members of a target audience must realize they have a genuine risk or problem and believe that the product is a feasible solution. Having a product that is both attractive and appealing to the target group increases effectiveness of the intervention. Efforts are placed on finding a niche for the product and identifying benefits so the product is perceived as more appealing than the competition. The product should address that question, "What's in it for me?"
Price	Price refers to the perceived costs of obtaining the social marketing product. The costs may be tangible (money) or intangible (time, effort, inconvenience, or risk of humiliation, or censure) in exchange for the product. Careful research must be done to determine the perceived costs and benefits so as to plan an intervention minimizing the costs and maximizing the benefits. Perception of greater benefits than costs is likely to encourage the consumer to "buy" the product.
Place	In commercial marketing, place often refers to the distribution channels in which customers are able to obtain the product. From a social marketing perspective, one could pose the question as: "Where is the behavior available to the target audience?" Nutrition messages should reach people in a place where they are making decisions about the behavior. For a program promoting low-sodium diets for hypertensive individuals, what are some venues to promote the message to the target audience? Perhaps viable settings would include doctors' office waiting rooms, Internet websites, shopping malls, grocery stores, television news programs, and radio talk shows.
Promotion	Promotion is an integral part of the marketing mix. It focuses on strategies to convey the message to persuade the target audience. Promotional strategies generally integrate the other Ps, such as defining the product, emphasizing the benefits, and distributing in appropriate places. Promotional messages need to match the "target population preferences and information processing styles."[34] There are a variety of styles to choose from including emotional or rational appeals, use of humor, social value

(continued)

Table 11.3   The Marketing Mix *(continued)*

Four Ps of the Marketing Mix

emphasis, to name a few. The objective is to create and sustain demand for the product. Promotion can involve various methods of communication including the following:
- Advertising (radio and television commercials, billboards, posters)
- Public relations (press releases, talk shows)
- Promotions (displays, coupons)
- Media advocacy (press events to promote policy change)
- Special events (health fairs)
- Entertainment (puppet shows, skits, concerts)

Additional Social Marketing Ps

Publics	Social marketing programs include both external and internal groups. Examples of the external publics are the target audience itself, family members, physicians, or even policymakers and media professionals. Internal publics refer to those involved in approval or implementation of the social marketing program including staff members and supervisors.
Partnership	Because of the complexity of social and health behavior change, collaboration with other organizations or groups within the community can increase resources and probability of a successful intervention.
Policy	A change in policy can be effective in supporting the desired behavior change. For example, legislative advocacy such as lobbying may be effective in promoting the sale of healthier foods in vending machines located in high schools.
Purse strings	Social marketing interventions may seek funding from a variety of sources such as foundations, donations, and governmental grants.

Application of Social Marketing

The Pawtucket Heart Health Program is an example of a notable short-term social marketing program focused on reducing blood cholesterol levels

---

EXERCISE 11.4   Application of Social Marketing

You are actively involved in a social marketing campaign aimed at lowering saturated fat consumption in middle-aged adults living in New York City. Work in groups to provide applications of social marketing principles by addressing the following questions.

1. What is the product?
2. What are potential costs and barriers that the target audience might associate with adopting a low saturated fat diet?
3. What are strategies to best position the product to minimize perceived costs associated with this dietary behavior change?
4. What are likely places where the message can be disseminated to the target audience?
5. What would be possible methods for promoting the health message among the adults?

---

in adult residents in Rhode Island.[41] Program staff developed a nutrition kit designed to help target members change their diets. Mass print campaigns were conducted at worksites, education, and religious institutions. Events were held in which screening, counseling, and referral evaluations were made available. The marketing strategy resulted in over 10,000 individuals having their blood cholesterol measured over a two year period and a significant reduction of blood cholesterol levels in the adult attendees.

KEYS TO SUCCESS #4—ESTABLISH GOALS AND OBJECTIVES

At this point in the planning process, you have a picture of the needs of your target population, a philosophy expressing your aims for an intervention, and a selection of one or more theories to guide your plan. You are now ready to set goals and objectives and develop **generalizations** for your nutrition education intervention. Clearly, they provide a road map for an intervention, driving the

Table 11.4 ABCs of Objectives

Example: "Upon completion of this program, older adult participants of a congregate meal site given four food labels will be able to identify high fiber foods using government standards."

Category	Definition	Example
Audience	"A" stands for audience and identifies who is the target for a nutrition education intervention. Clearly define the learner.	In the example, the target audience (who) are older adult participants at a congregate meal site.
Behavior	"B" stands for behavior. What measurable and observable behaviors (what and how much) will the learners be able to do as a result of the program?	The behavior ("what") is to identify high fiber foods and the "how much" is four food labels.
Condition	"C" refers to the condition (when or where) by which the learners' performance will be assessed. These conditions may be the resources used, the location, limitations, or time imposed.	The condition "when" is after completion of the program. In the example, "when" adequately defines the condition.

overall direction of the process, the instructional strategies employed, resources used, and evaluation methods selected. They also help the learner focus and set priorities.

Goals

Goals can be referred to as broadly stated learner outcomes. They reflect global learner outcomes or the overall intent of an intervention or program and generally include "who" and "what." For example, a nutrition educator might describe the goal of a wellness seminar in the following way: "Participants of this course will learn effective dietary methods for weight management." In this broadly stated goal the "who" are the course participants, and the "what" is learning effective dietary methods for weight management. As compared to objectives, goals often express a long range purpose and are more general. They provide overall direction, include all or most aspects of a program, and are usually not time-bound, measured or observed.

Objectives

Objectives are specifically stated learner outcomes or descriptions of what the learner will be able to do after participating in a learning experience. These objectives are helpful in planning nutrition education, even for day-to-day instruction. Use the mnemonic SMART to define objectives: Specific, Measurable, Attainable, Rewarding, and Time-bound.

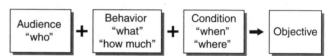

Figure 11.2 Components of Writing an Objective
Source: Designed by author.

Writing meaningful objectives is an important skill of every educator. The ABCs of Objectives are a helpful guideline for writing clear objectives.[42] See Table 11.4 and Figure 11.2. Be sure your objectives contain all three of these components. Incomplete objectives lack clarity and fluency, resulting in nebulous and confusing road maps.

Verbs

Goals and objectives should be worded in a clear, concise, and realistic manner. They can incorporate behavioral (measurable) and non-behavioral (non-measurable) verbs. Goals are more likely to include verbs that cannot be measured quantitatively, such as understood, know, appreciate, value, believe, and learn. Objectives are likely to contain behavioral verbs such as: identify, recite, differentiate, classify, construct, write, and compare. These verbs reflect behaviors that can be observed, measured, and evaluated.

Types of Objectives

We are now going to further develop the process of writing objectives by exploring three domains

---

EXERCISE 11.5   Evaluate an Objective

Identify the major components of the following objective:

After completion of this seminar, college students will correctly identify three foods containing saturated fats on an evaluation form.

1. Audience _____

2. Behavior _____

3. Condition _____

---

of learning, sometimes referred to as categories of learning outcomes.[43] As you develop your objectives and learning plan, the type of learning domain you choose will guide you in selecting learning activities and assessments. In order to provide a comprehensive approach, aim to address all three domains:

- **Cognitive:** mental skills, such as *knowledge* or *think*
- **Affective:** feelings or emotions, such as *attitude* or *feel*
- **Psychomotor:** manual or physical *skills*, such as *do*

This taxonomy has given rise to short-hand versions of the three categories known as KSA (Knowledge-Skills-Attitude) or Do-Think-Feel. However, a full understanding of the domains requires an investigation of subcategories of the three domains

of learning. The subcategories are arranged in a hierarchy starting with the simplest and moving to the complex, illustrating growth of learning as individuals advance in their abilities.

Cognitive Domain

Cognitive domain is the "head" of a body which emphasizes thinking, knowing, understanding, and comprehending. Any descriptions related to cognitions reflect this domain, which may entail recalling, explaining, reasoning, creating, or making a judgment. The cognitive domain has six major levels as identified in Figure 11.3 and described in Table 11.5.

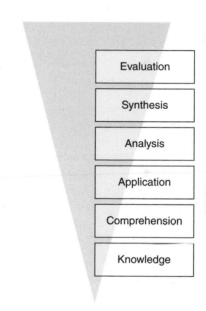

Figure 11.3 Levels of Cognitive Domain
Source: Designed by author.

Table 11.5 Cognitive Domain and Levels of Learning (Focus on Thinking)[43]

Level and Description	Examples* and Key Verbs
Knowledge: Recalling, remembering, and recognizing data or information.	**Examples:** Name the fat-soluble vitamins. Match nutrient deficiency disease with nutrient. Identify foods high in saturated fats. Recite a policy. **Key Verbs:** cite, define, identify, label, list, match, memorize, name, recall, recite, recognize, reproduce, state.
Comprehension: Lowest level of understanding; explaining; stating a problem in own words.	**Examples:** Describe the role of vitamin K in blood clotting. Give an example of a high fiber food. Explain functions of vitamin C. Report on the effects of a high fat diet. **Key Verbs:** classify, describe, discuss, estimate, explain, give an example, infer, illustrate, interpret, paraphrase, reiterate, review, report, reword, summarize.

(continued)

Table 11.5   Cognitive Domain and Levels of Learning (Focus on Thinking) *(continued)*

Level and Description	Examples* and Key Verbs
Application: Using ideas, information, and principles in specific situations; applying knowledge to solve a problem.	**Examples:** Use an equation to calculate body mass index. Modify a recipe to lower caloric content. Demonstrate cooking techniques. **Key Verbs:** apply, change, compute, construct, demonstrate, discover, implement, manipulate, modify, operate, predict, prepare, perform, produce, relate, respond, role-play, show, solve, use.
Analysis: Dissecting information into basic elements and organizing principles; reasoning; clarifying hidden meaning, distinguishing fact and opinion, and assessing degree of consistency.	**Examples:** Compare and contrast health disparities of two cultural groups. Analyze a health claim. Differentiate between microcytic and macrocytic anemia. **Key Verbs:** analyze, associate, break down, compare and contrast, deconstruct, determine, differentiate, discriminate, distinguish, experiment, plot, relate.
Synthesis: Re-assembling component parts for new meaning; building a structure or pattern from diverse elements.	**Examples:** Design a weight reduction manual. Propose a nutrition policy. Develop a behavior change model. Summarize research findings of a nutrition education intervention. Create a team to develop a new curriculum. **Key Verbs:** build, create, combine, compile, compose, develop, design, devise, integrate, formulate, modify, organize, plan, propose, revise, summarize.
Evaluation: Making a judgment, appraising the value of information, methods, or materials against internal and external standards.	**Examples:** Justify a new budget. Critique a research proposal. Evaluate the outcomes of a nutrition education intervention. **Key Verbs:** make a judgment, appraise, assess, compare, conclude, contrast, criticize, critique, defend, evaluate, justify.

*Note these are examples of how to use the verbs and are not fully developed objective statements.

Source: Adapted from: Bloom B. Taxonomy of Educational Objectives, Handbook 1: Cognitive Domain. New York: David McKay; 1956.

Affective Domain

The affective domain emphasizes how a learner feels or values a particular entity. It can be likened to the heart, reflecting one's attitude, feelings of acceptance or rejection, and levels of appreciation and valuing. The five categories of the affective domain are identified in Figure 11.4 and described in Table 11.6.

Psychomotor Domain

The psychomotor or behavioral domain involves action, control, and movement of the body. It is analogous to the "hands" of the body, relating to skill development and mastery. Such skills may include applying insulin injections, sautéing fresh vegetables, or even conducting a food experiment. The five subcategories of the psychomotor domain are found in Table 11.7 and Figure 11.5.[44]

GENERALIZATIONS

A tool used as an aid for organizing instruction is identifying meaningful generalizations. A **generalization** is a clear statement that is universally

true. Generalizations are written as complete statements, representing universal truths that the learners will need to grasp and comprehend for an effective intervention. In our instructional plan, we will be relating generalizations to objectives. See Table 11.8

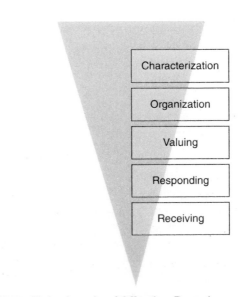

Figure 11.4   Levels of Affective Domain
Source: Designed by author.

Table 11.6 Affective Domain and Levels of Learning (Focus on Feeling)

Level and Description	Examples* and Key Verbs
Receiving: Willing to hear or experience; attending; and becoming aware.	**Examples:** Listen to a presentation with respect. Concentrate while receiving directions. Read and take notes during a presentation. Notice the printed food labels. Acknowledge the ingredients list. **Key Verbs:** accept, acknowledge, be alert, concentrate, focus, follow, hear, listen, notice, perceive, read.
Responding: Attending and reacting to the phenomenon; desire to be engaged in a subject or activity.	**Examples:** Contribute to class discussions. Practice safety standards in laboratory activities. Label toxic materials in laboratory. Seek clarifications of new food labeling guidelines. **Key Verbs:** agree to, answer freely, assist, care for, comply, contribute, cooperate, help, label, practice, react, read, recite, respond, seek clarification, tell.
Valuing: Developing attitudes; believing that the information or behavior is worthwhile based on an internal assessment and commitment; expressing personal opinion.	**Examples:** Propose a plan to reduce health disparities. Justify ruling to stop soft drink sales in high schools. Argue for calorie labeling in fast food restaurants. Express a need for physical training. Commit to the dietary regimen. Prefer low-fat dairy products. **Key Verbs:** adopt, argue, challenge, choose, commit, complete, confront, criticize, debate, desire, exhibit loyalty, express, follow, initiate, invite, join, justify, prefer, persuade, propose, select, share, show concern.
Organization: Arranging values into priorities by contrasting them, reconciling conflicts between them and creating an internal value system to guide behavior.	**Examples:** Adapt to a low sodium diet. Reveal interest in consuming whole grain foods. **Key Verbs:** adapt, alter, adhere, adjust, arrange, classify, group, compare, contrast, defend, formulate, generalize, integrate, modify, order, organize, prioritize, rank, reconcile, relate, synthesize, reveal
Characterization: Internalizing a set of values; controlling behavior in a consistent and predictable manner.	**Examples:** Advocate principles of sustainable agriculture. Show consistent devotion in mentoring children. **Key Verbs:** act upon, advocate, influence, defend, display, influence, maintain, qualify, serve, show consistent devotion to, verify

*Note these are examples of how to use the verbs and are not fully developed objective statements.

Source: Krathwohl DR, Bloom B, Maisa BB. Taxonomy of Educational Objectives, Handbook 2: Affective Domain. New York: Longman, 1964.

Table 11.7 Psychomotor Domain and Levels of Learning (Focus on Action)[44]

Level and Description	Examples* and Key Verbs
Perception: Using sensory cues, hearing, seeing, tasting, touching, and smelling to guide physical behavior.	**Examples:** Observe a food demonstration. Feel the texture of the fruit. Detect non-verbal communication cues. **Key Verbs:** detect, feel, hear, listen, observe, see, sense, smell, taste
Set: Becoming ready to act; reproducing an action through imitation or memory.	**Examples:** Assume a body stance for stretching. Station oneself besides the mixer. **Key Verbs:** achieve a posture, assume a body stance, place, hands, arms, feet, position the body, sit, stand, station
Guided response: Imitating and practicing a complex skill via trial and error. Performance of the task is usually imperfect.	**Examples:** Imitate dance movements. Repeat sanitizing procedures. **Key Verbs:** copy, duplicate, imitate, manipulate with guidance, practice, repeat, try
Mechanism: Increasing efficiency of learned responses; becoming habitual and proficient.	**Examples:** Demonstrate proper hand washing procedures. Show dexterity in mincing foods. **Key Verbs:** complete with confidence, conduct, demonstrate, execute, pace, produce, show dexterity

(continued)

Table 11.7   Psychomotor Domain and Levels of Learning (Focus on Action) *(continued)*

Level and Description	Examples* and Key Verbs
Complex overt response: Performing skill or task automatically without flaw; marked by accuracy, speed, and control.	**Examples:** Master the art of French cooking. Excel in operating computer applications and software. **Key Verbs:** act habitually, control, excel, guide, master, perfect, perform automatically

*Note these are examples of how to use the verbs and are not fully developed objective statements.

Source: Simpson, E. The classification of educational objectives, psychomotor domain. Illinois Teacher. 1967;10:110–145.

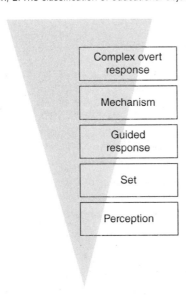

Figure 11.5   Levels of Psychomotor Domain
Source: Designed by author.

for guidelines for writing generalizations. There are three categories of generalizations:

- **Level 1:** The first level may be a statement of fact, definition, identification, or simple description. *Example: Cancer is a chronic disease.*
- **Level 2:** A second-level generalization makes comparisons or shows relationships among ideas, expressing increased scope of the subject matter. *Example: Cancer risk may be affected by environmental exposure to toxins.*
- **Level 3:** A third-level generalization is characterized by some form of explanation, prediction, or interpretation. *Example: An individual's intake of dietary fats may predict serum triglyceride levels.*

EXERCISE 11.6   Classify Objectives

For the following, identify the domain of learning (cognitive, affective, and psychomotor) and the level within each domain.

Example: Upon completion of the course, participants will be able to identify the physiological functions of insulin on an objective evaluation.

(Cognitive domain; knowledge level)

1. After completing the carbohydrate unit, students will be able to explain two functions of glycogen.
2. Student counselors will listen attentively to volunteer counselees during a mock counseling session.
3. After completion of the seminar, the participants will advocate lifestyle changes to promote cardiovascular health to a community group.
4. During the diabetes learning unit, students will cooperate with group members to complete a project.
5. Older adults will demonstrate knowledge of high fiber foods by placing four pictures of food products from lowest to highest amount of fiber.
6. Upon completion of the course, students will be able to formulate an effective nutrition education intervention, by using the Keys to Success model.
7. Based on concepts discussed in class, students will imitate proper hand-washing techniques before starting the cooking lab.
8. During a presentation, students will advocate for either commercial or sustainable agriculture.
9. In a written report, students will analyze the validity of a popular weight loss diet.
10. During a mock counseling session, students will execute effective nonverbal behavior.
11. Ninety percent of kindergarteners will be able to state examples of green vegetables served in the school cafeteria.

Table 11. 8 Guidelines for Writing Generalizations

- Aim for complete sentences of a maximum number of 20–22 words.
- Express only one key idea or universal truth. Avoid using colons or semi-colons.
- Use clear words that are technically and grammatically correct, free of ambiguous language.
- Avoid using value judgments with words such as "must", "should", "ought to be done."
- Use phrases that facilitate the expression of statements that show relationships. Some examples are as follows: "contributes to", "is promoted by", "is related to", "is an integral part of", "is influenced by", "may be associated with", "may be determined by", and "may be developed by".

EXERCISE 11.7   Identify Level of Generalization Statement

Identify the following as a Level 1, Level 2, or a Level 3 generalization statement.

1. Obesity is a problem in the United States because of overconsumption of calories.
2. Potassium is an intracellular cation.
3. Obesity may contribute to type 2 diabetes.
4. Sucrose is a type of simple sugar.
5. Cigarette smoking is a major risk factor for developing coronary heart disease.
6. Consumption of high levels of sodium is a risk factor for high blood pressure.

EXERCISE 11.8   Case Study Keys to Success #4 Related Activity

Go to the national heart lung and blood institute site and click on In Brief: Your Guide to Lowering Your Blood Pressure with DASH http://www.nhlbi.nih.gov/health/public/heart/hbp/dash/dash_inbrief.htm. Review the DASH eating plan and record two findings in your journal.

INSTRUCTIONAL PLAN

A written instructional (lesson) plan provides a roadmap for the educational session. See Table 11.9 for an example of a lesson plan. Without this

---

CASE STUDY   Keys to Success #4 –Establish Goals and Objectives

Goal: The broad based goals for this nutrition education are to implement behavior change strategies and increase awareness of the role of dietary factors influencing blood pressure.

Objective	Domain and Level	Generalization
1. Upon completion of this program, older adult participants will identify the negative health consequences of uncontrolled hypertension.	Cognitive Domain Knowledge Level	Stroke is a negative health consequence of uncontrolled hypertension. Level 1 Generalization
2. Upon completion of this program, older adult participants will perceive the health benefits of following the DASH Food Plan.	Affective Domain Receiving Level	The DASH Food Plan contributes to the management of high blood pressure in older adults. Level 2 Generalization
3. Upon completion of this program, 80% of older adult participants will commit to making two dietary changes recommended by the DASH Food Plan.	Affective Domain Valuing Level	Developing positive attitudes toward the DASH Food Plan may be associated with the commitment to consume these foods. Level 2 Generalization
4. Upon completion of this program, older adult participants will express confidence in their ability to prepare DASH foods at home.	Affective Domain Valuing Level	Confidence in one's ability to prepare DASH foods at home may be promoted by repetition in food preparation activities. Level 2 Generalization

Table 11.9   Sample Lesson Plan Using Constructs from Social Cognitive Theory

Simple Versus Complex Carbohydrates

Duration: One hour

Target Group: High school freshmen enrolled in a science class

Overall Goal: To increase knowledge of the functions of simple and complex carbohydrates

Major Concepts:
- Simple carbohydrates
  - Monosaccharides
  - Disaccharides
- Complex carbohydrates
  - Polysaccharides
  - Dietary fiber

Icebreaker or Attention Grabber: (5 minutes)
- Ask students to name food groups containing carbohydrates.
- As students name a food, write the name of the corresponding food group on the board: grains, vegetables, fruits, nuts, seeds, dairy, other (high sugar foods, soft drinks, candy).
- Have a sample or food model of each food group in a paper bag. As the food group is identified, take the corresponding item out of the bag and place on a table.
- Show a poster of the MyPyramid guidelines. Point out the carbohydrate-containing food groups.

Objectives and Learning Domains; Generalizations and Learning Experiences

1. The students will state examples of simple and complex carbohydrates commonly found in supermarkets.

Domain: Cognitive domain—knowledge

Generalization: Complex carbohydrates are composed of many glucose molecules joined together.

Learning Experiences: (15 minutes)

Provide a PowerPoint presentation covering the following concepts:

Define Simple Carbohydrates:
- A monosaccharide is a single glucose, fructose, or galactose molecule. A disaccharide is composed of two monosaccharides joined together.
- Foods high in simple sugars give quick energy, but usually do not contain other nutrients or fiber.
- Simple sugars include glucose, fructose, high fructose corn syrup, maltose, and lactose.
- Foods high in simple sugars include table sugar, soft drinks, desserts, and fruit juice.

Define Complex Carbohydrates:
- Complex carbohydrates are composed of many simple sugars (polysaccharides or starch and fiber).
- Foods high in complex carbohydrates supply energy and often other nutrients and fiber.
- Foods high in starch include bread, cereal, potatoes, pasta, rice, and legumes (dried peas and beans).
- Fiber is the indigestible parts of plant foods.
- Fiber is needed by the body to aid in the prevention of constipation, reduction of blood cholesterol, assist in weight loss.
- Foods high in fiber include bran, whole grain foods, vegetables, fruit, legumes, nuts, seeds and popcorn.

2. The students will observe a food demonstration involving the preparation of whole-grain pasta salad.

Domain: Psychomotor domain—perception

Generalization: Whole-grain pasta is prepared from natural, unprocessed ingredients consisting of vegetables and whole wheat products.

Learning Experiences: (15 minutes)
- A hands-on food demonstration by a peer leader (*observational learning*) will be presented involving the use of whole-grain ingredients in the preparation of a pasta salad.
- Step by step procedures will be outlined to increase *self-efficacy* and *behavioral capability* among the students.
- On a visible table, place one large bowl for mixing and smaller prepared bowls of food ingredients, (whole wheat pasta, canned legumes rinsed to remove salt, cooked and raw vegetables) and low-sodium salad dressing.

(continued)

Table 11.9 Sample Lesson Plan Using Constructs from Social Cognitive Theory *(continued)*

Objectives and Learning Domains; Generalizations and Learning Experiences

- Place ingredients in the bowl one at a time. First show the original container and then place each item in the bowl.
- A taste testing of the whole-grain pasta salad will be provided.

3. Students will value the consumption of food items high in whole grains and fiber.

Domain: Affective domain–valuing

Generalization: A favorable attitude towards consuming whole-grains and fiber-rich foods may be associated with increased dietary intake of these foods.

Learning Experiences: (10 minutes)

Provide a list of healthy selections of carbohydrates (baked beans, lentil soup, low-fat dairy, whole-grain bread, whole-wheat pasta) with a Likert scale for students to rate their overall attitude towards consuming or trying the food items. Words or pictures of faces can be used to determine favorable versus unfavorable attitudes toward each food. For example:

Circle the picture that best describes your attitude toward consuming each food.

Whole-grain bread

Teaching Aids and Materials:
- Paper bag containing food models or food items representing food groups containing carbohydrates
- Poster of MyPyramid
- PowerPoint projector and screen
- Survey of attitudes toward healthy selections of carbohydrates
- Prepared whole-grain ingredients for pasta salad including cooked whole wheat pasta, legumes, vegetables
- Low-sodium salad dressing and measuring cup
- Clear glass or plastic mixing bowls, one large and smaller ones for ingredients
- Salad mixing fork and spoon
- Food items in original packaging for demonstrations
- Small containers and spoons for tasting
- On-line or physical jeopardy board with review questions of learning activities

Summary: Carbohydrates are the body's primary source of energy. Simple carbohydrates include glucose, galactose, and fructose. Disaccharides are made up of two simple sugars joined together, such as sucrose or table sugar. Polysaccharides or complex carbohydrates include various sources of food such as bread, pasta, brown rice, and whole grains. Another example of polysaccharide is fiber or the indigestible part of plant foods. Fiber is needed by the body to aid in the prevention of constipation, and may help in the reduction of blood cholesterol and weight loss. The recommended dietary fiber intake is at least 25 grams per day. Whole-grain foods contain the three key ingredients of cereal grains—bran, endosperm, and germ.

Evaluation: (15 minutes)

The students will be divided into two opposing groups. An interactive jeopardy game will be presented at the end of the session to assess students' knowledge of the information presented. For example, "This is the name of a simple sugar that is commonly found in fruit" and "This disaccharide consists of two glucose molecules joined together." (10 minutes)

Assignment: An evaluative instrument on daily fiber intake will be distributed for students to monitor their intake on a daily basis for a period of 3 days (*self-monitoring*). Students will be asked to complete this instrument and present their findings in one week. The instrument will also include a list of commonly consumed foods with the nutritional information on actual fiber content.

detailed plan, the educator may not remember to use all the teaching techniques or review the necessary concepts, making last-minute adjustments difficult. In addition, the plan should be a clear outline for another instructor to implement. If you are working for an organization with multiple sites, there may be several educators implementing the instructional plan. Your instructional plan should provide answers for the following: Where are your learners going? How will they get there?

How will you know when they have arrived?[45] A good plan should include the following essential components:

- **Target Audience:** This section should inform the reader the intended developmental level of the instructional plan. Descriptive information could include age range or category; grade levels; ability levels; interests; attention spans; ability to work together in groups; prior knowledge and learning experiences; special needs or accommodations.

- **Specific Goals, Objectives, and Generalizations:** Aim to include multiple levels of domains of learning and generalizations. Lesson plans for school districts need to adhere to state or national curriculum standards. Enrich the lesson by incorporating cognitive, affective, and behaviorally-based (psychomotor) objectives.

- **Prerequisites:** Depending upon the intended audience, this may need to be specified in order to meet lesson objectives. For school based settings, this may mean advanced math skills or basic chemistry principles. For community programs, you may need to specify certain abilities, such as mobility or visual acuity.

- **Duration:** How much time will be needed to complete the instructional plan? You may also wish to include estimated set-up time.

- **Teaching Materials and Resources:** Include a list of teaching materials needed for the lesson such as markers, posters, handouts, food models, PowerPoint presentation, Internet access, and projector. This will also help instructors estimate time needed for organization based on their individual access to resources. These should be divided according to learning activity to ensure clarity.

- **Lesson Description:** This section includes a general overview of the lesson including core subject, activities, and purpose. Questions that could be answered in this section include: What is distinctive about this lesson? Did your students like it? What categories and levels of Bloom's Taxonomy are addressed?[43]

- **Lesson Procedure:** This is divided into two sections: Introduction and Learning Experiences

Introduction or Establishing Set: Make a good first impression! You want to gain attention of participants and encourage motivation to learn. This introduction should be a well-prepared, age-appropriate, and include a stimulating icebreaker that is relevant to the main theme of the program. Introduce yourself and the major goals of the educational session. Tie into past learning experiences, if appropriate. Some creative icebreaker activities may include telling a captivating anecdote, providing a brief self-assessment quiz, sharing a novel fact, statistic, or even a humorous cartoon.

Learning Experiences: Learning experiences are the activities in which learners participate to achieve the specified behavioral objectives. They guide the educator with the implementation of the lesson plan and to inform as to what participants need to do. Be creative and use varied, multi-sensory learning experiences that are age-appropriate and culturally sensitive. Include a time frame for each activity. This area should be clearly written so that any instructor can easily follow the flow of the experiences. El-Tigi [45] provides a Rule of Thumb # 1 about this portion of the lesson plan.

Rule of Thumb # 1:

Take into consideration learning objectives and content (a new skill, a rule or formula, a concept, fact, or idea, an attitude, or a value). Use the following to guide selection of techniques for planning your lesson and to complement your objectives:

Demonstration ⟹ list sequence of the steps in detail to be performed

Explanation ⟹ outline the information to be explained

Discussion ⟹ list of key questions to guide the discussion

Source: Adapted from: Manal El-Tigi, Write a Lesson Plan Guide. http://www.eduref.org/Virtual/Lessons/Guide.pdf. Accessed July 16, 2010.

Rule of Thumb # 2:

Be sure to provide students with opportunities to practice assessment material. Do not introduce new material during assessments. Also, do not ask higher level thinking questions if students have not had opportunities to engage in such activities during the lesson. For example, if you expect students to *apply* knowledge and skills, they should first be provided with the opportunity to practice application.

Source: Manal El-Tigi, Write a Lesson Plan Guide. http://www.eduref.org/Virtual/Lessons/Guide.pdf. Accessed July 16, 2010.

- **Summary and Closure:** This summative portion of the lesson brings closure by formulating key generalizations about the content that was addressed. This closure allows the educator to reiterate major concepts and evaluate student learning. Recalling original learner objectives and intended outcomes can help bring closure.

- **Evaluation:** This is a vital component to assess whether or not the program met the outlined objectives. Were the goals and objectives fulfilled at the end of the session? Appropriate evaluation tools for nutrition education may include pre- and post-tests, interviews with the target audience, games to assess knowledge gained, biochemical or anthropometric measurements, or nutrient analyses of food records.

- **Assignment:** Assignments may be used as a method for evaluating learner achievement of

---

EXERCISE 11.9   Survey Gateway to 21st Century Skills

Go to The Gateway to 21st Century Skills website sponsored by the National Education Association. Users have access to over 50,000 teaching resources including lesson plans. Although individual entries are not peer reviewed, the sponsoring organization supplying resources must become part of the Gateway Consortium, which requires a review process. Explore this website by going to http://www.thegateway.org/ Type in a topic of interest and review the resources that are displayed. In your journal, record two impressions, observations, or facts regarding your investigation.

---

program objectives, provide an opportunity for enriching the learning experience, and reinforce concepts covered in the lesson. Possibilities for assignments can include a small group project, experiment, observation, survey, or a written report on a relevant topic.

REVIEW QUESTIONS

1. What settings are relevant for nutrition education?

2. Describe the four needs assessment categories.

3. Compare and contrast quantitative and qualitative research methods for data collection.

4. Explain three types of nutrition education approaches used by various agencies.

5. Describe the importance of using theory in designing nutrition education interventions.

6. What are the major principles of social marketing?

7. What is the difference between goals and objectives as learner outcomes?

ASSIGNMENT

Design A Nutrition Intervention Choose and describe a target audience to design a nutrition intervention. Conduct a needs assessment using two different methods of data collection. Select a nutrition education approach and a theory congruent with your target audience. Formulate goals and objectives for a one hour nutrition intervention.

Answers to Chapter Exercises

Exercise 11.3 The HBM was chosen because hypertension is prevalent and may lead to stroke and cardiovascular disease. Case study participants perceive high blood pressure as a health threat impacting their physical and social livelihoods. Additionally, the HBM centers on perceptions of benefits and barriers to adopting health behaviors, which are useful constructs for addressing the beliefs and concerns of the participants.

Exercise 11.5 1 = college students; 2 = identify three foods containing saturated fats; 3 = after completion of the seminar

Exercise 11.6 1 = Cognitive- comprehension, 2 = Affective-receiving, 3 = Affective-characterization, 4 = Affective-responding, 5 = Cognitive-Application, 6 = Cognitive-synthesis, 7 = Psychomotor-guided response, 8 = Affective-characterization, 9 = Cognitive-analysis, 10 = Psychomotor-mechanism, 11 = Cognitive-knowledge

Exercise 11.7 1 = Level 3, 2 = Level 1, 3 = Level 2, 4 = Level 1, 5 = Level 2, 6 = Level 2

SUGGESTED READINGS, MATERIALS, AND INTERNET RESOURCES

Social Marketing: Influencing Behaviors for Good. 3rd Edition. P. Kotler & N.R. Lee. Thousand Oaks, CA: Sage Publications; 2008.

http://www.fruitsandveggiesmatter.gov/
Provides curricula and teaching tools, worksite wellness ideas, downloadable signs, and posters

http://collaborate.extension.org/wiki/Main_Page
Run by Cooperative Extension, one-stop web page for sharing nutrition education resources, how to use YouTube

http://www.nutritionquest.com/
Dietary survey questionnaire development and validation

http://medlineplus.gov/
National Nutrition Summit Information Resources

http://www.cdc.gov/nccdphp/dnpao/socialmarketing/index.html
Center for Disease Control and Prevention/ Social Marketing Resources, includes a course on how to use social marketing

http://www.eduref.org/Virtual/Lessons/Guide.shtml
Information Institute of Syracuse, Write a lesson plan guide.

http://www.sne.org
Society for Nutrition Education

REFERENCES

[1] International Food Information Council. *2009 Food & Health Survey: Consumer Attitudes toward Food, Nutrition & Health.* 2009. Available at: http://www.foodinsight.org/Resources/Detail.aspx?topic=2009_Food_Health_Survey_Consumer_Attitudes_toward_Food_Nutrition_and_Health. Accessed May 27, 2010.

[2] Bassett MT, Dumanovsky T, Huang C, et al. Purchasing Behavior and Calorie Information at Fast-Food Chains in New York City, 2007, *Am J Public Health.* 2008; 98: 1457–1460.

[3] Roberto CA, Larsen PD, Agnew H, Baik J, Brownell KD. Evaluating the Impact of Menu Labeling on Food Choices and Intake, *Am J Public Health.* 2010; 100: 312–318.

[4] Rabin RC. How posted calories affect food orders. *New York Times.* 2009 Accessed July 9, 2010. http://www.nytimes.com/2009/11/03/health/03nutrition.html?_r=1&fta=y.

[5] Lusk SL, Raymond DM. Impacting health through the worksite. *Nurs Clin North Am.* 2002; 37:247–56.

[6] Soler RE, Leeks KD, Razi S, et al. Task Force on Community Preventive Services A Systematic Review of Selected Interventions for Worksite Health Promotion: The Assessment of Health Risks with Feedback *Am J Prev Med.* 2010; 38: S237–S262.

[7] Milani, RV, Lavie CJ. Impact of worksite wellness intervention on cardiac risk factors and one-year health care costs. *Am J Cardiology.* 2009; 104 (10):1389–1392.

[8] Position of the American Dietetic Association Society for Nutrition Education, and American School Food Service Association—Nutrition services: An essential component of comprehensive school health programs. *J Am Diet Assoc.* 2003; 103:505–514.

[9] Franks AL, Kelder SH, Dino GA, Horn KA, Gortmaker SL, Wiecha JL, et al. School-based programs: Lessons learned from CATCH, Planet Health, and Not-On-Tobacco. *Prev Chronic Disease.* 2007 Available from: http://www.cdc.gov/pcd/issues/2007/apr/06_0105.htm Accessed July 15, 2010.

[10] Contento IR, Balch GI, Bronner YL, Lytle LA, Maloney SK, White SL, Olson CM, Swadener SS, Randell JS. The effectiveness of nutrition education and implications for nutrition education policy, programs, and research: A review of research. *J Nutr Educ.* 1995; 27:279–418.

[11] Simons-Morton BG, Greene WH, Gottlieb NH. *Introduction to Health Education and Health Promotion.* Prospect Heights, IL: Waveland Press, Inc.; 1995.

[12] Hitch EJ, Youatt JP. *Communicating Family and Consumer Sciences.* Tinley Park, IL: The Goodheart-Willcox Company, Inc.; 2002.

[13] Knowles MS, Holton EF, Swanson RA, Holton E. *The Adult Learner: The Definitive Classic in Adult Education and Human Resource Development.* 6th ed. Oxford, England: Butterworth-Heinemann; 2005.

[14] Walker EA. Characteristics of the adult learner. *Diabetes Educator.* 1999; 25:16–24.

[15] Kettner PM, Moroney RM, Martin LL. *Designing and Managing Programs: An Effectiveness-Based Approach.* Newbury Park, CA: Sage Publications; 1990.

[16]Liou D, Bauer K. Obesity Perceptions among Chinese Americans: The Interface of Traditional Chinese and American Values. *Food, Culture and Society*. 2010; 13(3):351–369.

[17]Chamberlain VM. *Creative Home Economics Instruction*. 3rd ed. New York: Glencoe McGraw-Hill; 1992.

[18]Moorman C. The effects of stimulus and consumer characteristics on the utilization of nutrition information. *J Consumer Res*. 1990; 17:362–374.

[19]Whitehead R. Nutrition education research. *World Rev Nutr Diet*. 1973; 17:91–149.

[20]Fleming, PL. Nutrition education and counseling. In: Paige DM, ed. *Clinical Nutrition*, 2nd ed. St. Louis: Mosby; 1988.

[21]Contento I. Review of nutrition education research in the Journal of Nutrition Education and Behavior, 1998 to 2007. *J Nutr Ed Behav*. 2008; 40:331–340.

[22]Green LW, Kreuter MW. Health Promotion Planning: An Educational and Environmental Approach. Mountain View, CA: Mayfield; 1991.

[23]Brownell KD, Warner KE. The perils of ignoring history: Big tobacco played dirty and millions died. How similar is big food? *Milbank Quarterly*. 2009; 87(1):259–294.

[24]Gilman S L. *Fat: A cultural history of obesity*. Cambridge, UK: Polity Press; 2008.

[25]Freire P. *Pedagogy of the Oppressed*. New York: Continuum; 1970.

[26]Baranowski T, Cerin E, Baranowski J. Steps in the design, development, and formative evaluation of obesity prevention-related behavior change. *Inter J Behav Nutr Phys Activity*. 2009; 6:6.

[27]Conner M, Armitage C.J. *The social psychology of food*. Buckingham, UK: Open University Press; 2002.

[28]Johnson DW, Johnson RT. Nutrition education: A model for effectiveness, a synthesis of research. *J Nutr Educ*. 1985; 17:S1–S44.

[29]Gillespie AH, Brun JK. Trends and challenges in nutrition education research. *J Nutr Educ*. 1992; 24:222–226.

[30]Glaser BG, Strauss A. *Discovery of Grounded Theory. Strategies for Qualitative Research*. Sociology Press; 1967.

[31]Glanz K, Rimer BK, Viswanath K. *Health Behavior and Health Education: Theory, Research, and Practice*. 4th ed. San Francisco, CA: Jossey-Bass; 2008.

[32]North Carolina Nutrition Network. Partners in Wellness (PIW) http://www.ces.ncsu.edu/Wellness/index.htm Accessed July 13, 2010.

[33]Kotler P, Zaltman G. Social marketing: An approach to planned social change. *J Marketing*. 1971; 35:3–12.

[34]Storey JD, Saffitz GB, Rimon JG. Social marketing. In: *Health Behavior and Health Education Theory, Research, Practice*. Glanz K, Rimer BK, Viswanath K, Eds. San Francisco, CA: Jossey-Bass, 2008.

[35]National Cancer Institute. 5 A Day for Better Health Program Evaluation Report. Updated 2006, Available at: http://cancercontrol.cancer.gov/5ad_exec.html Accessed October 24, 2010.

[36]Center for Disease Control, About the National Center for Health Marketing, 2008, Available at: http://www.cdc.gov/healthmarketing/ Accessed July 11, 2010.

[37]Social Marketing National Excellence Collaborative. *Social Marketing: A Resource Guide*, First in a series of Turning Point resources on social marketing, 2002 Available at: http://www.turningpointprogram.org/Pages/pdfs/social_market/social_marketing_101.pdf Accessed July 11, 2010.

[38]Petty RE, Cacioppo JT. Communicator and Persuasion: Central and Peripheral Routes to Attitude Change. New York: Springer-Verlag; 1986.

[39]Andreasen, AR. *Marketing Social Change: Changing Behavior to Promote Health, Social Development, and the Environment*. San Francisco, CA: Jossey-Bass; 1995.

[40]Weinreich NK. *Hands-On Social Marketing*. Thousand Oaks, CA: Sage Publications, Inc.; 1999.

[41]Lefebvre RC, Lasater TM, Carleton RA, Peterson G. Theory and delivery of health programming in the community: The Pawtucket Heart Health Program. *Preventive Medicine*. 1987; 16:80–95.

[42]Hitch EJ, Youatt JP. *Communicating Family and Consumer Sciences*. Tinley Park, IL: The Goodheart-Willcox Company, Inc.; 2002.

[43]Bloom B. *Taxonomy of Educational Objectives, Handbook 1: Cognitive Domain*. New York: David McKay; 1956.

[44]Simpson E. The classification of educational objectives, psychomotor domain. *Illinois Teacher*. 1967; 10:110–145.

[45]El-Tigi, M. Teacher Education Module Series. Develop a Lesson Plan, Module B-4 of Category B—Instructional Planning (1977). Ohio State Univ., Columbus. National Center for Research in Vocational Education. ED149062 - An ERIC Document. 2003, http://www.eduref.org/Virtual/Lessons/Guide.pdf. Accessed July 16, 2010.

12
Educational Strategies, Mass Media, and Evaluation

© David Pogue/The New York Times

Self-initiated learning, once begun, develops its own momentum.

—RAY HARTJEN

Behavioral Objectives

- Describe key elements for delivering effective oral presentations.
- Identify steps for planning a demonstration.
- Compare and contrast action-oriented learning activities.
- Understand the process of developing mass media educational tools.
- Describe selected examples of emerging technologies used in education.
- Describe the basic components of nutrition education evaluations.

Key Terms

- **Teaching Techniques:** means in which educational objectives are achieved to create meaningful learning experiences.
- **Action-oriented Techniques:** strategies such as debates which allow individuals to exercise a level of control of what is learned.
- **Formative Evaluation:** systematic assessment occurring before or during a learning activity to improve the educational process.
- **Summative Evaluation:** systematic assessment at the conclusion of a course, program, or learning activity.

INTRODUCTION

In Chapter 11, we reviewed the Keys to Success Nutrition Education Process Model and covered the first four keys setting the foundation for instructional planning and developing educational materials. We discussed the need to know your audience, determine an educational philosophy that fits their needs, select a theory or components of a theory that would best guide an intervention, and then set goals and objectives for the intervention. In this chapter, we review a variety of educational strategies that can be used to meet these needs. A good educational plan needs to integrate an evaluation component to monitor progress, revise strategies, and evaluate outcomes. The basics of the evaluation process are covered at the end of this chapter.

KEYS TO SUCCESS # 5–PROVIDE INSTRUCTION PLANNING AND INCORPORATE LEARNING STRATEGIES

Several factors need to be considered when selecting learning strategies. In addition to the educational factors related to your audience discussed in previous keys, there are logistical factors that will influence your choice of strategies, such as the size of the group, number of meetings, location, and resources. For example, planning for an on-line virtual audience as compared to an in-person meeting will likely promote certain choices over others. No matter what techniques you employ, emphasis should be placed on engaging your audience by building an environment conducive to hearing your messages. If possible, consideration should be given to providing active learning opportunities. As illustrated in Figure 12.1, the impact of

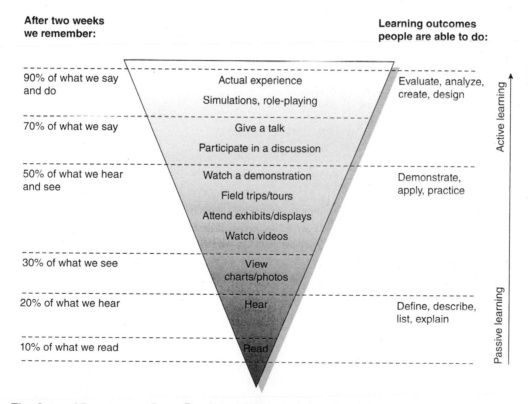

Figure 12.1 The Core of Experience: From Passive Learning to Active Learning
Source: From BOYLE/HOLBEN. Community Nutrition in Action, 5E. © 2010 Brooks/Cole, a part of Cengage Learning, Inc. Reproduced by permission. www.cengage.com/permissions

the learning experience provides better and more lasting outcomes as the audience becomes actively engaged. The following provides a synopsis of commonly used **teaching techniques**.

Presentation

Lecturing may be the oldest method of communicating information to groups and can be used in a wide variety of settings. Effective public speakers develop skills based on a solid foundation of knowledge, enthusiasm for the subject matter, technique development, and practice. Work at delivering a lecture with confidence and animation to generate enthusiasm for the presentation. Some helpful pointers for organizing a presentation are given in Table 12.1 and for effectively delivering a presentation in Table 12.2. A list of common presentation behaviors that detract from the impact of the intervention are listed in Table 12.3.

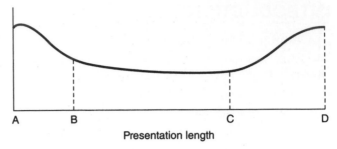

Figure 12.2   Presentation Learning Curve
Source: Morrisey GL, Sechrest TL, Warman WB. (1997). *Loud and Clear: How to Prepare and Deliver Effective Business and Technical Presentations.* Reading, MA: Addison Wesley Longman, Inc.

Both short and long presentations have a learning curve. In the beginning (A–B) curiosity enhances attention and retention, in the middle (B–C) retention will drop, and at the end (C–D) a summary warning, "Note that we are coming to the end," will encourage an upward swing in retention.

Table 12.1   Organization of a Presentation

Category	Description
Analyze your audience	As part of a comprehensive nutrition intervention, you conducted a needs assessment. This assessment should provide you with a wealth of information, such as demographics, learning interests, and concerns about the topic you are presenting.
Tailor presentation to the audience	Narrow the scope of your topic. Be specific. For example: "Six Food Behaviors to Lower Cholesterol." "Herbs–Fountain of Youth in the Cupboard?"
Be organized and focused	Have a clear goal and 3 to 5 defined objectives for conveying information. Information intense programs will overwhelm the listener and reduce the learning impact. Be sure to emphasize important points at least 3 times, especially in the beginning and at the end. See the presentation learning curve in Figure 12.2. Include an introduction, body, and concluding remarks. Be sure to include appropriate transitions between major sections of the presentation. Stick to your outline.
Have an attention-grabbing introduction	You have two minutes to capture the attention of the attendees. Possible attention getters include the following: • Provide a startling fact • Involve the audience • Ask a question • Recite a quote • Provide a demonstration • Tell an anecdote or story • Comment on a current or local event
Answer two questions in the introduction	*Why are you up there?* "I'm here to help you find workable ways to lower your cholesterol." *What's in it for the audience? (What are your credentials?)* "As someone who has lowered her own cholesterol levels and helped numerous others to do the same, I can help you."

(continued)

Table 12.1   Organization of a Presentation *(continued)*

Category	Description
Effective body	The goal (main idea) of your presentation should be supported with one or more of the following: • Provide examples–offer actual samples of people, places, objects, actions, conditions, or experiences • Give definitions–clarify an unfamiliar word or phrase • Furnish comparisons–show similarities, such as increases in obesity rates around the world • Provide contrasts–show differences, such as health disparities for various cultural groups • Present statistics–use charts or graphs to illustrate numbers • Provide testimony–use the words of a client or a renowned expert • Give research findings–provide results of research studies
Add "sparkle" periodically	Keep audience interest by adding a "dash of spice" every few minutes to generate learner interest and retention. Possibilities include the following: • Ask questions • Provide anecdotes • Cartoons • Jokes • Flags–"If you remember one thing from this presentation, I hope it will be . . ." • Props or costumes • Activities • "Real people"–videos, pictures, or quotes Using humor can promote a relaxed atmosphere for learning, thereby, creating a feeling of well-being and likelihood of retention of information.[1] However, care should be given not to use inappropriate jokes and not to overdo the use of humor so as not to make your topic appear irrelevant.
Memorable ending	• Alert your audience that that you are ending the presentation. Summarize objectives (key points). For example, "Now that we are coming to an end, let's review the main points covered." • Encourage a "call to action." Give clear mental or behavior action steps. • End with an impressive quote, anecdote, or personal observation.

Adapted from Source: International Food Information Council Foundation. Sharpen your skills, the second P: Presenting. Foodinsight, www.foodinsight.org. 2009. Accessed October 16, 2010.

Table 12.2   Pointers for Delivery of a Presentation

Category	Description
Use vocal variety	A dry monotone delivery will dampen the most interesting topics. Minds will wander. Be lively and animated to spark interest. Your presentation should convey to the attendees that you are passionate about the topic. Use vocal variety to capture attention. Vary your pitch and volume. Everyone has a range in pitch. Women tend to speak in a high range and men in a low range. If there is little variation, the result will be a monotone delivery. You can also add interest by altering the volume of your voice. Both speaking softly or loudly can add emphasis to a point you wish to make. The tone of your voice should be in harmony with the emotional message you wish to convey. If a presentation lacks energy, a critic is likely to label the presentation as flat. The speed or rate of your speech can also vary to place emphasis. A faster rate illustrates that you have something exciting to say. A slower rate and a pause signals that the topic is important, that is, something you want to remember.
Speak naturally and clearly	Pay attention to diction, pronounce words carefully, and avoid slang words, unless appropriate for your population group. Avoid vocal fillers such as, "ums," "uhs," "likes," and "you knows."

(continued)

Table 12.2   Pointers for Delivery of a Presentation *(continued)*

Category	Description
Use engaging language and body behavior	A popular song title states, "When You're Smiling the Whole World Smiles With You." Attendees at a presentation are more likely to feel good about the experience if the speaker has a smile, conveys a positive attitude, and displays confident body language. Using the inclusive word "we" rather than "I" also tends to engage attendees.
Maintain eye contact	In Western societies, eye contact is a vital element in capturing attention and interest of learners. Your eye contact should be approximately 85% of the time.[2] Looking away or focusing on lecture notes will distance you from the audience. If maintaining eye contact is uncomfortable for you, try establishing "nose contact," or briefly looking at individuals' noses instead of their eyes.[3]
Move	Gestures such as hand movements and facial expressions provide visual stimulation for learners. Physical movement and mobility of the educator in a classroom setting has been found to elicit desirable behavior among learners.[4] Learners tend to be less distracted and focus on the educator when effective gestures are used.
Eliminate distracting mannerisms	Be careful that gestures and movements are used appropriately and creatively instead of becoming a distraction. Avoid annoying mannerisms and nervous habits such as gesturing wildly with the hands, rattling coins in pockets, or rocking on your heels.
Dress professionally	Your attire should not attract undue attention. Avoid clothing with unusual designs and accessories that are too bright or extraordinary.
Practice	Practice at least three times using a watch to time yourself. Practice in front of a mirror, in front of one other person, and then in front of several people. Ask for feedback.

Table 12.3   Presenter Never–Evers

Never give an apologetic beginning	For example, "I'm not sure I am the most qualified person to give this presentation." "I am a bit nervous, so I hope you will bear with me." "I didn't have time to prepare."
Never use wrong names	If you are thanking an individual or an organization, be sure you have correct names.
Never use gimmicks	For example, writing SEX in big letters on the board and then saying, "Now that I have your attention, we can begin."
Never ridicule	Do not denigrate participants for their lack of knowledge or experience. Each person's background should be appreciated. They should be encouraged to build upon their knowledge base to integrate your presentation concepts.
Never say you compressed your presentation to save time	Your presentation should be designed for your audience taking into consideration the amount of time allotted. If time runs short, simply provide what you can.
Never say you would have brought more materials if possible	Do not give excuses and indicate to your audience that they are missing out on something interesting. This will not please or inspire participants.
Never tell participants what you have forgotten	Participants will have no idea that you have forgotten something. Hearing this information will lead participants to think you are disorganized and encourage them to wonder if their attendance at your presentation is worthwhile.
Never read from a lengthy prepared text	Few people can read lengthy material and engage an audience. A particularly poignant short passage or poem would be appropriate. If participants need verbatim material, provide copies or a website to locate the information.

(continued)

Table 12.3 Presenter Never–Evers *(continued)*

Never supply sloppy handouts	Handouts should be clear, concise, and legible. Confusing or difficult to read handouts are not useful and are not likely to be read.
Never share a schedule that is not likely to be completed	If a program has been tightly scheduled and you are not likely to get to all the topics, give only a broad scope of what will be covered. Sparking interest in a particular topic leads to disappointment if time runs out and the program was not completed.
Never go past the scheduled time	If a lecturer goes over time, participants will start worrying about the end of the presentation rather than focusing on the presentation.

Source: Adapted from: Sharp P. The never evers of Workshop Facilitation. Tools for Schools. National Staff Development Council; December/January, 2000.

Discussion

A discussion is a goal-oriented interaction or conversation between individuals on a particular topic. When discussion is used for instructional purposes, it can lead to critical thinking, skill development, problem solving, and articulation of perceptions and opinions in a logical manner. Good discussions stem from careful use of questions. See Chapter 10 for a review of effective questions and group facilitation methods.

Demonstration

Demonstrations combine telling with showing. They can visually illustrate procedures, show techniques, or provide symbolic representations. A small demonstration may be inserted into a lecture, such as, having a percentage of the audience stand to represent the percent of people in the world who do not have access to safe water. For demonstrations showing techniques, learners are often expected to imitate the procedures viewed or adapt it to a specific situation. Table 12.4 provides a guide for planning more complex demonstrations.

Visual Aids

According to an ancient Chinese proverb, "one picture is worth more than 1,000 words." Visual aids provide clarity and add vitality to a presentation. Compare these two approaches: (1) hearing about a Japanese tea ceremony or (2) hearing about a Japanese tea ceremony, watching a video clip, and seeing an actual Japanese teacup. You probably imagine the second approach to be more

Simon Curtis/Alamy

appealing. Visual aids also enhance an educator's credibility. As the saying goes, "seeing is believing." You may have a hard time believing that people really enjoy eating chocolate covered mealworms or crickets unless you watch an episode of *Extreme Sweets*. Visual aids also promote learning by improving memory through visual stimulation. As illustrated in Figure 12.1, seeing enhances learning outcomes and retention. Table 12.5 provides tips for using visual aids effectively in presentations.

The following lists some commonly used visual aids to enhance learning:

- **Blackboard or whiteboard:** The use of a blackboard is the most conventional way to provide written and visual aids in learning. More recently, whiteboards with a plastic or ceramic and magnetic surface have replaced the traditional blackboard because of

---

EXERCISE 12.1 **Evaluate Your Presentation**

Arrange to give an eight minute presentation to a group and videotape the proceedings. Use the lesson plan guidelines in Chapter 11 to design the presentation. Review the videotape and evaluate the impact of your presentation.

	Good	Average	Needs Work
1. Introduction captured interest.	❏	❏	❏
2. Used illustrations or examples to explain concepts.	❏	❏	❏
3. Memorable ending, summarized main points, impressive conclusion.	❏	❏	❏
4. Connected to the audience: smiled, good eye contact, used "we" instead of "I".	❏	❏	❏
5. Fluent, no distracting "and then" and " um".	❏	❏	❏
6. Effective body language: no distracting habits, showed confidence, creative gestures.	❏	❏	❏
7. Added "sparkle", used humor.	❏	❏	❏
8. Demonstrated enthusiasm.	❏	❏	❏
9. Varied pitch of voice.	❏	❏	❏
10. Varied volume of voice.	❏	❏	❏
11. Tone of voice conveyed appropriate emotion.	❏	❏	❏
12. Rate of speech was effective.	❏	❏	❏

❏ Explain what you believe was the most effective component of your presentation.
❏ What would you like to do differently next time?

EXERCISE 12.2 **Examine I Have a Dream**

Martin Luther King's "I Have a Dream" speech on August 28, 1963, in Washington DC has often been credited with being the most effective presentation of its kind. Go to You Tube, http://www.youtube.com/watch?v=PbUtL_0vAJk, and watch the 17 minute speech and evaluate why this claim has been made. Comment on the following:

1. Voice quality: pitch, tone, volume, rate
2. Content: Phrasing
3. What do you believe was the most effective component of his presentation?
4. Is there anything about his presentation that you would like to emulate in your lectures?

their versatility for displaying charts, visual messages, and highlighting key points or concepts. A SMART Board is an electronic whiteboard that interfaces with a computer. Computer images can be displayed on the board via a digital projector, and may be further modified by the educator. Because the whiteboard is touch-sensitive, notations, drawings, and text can be written in digital ink and saved on the computer and printed.

• **Overhead Transparency Projector:** The beneficial features of teaching with an overhead projector include the ability to save and store materials on transparencies, the versatility of being able to teach and face the audience, and

Table 12.4 Steps for Planning a Demonstration

Category	Description
Define a topic	The topic should lend itself to be broken down into sequential steps.
Develop an outline	Include an introduction, major concepts to be covered, and concluding remarks. Consider audience participation, if appropriate. For example, if you are demonstrating food art with cabbage and carrots, give these items to all participants to prepare an individual food art as you demonstrate. If the program is geared to children, having them participate will increase enthusiasm and the likelihood of trying new foods.
Develop a sequential plan	Sequence the presentation to show the most logical way of demonstrating the procedures. Each step of your plan should indicate the amount of time and resources needed for the identified task.
Pre-preparation activities	Your plan should include a list of all resources that need to be gathered and activities that must be done before the actual demonstration. For example, cutting and mixing of ingredients often needs to be done ahead of time.
Practice	Practice and perfect the demonstration by ensuring all equipment is in optimal working condition and properly positioned for audience viewing. Food laboratories may include an overhead mirror or projection to enable all learners to observe the entire demonstration. Have others watch your practice demonstration to comment on the clarity of your explanations and effectiveness of your visuals. For food demonstrations, use clear bowls and have samples of the ingredients in their packaging or fresh produce on display as props.
Plan an attractive ending	Arrange for an attractive final product. For food demonstrations, provide samples of the final product for the audience to taste.

Table 12.5 Tips for Using Visual Aids

Category	Description
Use the visual for clarity	Use the visual to enhance an objective of your presentation. Do not get side-tracked into dwelling on various components of the visual aid. Speak to the audience, not the prop.
Integrate into presentation	Do not wait until the end of the presentation to show visuals. Integrate them into the body of your presentation to stimulate interest and enhance understanding of concepts. Normally keep them hidden until ready to use.
Be sure the visual is viable	Consider all of the support you need to use the visual aid, such as refrigeration, electricity, and assistance of someone else.
Use only a few best examples	Too many visuals will overstimulate and distract focus from the main concepts.
Test electronics	Test electronics at least a day beforehand so you know how to use the equipment and there is enough time to make alterations if needed.
Order visuals	If you have a number of visual aids, put them in order. Consider labeling them with numbers.
Rehearse with an assistant	If using your visual aid requires assistance from another individual, rehearse and provide clear instructions with a cue sheet.

Source: Write-out-loud.com. How to Use Props http://www.write-out-loud.com/howtouseprops.html Accessed October 24, 2010.

portable projectors can be used easily in most locations. The contents of parts of a transparency can be covered and revealed as needed by the lecturer. Follow the 6 × 6 rule: maximum 6 lines, maximum 6 words per line. Additional lines can be used if the number of words is reduced. Another option is writing directly on a transparency to jot down key points. Using colored markers increases the visual interest of this procedure. Carry extra bulbs, an extension cord, and a three-prong adapter.

- **Flipcharts:** A flipchart consists of a large paper pad that is either propped on an easel or mounted on a wall. This device is especially useful when a whiteboard or overhead projector is not available. A flipchart is useful for brainstorming to record audience ideas. Instructors can either write notes on the flipchart during lectures or reveal a previously prepared sequence chronologically during a presentation. These techniques would be useful for explaining something such as the digestion of carbohydrates within the gastrointestinal tract.

- **PowerPoint Presentations:** PowerPoint presentations provide numerous visual enhancement options. This method has quickly replaced earlier lecture aids because of its professional appeal, versatility, and relatively convenient access. Esthetically pleasing slides can easily be designed on computers that include script, graphics, digital pictures, sound effects, and video clips. PowerPoint presentations can be saved, duplicated, or modified easily. In addition, individual slides can be effortlessly reordered to meet changing needs. Avoid information overload on any single slide. Use the KISS principle–keep it simple and straightforward. Also, follow the 6 × 6 rule described previously. Incorporate key phrases or words on a given point as opposed to writing lengthy sentences. Always ensure that the font size and style is visible to the last row of the audience. Use a simple font, Times New Roman or Arial, and no less than size 24. Do not use more than two font sizes per slide. Color combinations and highlighting should be visible

and aesthetically pleasing. A mix of text and graphics create visual appeal. Avoid too many special effects–animations, colors, and sounds. These can be distracting. Do not use more than three colors per slide and keep your theme consistent throughout your presentation. When lecturing with PowerPoint presentations, be sure to provide adequate time for viewers to read or jot down slide information.

- **Videotapes (CDs, DVDs):** Videotape, CD, or DVD recordings can be used to add interest to presentations. Relevant movies, news reports, or musical recordings can be played during educational sessions to highlight certain points. For example, clippings of a documentary can be shown during a nutrition class to depict the potential impact of fast food eating on the health of Americans.

Action-Oriented Techniques

When planning an education intervention, consideration should always be given to possible ways to actively engage the audience. As indicated in Figure 12.1, participation and interaction stimulate interest, increasing acquisition and recall of content.[5] **Action-oriented techniques** encourage learners to take responsibility for their own learning by exercising a level of control over what is learned. Selected examples of action-oriented learning activities are reviewed in this bulleted list.

- **Debate:** A debate occurs when two opposing groups present affirmative and negative perspectives on a controversial issue. For example, college students may debate the issue of genetically modified foods as it pertains to the safety of consumers and the environment. The critical analysis deepens understanding of the subject and is likely to influence attitude and possibly behavior. A number of useful strategies have been developed for including informal debates in the classroom. See resources at the end of the chapter. In a formal debate, prescribed rules and procedures are followed that dictate the frequency, length of time, and actual members that may speak.

- **Role-playing:** In role-playing, participants engage in a spontaneous acting out of a scenario, displaying the emotional reactions of individuals in a particular situation. For example, students can engage in a role-play of peer pressures that exist when socializing at parties. Participants are provided brief descriptions of the roles and how they should be played without using a script or rehearsing lines. A debriefing should follow immediately after a role-play in order to enable learners to analyze what transpired.

- **Educational Games:** Games used in an educational setting are designed to teach, reinforce or introduce specific content, or develop a skill.[6] The outcome of this activity should depend largely on the knowledge or skill a learner is expected to achieve. Younger and older audiences generally can benefit from this teaching technique, adding excitement to learning by stimulating competition among members. Board games, puzzles, and simulated game shows such as "Nutrition Jeopardy" can be incorporated into educational settings. Textbook publishers and on-line educational and commercial sites offer numerous web-based games to reinforce learning such as Club Penguin, Webkinz, and PBS Kids.

- **Simulations:** Simulations contain components of both role-playing and games. These activities are a symbolic representation of a particular life experience. Learners are provided a role and interact in a scenario under specific guidelines. Participants are required to make decisions which will, in turn, influence the entire system. Simulations provide learners the opportunity to experience the complexities and emotions of real-life in an environment guarded from actual risks and consequences. A popular simulation used to explore intercultural communication difficulties is Barnga. This is a card game in which participants unknowingly play with different rules. Increasingly computer and on-line simulations are offering real world learning experiences.

- **Laboratories:** In a laboratory, learners are guided through a planned, supervised practice experience. Students are provided the opportunity to apply a principle, investigate a phenomenon, or to practice a process or skill. For example, a food laboratory lesson could include cooking with lower sodium via the use of herbal seasonings.

Technology-Based Techniques

The quickly changing world of technology has created a multitude of exciting options and challenges for educators.[7] With the availability of computers, wikis, blogs, Twitter, YouTube, virtual worlds, mobile devices, and electronic books, there appears to be a learning revolution and the opportunities sometimes seem endless. Taking courses "on-line" has become commonplace and a variety of software is available for developing and managing web-based courses. The challenge is to stay abreast of technology changes, keep up with technologies commonly used by the Net Geners (born in the 1980s) and iGeners (born in the 1990s), and to find ways to effectively incorporate them into learning experiences. The most recent technology savvy generation has "i" in its title because many of the most recent devices and websites have this letter in their names, such as iPod, iTunes, Wii, iChat, iHome, and iPhone. Your age is likely to reflect your selection of technologies and how they are used. For example, iGeners are more likely to communicate with cell phones for text messaging, instant messaging, accessing Facebook and other social networks, and video conferencing than by talking.[7] Increasingly educational programs are using these resources. In addition to an abundance of technology opportunities, there are many easily available educational resources on the Internet, termed "fingertip knowledge."[8] At no other time in the history of the world has there been easy access to so much information. As a result, there is less need to memorize and a greater need to guide students in finding and evaluating credible resources.

Incorporating web-based strategies and mobile technologies to promote nutrition education has become increasingly popular in community, clinical, and academic settings. See Table 12.6 for examples of technology-based resources and Table 12.7 for examples of web-based nutrition education activities.

Learning Domains and Strategies

An evaluation of a comprehensive review of the literature shows that there are advantages to using a variety of educational strategies as opposed to only one method.[14] Your lesson plan should use a variety of approaches to accommodate diverse learning styles and to influence all three learning domains discussed in Chapter 11. All the educational strategies discussed in this section address educational objectives of the cognitive domain and some also attend to affective and psychomotor domains. In targeting the affective domain, educators should seek to influence learner's attitudes, beliefs, and values. Particularly useful are interactive

Table 12.6 Examples of Technology-Based Education Resources

Video clips	The Internet is full of free video clips that can be used to demonstrate an activity or a concept.
WebQuest model[9]	A WebQuest is an inquiry-oriented lesson format in which learners actively use the Internet as a resource to gain different knowledge from tasks and subtasks designed around the major themes of a topic. Educators can use this technology as an assessment tool of students' acquisition of knowledge and their ability to apply higher-order thinking skills.
Podcasting	Podcasts are a series of audio or video files that can be downloaded on a computer or a digital music player (MP3 player) and listened to at a convenient time. A study showed that using podcasting with specific educational goals focusing on self-efficacy was an effective method to encourage weight loss and higher intakes of fruits and vegetables among overweight adults.[10]
Wireless text messaging	Wireless text messaging has been used to provide reminders. In a randomized controlled trial, wireless text messaging reminders improved adherence in taking daily vitamin C tablets.[11]
Virtual life sites	Virtual life sites such as Second Life in Education or Active Worlds Education Universe offer on-line interactive discovery and problem solving opportunities in realistic settings that can be done by individuals or groups. We are likely to see an expansion of opportunities to use these resources in nutrition counseling and education in the future. As of 2009 there were over 300 colleges and universities in Second Life with expectations for continued growth.[12] The possibilities are limited only by your imagination. For example, imagine a group of individuals who are attempting to make a behavior change, such as losing weight. Each person selects and controls an avatar. They meet weekly at a designated time in a plush office in the virtual world with a counselor who guides a group meeting. There could be an auxiliary site that encourages discussion and the posting of progress in meeting goals. The counselor could take a group on a virtual grocery store tour or prepare a meal. Because this is all online, the group could be made of individuals from anywhere in the world. A study at Penn State University had two teams attempt to solve a problem. One group worked face-to-face and the other group worked as avatars on Second Life. The latter group provided the most accurate solutions.[13]
Social networks	Social networks such as Facebook and LinkedIn provide opportunities for group learning and communication with individuals who may have knowledge about resources on a topic a person is exploring. For example, while interacting with others on a social network site, an individual may be directed to web resources to aid in memorizing the glycolysis pathway, such as a YouTube video of a rap song of the pathway, a parody of the pathway using the song "Sugar Sugar," or an open university website that provides a lecture about the pathway.

Table 12.7 Examples of Web-Based Nutrition Education Activities

Goal	Description
Provide a global perspective	The Internet provides a vast number of global resources. Projects could be designed to take advantage of exploring them. For example, a project could be to compare and contrast dietary analysis and dietary guidelines of various countries, such as the United States MyPyramid, Health Canada, Dietary Guidelines for Australians, and Japanese Dietary Guidelines.
Provide a real audience	Motivation to produce high-quality projects increases when there is expectation that the project will be placed on an actual on-line site, such as YouTube. Include opportunities on these sites for global feedback regarding these projects, such as comments and discussion boards.
Encourage the use of multiple mediums	Allow students to create projects using a variety of media including video, music, images, art, collages, dioramas, and PowerPoint as well as written stories. For example, an audio podcast could be combined with downloaded visual images of a topic to include in a presentation.
Consider online journaling	In classrooms, online journaling can often be done through course support websites such as Blackboard. Individuals or groups working together to make a common behavior change could use a social network site to document their progress, pose problems, and offer support to each other.

EXERCISE 12.3   Selecting Appropriate Teaching Techniques

You are approached by the dean of a local community college to design a 1-hour nutrition workshop targeting healthy snack consumption among 100 college freshmen. You are provided access to a large classroom facility equipped with the state-of-the-art technology. Work with a small group of colleagues and select three teaching techniques to maximize effectiveness in reaching this target audience. Be sure to address all three learning domains–cognitive, affective, and psychomotor. Note the following in your journal:

❑ Describe each strategy.
❑ Identify what objective or objectives are being addressed for each strategy.
❑ Identify what domain or domains of learning are being targeted for each strategy.

---

strategies such as debates or role playing that may lead to greater self-awareness and commitment to values and evaluation of beliefs. Choose activities that include opportunities to verbalize opinions, justify actions of others, or prioritize values. Keep in mind that attitudinal changes usually require longer periods of time and are not likely to alter over the course of a single educational intervention. In addressing behaviorally based or psychomotor objectives, demonstrations and experiments that include hands-on activities for the learner provide strategies for achieving desired outcomes in this domain.

KEYS TO SUCCESS # 6–DEVELOP APPEALING AND INFORMATIVE MASS MEDIA MATERIALS

Mass media can rapidly deliver persuasive and powerful health messages to a large audience. The sponsors of the messages are often commercial enterprises, news programs, consumer groups, and government, or professional organizations. Information from mass media is often the public's major source for acquiring knowledge about health issues. However, the desire of news programs to entertain may encourage selection of novel issues that apply to a limited section of the population. In addition, the need to rapidly transmit news often does not allow full coverage of the complexity or controversial aspects of an issue. Non-profit organizations use mass media to deliver nutrition information, but the impact tends to be limited due to restricted resources. The nutrition information from commercial advertising tends to be heavily biased with the primary goal of making a profit.[15] Children and adults are bombarded with numerous media messages every day. In order to create a balance, schools have begun to include media literacy in their curriculum providing guidance for analyzing, evaluating, and creating media.

Nutrition professionals can also help bring credible information to the public. Opportunities may arise involving both traditional print, audio, and audiovisual media as well as newer web-based venues. Federal agencies and national voluntary organizations may use public service campaigns to promote healthful behaviors. Many health campaigns use multiple outlets to deliver messages including social media venues, websites, television and radio public service announcements (PSAs), public service transit ads and billboards, posters, pamphlets, and special events. PSAs are usually 60-, 30-, or 10-second radio or TV announcements that provide novel information promoting activities or programs geared towards community interests and needs. They are most useful for creating public awareness and sensitivity to a health problem and for reinforcing a new health behavior.

Developing Audio and Audiovisual Messages

If you are involved in a large campaign, there is likely to be a team member who is a communication and media specialist. If this is the case, understanding the basics of developing media messages is likely to make your involvement more effective. On a smaller scale, you may have a greater portion of the responsibility developing audio messages or audiovisuals for a facility website or an in-house education program, such as health education videos that run continually in the lobby. Note that at this stage in planning a nutrition intervention, you have already defined your education philosophy, goals, and objectives. If you decide that audio or audiovisual materials are appropriate and realistic for your intervention, use the Message Development Model, presented in Figure 12.3 and outlined in Table 12.8 to guide the process.

Developing Print Materials

Good writing skills are an asset for nutrition professionals and can enhance your credibility and visibility. Opportunities are likely to arise requiring high quality written materials to influence target audience behavior. You may have opportunities to write for newspapers, magazines, or websites. Nutrition educators are likely to be involved in the

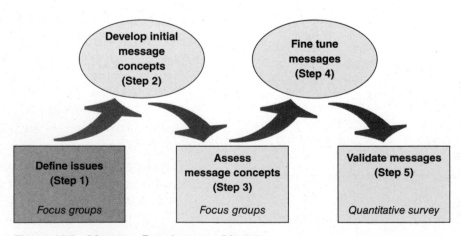

Figure 12.3   Message Development Model
Source: The consumer message development model. Adapted from Wirthlin Worldwide, Chicago, Ill; 1998.

Table 12.8 Five Basic Steps of the Message Development Model

Step	Description
1. Define the issue	• Define the central issue (information) you want to communicate. • What are the characteristics of the target audience? Find out what motivates your audience. Focus on demographics, family structure, hobbies and interests, hopes for the future, concerns and biases, life goals, and preferred recreational activities. • Is your target audience aware of the issue? Why or why not? What do they think, feel, and believe about the issue? Are their perceptions and attitudes towards the issue positive, negative, or neutral? Why? How does your target audience respond to the issue? Are they doing anything about it? Why or why not? • One-on-one conversations or focus groups are often used to obtain this information.
2. Develop initial message concepts	• Analysis of material received in Step 1 should provide clues regarding behaviors to be encouraged to address the health issue and how to approach your audience. • What specific action(s) or behavioral change(s) do you want your target audience to implement? • What potential incentives may inspire your target audience to adopt the behavior change(s)? • What potential barriers could inhibit your target audience from adopting the behavior change? • Develop the message. See Table 12.9 for presentation styles and Exhibit 12.1 for factors to consider for developing audio and audiovisual materials. • Does the language carry a positive tone? Is it "empowering" for the audience to make changes?
3. Assess message concepts	• Pretest message concepts with selected members of the target audience. Assess comprehension, personal relevance, appropriateness, and weak and strong points. In focus groups or personal conversations ask questions such as: ❑ "What does this message mean to you?" ❑ "Do you find this message motivating?" ❑ "Does this message speak to what you value in life?" ❑ Assess self-efficacy. Do you believe you can make this behavior change? What would you need to make this behavior change?
4. Fine tune message	• If your target audience did not interpret your message the way you intended, go back to Step 2. However, you will likely need to make only a few changes. Use the following questions to evaluate your message: ❑ How receptive was your test audience to the message? (Positive, negative, or neutral reaction–to what degree?) ❑ Did the test audience interpret the message the way you intended? Why or why not? What are the connotations they derived from the message? ❑ Which part of the message was clear or unclear? Will minor changes improve the meaning or should the message be completely rewritten?
5. Validate message	• Same step as four but apply to a wider audience. In advertising and marketing, telephone surveys and questionnaires are often used to reach a large number of individuals. • However, if your resources do not permit a large quantitative survey, you can use an informal survey to assess the impact of your message.

Source: Adapted from International Food Information Council Foundation. Message making 101: creating consumer-friendly messages. Food Insight, www.foodinsight.org. 2009. Accessed October 16, 2010.

development of fact sheets, pamphlets, direct mailings, or brochures. Few people are born with innate writing skills. With practice you can sharpen your skills to write communications that are "ICIC," which stands for Interesting, Clear, Informative, and Concise. Sounding out this acronym provides

Table 12.9 Categories of Intervention Presentation Styles

Category	Description
Testimonials	Use of a credible spokesperson or a celebrity may be attention-getting. Consider using physicians, organization leaders, or credible role models to convey the importance of your health message.
Slice of Life	This is a dramatization or a simple story within an everyday setting that may help the audience to relate to your message.
Demonstration	Provide an audiovisual demonstration of the desired health behavior, especially if skills must be taught.
Animation	Animation is eye-catching for young children and adults alike. Cartoon characters can demonstrate desired behaviors (for example, eating fruits and vegetables), and present abstract or sensitive issues (for example, eating disorders) in a non-threatening manner.
Humor	Proper use of humor can be memorable and heart-warming. Beware of offensive, stale, or corny jokes that serve no purpose.
Emotion	Emotional approaches to message presentation can range from warm and caring to fear and anxiety arousing. Be careful to pretest your messages and product so that the appropriate emotional tone is conveyed.
Use of music	Including a musical clip or overture can depict a mood you are trying to create. Use music judiciously so that it will not be distracting or compete with the message.

Source: U.S. Department of Health and Human Services. Making Health Communication Programs Work. NIH Publication No. 04-5145; 2001.

EXHIBIT 12.1 Tips for Developing Audio and Audiovisual Messages

❏ Present information in a direct manner by using relevant and compelling language.
❏ Use an attention-getter and identify the main issue in the first 10 seconds.
❏ Recommend a practical, easy-to-implement strategy and if possible, demonstrate the health behavior.
❏ Provide specific meaningful reasons for changing behaviors. For example, indicate benefits of taste, convenience, fun, culture, health benefits, or feeling good.
❏ Keep messages simple with one or two key points.
❏ Use a memorable slogan, theme, or sound effects to aid recall.
❏ Select an appropriate presentation style (for example., slice of life, testimonials, etc.).
❏ Generally a positive rather than negative appeal is more effective.
❏ Offer choices for making behavior changes. Choosing an option is motivational.
❏ If there is an action recommended, show the telephone number, website, or address on the screen for at least 5 seconds, and provide verbal reinforcement.

Source: U.S. Department of Health and Human Services. Making Health Communication Programs Work. NIH Publication No. 04-5145; 2001.

EXERCISE 12.4 Evaluate Public Service Audiovisuals

Go to YouTube to view New York City's "Pouring on the Pounds Campaign" videos: Man Eating Sugar and Man Drinks Fat. In your journal, answer the following:

❏ Describe your reactions to the videos.
❏ How effective do you believe these audiovisuals are in educating and altering behavior?
❏ Which categories of presentation styles presented in Table 12.9 were used in the audiovisuals?
❏ What tips for developing audiovisuals in Exhibit 12.1 were evident in these audiovisuals?

---

what you hope your readers feel after reading your material, "I see, I see!"[16] There are four steps for writing "ICIC" communications:

STEP 1: ASOAP Analysis Begin your preparation with an ASOAP analysis. ASOAP is an acronym for Audience, Subject, Objective, Angle, and Publication and provides a clear direction for your writing.

- **Audience.** Based on your needs assessment, you should have an understanding of your target audience's demographic profile

---

EXERCISE 12.5   Investigate Message Development Model

Read the following articles focusing on the Message Development Model:

Borra S, Kelly L, Tuttle M, Neville K. Developing actionable dietary guidance messages (Dietary fat as a case study). *J Am Diet Assoc.* 2001;101:678–684.

Hoffman EW, Bergmann V, Shultz JA. Application of a five-step message development model for food safety education materials targeting people with HIV/AIDS. *J Am Diet Assoc.* 2005;105: 1597–1604.

In your journal, provide the following:

❏ Select one quote from each article related to the Message Development Model. Write the quotes in your journal; identify the source and page number.
❏ For each quote, write how this information was significant for developing a consumer message.
❏ Explain how you may be able to apply what you learned to developing your own consumer messages.

(age, gender, ethnicity, religious affiliation, socioeconomic status, etc.) and psychographic profile (lifestyle, goals, values, beliefs, biases, etc.). Your investigation should have led you to understand what your target audience needs to know, what they want to know, and what type of message is likely to inspire them.

- **Subject.** You need to thoroughly research the subject matter. Identify reliable resources, and be sure to ask a colleague familiar with the topic to review your draft.
- **Objective.** You should clearly understand what you are trying to accomplish. Readers generally expect answers to the following questions: "Why are you giving this to me?" "How does this affect me?" "What am I supposed to do?" Is your primary objective to inform, stimulate interest, or change behavior? An evaluation of the stages of change of your target audience can indicate the motivational stage of a majority of your target population. If you found a large portion in the precontemplation stage, your objective could be to stimulate interest with a novel or persuasive message.[17]
- **Angle.** Your needs assessment is likely to provide guidance as to what are motivating factors for your target audience. Taping into motivating factors has been shown to increase attention and integration of reading about nutrition information.[18]
- **Publication.** Develop printed materials that are intended for the audience to understand, accept, and use.[15] The publication source may define a particular style and format. In some instances government guidelines or institution policy may require a certain template.[19] Choose an appropriate serious or light tone based on your audience norms and expectations.

STEP 2: Outline and Collect Resources An outline is a roadmap for writing your paper. It provides guidance for keeping you on target and insures that you do not leave out important information. You should not consider your outline written in stone. As you begin to write your draft and review your resources more closely, you may find the need to make alterations.

STEP 3: Write the First Draft You should not expect your first writing to be a final copy. Table 12.10 contains factors to consider for effective writing.

STEP 4: Polish Your Paper Depending upon your resources, there are various degrees of evaluation that can be done to assess effectiveness and readability. Ask colleagues to evaluate the document for the following:[21]

❏ Spelling, grammar, and punctuation
❏ Appropriate dating, numbering, and consistency
❏ Visual appeal
❏ Effectiveness and consistency of text enhancements
❏ Odd breaks or anything that reduces clarity

Ask individuals of your target audience to evaluate your document for comprehension,

Table 12.10   Factors to Consider for Effective Writing for the Public

Sentence Structure	
Unnecessary words	Unnecessary words clutter your message and reduce comprehension.[20] Limit most sentences to 8 to 10 words. Examples: Less effective: Past experience with high fiber diets has shown us that higher levels of fiber in the diet may help reduce cholesterol levels. More effective: Dietary fiber may help reduce cholesterol levels.
Simple	Simple sentences rather than complex increase health information comprehension, especially if loaded with technological jargon.[19] Examples: Complex: Increased consumption of cholecalciferol has been shown to decrease risk of osteopenia and osteoporosis and reduce the risk of falls incapacitating an individual. Simple: Higher intakes of vitamin D reduce risk of bone loss and falls.
Active	Use an active voice rather than a passive voice whenever possible. Examples: Passive: The athletes were taught the basics of carbohydrate loading by the nutritionist. Active: The nutritionist taught the athletes the basics of carbohydrate loading.
Personal and direct	Address your audience directly. Use *we, you,* and, *us* to provide a personal message. A conversational style is easier to read. Examples: Indirect: Reducing your cholesterol can be accomplished. Direct: You can reduce your cholesterol.
Specific	When giving advice, be clear and specific, not general. Avoid inconsistencies within the messages caused by controversies among scientists, government agencies, and industry groups. Examples: Indirect: Eat more fiber. Direct: Eat three servings of whole grains each day.
Analogies	Use comparisons familiar to your audience. Examples: Less effective: One portion of meat is approximately four ounces. More effective: One portion of meat is approximately the size of a deck of cards.
Readability	Reading level is defined as the number of years of education required for a reader to understand a written passage. Generally information for the public should be written at the fourth- to eighth-grade level.[21] A third- to a fifth-grade level is more appropriate for low-literacy readers. Too often this factor has not been taken into consideration when developing health education materials, even the 1990 Dietary Guidelines.[22,23] A number of readability formulas can be found on the Internet or word processing programs such as Microsoft Word. Exhibit 12.2 and Table 12.11 review SMOG procedures, a commonly used readability formula.
Paragraphs	
Main idea first	Readers expect the main idea to be the first sentence of a paragraph. Reading comprehension of health information increases if paragraphs follow expectations.[19]
Organization	The organization should follow a logical sequence. Limit paragraphs to three to five sentences.
Document Design	
Use effective text signaling	Bold text, underlines, and capitals are all examples of text signaling. Overuse or ineffective use of these methods decreases reading comprehension. See Figure 12.4 for an example of effective use.
Use engaging titles	Section titles should be stimulating and reflect the main theme.
Use effective visuals	Do not use decorative visuals that are abstract. They tend to take away from the text. Show images of what to do, rather than what not to do. For example, show fruits and vegetables, rather than soft drinks and candy.

(continued)

Table 12.10   Factors to Consider for Effective Writing for the Public *(continued)*

Use readable fonts	Use font sizes between 12 and 14 points. The heading should be 2 font sizes larger.[21] All caps, white on black, and italics are more difficult to read. Use fonts with serifs, letters with feet, such as Times New Roman, for the main text. For titles, use fonts that are sans serif (no feet) such as Calibri.
Provide visual appeal	The following list contains general suggestions for providing pleasing useable visual appeal. Not all pointers are applicable for every written document. • Select compatible colors and typefaces. • As a rule of thumb, select two to three colors. • Avoid light colors for text, because they reduce visibility. • Provide adequate white space for writing personalized messages. • Use the best quality paper the budget allows. • Use text boxes and borders to highlight particular sections.
Content Issues	
Accuracy	Check scientific accuracy. If in doubt, ask colleagues to review.
Cultural sensitivity	Be sensitive to the cultural and regional practices and taboos. For example, if designing materials for an Indian Hindu population on low fat foods, giving examples of low fat cuts of beef would not be appropriate. In another case, if you were designing a fact sheet on healthful food selections for Mexican Americans, choose colors and content familiar for this group.
Resources	Include resources for readings and educational websites.

EXHIBIT 12.2   Procedure for Calculating SMOG Reading Level

1. Count 10 sentences at the beginning, middle, and near the end of the document. If there are fewer than 30 sentences, use all that are provided.
2. Using the 30 sentence sample, circle all the words containing three or more syllables (polysyllabic).
   ❑ Include repetitions of the same word and numbers that are spelled out.
   ❑ Hyphenated words are considered one word.
   ❑ Abbreviations should be read as unabbreviated.
3. Total the number of circled words.
4. Use the SMOG conversion table to determine approximate grade level. See Table 12.11.[29]

Table 12.11   SMOG Conversion Table

Total Polysyllabic Word Counts	Approximate Grade Level (± 1.5 Grades)
0–2	4
3–6	5
7–12	6
13–20	7
21–30	8
31–42	9
43–56	10
57–72	11
73–90	12
91–110	13
111–132	14
133–156	15
157–182	16
183–210	17
211–240	18

Source: Adapted from: McLaughlin G. SMOG grading: A new readability formula. *J Reading.* 1969;12(8):639–646.

appropriateness, and readability. A formal protocol testing requires interviewing from three to nine people and asking them the meaning of each sentence of the document.[21]

Application of Emotion-Based Approach

"Any training that does not include the emotions, mind and body is incomplete; knowledge fades without feeling."

—Anonymous

Emotion-based messages have been found to be particularly persuasive for influencing nutrition behaviors. Advertising and marketing research has shown that people are more likely to make behavioral decisions in response to emotions rather than rational thought.[24,25] The Women, Infants and Children (WIC) *Touching Hearts, Touching*

Original Fragment	Revised Fragment
The doctor decides the appropriate dose, taking into account the nature of the complaints. In cases of anxiety and tension, the usual dose is one 10 mg tablet, taken 3 to 4 times a day. In serious cases it may be necessary to increase the dose to 150 mg a day with a maximum of 300 mg. –*In case of sleeping problems* 20 to 50 mg, to be taken at least one hour before going to bed. One should start with the lowest dose, as the risk of side effects increases with higher doses. A lower dose is prescribed for elderly patients, children and patients suffering from liver or kidney problems or from a chronic respiratory disease called hypercapnia. Take the tablets with water.	**How and when should you take Oxazepam?** Take the tablet with water. Swallow it whole with a glass of water. Do not dissolve the tablet in water and do not chew on it. For sleeping problems you should take Oxazepam at least one hour before going to bed. **How much Oxazepam should you take?** • In case of anxiety or tension, the usual dose is one 10 mg tablet, taken 3 to 4 times a day. In serious cases it may be necessary to increase the dose to 150 mg a day with a maximum of 300 mg. • In case of sleeping problems, the usual dose is 20 to 50 mg. You will be given the lowest dose to start with, as the risk of side effects increases with higher doses. Furthermore, you will be given a lower dose when you belong to one of the following groups: • elderly • children • patients suffering from liver problems • patients suffering from kidney problems • patients suffering from a chronic respiratory disease called hypercapnia

Figure 12.4 Example of Effective Use of Text Signaling

Source: Reprinted from Patient Education Counsel 80:113–119. Maat H.P., Lentz L., Improving the usability of patient information leaflets, page 115, 2010, with permission from Elsevier.

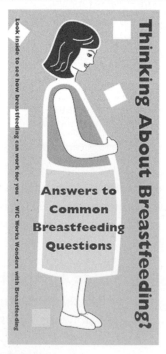

Figure 12.5 Comparison of Traditional WIC Material and Emotion-Based Material

Source: Massachusetts WIC Nutrition Program, Massachusetts Department of Public Health. Used with permission.

Minds developed educational materials and counseling approaches geared towards parent-identified emotional "pulse points" to guide clients towards making behavior changes to improve eating and physical activity behaviors.[26] The new emotion-based printed materials used photographs instead cartoon-style graphics focusing on messages that healthy nutrition behaviors leads to feelings of joy and pride. See Figure 12.5 for an example of their emotion-based material as compared to traditional WIC materials. Evaluation of the printed materials found that including a colorful photo with an emotion-based message, personal testimonials, cooking and snacking tips, and recipes were best received.

EXHIBIT 12.3   Program Evaluation Questions

Did the intervention reach the target population?

Which participants benefitted the most from the program?

Which participants benefitted the least from the program?

Was the program implemented as planned?

Was the original program plan effective?

How much did the program cost?

Was the program cost-effective?

Source: Adapted from: Boyle MA, Holben DH. Program planning for success, Belmont, CA: Wadsworth Cengage Learning; 2010, p. 115.

KEYS TO SUCCESS # 7–CONDUCT EVALUATIONS

However beautiful the strategy, you should occasionally look at the results.

—Winston Churchill

Evaluations are needed to determine the effectiveness of nutrition education interventions. Evaluation was discussed in Chapter 5 as part of the Nutrition Care Process, and as part of a nutrition counseling intervention in Chapter 7, and Chapter 11 addressed evaluation in setting program objectives and in order to conduct a needs assessment of a target audience. Components of all the previous discussions are applicable to the present review; however, we will focus on special aspects particularly significant in nutrition education programs. In this case, evaluations may be needed to provide information regarding distribution of resources, altering program delivery, continuing a program, or meeting funding requirements. For successful programs, having data to support the quality of a program provides useful publicity information for the community and policy makers.

Planning for an Evaluation

The framework for an evaluation needs to be arranged during the planning phase of an intervention since evaluation methods should be closely integrated with the design of educational strategies.[27] The exact form an evaluation takes depends on the scope of the intervention, such as an individual lecture at a health care facility or a comprehensive program for a multicenter organization. An evaluation may focus on elements of an intervention, such as the quality of a facility, audiovisuals, or handouts. Effectiveness of an intervention may assess knowledge, skills, attitudes, perceptions, or adoption of new behaviors by the participants. Also, an evaluation needs to measure the perceptions of the organizers and implementers of an intervention. A comprehensive evaluation will take into account all of these elements, the total design of the program, and all components of the process of implementation. Exhibit 12.3 lists questions to ask when planning an evaluation of outcomes. Answers to these questions will help determine if the plan includes appropriate goals and objectives or if they need to be altered.

After determining what factors should be evaluated, the plan needs to determine appropriate instruments and methods for the assessment. There are a variety of assessment procedures to choose among, including focus groups, questionnaires, interviews, biochemical analysis, or nutrient intake assessments of food records. An evaluation should be built into each phase of the operation. In addition, instruments should meet validity and reliability standards. Validity addresses the question of whether an instrument truly measures what it purports to be measuring. Reliability refers to the question of

---

EXHIBIT 12.4   Data Analysis Questions

Who will store the data?

Who will analyze the data?

Who will interpret the findings of the analysis?

Who will justify conclusions?

What are your plans for using evaluation findings?

Source: Adapted from: Centers for Disease Control (CDC). Evaluation Plan Template. www.cdc.gov/tb/programs/evaluation 2010, Accessed October 19, 2010.

---

whether the outcomes of the evaluation are reproducible, repeatable, or consistent.[28] For example, the validity of food frequency questionnaires needs to be established if the purpose is to measure usual intakes of vitamin A over a specific time frame. Reliability of skinfold evaluations is established when results are similar every time measurements are taken. Whether you borrow a previously designed instrument or develop a new one, extensive pilot testing is essential. Instruments or surveys need to be tested with your target audience.

The plan also needs to take into consideration who will administer the instruments or evaluation methods and when the assessments will be done. Exhibit 12.4 includes questions to consider regarding handling the analysis of data. Conducting evaluations can be extremely useful in detecting program deficiencies and strengths.

Formative Evaluations

Formative evaluations often involve qualitative data collection via observation, interviewing, and structured discussions. An evaluation may be conducted before a program begins to assess certain design elements to be sure they will be effective and to determine what theory variables should be included in an intervention. For example, a brochure may be evaluated for accuracy, appropriateness, and readability. At the beginning of a program, an evaluation may be conducted to provide a baseline to measure the impact of an intervention. For on-going projects, formative evaluations are implemented periodically to assess progress and to make necessary adjustments to improve methodology as indicated by the analysis.

Summative Evaluations

At the conclusion of a program or learning activity, a **summative evaluation** is conducted to assess outcomes. You want to know whether anticipated changes occurred in relation to the nutrition education intervention. Summative evaluations often include a quantitative evaluation including performance tests, observations, surveys, biochemical and anthropometric measurements, and self-assessment tools.[6] Careful assessment of summative evaluations help program managers determine whether a nutrition education program actually accomplished what it was designed to do. Based on these results, consideration can be given to modifying program goals and objectives for future interventions.

CASE STUDY   Presentation to High School Class

The school principal offers you the opportunity to speak to a group of high school seniors on the importance of healthy snack options for National Nutrition Month. There are approximately 30 teenagers, 13 females and 17 males. Some of the individuals participate in sports, but by appearance you estimate that about one-third of the students are overweight. You will provide a 30-minute presentation in a mediated classroom. A previous investigation indicated that the target audience frequently skips breakfast and snacks on high-fat and high-sugar snack foods purchased in vending machines. For your presentation, you are considering analyzing a snack food commercial; taste testing; video clip of a sports figure emphasizing the importance of eating well; demonstration of the sugar, fat, and salt content of vending machine snacks; a game; or discussion of "What does it take to get me to eat breakfast?"

1. Analyze the teaching techniques under consideration for effectively reaching this group of high school seniors. What are the advantages and disadvantages of each?

2. Brainstorm some titles and attention-grabbing introductions for your presentation.

3. Which teaching techniques would not be appropriate for this teenage audience? Explain.

4. What would be your call to action at the end of the presentation?

REVIEW QUESTIONS

1. When presenting a lecture, what are techniques for holding the attention of the listeners?

2. What key elements need to be considered when planning a food demonstration?

3. Under what conditions would the following methods be useful teaching techniques: role-playing, debates, simulations, and laboratories?

4. Identify and explain four possible ways to integrate web-based activities into nutrition education.

5. Explain the five basic steps of the Message Development Model.

6. Identify and explain the four steps of writing "ICIC" communications.

7. Define and compare formative and summative evaluations.

ASSIGNMENT—DEVELOP A TV PUBLIC SERVICE ANNOUNCEMENT

Work with a group to create a 60-second TV public service announcement using the Message Development Model in Figure 12.3 and described in Table 12.8 to guide the process. Choose a population group to target with an audiovisual message advising individuals to increase vegetable consumption. Possible population groups could be adult men and women with risk factors for cardiovascular disease, female college athletes, or working mothers.

STEP 1: Define the Issue

Your group should interview five people in your target population.

- Write a clear statement of the issue and 10 questions that will be used to conduct the interview using the questions in Table 12.8 to guide the process.

- Ask each interviewee to sign the Interview Agreement, LMF 5.7 in Appendix D.

- Conduct the interviews and record their answers.

STEP 2: Develop Initial Message Concepts

- Review material found in Step 2 of Table 12.8. Compare and analyze answers received in Step 1. Answer Step 2 questions in Table 12.8, and decide on a major motivating factor for behavior change for your target population.

- Based on the selected motivating factor, choose a presentation option (for example, testimonials, slice of life, animation, etc.) and develop an appealing script to support the central theme. See Table 12.9.

  1. Plan your video theme. Decide how you want to convey the central message of your video.

  2. Choose actors or create animations to convey the importance of this health message.

  3. If music is appropriate, choose an existing tune or write original music with lyrics that fits the central message.

  4. Create a storyboard. A storyboard is a set of drawings or pictures detailing how the video will unfold. Start with one central idea and build on it.

  5. If using actors, rehearse each scene. Rehearse each storyboard scene a few times before videotaping. This will help the camera person to anticipate movements of the actors and allow experimentation with shooting angles, lighting, and focusing.

  6. Shooting and editing. Shoot each scene and edit. Include optional background music or supporting captions.

STEP 3: Assess Message Concepts

- Pretest your 60-second TV public service announcement with three individuals of your target audience.

- Write four questions to ask your target population. Use the questions in Table 12.8 to guide the process.

- Record and analyze answers to the questions.

STEP 4: Fine Tune Your Message

- Answer all three questions in Step 4 of Table 12.8.
- Make any needed changes to your message.

STEP 5: Validate Message

- Place your presentation on a public or classroom access site such as YouTube or Blackboard that allows comments.
- Ask your target audience to go to the site for an evaluation.
- Develop a short quantitative questionnaire to email or give to 10 members of your target audience. Possible questions include those in Table 12.12.

STEP 6: Write a Report

Title each section of your report. Under each section, write the following:

Step 1: Define your target population. Provide a clear statement of the issue and the 10 questions used to conduct the interview. Under each question, write the responses received during the interviews. Designate each interviewee with a letter, such as G, H, L, etc.

Step 2: In your report, rewrite questions found in Step 2 of Table 12.8 and answer the questions based on the responses to the interviews. Explain your selection of the motivating factor to guide the development of your audiovisual.

Step 3: Provide the four questions used to assess the pretest. Under each question write the response of your interviewees designating each interviewee with a letter.

Step 4: Rewrite the questions in Step 4 of Table 12.8 and provide answers to these questions. As a result of these responses, did you make any changes to your audiovisual? Explain.

Step 5: Provide a list of comments you received from the public access site and a list of the four questions you used for the quantitative survey. Give a summary of the responses for each question.

- **Final evaluation of your participation in this project:** Each person in the group should answer the following questions:
  1. What were the best learning experiences for you?
  2. What would you do differently if you could redo the project?
  3. Think about what you learned. Explain how this knowledge or skill could be applied elsewhere in your nutrition education endeavors.

- **Hand in:** In addition to your report, hand-in the signed Interviewee Agreements from Step 1 and the storyboard.

SUGGESTED READINGS, MATERIALS, AND INTERNET RESOURCES

Books

U.S. Department of Health and Human Services. *Making Health Communication Programs Work.* National Cancer Institute Publication No. 04-5145; 2001.

Table 12.12 Evaluation of 60-Second TV Public Service Announcement

After watching the video, answer the following questions	Agree	Neutral	Disagree
1. I found the message appealing.	❏	❏	❏
2. I have a greater commitment to eat more vegetables.	❏	❏	❏
3. I am more likely to eat more vegetables.	❏	❏	❏
4. I am inspired to find out more information about eating vegetables.	❏	❏	❏

Brookfield SD. *The Skillful Teacher: On Technique, Trust, and Responsiveness in the Classroom.* Jossey-Bass, 2nd Edition; 2006.

Galbraith, MW (Editor). *Adult Learning Methods: A Guide for Effective Instruction.* 3rd Edition. Malabar, FL: Krieger Publishing Company, Inc.; 2003.

Richardson, W. *Blogs, Wikis, Podcasts, and Other Powerful Web Tools for Classrooms.* 3rd Edition. Corwin Press, 2010.

Educational Strategies Websites

www.speaking-tips.com
Information and articles on public speaking

www.write-out-loud.com/index.html
Numerous ideas for presentations

www.powerfulpresentations.net
Information for creative presentations

www.TeamNutrition.usda.gov
Information on nutrition education for schools

www.webQuest.org
Web-based inquiry-oriented lesson format

www.nutritionexplorations.org
Nutrition lessons and activities for all grade levels

www.educationworld.com/
Search: It's Up for Debate

Media Technique Websites

www.justthink.org
Just Think Foundation; information to promote media literacy and build skills in creative media production.

www.foodinsight.org/food-research.aspx
Tools for Effective Communication. The International Food Information Council Foundation provides professionals with numerous resources and suggestions for formulating effective nutrition education messages.

http://plainlanguage.nih.gov/
NIH Plain Language Online Training

www.cdc.gov
Simply Put. Developed by the Centers for Disease Control and Prevention (CDC)

www.hsph.harvard.edu/healthliteracy/
Harvard Health Literacy Studies, numerous strategies and tools

www.harrymclaughlin.com
SMOG Readability Calculator

Evaluation Website

www.cdc.gov/eval/resources.htm
Evaluation Working Group at CDC

REFERENCES

[1] Powell JP, Andresen LW. Humour and teaching in higher education. *Studies in Higher Education.* 1985; 10:79–90.

[2] International Food Information Council Foundation. Tools for effective communication, the second P: Presenting. *Food Insight,* www.foodinsight.org. 2009. Accessed October 16, 2010.

[3] Schloff L, Yudkin M. *Smart Speaking.* New York: Penguin Putnam Inc.; 1992.

[4] Fifer F. Teacher mobility and classroom management. *Academic Therapy.* 1986;21:401–410.

[5] Yelon SL. *Powerful Principles of Instruction.* White Plains, NY: Longman Publishers; 1996.

[6] Hitch EJ, Youatt JP. *Communicating Family and Consumer Sciences.* Tinley Park, IL: The Goodheart-Willcox Company, Inc.; 2002.

[7] Rosen LD. *Understanding the iGeneration and the Way They Learn.* New York, NY: Palgrave Macmillan; 2010.

[8] Bonk CJ. *The World is Open How Web Technology Is Revolutionizing Education.* San Francisco, CA: Jossey-Bass; 2009.

[9] Dodge, B. Five rules for writing a great WebQuest. *Learning and Leading with Technology.* 2001;28:6–10.

[10] Turner-McGrievy GM, Campbell MK, Tate DF, et al. Pounds Off Digitally Study: A randomized podcasting weight loss intervention. *Am J Prev Med.* 2009; 37: 263–269.

[11] Cocosila M, Archer N, Haynes RB, et al. Can wireless text messaging improve adherence to preventive activities? Results of a randomized controlled trial. *Intern J Med Informatics.* 2009; 78:230–238.

[12] Papp R. Virtual worlds and social networking: reaching the millennials. *J Technol Research.* 2009; 2. Available at: http://www.aabri.com/manuscripts/10427.pdf Accessed October 21, 2010.

[13]Spinelle J, Messer A. Virtual world offers new locale for problem solving. *Penn State University Press Release.* 2008, (September 29). Available at: http://live.psu.edu/story/34908, Accessed October 21, 2010.

[14]Contento IR, Balch GI, Bronner YL, et al. The effectiveness of nutrition education and implications for nutrition education policy, programs, and research: A review of research. *J Nutr Educ.* 1995; 27:279–418.

[15]U.S. Department of Health and Human Services. *Making Health Communication Programs Work.* NIH Publication No. 04-5145; 2001.

[16]International Food Information Council Foundation. Tools for effective communication, the first P: Publishing (a.k.a. writing). *Food Insight,* www.foodinsight.org. 2009. Accessed October 16, 2010.

[17]Prochaska JO, DiClemente CC. Stages and processes of self-change of smoking: Toward an integrative model of change. *J Consul Clin Psych.*1983; 51(3):390–395.

[18]Miller LMS, Gibson TN, Applegate EA. Predictors of nutrition information comprehension in adulthood. *Pt Educ Counsel.* 2010; 80:107–112.

[19]Maat HP, Lentz L. Improving the usability of patient information leaflets. *Pt Educ Counsel.* 2010; 80:113–119.

[20]Peregrin T. Picture this: visual cues enhance health education messages for people with low literacy skills. *J Am Diet Assoc.* 2010; 110:500–505.

[21]National Institutes of Health. The plain language initiative. 2009. Available at: http://execsec.od.nih.gov/plainlang/index.html. Accessed October 17, 2010.

[22]Busselman KM, Holcomb CA. Reading skill and comprehension of the Dietary Guidelines by WIC participants. *J Am Diet Assoc.* 1994; 94:622.

[23]Dollahite J, Thompson C, McNew R. Readability of printed sources of diet and health information. *Pt Educ Counsel.* 1996; 27:123.

[24]Bagozzi RP, Gopinath M, Nyer P. The role of emotions in marketing. *J Acad Market Sc.* 1999; 27:184–206.

[25]Bagozzi R. The poverty of economic explanations of consumption and an action theory alternative. *Managerial and Decision Economics.* 2000; 21:95–109.

[26]Colchamiro R, Ghiringhelli K, Hause J. Touching hearts, touching minds: Using emotion-based messaging to promote healthful behavior in the Massachusetts WIC Program. *J Nutr Educ Behav,* 2010; 42:S59–S65.

[27]Contento IR. *Nutrition Education: Linking Research, Theory, and Practice.* 2nd ed. Sudbury, MA: Jones and Bartlett Publishers; 2011.

[28]Vockell EL, Asher JW. *Educational Research.* 2nd ed. Englewood Cliffs, NJ: Prentice-Hall, Inc.; 1995.

[29]McGraw HC. SMOG Conversion Table. Office of Educational Research. Baltimore County Public Schools. Towson, MD.

13

Professionalism and Final Issues

*Being a professional is doing all the things you love
to do on the days when you don't feel like doing them.*
—JULIUS ERVING

Behavioral Objectives

- Describe professionalism.
- Explain the three blocks of the scope of the dietetic practice framework.
- Describe four moral principles of biomedical ethics.
- Describe steps to take for ethical decision making.
- Use ADA Code of Ethics to evaluate professional behavior.
- Identify boundaries between nutrition counseling and psychotherapy.
- Describe factors that need to be considered for starting a private practice.
- Explain social media marketing.
- Explain professional concerns regarding interactions with clients of Web based platforms.

Key Terms

- **Code of Ethics:** ethical standards published by a professional organization.
- **Standards of Professional Practice:** components of high-quality dietetic practice.
- **Biomedical Ethics:** considering autonomy, nonmaleficence, beneficence, and justice when providing care.
- **Client Rights:** clients have a right to confidentiality, clarification of procedures, and goals, and qualifications and practices of the health practitioner.
- **Nutrition Therapist:** nutrition counselor who incorporates the dynamics of the counseling relationship for helping clients make behavior changes.
- **Business Roadmap:** a formal business plan.
- **Social Media Marketing:** uses social media to influence consumers to purchase products and services.

INTRODUCTION

In this chapter, we review the basic components of professionalism as they relate to nutrition counseling and education practitioners. First, we examine the three-block framework of the dietetics profession as established by the American Dietetic Association. Then we explore ethical behavior in the context of all human behavior with special emphasis on ethical decision making, client's rights, boundaries between nutrition counseling and psychotherapy, and when referrals are in order. For those interested in starting a private practice or small business, important factors to consider are presented including a discussion of marketing in general and **social media marketing** in particular.

PROFESSIONALISM

The Merriam-Webster dictionary defines professionalism as "the conduct, aims, or qualities that characterize or mark a profession."[1] The American Dietetic Association (ADA), the largest organization of food and nutrition professionals, defines the dietetics profession as, "the integration and application of principles derived from the sciences of food, nutrition, management, communication, and biological, physiological, behavioral, and social sciences to achieve and maintain optimal human health."[2] In order to implement these roles and responsibilities, ADA designed a three-block framework with flexible boundaries for the profession. See Table 13.1.

Table 13.1 Scope of Dietetics Practice Framework

Block One: Foundation Knowledge

Five Characteristics of the Profession	Professionals Who Demonstrate This Characteristic	Core Professional Resources	
Code of Ethics	Follow a Code of Ethics for practice	Code of Ethics	Ethics Opinions
Body of Knowledge	Possess a unique theoretical body of knowledge and science-based knowledge that leads to defined skills, abilities, and norms	Philosophy and Mission: Research Philosophy and Diagram	Research, Position Papers, Practice Papers, Published Literature
Education	Demonstrate competency at selected level by meeting set criteria and passing credentialing exams	CADE (Core Competencies and Emphasis Areas)	CDR Certification (RD, DTR)
Autonomy	• Are reasonably independent and self-governing in decision-making and practice • Demonstrate critical thinking skills • Take on roles that require greater responsibility and accountability both professionally and legally • Stay abreast of new knowledge and technical skills	The CDR Professional Development Portfolio Process offers a framework for credentialed professionals to develop specific goals, identify learning needs, and pursue continuing education opportunities. This may encompass certificates (such as weight management), specialty certificates, advanced practice certification, or advanced degrees.	
Service	Provide food and nutrition care services for individuals and population groups and other stakeholders; manage food and other material resources; market services and products; teach professionals and students; conduct research; manage human resources; manage facilities	Nutrition Care Process and Model Practice Based Evidence • Dietetics Practice Outcomes Research • Dietetics Practice Audit	Nationally Developed Guidelines ADA Evidence-Based Guidelines for Practice ADA Nutrition Care Manual

Table 13.1 Scope of Dietetic Practice Framework *(continued)*

Block Two: Evaluation Resources

The evaluation resources are intended to be used in conjunction with relevant state, federal, and licensure laws. They serve as a guide for ensuring safe and effective dietetics practice. They can be used to determine whether a particular activity falls within their legitimate scope of practice, evaluate their performance, make hiring decisions, and initiate regulatory reform.

Core of Ethics	DTR Standards of Practice in Nutrition Care RD Standards of Practice in Nutrition Care ↓ RD Specialty or RD Advanced Standards of Practice	Standards of Professional Performance for Dietetics Professionals ↓ RD Specialty or RD Advanced Standards of Professional Performance

Block Three: Decisional Aids

Because health care environments are highly diverse and evolving, resources are provided to respond to new demands and to consider whether a new role or activity falls within legitimate scopes of practice.

Decisional Analysis Tool	Decision Tree	Definition of Terms

Supporting Documentation for use with Decision Tree and Decision Analysis Tool

Licensure or Certification Credentials

Organizational Privileging

Individual CDR Professional Development Portfolio

Best Available Evidence
➤ ADA's Evidence Library, ADA Position and Practice Papers, Ethics Opinions Published Literature and National Evidence Database

Practice Guidelines
➤ Nationally-Developed Guidelines and ADA Guides for Practice

Practice Based Evidence
➤ Dietetics Practice Outcomes Research

Source: Adapted from American Dietetic Association Scope of Dietetics Practice Framework. © 2005 American Dietetic Association. Available at: www.eatright.org/scope.

Block One: Foundation Knowledge

Foundation of knowledge is divided into five categories: **Code of Ethics**, Body of Knowledge, Education, Autonomy, and Service. A discussion of ethics can be found later in this chapter. The body of knowledge needed to perform effectively has to be based on sound research as published in referred journals and practice guidelines provided by professional organizations. Accredited dietetic education programs need to meet educational standards and perform regularly scheduled self studies and evaluations. Dietetic professionals demonstrate their acquisition of knowledge by passing credentialing exams. Dietetic professionals are encouraged to be autonomous and continually assess their competencies, update skills, and to assume positions of increasing responsibility. A nutrition professional is expected to provide quality service for clients following best evidence practices and critical thinking skills. Because clients are often not in the best position to evaluate services of a professional,

they depend on the integrity of the professional to make the right decision on their behalf.[3]

Block Two: Evaluation Resources

The American Dietetic Association provides two categories of standards for assessing performance of registered dietitians (RDs) and dietetic technicians, registered (DTRs). The Standards of Practice (SOP) in Nutrition Care are based on the four steps of the Nutrition Care Process and Model: assessment, diagnosis, intervention, and monitoring and evaluation. SOP provides a guide for evaluation for dietetic professions working in patient and client care. The Standards of Professional Performance (SOPP) apply to dietetic professionals working in all settings including non-direct care roles. They are based on the six domains of professional behavior described in the Commission on Accreditation for Dietetic Education Educational core competencies: provision of services, application of research, communication and application of knowledge, use and management of resources, quality in practice, and competence and accountability.[4] Additional advanced standards have been developed for specific practice areas. The various standards are periodically reviewed and revised as the needs of the dietetic profession change.

Block Three: Decisional Aids

Since health services are a dynamic field, the dietetic practice framework provides several easy to use decisional aids to determine if a new service is within your range of practice. These aids guide you through factors to consider before you embark on a new venture, such as reviewing state or regulatory agency guidelines, receiving core training, and additional responsibilities.

> *Our very lives depend on the ethics of strangers, and most of us are always strangers to other people.*
> **—Bill Moyers**

ETHICS

Ethics refers to "well-founded standards of right and wrong that prescribe what humans ought to do, usually in terms of rights, obligations, benefits to society, fairness, or specific virtues."[5] Ethical behavior can be analyzed in the context of all human behaviors. See Figure 13.1. Two factors about human behavior are self-evident. Humans are social animals and capable of exhibiting a broad range of

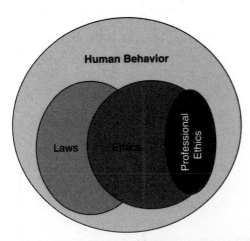

Figure 13.1 Ethics As a Component of Human Behavior
Source: Doris Derelian, Phb, jD, Rd, FADA Professor and Head, Food Science and Nutrition Department, California Polytechnic State University, San Luis Obispo, CA.

behaviors. Our feelings and desires can be selfish and infringe on the rights of others. Our ancestors must have quickly realized that society needed to provide some guidelines and control over behaviors in order for society to function effectively and comfortably. Societal pressures influencing our conduct come from a variety of sources including laws, family values, educational principles, religious morals, and social agreements, such as ethical standards of a professional organization.[3] However, none of these influencing factors has consistently advocated ethical behavior. For example, there have been times when societal organizations universally supported slavery or other human rights violations. As a result, in order to behave in an ethical manner, self-analysis of our own moral beliefs and moral conduct is required.[6]

To help influence the manifestation of the best ethical behavior, professional organizations develop and publish ethical standards. The American Dietetic Association, for example, publishes the Code of Ethics for the Profession of Dietetics. The code, found in Exhibit 13.1, "provides guidance to dietetics practitioners in their professional practice and conduct."[6] There are times when taking a course of action may be unclear, especially if the ethical dilemma is subtle. Ethical principles and procedural models are available to assist decision making.[7-9] See Exhibit 13.2 and Exhibit 13.3.

CLIENT RIGHTS

Clients are entitled to know their rights and options during the course of an ongoing counseling relationship. Cormier and Cormier[10] point out that ethical counselors should provide their clients with enough information about the counseling process to enable them to make informed choices (also known as empowered consent). This includes (1) confidentiality, (2) the procedures and goals of counseling, and (3) the counselor's qualifications and practices. In addition, long-term care residents and acute care patients have established rights that allow for personal choices to be addressed.

Confidentiality

A breach of confidentiality can do irreparable harm to a counseling relationship because it undermines

---

EXHIBIT 13.1   Code of Ethics for the Profession of Dietetics

The American Dietetic Association and its credentialing agency, the Commission on Dietetic Registration (CDR), believe it is in the best interest of the profession and the public it serves to have a Code of Ethics in place that provides guidance to dietetics practitioners in their professional practice and conduct. Dietetics practitioners have voluntarily adopted a Code of Ethics to reflect the values and ethical principles guiding the dietetics profession and to set forth commitments and obligations of the dietetics practitioner to the public, clients, the profession, colleagues, and other professionals.

The Code of Ethics applies to the following practitioners: (a) In its entirety to members of ADA who are Registered Dietitians (RDs) or Dietetic Technicians, Registered (DTRs); (b) Except for sections dealing solely with the credentials, to all members of ADA who are not RDs or DTRs; and (c) Except for aspects dealing solely with membership, to all RDs and DTRs who are not members of ADA. All individuals to whom the Code applies are referred to as "dietetics practitioners," and all such individuals who are RDs and DTRs shall be known as "credentialed practitioners." By accepting membership in ADA or accepting and maintaining CDR credentials, all members of ADA and credentialed dietetics practitioners agree to abide by the Code.

PRINCIPLES
Fundamental Principles
1. The dietetics practitioner conducts himself or herself with honesty, integrity, and fairness.
2. The dietetics practitioner supports and promotes high **standards of professional practice.** The dietetics practitioner accepts the obligation to protect clients, the public, and the profession by upholding the Code of Ethics for the Profession of Dietetics and by reporting perceived violations of the Code through the processes established by ADA and its credentialing agency, CDR.

Responsibilities to the Public
3. The dietetics practitioner considers the health, safety, and welfare of the public at all times.
4. The dietetics practitioner complies with all laws and regulations applicable or related to the profession or to the practitioner's ethical obligations as described in this Code.
5. The dietetics practitioner provides professional services with objectivity and with respect for the unique needs and values of individuals.
6. The dietetics practitioner does not engage in false or misleading practices or communications.
7. The dietetics practitioner withdraws from professional practice when unable to fulfill his or her professional duties and responsibilities to clients and others.
8. The dietetics practitioner recognizes and exercises professional judgment within the limits of his or her qualifications and collaborates with others, seeks counsel, or makes referrals as appropriate.
9. The dietetics practitioner treats clients and patients with respect and consideration.
10. The dietetics practitioner protects confidential information and makes full disclosure about any limitations on his or her ability to guarantee full confidentiality.
11. The dietetics practitioner, in dealing with and providing services to clients and others, complies with the same principles set forth above in "Responsibilities to the Public" (Principles #3–7).

Responsibilities to the Profession
12. The dietetics practitioner practices dietetics based on evidence-based principles and current information.
13. The dietetics practitioner presents reliable and substantiated information and interprets controversial information without personal bias, recognizing that legitimate differences of opinion exist.
14. The dietetics practitioner assumes a life-long responsibility and accountability for personal competence in practice, consistent with accepted professional standards, continually striving to increase professional knowledge and skills and to apply them in practice.
15. The dietetics practitioner is alert to the occurrence of a real or potential conflict of interest and takes appropriate action whenever a conflict arises.
16. The dietetics practitioner permits the use of his or her name for the purpose of certifying that dietetics services have been rendered only if he or she has provided or supervised the provision of those services.
17. The dietetics practitioner accurately presents professional qualifications and credentials.

(continued)

EXHIBIT 13.1 Code of Ethics for the Profession of Dietetics *(continued)*

18. The dietetics practitioner does not invite, accept, or offer gifts, monetary incentives, or other considerations that affect or reasonably give an appearance of affecting his/her professional judgment.

Responsibilities to Colleagues and Other Professionals

19. The dietetics practitioner demonstrates respect for the values, rights, knowledge, and skills of colleagues and other professionals.

Source: American Dietetic Association/Commission on Dietetic Registration Code of Ethics for the Profession of Dietetics and Process for Consideration of Ethics Issues. *J Am Diet Assoc.* 2009; 109(8):1461–1467.

EXHIBIT 13.2 Four Moral Principles of Biomedical Ethics

1. Autonomy: Client's right to self-determination, decision making, and to choose a given path of health care.
2. Nonmaleficence: Health care professionals have an obligation to not intentionally inflict harm.
3. Beneficence: Positive steps need to be taken to benefit clients, balancing and assessing benefit, risk, and cost to produce the best overall result.
4. Justice: Clients should expect fair, equitable, and appropriate treatment.

Source: Beauchamp TL, Childress JF. Principles of Biomedical Ethics. 6th ed. New York, NY: Oxford University Press; 2009.

EXERCISE 13.1 Working with Ethics

Revisit the case study about John in Chapter 2. Review the American Dietetic Association Code of Ethics in Exhibit 13.1 and the moral principles of **biomedical ethics** in Exhibit 13.2. In your journal address each of the following:

❏ Use the American Dietetic Association code as a general guide for professional behavior. Identify by number and explain specific ethical issues of concern in the Chapter 2 case study exhibited by the nursing home staff.

❏ List each moral principle identified in Exhibit 13.2 and give your impression of how each is reflected in the case study about John.

EXHIBIT 13.3 Steps for Making Ethical Decisions

1. Identify the problem or dilemma. Gather information that will shed light on the nature of the problem. This will help you decide whether the problem is mainly ethical, legal, or moral.
2. Identify the potential issues. Evaluate the rights, responsibilities, and welfare of all those who are involved in the situation. Consider how opposing sides view the dilemma.
3. Brainstorm how culture may be influencing the decision making process. Possible influencing factors may include religious beliefs, language differences, food customs, and societal customs.
4. Look at the relevant ethical codes for general guidance on the matter. Consider whether your own values and ethics are consistent with or in conflict with the relevant guidelines.
5. Consider the applicable laws and regulations, and determine how they may have a bearing on an ethical dilemma.
6. Seek consultation from more than one source to obtain various perspectives on the dilemma, and document in the client's record what suggestions you received from this consultation.
7. Brainstorm various possible courses of action. Continue discussing options with other professionals. Include the client in this process of considering options for action. Again, document the nature of this discussion with your client.
8. Enumerate the consequences of various decisions, and reflect on the implications of each course of action for your client. Consider the ethical principles of autonomy, beneficence, nonmaleficence, and justice as a framework for evaluation of the consequences of a given course of action. See Exhibit 13.2.
9. Decide on what appears to be the best possible course of action. After the course of action has been implemented, follow up to evaluate the outcomes and to determine if further action is necessary. Document the reasons for the actions you took as well as your evaluation measures.

Source: From COREY. Theory and Practice of Counseling and Psychotherapy, 8E. © 2009 Wadsworth, a part of Cengage Learning, Inc. Reproduced by permission. www.cengage.com/permissions.

EXERCISE 13.2   Use Ethical Theory and Decision Making Guidelines

Consider the following scenario: A sick child is in a hospital room and is likely to be there for an extended period of time. The large family of the child is part of a cultural group that believes they need to help the child get well by performing loud, religious ceremonies that involve the whole family. This is against hospital policy, but the family refuses to abide and their activities are disturbing other patients and staff. With a colleague review the steps of ethical decision making identified in Exhibit 3.3 and discuss influencing factors for each step.

---

the essential component of trust. Generally, counselors are not free to disclose information about their clients unless they first receive written permission.[10] A discussion regarding the confidential nature of the sessions should be included in the first counseling session with your client. The American Dietetic Association Professional Code of Ethics states, "The dietetics practitioner protects confidential information and makes full disclosure about any limitations on his/her ability to guarantee full confidentiality."[6]

Procedures and Goals of Counseling

The procedures and goals of the counseling program should be discussed with your client during the first session, including a clarification of any fees and a time frame for payment. Depending on the setting of the counseling intervention, there may be an institutional form that clients will be asked to sign. If you are working in a doctor's office or a private practice office, you should design an appropriate form. Appendix D includes a sample agreement form for a student working with a volunteer. See Lifestyle Management Form 14.2, Student Nutrition Counseling Agreement.

EXERCISE 13.3   Personal Inventory of Attitudes Relating to Ethical Issues

This inventory is designed to assess your attitudes and beliefs on specific ethical issues common to all counselors or particularly relevant to nutrition counselors. Select the response that comes closest to your position, or write your own response in *e*. There are no right or wrong answers. Discuss your selections with your colleagues.

1. A counselor's primary responsibility is to
   a. the client.
   b. the counselor's agency.
   c. society.
   d. the client's family.
   e. _____

2. Regarding confidentiality, my position is that
   a. it is never ethical to disclose anything a client tells me under any circumstances.
   b. it is ethical to break a confidence when the counselor deems that the client might do harm to himself or herself or to others.
   c. personal information can be shared with the parents of the client if the parents request it.
   d. it applies only to licensed therapists.
   e. _____

3. Concerning the issue of physically touching clients, my position is that
   a. touching is an important part of a helping relationship.
   b. touching a client is not wise.
   c. touching a client is ethical when the client initiates physical closeness with the counselor.
   d. it should be done only when the counselor feels like doing so.
   e. _____

(continued)

EXERCISE 13.3   Personal Inventory of Attitudes Relating to Ethical Issues *(continued)*

4. The way I can best determine my level of competence in working with a given type of client is
   a. by having training, supervision, and experience in the areas in which I am practicing.
   b. by asking my clients whether they feel they are being helped.
   c. by possessing an advanced degree and a license.
   d. by relying on reactions and judgments from colleagues who are familiar with my work.
   e. _____

5. Regarding the ethics of social and personal relationships with clients, it is my position that
   a. it is never wise to see or to get involved with clients on a social basis.
   b. it is an acceptable practice to strike up a social relationship after the counseling has ended if both parties consent.
   c. with some clients a personal and social relationship might well enhance the therapeutic relationship by building trust.
   d. it is ethical to combine a social and counseling relationship if both parties agree.
   e. _____

6. If I am counseling individuals who are engaging in a cultural practice that is morally repugnant to me (for example, the sacrifice of a dog to achieve healing may be repugnant to many Westerners), I believe it is my responsibility to
   a. learn about their values and not impose mine on them.
   b. encourage them to accept the values of the dominant culture for survival purposes.
   c. modify my counseling procedures to fit their cultural values.
   d. end the counseling relationship because I cannot accept their values.
   e. _____

7. When working with an overweight client who has a long history of losing weight and gaining back more weight than lost, the focus of my counseling should be to
   a. encourage the client to accept his or her present weight.
   b. encourage the client to join a self-help group.
   c. put the client on a strict calorie controlled diet.
   d. encourage the client to set weekly goals to improve the quality of his or her diet.
   e. _____

8. For clients who have minimal financial resources,
   a. it is acceptable to file false claims for services.
   b. refuse to take them as clients.
   c. it would be appropriate for me to charge no fee or less than I charge other clients.
   d. refer clients who cannot pay to self-help groups.
   e. _____

9. When working with sales people,
   a. it would be appropriate to accept expensive perks, such as a trip to a resort.
   b. it would be acceptable to accept modest perks, such as a fruit basket.
   c. it would not be acceptable to accept any presents.
   d. it would not be appropriate for me to continue working with a sales person who offered me perks.
   e. _____

10. When counseling a client referred to me by a doctor who has given incorrect nutrition advice (such as "Don't eat fruit. Pregnant women should not eat fruit. It holds water"), I should
    a. tell the client it is OK to eat fruit.
    b. tell the client that you will talk to her doctor about that and get back to her.
    c. avoid the topic and talk about what other foods she should eat.
    d. agree with the statement.
    e. _____

Qualifications and Practices of the Counselor

Your clients should know what you can and cannot do for them. For example, clients may be coming to you hoping to obtain a diagnosis for their condition. The counseling agreement form should contain information about your credentials and something about the scope of your practice. During your first session, ask your clients what they are expecting to get out of the counseling intervention with you. At that point, any discrepancies should be clarified.

BOUNDARY BETWEEN NUTRITION COUNSELING AND PSYCHOTHERAPY

According to the American Dietetic Association Professional Code of Ethics, "The dietetic practitioner recognizes and exercises professional judgment within the limits of his/her qualifications and collaborates with others, seeks counsel, or makes referrals as appropriate."[6] After reviewing various theoretical approaches to counseling, a beginning nutrition counselor may well feel overwhelmed and wonder what the boundaries are between counseling and psychotherapy. Expanding your skills in psychotherapy can in fact be useful if you decide to specialize in nutrition counseling. You may decide to obtain additional education in areas such as addiction or family counseling. Understanding that there is overlap between the role of a psychotherapist and nutrition counselor, some have advocated a client-centered nutrition therapy paradigm where there is exploration of a client's personal issues related to food behavior. For a counselor who appreciates the dynamics of the counseling relationship and factors involved in healing, the term **nutrition therapist**

> One time when I was counseling a middle-aged man with high blood pressure and elevated serum cholesterol, he came to our weekly session distraught over a decision made by his unmarried teenage daughter to have an abortion. It was against his moral beliefs, and he was having trouble functioning. He said I was a counselor and maybe I could give him some advice. I told him I sympathized with his dilemma, I understood that being a parent is sometimes a heart-wrenching task, and I wished I could change things for him. Because there did not seem to be anything he could do about the abortion, I suggested that he consider individual or family counseling with a psychotherapist so at least he could better cope with the issue.

> I had a client who appeared to be playing unusual games with food. She was forever "finding" food hidden deep in her closet or buried under her bed. After several weeks of counseling, it appeared to me that her eating problems had deep-rooted emotional origins, and I suggested that she seek the help of a mental health professional.

appears more appropriate.[11] However, some authorities have addressed the boundary issue. Saloff-Coste, Hamburg, and Herzog[12] cover boundaries between psychotherapy and nutrition counseling in their article on eating disorders. They believe the nutrition counselor's territory includes almost any issue related to food, weight, eating patterns, and body image. King[13] emphasizes the need to maintain healthy boundaries in a counseling relationship; otherwise, clients become confused as to what you can do with them and for them. The counselor does this by setting an example in his or her discussions of what can be handled in the nutrition counseling session. The nutrition counselor needs to be clear that the psychotherapist's work is based on feelings and the nutrition counselor's work is based on food. Should feeling issues arise that do not fall into a nutritionist's scope of practice, the nutrition counselor should be cautious about giving advice. It would be appropriate to acknowledge those feelings, provide any nutrition information related to the issue, and suggest the client receive assistance from a therapist qualified to deal with his or her problem.[13]

REFERRALS

At times you will want to suggest to your clients that they seek additional assistance from a person qualified to work with them regarding their feelings. In some clinical situations, you will be working with a medical team, and your client will already be receiving help from a psychotherapist. If this is the case, then you can suggest your client discuss the issue in question with his or her therapist. Helm[14] states that referrals to a mental health professional should be made when a client discloses information such as suicidal tendencies, physical abuse, severe marital difficulties, feelings of depression, unresolved sexual abuse, recurring self-destructive behaviors, eating disorders, and

---

EXERCISE 13.4   Evaluate Counseling Effectiveness and Professional Behavior

Read the following scenario, and identify what the counselor could have done or said differently to have a more effective professional encounter. Record your ideas in your journal, and discuss them with your colleagues.

ANITA   Hi, Nancy. Please come in. Have a seat and relax. I am just finishing up a few things. (Four minutes later . . .)

ANITA   OK. Let's see if we can get to the root of your problem here. How long have you had a weight problem?

NANCY   Well, when I was pregnant in 1997 I gained fifty-three pounds. I had only lost about twenty pounds when I realized that I was pregnant again. Unfortunately, I gained another thirty-five pounds with the second pregnancy.

ANITA   Has your doctor said anything to you about it?

NANCY   Not really.

ANITA   OK. Tell me about yourself.

NANCY   Well, I am forty-two years old, married, with two children. I work at a nursing home from 11 P.M. to 7 A.M. as the charge nurse. I am an LPN. I have a lot of stress in this job because I am the only nurse on duty. Whenever anything goes wrong, I have to make the decision as to what to do all by myself. Do I call the doctor and wake him up; do I send the patient straight to the emergency room; do I wait until morning; what if the patient dies by then? To cope with my anxieties during my night shift, I eat.

ANITA   What types of foods do you eat?

NANCY   Usually families are trying to be nice, and they bring in cookies, donuts, or chocolates. These things are always at the nurse's station.

ANITA   Do you eat any meals?

NANCY   Yes, the kitchen sets us up with a hot meal, which they leave in the refrigerator, and we microwave it when we are ready to eat. It is the same dinner the patients eat.

ANITA   Do you eat when you go home?

NANCY   Usually I stop at one of the fast-food restaurants on Route 5. I have a croissant sandwich or a biscuit with an egg and Taylor ham.

ANITA   Well, we will have to change that!

NANCY   I have that and then in about an hour or so I go to bed. I sleep most of the day, which is a problem because of my daughter.

ANITA   What about your daughter?

NANCY   She is in high school, and because I usually sleep until 6 or 7:00 P.M., she goes unsupervised after school. She has gotten to be a handful.

ANITA   What about your husband—isn't he available to watch her?

NANCY   My husband works from 3:00 P.M. to 1:00 A.M. But he really doesn't care anymore. He feels that Marie's problems are a direct result of my inability to control her and blames me for everything. He has little respect for me, especially with me being so overweight.

ANITA   Well, from the Client Assessment Questionnaire you completed, it seems to me that you are about a hundred pounds overweight. Therefore, we are talking about a long-term lifestyle change for you to get to your goal weight of 120 pounds. You need first of all to get your husband to be supportive of you! He should be happy with your efforts at improving yourself.

NANCY   I guess I can try.

ANITA   Is there any chance you could get another job? You are so sedentary in what you are doing.

NANCY   I have a pretty good pension and seniority. I really can't change that part of my life.

ANITA   Well, OK, let's see if we can design a diet for you that will work. But I must tell you that to be successful, you are going to have to join a gym.

---

other severe problems that are beyond the scope of nutrition practice.

Referrals are in order whenever the needs of the client are outside the scope of a particular counselor's expertise. It is useful to have on hand a list of professionals, such as social workers, physical therapists, psychologists, or physical trainers.

PROPER DRESS ATTIRE

Proper dress attire is a component of creating a professionalism image. A CareerBuilder.com survey found that 41 percent of employers felt that people who dress better or more professionally tend to be promoted more often than others in their organization.[15] Dressing for a successful nutrition intervention means to dress in a manner that does not create discomfort for the clientele. Clean, neat, and modest clothing and jewelry are suitable for most interventions. Expensive suits and jewelry would not be appropriate for working in poverty programs. Tight-fitting, revealing clothes are probably not suitable for any setting, but would be particularly inappropriate in an obesity clinic. Strong scents should be avoided because the odor can be irritating, and some people are allergic to them. Dangling jewelry or any clothing item that could be distracting should not be worn. Often nutrition counselors in health care settings wear a white lab coat with a nametag over professional attire.

STARTING A PRIVATE PRACTICE

Having your own business is appealing for individuals who have an entrepreneurial spirit and desire control over their career destiny. Table 13.2 provides a list of personal qualities that enhance the likelihood of career success. Starting a private practice provides you with opportunities to work on projects for which you have a passion and developing a creative enterprise can be rewarding. However, having total responsibility for the progress of your business can be challenging. You will be using all the traits you inherited and the skills you acquired throughout your life to find, develop, and organize resources. When making a decision to start a business, consider the following: Do you possess self-motivation and self-confidence; are you a risk taker, tenacious, disciplined, and adaptable; have the ability to roll with inevitable down turns, critically evaluate difficulties, and analyze and build upon successes? If you answered yes to all of these questions, you are probably a good candidate for starting a private practice. However, the Bureau of Labor Statistics indicates that about one-third of small businesses fail within the first two years so you should minimize the risk by careful planning.[16] Successful entrepreneurial dietitians generally earn more money than those working for an employer, but most small businesses take from three to five years to turn a profit.[17] You should plan accordingly; for example, you may wish to keep a part-time job while starting your business. The following provides an overview of factors to address when starting a private practice.[18]

Define a Focus

Clarify a vision of what you want for yourself and your career. You are not likely to hit a target if you do not know where to aim. What are you passionate about? To what are you willing to devote many hours of your life? Critically analyze whether your interest will produce a viable business. Becky Dorner, a successful long-term care nutritionist who owns two companies and has 25 employees, stated that her first attempts in the private practice arena did not go well.[19] Wellness was her first focus area, but she was not able to make that specialty area work financially so she changed her attention to another area of interest, long-term care. An investigation of market possibilities can help you to define a focus area. Finding a niche allows you to master skills in a particular area of nutrition. By spreading yourself too thin, you run the risk of being a "jack of all trades, but a master of none."[20]

Professional Credentials and Achievements

Your area of interest will guide the credentials you wish to obtain. If you wish to provide direct nutrition counseling services to clients, there will be more opportunities if you are a registered dietitian and obtain state licensure or certification credentials. The latter differs by state, and there may be defined legal guidelines as to who can provide counseling services. There may be useful degrees or credentials in other specialty areas, such as business management, food science, personal training, or diabetes education. By keeping a record of your education credits and a portfolio highlighting your statistics, business customers, positive evaluations, patient testimonials,

EXERCISE 13.5   Are You Ready to Start a Small Business?

Take the entrepreneurial aptitude test in Table 13.2. In your journal write the following:

❑ What was your score on the aptitude test?
❑ Do you agree with the number you received? Explain.

references, and achievements, you will be ready for an interview with a potential business client.

Learn and Connect

Continuing your education either formally or informally through webinars, conferences, and workshops can improve your skills. Consider possible mentors in your education or social network. Joining and taking an active role in professional groups will allow

Table 13.2   Entrepreneurial Aptitude Evaluation

After reading each question, circle the number that indicates your agreement with the statement. The number 5 indicates total agreement, and number 1 indicates no agreement.

1. In the games I play, I play harder when I fall behind.	5 4 3 2 1
2. When I go to a sports event or concert, I often try to figure out the promoter's or the owner's gross revenues.	5 4 3 2 1
3. When things take a serious turn for the worse, my first impulse is to look for alternatives and solutions, not for someone to blame.	5 4 3 2 1
4. Using my friends and co-workers as a barometer, my energy level is high.	5 4 3 2 1
5. I often daydream about business opportunities while commuting to work, flying on an airplane, waiting in the doctor's office, or other quiet times.	5 4 3 2 1
6. Looking back on significant changes I have made in my life, such as schools, jobs, relocations, and relationships, I have looked forward to them with excitement and been able to make tough decisions after doing some research.	5 4 3 2 1
7. My first consideration of any opportunity is always the upside, not the downside.	5 4 3 2 1
8. I am happiest when I am busy, not when I have nothing to do.	5 4 3 2 1
9. As an older child or young adult, I often had a job, plan, or idea to make money.	5 4 3 2 1
10. As a youth, I worked part-time and had summer jobs, rather than spending little or no time working. I did not primarily spend my time participating in recreation and enjoying a total break over the summer.	5 4 3 2 1
11. My parents worked many years owning a small business.	5 4 3 2 1
12. I have worked for a small business for more than one year.	5 4 3 2 1
13. I really enjoy being in charge, in control, and at the center of attention.	5 4 3 2 1
14. I am comfortable borrowing money to finance an investment, such as buying a home.	5 4 3 2 1
15. I am extremely creative.	5 4 3 2 1
16. When I balance a checkbook, "close" is good enough; it does not have to be to the penny.	5 4 3 2 1
17. When I fail at a project or task, I am inspired to do it better the next time; it will not scar me forever.	5 4 3 2 1
18. When I truly believe in something, whether it's an idea, a product, or a service, I am able to sell it.	5 4 3 2 1
19. In my current social and business environment, I am most often a leader, rather than a follower.	5 4 3 2 1
20. I am good at keeping my New Year's resolutions.	5 4 3 2 1

Scoring the test

80–100:   GO FOR IT. . . . YOU SHOULD BE A SUCCESSFUL ENTREPRENEUR.

60–79:   You probably have what it takes, but review the statements with the lowest scores to determine if there are trends.

40–59:   Too close to call. Seriously look at the low scores, and see if there is something you can learn to tilt the scales in your favor.

0–39:   Tests could be wrong, but you are probably better off working as an employee.

Source: Adapted from Tyson E, Schell J. Small Business for Dummies, 3rd ed. Hoboken, NJ: Wiley Publishing, Inc.; 2008.

you to make important networking connections to learn about opportunities in your interest area. Also, networking with marketing, communications, or consulting organizations can be valuable, particularly if you are planning to work with corporate organizations. There are several resources available to informally learn business skills. The United States Small Business Administration has a mission to help build and grow small businesses.[21] The Administration website has a great deal of helpful information, and most states and cities have regional offices that offer classes. Another helpful organization is SCORE (Service Corps of Retired Executives).[22] Volunteer mentors offer free online or in-person advice to small business entrepreneurs.

Create a Business Roadmap

You may consider starting with a vision (goal) board, depicting what you want to keep and expand in your life. As the saying goes, "What you focus on expands." This can be done as a tactile collage with drawings or pictures or it can be done on a computer. Some people respond to a visual representation to help keep their lives focused on their dreams and goals.[19]

A formal business plan can take several forms, such as a one page concept summary or a comprehensive analysis including an executive summary, mission statement, description of product or service, goals for your company, target market, implementation milestones, management team, competition, and financial projections.[18,23] There are small business planning computer templates available online at inc.com and entrepreneur.com. If you are looking to obtain outside financing, a detailed plan is essential.

Nutrition professionals working in private practice often wear many hats. For example, they may see private clients for diet consultation, give presentations to the public and professional groups, create educational materials, develop recipes or menus, give supermarket tours, or write a book. The most lucrative appear to be working as a media spokesperson, consulting for businesses, and writing books.[24]

Professional Support Systems

Setnick[25] reported that her life was hectic and she felt over-worked running a private business. After she hired a support team including an office manager, a computer consultant, a webmaster, and a book assistant, her life changed. Despite the salaries and consultant fees, her net income increased and she had more time to devote to her personal life. Time is money, and you cannot be an expert at all aspects of managing a business. Knowledgeable professional support can help your business to grow and be more efficient. For example, accountants design accounting and payroll systems; corporate attorneys create appropriate business structures; marketing consultants establish workable marketing plans; and technology consultants are able to integrate the latest technology resources into various components of a business structure.[18]

Business Basics

Although a professional support system is helpful, you need some basic business knowledge and skills. Financial planning is essential for making a profit. You have fixed costs, such as rent and malpractice insurance; variable costs, such as printing; program costs, such as travel expenses; and wages. Account for the amount of unpaid time you spend marketing; interacting with therapists, physicians, and corporations; and responding to email. Use a variety of resources to determine fees including an investigation of competition prices, discussions with colleagues, and review published formulas.[26,27] Also explore the possibility of becoming an in-network provider for individual health plans. You will be listed as a referral resource and billing and payments will be done through the health care provider.

MARKETING BASICS

Whether or not you are starting a private practice, you should consider your "brand image", that is, how you are perceived as others. Having a positive image will likely provide you with more opportunities for advancement and greater satisfaction during your career. See Table 13.3 for a list of habits that help people find success by developing a positive brand image.

One place to start a self-promotion campaign is developing a sound bite to describe your work.[28] How many times are you asked what do you do? You want to have a clear, concise answer that sounds interesting and rewarding. A general formula includes the following: (1) offer a general sentence about where you work and your area of expertise; (2) provide more depth about what you do, possibly give an example of how you contribute to the organization; and (3) give your title.[28] For example, "I market probiotics. I write articles and maintain a website to help people have better digestion. I own my own company." The following discussion focuses on marketing for starting a private practice, but even if that is not your objective, you are likely to find pointers for building a positive personal identity.

Table 13.3 Habits That Help You Strive for Success

Habit	Description
Be Proactive	Behavior is a function of our decisions, not our conditions. We have the initiative and responsibility to take action to make things happen.
Begin with the End in Mind	Know where you are now and where you are going with a clear vision of your destination.
Put First Things First	Manage your behavior to accomplish a desired purpose–doing what needs to be done, whether or not the task is viewed as enjoyable.
Think Win-Win	Search for mutually beneficial interactions, solutions, or agreements.
Seek First to Understand . . . . Then to be Understood	Empathic listening is essential, that is, with the intent to understand another person emotionally and intellectually. You see the world through the eyes of another person.
Synergize	Growth, creativity, and excitement occur when there is mutual appreciation for abilities, learnings, and insights of all involved.

Source: Adapted from Covey SR, The 7 Habits of Highly Effective People. New York, NY: Free Press; 2004.

EXERCISE 13.6   Apply Characteristics That Help You Strive for Success in Nutrition Counseling

Review the list of characteristics in Exhibit 13.3 that can aid in the pursuit of a successful career. In your journal, indicate how you believe they apply to a career in nutrition counseling; discuss your answers with your colleagues.

EXERCISE 13.7   Write a Self-Promotion Sound Bite

Follow the guidelines for writing a sound bite to describe your work. Write this statement in your journal.

We are advertis'd by our loving friends.
—William Shakespeare

Marketing Plan

A marketing plan is essential for finding clients. A marketing consultant can help develop a comprehensive plan which may include identifying a brand image, creating marketing materials, identifying referral resources, and developing advertising materials. Consider conducting a focus group or surveying potential customers regarding their knowledge, beliefs, and attitudes. Some of the most successful marketing strategies identified by nutrition entrepreneurs include word-of-mouth from satisfied customers, networking, Internet (website and social marketing), and personal meetings or phone calls to referral resources.[24] In order to evaluate the best marketing approaches for your audience, you need to keep track of the responses. Ask clients how they heard about your services. Telephone, email, or send a note to thank referrals.[18]

Web-Based Marketing

Opportunities abound for marketing on the Internet as shown by statistics: 260 million people in the United States use the Internet on a regular basis, on average, 29 hours per month.[29] As of December 2009, 74 percent of adults use the Internet and 61 percent look for health information online.[30] Web-based advertising has been steadily replacing traditional marketing strategies. About three-fifths of the small business owners surveyed by Vistaprint, believed that the Internet opened new markets, increased their competitiveness, and lowered costs.[31]

Having a creative, easy to navigate website is essential for all businesses. Create original content and update as often as possible. This will encourage readers to regularly return to your site and keep search engines indexing your site. Providing opportunities for your audience to make comments helps you to better understand the concerns of your target population and may give you ideas regarding additional content to post.[32] See Table 13.4 for suggestions to draw attention to your site.

Table 13.4 Techniques to Draw Attention to Your Website

Category	Description
Register with search engines	Major search engines allow you to register your website.
Trade links	Exchange links with complementary websites.
Ping your website	Ping-O-Matic (http://pingomatic.com) informs search engines when you have updated your site.
Provide website address on printed material	Include your website address on letterheads, brochures, business cards, and so on.
Post on ADA's National Nutrition Network	If you are an ADA member, you can post website and blog information on the ADA Network.
Mention in presentations	When providing presentations, mention your website three times, at the beginning, middle and end of your talk. Be sure there is printed material with your web address for audience members to take.
Write articles	Include your web address in any articles you write.
Write newsletters	Provide an option to receive a periodic newsletter via email. The newsletter should contain useful information, not just advertising.

Source: Switt JT. Drawing attention to your web site. *J Am Diet Assoc.* 2008; 109:20.

Social Media Marketing

Using social media marketing can create interest in your business and bring potential customers to your website. In the past few years, there has been tremendous growth in consumers using social media sites. For example, in April 2009, 13.9 billion minutes were spent on Facebook. Social media marketing uses social media to influence consumers to purchase products and services. Social media is defined as "content created by everyday people using highly accessible and scalable technologies such as blogs, message boards, podcasts, microblogs, bookmarks, social networks, communities, wikis, and vlogs."[29]

Using social media to influence consumers provides a number of advantages to small business owners.[33] If used properly, social media marketing can be directed towards your target audience. For example, a study of ad options on social sites found that individuals were most likely to buy and recommend products from a brand with a profile with fans.[34] The skills to use social media sites are not complex. If someone knows how to use the Internet, the learning curve will be fast. This type of marketing has a high return on investment; in fact there are often no costs at all. Consumers are leery of money backed on-line advertising links or banners. After family and friends, online reviews and feedback from social media has the greatest impact on consumer purchasing decisions and beliefs about a product or service.[35] To help guide the decision-making process, potential consumers want to read reviews written by previous customers.

You will need to make decisions regarding the social media sites most useful for your business and periodically monitor changes in trafficking for your target audience. There are several free online tools for measuring social activity on the Web.[29] As you can see from Figure 13.2, visitors to Myspace surpassed Facebook in 2007, but only two years later Facebook clearly had the lead in activity. In 2003, Friendster was a social network leader, but now many are not aware of its existence. At the present time, the fastest growing social platform is Twitter, with a growth of 1,382 percent from February 2008 to February 2009 and increases in growth every month.[36] However, the

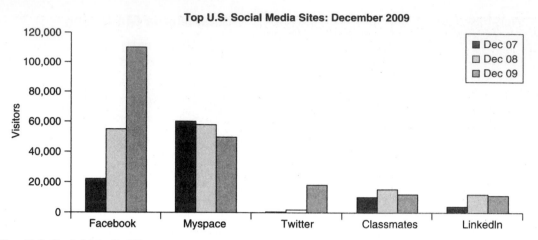

Figure 13.2 Top U.S. Social Media Sites

Source: Nielsenwire.com. Led by Facebook, Twitter, Global Time Spent on Social Media Sites up 82% Year over Year, January 22, 2010, Available at: http://blog.nielsen.com/nielsenwire/global/led-by-facebook-twitter-global-time-spent-on-social-media-sites-up-82-year-over-year/ Accessed November 2, 2010.

social media audience is fluid, so marketers need to keep track of changes in social media activity. "The one thing that is clear about social networking is that regardless of how fast a site is growing, or a how big it is, it can quickly fall out of favor with consumers," said Jon Gibs, vice president, media and agency insights, Nielsen Online.[37] Table 13.5 lists popular social media sites that may be considered for an advertising campaign.

Many health professionals have embraced the new web technologies, but others are cautious and question the legal ramifications, particularly for health professionals providing direct patient services.[38] Some of these issues are likely to be played out in the courts in the future. For example, what is a health provider's responsibility if a Facebook friend posts incorrect health information? Does the disclosure of a patient's medical condition by the patient on a social network site by the patient violate the Health Insurance Portability and Accountability Act (HIPPA)? Although there are discussions about e-professionalism, so far health professional organizations have not offered guidance regarding

Table 13.5 Popular Social Media Sites

Name	Description
Facebook	Popular for organizing and communicating with social groups easily; largest segment of users 25-49 years old.[37]
MySpace	Leads the industry in video, music, and audio material; audience tends to be young, 12-24 years old.[37]
Twitter	A microblogging social platform, allows 140 characters per message (tweet); audience tends to be older; sign-up with your brand or company name as the handle (hopefully no one else has it).
LinkedIn	A professional site; has several dietetic groups; has a slide share feature allowing individuals to share slide presentations. As of 2008, the average age was 35 years old and average income was $140,000.[29]
YouTube	Can be used to upload videos and link to social networking sites.

professional social media interactions. However, everyone agrees that there is no going back to pre-digital days.

CASE STUDY   Interactive Personal Case Study

The objective of this experience is to use theories, strategies, and skills learned throughout the course of your nutrition counseling and education studies and apply them to yourself. Think of yourself as the client.

PART I. Think of a time in your life when you wanted to make a lifestyle change. (You could also choose a change you are currently attempting to establish. In that case, all the following questions would be posed in the present tense, rather than the past.) For this activity, it does not matter to what degree you feel your efforts were successful. Analyze

(continued)

CASE STUDY   Interactive Personal Case Study *(continued)*

and reflect on the experience in as much detail as possible in a case study format. Number and answer each of the following questions:

1. Describe the issue:
   a. What was the problem and why were you inspired to try to change?
   b. Label the feelings you experienced at the time.

2. Provide background and depth to the case:
   a. How did the issue impact your life and the life of those close to you? What difficulties or potential problems were related to the problem?

3. Identify your motivation level at the time:
   a. Describe your readiness for change, the degree of importance, and confidence to succeed using the Assessment Graphic, Lifestyle Management Form 4.1 in Appendix D.

4. Identify your goal or goals.

5. Analyze your approach or intervention
   a. What approaches or interventions did you use?
   b. Were they appropriate for your motivational level?

6. Describe the difficulties you encountered:
   a. What were your cues?
   b. What were your barriers to making a behavior change?
   c. What cognitive distortions may have hampered your progress?
   d. What prompted lapses or a relapse?

PART II. Now imagine yourself as a counselor working with a person just like yourself at the time you were attempting the behavior change. Considering what you have learned during the course of your counseling studies, answer the following questions:

1. Which theory or theories in Chapter 2 would you want to guide your counseling approach? Explain.

2. How would you view your role as a counselor?

3. What behavior change strategies do you believe would be effective? Explain.

4. How would you assist your client in preparing for a lapse?

REVIEW QUESTIONS

1. Explain the three blocks of the Scope of Dietetic Practice Framework.

2. What are the four moral principles of biomedical ethics guiding individuals to behave in an ethical manner?

3. Why do professional organizations develop codes of ethics?

4. Identify and explain three factors clients have a right to know before engaging in a counseling relationship.

5. Why is having a well-defined focus important before starting a private practice?

6. What is a vision board and a roadmap?

7. What is a brand image?

8. What is social media marketing?

ASSIGNMENT—EVALUATE YOUR COUNSELING EFFECTIVENESS

Audio- or videotape a counseling session with a volunteer client following the counseling format outlined in Chapters 4 and 5, or use one of the checklists for a counseling session found in Chapter 14. Review the tape carefully, and complete the LMF 7.5 Interview Assessment Form, and the LMF 7.6 Counseling Response Competency (Appendix D). Answer the following questions:

1. Explain what materials you reviewed and what educational materials or activities you prepared before meeting your client.

2. What verbal facilitation techniques and relationship-building responses appeared to work most effectively with your client? Explain why.

3. What behavior change techniques did you use?

4. How effective were these methods?

5. Are there cues or messages that you missed or interpret differently after listening to the tape?

6. Do you believe you focused on the main life-style issues that concerned your client?

7. Did you keep the focus of the session on your client's main issues? If yes, what did you do to keep your client on track? If not, what could you have done differently?

8. What were your impressions of the emotions expressed by your client while you listened to the tape? Were these impressions different than what you understood at the time of the counseling session?

9. What was your emotional state at the time of the session? How do you believe this impacted the course of the session?

10. What is your overall impression of how your client responded? Explain.

11. If you could redo the counseling experience, what would you do differently?

12. What did you learn from this experience?

Additional Considerations of Videotape Observations

To focus on and assess the visual aspects of the counseling experience, watch your video with the sound turned off.

1. Describe your body language during the course of the counseling session. What messages do you believe your body behavior conveyed to your client?

2. Describe your client's body language during the course of the counseling session.

3. After analyzing the nonverbal behavior, did you change any of your impressions regarding the counseling encounter?

SUGGESTED READINGS, MATERIALS, AND INTERNET RESOURCES

Websites

Working with the Media: A Handbook for Members of the American Dietetic Association
www.eatright.org

Nutrition Entrepreneurs, ADA DGP
www.nedpg.org

Voices.com
www.voices.com/podcasting/how-to-create-a-podcast.htm
A tutorial for creating your own podcast.

AllAboutYourOwnWebsite.Com
www.allaboutyourownwebsite.com
A seven step guide to creating a website

Tech Soup's Web Building Page
www.techsoup.org/learningcenter/webbuilding/index.cfm

Blogger
www.blogger.com/start
You can learn blogging basics, including how to create your own blog in minutes on this site.

Building a private practice
www.UnderstandingNutrition.com

Books

Mitchell FB, Silver AM. *Making Nutrition Your Business: Private Practice and Beyond*. American Dietetic Association; 2011.

REFERENCES

[1]http://www.merriam-webster.com/dictionary/professionalism Accessed November 1, 2010.

[2]Maillet JO, Skates J, Pritchett E. American Dietetic Association: Scope of dietetics practice framework. *J Am Diet Assoc.* 2005; 105:634–640.

[3]Derelian D. Checking Your Ethics IQ. American Dietetic Association Food and Nutrition Conference and Expo Presentation, October 19, 2009.

[4]American Dietetic Association Quality Management Committee. American Dietetic Association revised 2008 standards of practice for registered dietitians in nutrition care; standards of professional performance for registered dietitians; standards of practice for dietetic technicians, registered, in nutrition care; and standards of professional performance for dietetic technicians, registered. *J Am Diet Assoc.* 2008; 108:1538–1542.

[5]Velasquez M, Andre C, Shanks T, Meyer MJ. What is ethics? Markkula Center for Applied Ethics, Santa Clara University, 2010. http://www.scu.edu/ethics/practicing/decision/whatisethics.html Accessed October 29, 2010.

[6]American Dietetic Association/Commission on Dietetic Registration Code of Ethics for the Profession of Dietetics and Process for Consideration of Ethics Issues. *J Am Diet Assoc.* 2009; 109(8):1461–1467.

[7]Corey G, *Theory and Practice of Counseling and Psychotherapy.* 8th ed. Pacific Grove, CA: Brooks/Cole, Cengage Learning; 2008.

[8]Fornari A. Approaches to ethical decision making. *J Am Diet Assoc.* 2002; 102:865.

[9]Beauchamp TL, Childress JF. *Principles of Biomedical Ethics.* 6th ed. Oxford University Press; 2008.

[10]Cormier S, Nurius PS, Osborn CJ. Interviewing Strategies for Helpers: Fundamental Skills and Cognitive Behavioral Interventions. 6th ed. Pacific Grove, CA: Brooks/Cole; 2008.

[11]Israel D, Moores S. *Beyond Nutrition Counseling.* Chicago, IL: American Dietetic Association; 1996.

[12]Saloff-Coste C, Hamburg P, Herzog D. Nutrition and psychotherapy: Collaborative treatment of patients with eating disorders. *Bulletin of the Menninger Clinic.* 1993; 57(4):504–516.

[13]King NL. *Counseling for Health & Fitness.* Eureka, CA: Nutrition Dimension; 1999.

[14]Helm KK. Group process. In: Helm KK, Klawitter B. eds. *Nutrition Therapy Advanced Counseling Skills.* Lake Dallas, TX: Helm Seminars; 1995:207–213.

[15]Haefner R. How To Dress For Success At Work. CNN.com website. July 30, 2008. Available at: http://www.cnn.com/2008/LIVING/worklife/07/30/cb.dress.for.success/index.html Accessed July 31, 2010.

[16]US Small Business Administration. Advocacy Small Business Statistics and Research http://web.sba.gov/faqs/faqindex.cfm?areaID=24 Accessed November 4, 2010.

[17]Rogers D. Compensation and benefits survey 2007: Above average pay gains seen for registered dietitians. *J Am Diet Assoc.* 2008:108:416–427.

[18]Gross M, Ostrowski C. Getting started in private practice: A checklist to your entrepreneurial path. *J Am Diet Assoc.* 2008; 108:21–24.

[19]Dorner D. *Entrepreneurship: Six Figures or Simply Surviving.* Food and Nutrition Conference and Expo, American Dietetic Association; 2009.

[20]Mathieu J. Consulting do's and don'ts. *J Am Diet Assoc.* 2008; 108:1827–1828.

[21]US Small Business Administration. www.sba.gov Accessed November 1, 2010.

[22]SCORE Counselors to America's Small Business. http://www.score.org/index.html Accessed November 1, 2010.

[23]Peregrin T. Business plan 2.0: putting technology to work. *J Am Diet Assoc.* 2008; 108:212–214.

[24]King K. *Entrepreneurship: Six Figures or Simply Surviving.* Food and Nutrition Conference and Expo, American Dietetic Association; 2009.

[25]Setnick J. *Everything I Needed to Know I Learned in Private Practice: How My Career Taught Me Life Balance.* 2009, www.understandingnutrition.com Accessed October 30, 2010.

[26]King K. *The Entrepreneurial Nutritionist,* 4th ed. Lippincott Williams & Wilkins; 2009.

[27]Litt AS, Mitchell FB. *American Dietetic Association Guide To Private Practice: An Introduction To Starting Your Own Business.* Chicago, IL: American Dietetic Association; 2004.

[28]McCaffree J. Promote yourself and your profession with sound bites. *J Am Diet Assoc.* 2009; 109:S4–S5.

[29]Singh S. *Social Media Marketing for Dummies.* Hoboken, NJ: Wiley Publishing; 2010.

[30]Rainie L. Internet, broadband, and cell phone statistics. Pew Internet and American Life Project. January 5, 2010 Available at: http://www.pewinternet.org/~/media//Files/Reports/2010/PIP_December09_update.pdf Accessed October 31, 2010.

[31]Kelly S. *Web Skills for Dietitians: A Guide to Creating and Evolving Your Website, Marketing Yourself, and Counseling Clients Online*, 2nd ed. Hermos Beach, CA: Skelly Publishing; 2010.

[32]Switt JT. Drawing attention to your web site. *J Am Diet Assoc.* 2008; 109:20.

[33]Hubpages.com. 5 Big Advantages to Social Marketing, Available at: http://hubpages.com/hub/5-Big-Advantages-to-Social-Marketing Accessed November 2, 2010.

[34]Evans D C, Epstein E. Comparing User Engagement across Seven Interactive and Social-Media Ad Types. 2010. Psychology of social media. Psychster Inc. Available at: www.psychster.com Accessed November 2, 2010.

[35]Nielsenwire.com. Friending The Social Consumer, June 16, 2010, Available at: http://blog.nielsen.com/nielsenwire/online_mobile/friending-the-social-consumer Accessed November 2, 2010.

[36]Nielsenwire.com. Led by Facebook, Twitter, Global Time Spent on Social Media Sites up 82% Year over Year, January 22, 2010, Available at: http://blog.nielsen.com/nielsenwire/global/led-by-facebook-twitter-global-time-spent-on-social-media-sites-up-82-year-over-year/ Accessed November 2, 2010.

[37]Nielsenwire.com. Time Spent on Facebook up 700%, but MySpace Still Tops for Video, June 2, 2009, http://blog.nielsen.com/nielsenwire/online_mobile/time-spent-on-facebook-up-700-but-myspace-still-tops-for-video Accessed November 2, 2010.

[38]Aase S. Toward e-professionalism: thinking through the implications of navigating the digital world. *J Am Diet Assoc.* 2010; 110:1442–1440.

14

Guided Counseling Experience

Digital Imagery/PhotoDisc, Inc.

Before everything else, getting ready is the secret of success.

—HENRY FORD

Behavioral Objectives

- Use standard counseling procedures.
- Employ interpersonal skills.
- Demonstrate use of basic assessment tools.
- Employ goal setting processes.
- Tailor educational interventions.
- Use standard documentation procedures.
- Evaluate counseling effectiveness.

Key Terms

- **Counseling Checklists:** step-by-step counseling guides.
- **Informed Consent:** sufficient information was supplied to a participant for making a decision regarding a course of action.

INTRODUCTION

This chapter provides a guided approach for conducting four one-hour individual counseling sessions with a volunteer adult client, who would like to lose weight or improve the quality of his or her diet. The sessions integrate material covered in all previous chapters. This guide follows the counseling protocol and the motivational counseling algorithm, Figure 4.2, presented in Chapter 4, and contains the basic elements of the counseling process generally accepted as standard. The application of commonly used nutrition assessment and counseling tools is integrated into the guide. This guided experience was designed to introduce basic nutrition counseling procedures to a college student with some knowledge of nutrition. It can be applied to other settings, but was not intended to cover the spectrum of all nutrition counseling experiences.

DEVELOPING A COUNSELING STYLE

The anticipation of meeting with your client for the first time can be exciting as well as stressful. Interviewing and counseling skills take time to master, but your competence will improve with each session. There is no one perfect counseling method; however, a good place for a novice counselor to begin is with a structured, well-defined protocol. We recommend that this structured program be used as a springboard on which to build and modify individual counseling styles that mesh with your disposition and personality.

Also, you will find that adjustments are necessary to meet your clients' specific needs. For example, in this guidebook we suggest setting explicit goals. This has been found to work well for most clients. However, our experience has shown that occasionally a client will rebel against this structure and prefer to set more general goals. Your counseling strategies and style will evolve over the course of your career as you gain experience, take continuing education courses, and read educational materials. If nutrition counseling becomes a major career goal, then you may consider obtaining additional counseling credentials and establishing a relationship with a mentor.

FINDING VOLUNTEER CLIENTS

Volunteer clients should be relatively healthy, and any expectations of weight loss should be modest. No client should be accepted into the program with severe medical problems. Any individual with a complicated medical condition should be reviewed with your instructor before proceeding. This program can be used with a friend or relative, but the impact of the learning experience is likely to be greater if you do not know the client. Established relationships may have patterns and issues that could interfere with the counseling relationship. The process of establishing a counseling relationship is a valuable learning experience that would be lost if the counselor were closely connected to the client. An additional concern is the strain such an experience can put on a relationship if a friend or family member is having difficulty meeting his or her obligations to take part in the program. An associate that has been coerced into the program may be a reluctant client and not have the motivation to make lifestyle changes.

In our clinic, we have been successful at finding volunteers by advertising on our university campus through flyers, school newspaper stories, school radio advertisements, listservs, and classroom announcements. Student counselors schedule a counseling room for four weekly one-hour sessions. Volunteers enroll in the program during a designated registration time. Most volunteers follow through on their commitment to complete the program. However, for those who do not, we keep a record of all the potential volunteers who we were not able to accommodate during registration. Usually we are able to fill a vacancy from the record. As a last resort, a counseling student may need to secure a friend or relative to be his or her volunteer.

GOALS OF THE GUIDED COUNSELING EXPERIENCE

The overall goals of the guided counseling experience can be divided into skill goals and attitudinal goals.

Skill Goals

This program addresses the following skill goals:

- Conduct a four-session counseling program with an adult volunteer client.
- Demonstrate use of standard nutrition assessment tools: client assessment questionnaire, food frequency form, usual diet form, and computerized analysis of three-day log.
- Use appropriate interpersonal skills for facilitation and relationship building.
- Demonstrate use of basic counseling responses.
- Tailor education approaches and intervention strategies to a client's motivational level and needs.
- Employ effective goal setting during counseling sessions.
- Effectively evaluate, document, and plan after each session.
- Provide self-assessment for counseling effectiveness.

Attitude Goals

Attitudinal goals for this intervention include the following:

- Unconditional positive regard for clients.
- Curious and nonjudgmental approach to clients.
- Desire to work in collaboration with clients.
- Willingness to learn from clients.

THE FOUR COUNSELING SESSIONS

The following guidelines contains a **counseling checklist** describing procedures and goals for four one-hour counseling sessions. Each session has a detailed checklist to aid in the flow and pacing of a counseling session. Time frames are given for the phases to help with pacing. You may need to modify some segments of the protocol to keep the sessions moving.

Student need and desire for structure varies. Many students welcome well-defined guidelines in order to visualize the flow of an intervention before meeting a client. Some students use the checklists as a reference and develop their own interview guide. The best approach is to simply have a short list of tasks on a card and take the checklist into the counseling session. Be careful not to let the tool or guide interfere with the development of a counseling relationship. Before meeting with your client, you should be so clear about your counseling approach and tasks that you only need to glance at your interview tool to stay on track. All components of the checklists are addressed in more detail in other chapters; however, those that require additional explanation for the intervention are highlighted here.

Preparation for Session 1

Certain preliminaries should be addressed before your first session. These include the following: completing a registration form. A sample registration form is found in Appendix D, LMF 14.1 Registration for Nutrition Clinic; getting your client's consent, and giving your client a welcome packet containing a LMF 5.1 Client Assessment Questionnaire, LMF 5.4 Food Frequency Questionnaire, and LMF 8.5 Medical Release to complete before the first session. During the registration process, obtain your volunteer client's consent to participate in the counseling program. Use Lifestyle Management Form 14.2, to obtain your client's **informed consent**. You need to thoroughly explain the content of the form. Adhere to the protocol in Exhibit 14.1.

If your client did not formally register for the program before your first meeting, you should call your client to arrange a registration meeting to give the future client a welcome packet and to complete the counseling agreement form, LMF 4.2 in Appendix D. If you were not able to meet with your client for registration before the first session, your first counseling session will probably require an extra twenty minutes.

If your client formally went through the registration process with someone else, you should contact your client to verify your meeting time and place. This can be done through email, but first try to call, since direct discussions are more personal and you can better answer questions through direct

EXHIBIT 14.1 **Informed Consent Protocol**

1. Obtain two copies of the nutrition counseling agreement form, Lifestyle Management Form 14.2 (LMF 14.2).
2. Give one copy of the agreement form to your potential client to read along with you.
3. Tell the potential volunteer that the purpose of reviewing LMF 14.2 is to help the individual decide whether or not to participate in the nutrition counseling program.
4. Read your copy of LMF 14.2 out loud.
5. Periodically check for understanding. Ask, "Do you have any questions about anything I have read?"
6. Tell the potential client that participation is totally voluntary and that another volunteer will be available if the individual would rather not participate in the nutrition counseling program.
7. After reading through the complete nutrition counseling agreement form, ask a final time, "Do you have any questions or concerns about participating in this program?"
8. If your client indicates that he or she would prefer not to participate, thank the individual for their interest in the program and assure the person that another volunteer will be available.
9. If your potential client clearly states an interest in participation, ask the individual to sign two copies of LMF 14.2.
10. Give one copy of the form to the volunteer, and give one copy to your instructor.

discussion. Your telephone conversation could go something like this:

Hello, Mrs. Jones. This is [give your full name]. I am a student at University X taking a nutrition counseling course. I understand that you have been assigned to my time slot for nutrition counseling. I called to verify our meeting on [give day of the week and date] at [give time] in [give location]. The session should take approximately one hour. Is this OK? Have you been able to look over the forms? Do you have any questions? Please try to fill them out to the best of your ability. We will be going over them during our session. Do you know where the [name location] is? Please give me a call if you have any problems keeping our meeting

time. Do you have my phone number? The best time to reach me is at [give time of day]. I look forward to meeting you.

Note that the session 1 checklist indicates that you should bring copies of the forms contained in your client's welcome packet in case the forms have been forgotten.

Session 1

The involving phase for session 1 is more extensive than for subsequent sessions. During the greeting, the tone of your voice and your body language should convey the message that you are happy to meet your client. Be sure to stand and smile during the welcome and invite your client to sit down. The following are examples of possible greetings:

Good morning, Mr. Gray. I am very happy to meet you. I am Sally Mason. I hope you managed to find my office easily. May I call you Jim?

Good afternoon, Mrs. Jones! I am Sally Mason. Please come in. It is great to finally meet you. How do you prefer to be addressed? Is calling you Mrs. Jones OK?

The amount of time you will need to devote to explaining the program will depend on how much of this topic was covered during the registration process. The following are some aspects of the program you may wish to emphasize.

Overview: The objective of the four sessions is for us to work in partnership to discover ways for you to make lifestyle changes to improve the nutritional quality of your diet and to achieve or maintain a desirable weight. This program is designed to achieve slow, steady changes.

Nutritional Assessment: To analyze the nutritional quality of your diet and make recommendations, it will be necessary to complete some forms. You have completed two of them already, client assessment questionnaire and the food frequency evaluation. They were in your welcome packet. We will go over them today. I will also ask you to describe what you eat on a usual day. In a few weeks I will ask you to record your food intake for three days in order to analyze your diet with the aid of a computer program.

Educational Component: Every session will include an educational component geared to your needs and interests. This could include short videos or reading materials.

Goal Setting: Each session, we will evaluate the possibility of setting goals or modifying old ones based on your food habit problems and strengths.

Food Management Options: You have a choice in how you would like to proceed with aid in making dietary changes. On your client assessment questionnaire, you indicated how much structure you believe would work for you.

Audiotape: I will be asking your permission to audiotape a session to evaluate my counseling skills.

To develop a better understanding of your client's issues, a typical day strategy (Exhibit 4.2 in Chapter 4) and usual diet method (Chapter 5) have been combined during the exploration phase. You will need to jot down notes on the usual diet form while your client is talking. Care should be taken not to let this process interfere with the flow of conversation. For example, do not ask your client to repeat in order to correctly record what is usually eaten for lunch. Instead, ask clarifying questions at the conclusion of your client's story.

The resolving phase intervention is geared toward your client's degree of readiness to change. For most clients, the sequencing of the questions and topics should work well. However, you should not feel tied to the protocol. If you believe certain questions or topics would be useful to cover with your client, cross-over is appropriate. The closing part of the interview is just as important as the opening. Each session should end on a positive note. The post session activities guide you to evaluate your client's needs, assess your intervention effectiveness, and prepare for the next session. To gain experience with two documentation methods, use both ADIME and SOAP formats for your first session. In subsequent sessions, only the ADIME format will be used for charting. See Chapter 5 for guidelines. Since Session 2 addresses physical activity, the guidelines include preparation tasks in this area.

You will also need to prepare an education intervention for your client that should last ten minutes or less. Your guide to selecting a topic will be your discussions during Session 1 as well as the areas of interest indicated in LMF 5.1, Client Assessment Questionnaire. In addition, one of the post session activities is to prepare a preliminary copy of the food management tool (detailed food plan, exchange system, food group plan, or goal setting) selected by your client in the client assessment questionnaire. If your client indicated a desire to only set goals each week, then a food management tool does not need to be developed.

Session 2

After the opening, one of your first objectives will be to review your client's progress since the first session, and then provide the tailored educational intervention you prepared for your client. The next counseling activity will be to review the preliminary food management tool you prepared or confirm that goal setting will be the foundation for behavior change during your session. For a food management tool you prepared, discuss with your client how the tool should be used and modified. Motivated clients may wish to implement the total food plan immediately; others may want to use the pattern as a picture of what the client is attempting to achieve through small steps.

Another issue to be addressed is the need to keep a food diary for at least three days to complete a computerized diet analysis. Review with your client procedures for keeping a food diary. See LMF 5.2 Food Record in Appendix D. The final part of the counseling session will be devoted to the protocol for physical activity counseling. See Chapter 8 for a review of this protocol.

Session 3

After the greeting and reviewing progress in fulfilling previously set goals, go over your client's three-day food record and verify portion sizes and preparation methods. If your client did not bring his or her records, then conduct a 24-hour recall. See Chapter 5 to review this procedure. Use LMF 5.3 24-Hour Recall, Usual Diet in Appendix D. In that case, you will do a computer analysis based on this recall rather than a three-day food record. Then proceed with the tailored education intervention you prepared for your client.

Your next activity will be to review your client's food management tool, if appropriate. Whether or not this topic needs to be addressed will depend on the outcome of your food management tool discussion during the previous session. The rest of the exploration-education phase will cover the role of behavior management and the importance of a support system for making a behavior change. The guidelines contain three behavior management strategies to discuss with your client. However, you may wish to address alternative strategies identified in Chapters 6 or 7 based on your client's needs.

Session 4

This is your final session. After greeting your client and setting the agenda, investigate your client's thoughts about his or her progress since the last session. Then assess all goals your client is currently pursuing and consider altering or continuing them. Present the computerized analysis to your client and review it point by point. Ask your client to give his or her thoughts about the feedback, and then summarize the discussion regarding the analysis.

Next, present the tailored education intervention that you prepared for your client. The last activity of the exploration-education phase is to explore some aspects of relapse prevention counseling. Specifically explain the spiral of change in Lifestyle Management Form 7.4 in Appendix D, and investigate high-risk situations that apply to your client. Discuss the concept of apparent irrelevant decisions, and identify those that could apply to your client's lifestyle change goals.

Continuing with relapse prevention counseling into the resolving phase, discuss coping strategies to deal with the high-risk situations and apparent irrelevant decisions that your client identified. The final counseling objective is to address cognitive restructuring. First discuss the concept with your client, and investigate possible dysfunctional thinking patterns your client may exhibit. If this is an issue, ask your client to consider the validity of the destructive self-talk and prepare some coping strategies, such as thought stopping and alternative responses. Imagery can be used to rehearse the use of the strategies.

Ending the relationship in a meaningful manner will have a significant impact on the counseling encounter. The guidelines contain a sequence of culminating points to address. At the end of the session, express your appreciation for your client's willingness to participate in the clinic, and present your client with a certificate of appreciation.

PREPARATION FOR SESSION 1 CHECKLIST

REGISTRATION

- ❏ Complete registration form—LMF* 14.1 Registration for Nutrition Clinic in duplicate.
  - ○ Give one copy to the client and one copy to the counselor.
- ❏ Complete the agreement form—LMF 14.2 Counseling Agreement in duplicate.
  - ○ Adhere to the protocol for obtaining consent in preparation for Session 1 guidelines in Exhibit 14.1.
  - ○ Both you and your client should sign each copy of the form.
  - ○ Give a copy to your client.
- ❏ Give or send the client a welcome packet containing the following:
  - ○ LMF 5.1 Client Assessment Questionnaire
  - ○ LMF 5.4 Food Frequency Questionnaire
  - ○ LMF 8.5 Medical Release

PHONE CALL

- ❏ Verify date, time, and place of counseling session.
- ❏ Remind client about forms and inquire if there were any questions.
- ❏ Verify how to get in touch with each other if the meeting needs to be postponed.
- ❏ Express desire to meet your client.

PREPARE FOR SESSION 1

- ❏ Review the following procedures/guidelines:
  - ○ Analysis and flow of a counseling interview/counseling session (Chapter 4)
  - ○ Nutrition Counseling Motivational Algorithm (Figure 4.2 in Chapter 4)
  - ○ Session 1 guidelines in Chapter 14
  - ○ "A typical day strategy" (Exhibit 4.2 in Chapter 4)
  - ○ Goal setting process (Chapter 5)
  - ○ MyPyramid website and DASH Diet (Appendix A)
  - ○ Review Session 1 checklist
- ❏ Bring copies of the following forms:
  - ○ LMF 4.1 Assessment Graphic
  - ○ LMF 5.1 Client Assessment Questionnaire
  - ○ LMF 5.3 24-Hour Recall/Usual Diet Form
  - ○ LMF 8.4 Physical Activity Par-Q Form
  - ○ Copy of the DASH Food Plan (Appendix A)
- ❏ Bring visuals to estimate portion size.
- ❏ Minimize distractions.
- ❏ Remind yourself of the six relationship-building responses: attending, reflection, legitimation, support, partnership, and respect.

*LMF = Lifestyle Management Form.

SESSION 1 CHECKLIST	
TASK	**POSSIBLE DIALOGUE**
Involving Phase (10–15 minutes)	

❏ Greeting
  ◯ Greet client verbally.
  ◯ Shake hands.
  ◯ Introduce yourself.
  ◯ Resolve how to address each other.
	Good morning. I'm very happy to meet you.
	
	How would you like to be addressed?

❏ Small talk	*How did you hear about our program?*

❏ Investigation of client's long-term objectives, expectations, needs, and concerns.
If appropriate, use
  ◯ Reflection statements.
  ◯ Legitimation statements.
  ◯ Respect statements.
❏ Summarize.
	Have you ever worked with a nutrition counselor before?
	
	What do you want to achieve in this program?

❏ Explain program and counseling process. ◯ Use partnership statement.	*I'd like to tell you about the design of this program.* *My hope is that we will work together to build on skills you already have to make dietary changes.*
❏ Review weight loss expectations, if appropriate.	
❏ Discuss monitoring of weight, if appropriate.	
❏ Set agenda.	*What we will do this session is . . . review the forms you completed, go over the foods you typically eat and set a goal, if you believe you are ready.*
Transition ❏ Transition statement	*Now that we have gone over the basics of the program, we will explore your needs in greater detail.*
Exploration-Education Phase (25 minutes)	
❏ Review completed Client Assessment Questionnaire, LMF 5.1. ◯ Clarify any highlights on the form.	*I am wondering what came to your mind as you were completing this form.* *What topics in this form do you think have particular importance for your food issues?*

❏ A Typical Day
  ◯ Interrupt as little as possible.
  ◯ Do not impose your ideas.
  ◯ Speed up the pace, if necessary.
  ◯ Summarize.
	Can you take me through a typical day so I can understand more fully what happens and tell me where eating fits into the picture?
	Start with when you get up.
	Is there anything else you would like to add?

Exploration-Education Phase (25 minutes) *(continued)*	

❑ Complete 24-Hour Recall, Usual Diet Form, LMF 5.3.
  ○ Fill in during a Typical Day review.
  ○ Clarify; ask questions after your client has completed his or her description.
	How was the chicken cooked?

❑ Review completed Food Frequency Questionnaire, LMF 5.4.
  ○ Clarify portion sizes; use visuals.
  ○ Clarify preparation methods.
  ○ If client did not complete, use only the 24-Hour, Usual Diet form, LMF 5.3, completed during the typical day discussion to provide feedback.
	Thank you for completing this questionnaire.
	What came to your mind as you were filling it out?
	Did you feel a need to clarify or expand on anything?

❑ Provide Feedback and Education
  ○ Review totals on the bottom of LMF 5.3, 24-Hour Recall, Usual Diet.
  ○ Compare LMF 5.3 and LMF 5.4 to recommended intakes for 2000 kcalorie diet; point by point, nonjudgmentally.
  ○ Clarify when needed.
  ○ Explain MyPyramid.
  ○ Ask opinion of comparison.
  ○ Give your opinion, if requested.
  ○ Summarize.
	What do you think about the comparison?

❑ What's next?	*How would you like to proceed?*

❑ Check readiness with Assessment Graphic, LMF 4.1.
  ○ Check importance.
  ○ Check confidence.
	To get a better idea of how ready you are to make a change, we will use this picture of a ruler. If 0 is not ready at all and 10 is ready, where are you?
	Using the same scale, how do you feel right now about how important this change is for you? How confident are you that you can make this change?

Transition to Resolving Phase ❑ Transition Statement	*Now I'd like to talk more about the possibility of changing your food patterns.*
Resolving Phase (15 minutes)	
Level 1—Not Ready (1–3 on LMF 4.1 Assessment Graphic)	
❑ Raise awareness. ❑ Personalize benefits. ❑ Request permission to discuss.	*Summarize benefits of following MyPyramid, DASH Food Plan, or goal setting.* *Because of (family history, past concerns, present medical problems), you would benefit from . . .* *Would you like to discuss the possibility of such a change?*

Resolving Phase (15 minutes) *(continued)*

❑ Ask key open-ended questions. 　○ Discuss importance. 　○ Identify motivating factors. ❑ Summarize. ❑ Ask permission to give advice. ❑ Express support.	*What do you believe will happen if you don't change?* *How come you picked 3 and not 1 on the graphic?* *What would have to happen for you to move up to the number 8?* *It's up to you—you know best—but small changes can make a difference.* *You probably need some time to think about this. . . .* *Do not hesitate to call me if you have any questions.*

Level 2—Unsure (4–7 on LMF 4.1 Assessment Graphic)

❑ Raise awareness. ❑ Ask key open-ended questions; promote change talk. 　○ Explore confidence. 　○ Explore barriers. ❑ Examine pros and cons. 　○ Summarize. ❑ Imagine the future. ❑ Explore past successes. ❑ Explore social supports. ❑ Summarize. ❑ Ask about next step—go to goal setting in Level 3, if appropriate	*Summarize benefits of following MyPyramid, DASH Food Plan, or goal setting.* *How come you rated your confidence as a 6 instead of a 1 on the ruler?* *What would you need to get to 10 on the graphic?* *What is preventing you from making changes?* *What do you like about your present diet? Dislike?* *Advantages of changing? Disadvantages?* *What would your life be like if . . .? What is the first thing you notice? How do you feel?* *Were you ever able to . . .?* *Do you have someone who could support you?*

Level 3—Ready (8–10 on LMF 4.1 Assessment Graphic)

❑ Praise positive behaviors. ❑ Explore change options to develop a broadly stated goal. 　○ Elicit client's thoughts. 　○ Make an options tool, if appropriate. 　○ Probe for concerns about the selected option. ❑ Explain goal setting basics. ❑ Identify a specific goal. 　○ Small talk 　○ Look to the past. 　○ Build on past. The stated goal is 　○ achievable. 　○ measurable. 　○ totally under client's control. 　○ stated positively.	*It is so good that you . . .* *Do you have an idea of what will work for you?* *This is an options tool. Let's brainstorm ideas, and we'll write them in the circles.* *This seems to be the best choice, but will it work for you?* *What is the smallest goal you believe is worth pursing?* *When have you eaten . . . before?* *Explain how it happened that you ate . . . before.* *Would this work for someone else?*

Resolving Phase (15 minutes) *(continued)*	

❑ Develop an action plan.
  ○ Investigate physical environment.
  ○ Examine social support.
  ○ Review cognitive environment. Explain positive coping talk, if necessary.
❑ Select tracking technique—chart, journal, etc.
❑ Ask your client to verbalize the goal.
❑ Write down goal on a card and give it to your client.
	Do you have everything you need?
	Is there anyone who could help you achieve your goal?
	What are you saying to yourself right now about this goal?
	
	Just to be sure we are both clear about your goal, could you please state the goal?

Closing Phase (5 minutes)	
❑ Support self-efficacy. ❑ Review issues and strengths. ❑ Use a relationship-building response (respect). ❑ Restate food goal. ❑ Give LMF 8.4 Physical Activity Par-Q Form. ❑ Review next meeting time. ❑ Set date and time to call or other method to confirm meeting. ❑ Shake hands. ❑ Express appreciation for participation. ○ Use support and partnership statement.	*I think we did a good job selecting a goal that will work for you.* *I am very impressed with . . .* *Thank you so much for your participation in this program. I look forward to working with you to implement some of the options we've identified.*

POST-SESSION I

Congratulations! You have just finished your first nutrition counseling session. It is now time to reflect, evaluate, document, and plan for the next session.

EVALUATE AND DOCUMENT SESSION 1

❏ Complete the following forms:
  ○ LMF 5.6 Client Concerns and Strengths Log
  ○ LMF 7.5 Interview Assessment Form
❏ Document the session twice by using SOAP and ADIME format.

PLAN FOR SESSION 2

❏ Prepare a preliminary copy of your client's food monitoring tool unless your client will only be setting goals
❏ Calculate your client's exercise target zone (Chapter 8).
❏ Prepare an education intervention according to the needs and desires of your client as indicated in the Client Assessment Questionnaire, LMF 5.1, and your discussions.

PREPARE FOR SESSION 2

❏ Review the following procedures/guidelines:
  ○ Session 2 guidelines and checklist found in the beginning of this chapter.
  ○ Physical Activity Algorithm, Figure 8.2, and protocols (Chapter 8)
  ○ Goal setting process (Chapter 5)
  ○ Food diary guidelines (Chapter 5)
  ○ SOAP and ADIME notes from Session 1
❏ Bring copies of the following forms:
  ○ LMF 4.1 Assessment Graphic
  ○ LMF 8.1 Benefits of Regular Moderate Physical Activity
  ○ LMF 8.2 Physical Activity Log
  ○ LMF 8.3 Physical Activity Options
  ○ LMF 8.4 Physical Activity Par-Q Form
  ○ LMF 8.6 Physical Activity Assessment and Feedback Form, 2 copies
  ○ LMF 5.2 Food Record
  ○ Preliminary copy of your client's food-monitoring tool
  ○ Calculated BMI
❏ Bring visuals to estimate portion size.
❏ Minimize distractions.
❏ Remind yourself of the six relationship-building responses: attending, reflection, legitimation, support, partnership, and respect.

SESSION 2 CHECKLIST	
TASK	**POSSIBLE DIALOGUE**
Involving Phase (5 minutes)	

❑ Greeting
   ○ Extend verbal greeting.
   ○ Shake hands.
	Good morning. It is nice to see you again.

❑ Set agenda.	*In the session today, I thought we would review how your goal worked out this week, go over [client's requested educational need], discuss the food management tool you indicated that you desired, review how to keep a food diary, and then discuss physical activity. How does this sound?*
Transition ❑ Transition statement.	*So let's move on. First I'd like to address last week's goal.*
Exploration-Education Phase (25 minutes)	

❑ Investigation of client thoughts about the week.
If appropriate, use
   ○ Reflection statements.
   ○ Legitimation statements.
   ○ Respect statements.
	How did your week go?
	Let's look at the assessment graphic. What number would you pick to describe how closely you have been following your plan?
	So you feel good about your diet but you are not sure if you can keep it going.

❑ Evaluate effectiveness of plan.
   ○ Identify barriers to goal achievement.
   ○ Clarify client strengths and weaknesses.
❑ Summarize—keep present goal or modify.
	So overall this week you . . .

❑ Tailor educational experience.
❑ Determine appropriate food management tool.
   ○ Show preliminary food pattern based on indicated pattern.
   ○ If appropriate, review each tool (detailed plan, exchanges, food groups, only goal setting).
   ○ Identify advantages and disadvantages, if appropriate.
	On your Client Assessment Questionnaire, you indicated that you wanted some structure but freedom to select foods from food groups. I prepared a preliminary food management tool based on the DASH Food Plan. I thought we could discuss the plan today to see whether this is truly what you think would be useful to you and adjust it according to what you believe would work for you. We could use this plan in two ways: (1) One would be a defined pattern that you would try to implement immediately. (2) It could also be used as a tool that we periodically review to identify what we are working toward.

❑ Instructions on use of tool (detailed plan, exchanges, food groups, only goal setting), if appropriate.	

Exploration-Education Phase (25 minutes) *(continued)*	

❑ Food Diary/Food Record.
  ○ Give instructions, LMF 5.2 Food Record
  ○ Select length of time (need 3 days for computer analysis).
  ○ State purpose of computer analysis.
	Keeping a food diary is an excellent way to monitor and influence your eating habits. I really encourage you to keep records periodically. For the purpose of the computer analysis, I hope I can get you to do the analysis for three days. What do you think?

❑ Provide physical activity feedback.
  ○ Discuss BMI and health.
  ○ Collect LMF 8.4 Physical Activity Par-Q Form.
  ○ Complete LMF 8.6 in duplicate. Give one to the client.
❑ Ask opinion of comparison.
❑ Give your opinion, if appropriate.
❑ Summarize.
	What do you think about this comparison?
	Does this information surprise you?

Transition to Resolving Phase ❑ Transition Statement.	*Now I'd like to talk about any changes you would like to make.*

Resolving Phase (25 minutes)

Level 1—Not Ready (1–3 on LMF 4.1 Assessment Graphic)

❑ Summarize benefits of physical activity; use LMF 8.1.
❑ Personalize benefits to health status.
❑ Request permission to discuss possibility of change.
❑ Ask key open-ended questions.
  ○ Discuss importance/reasons to be physically active.
  ○ Elicit barriers to physical activity.
❑ Summarize.
❑ Ask permission to give advice, if appropriate.
❑ Give advice.
❑ Support self-efficacy.
	Would you be willing to discuss the possibility of a change in your physical activity patterns?
	What benefits of physical activity do you believe most likely apply to you?
	How come you picked 3 and not 1 on the assessment graphic?
	What would have to happen for you to move up to the number 8?
	It's up to you—you know best—but small changes can make a difference.
	I really admire how you. . . . Look how capable you are when you. . . . You have the resources to be physically active. When you are ready, you will be able to increase your physical activity.

Resolving Phase (25 minutes) *(continued)*	
Level 2—Unsure (4–7 on LMF 4.1 Assessment Graphic)	

❑ Raise awareness. ❑ Ask key open-ended questions to explore ambivalence. ○ Identify disadvantages of changing. ○ Explore consequences of inactivity. ○ Identify anticipated benefits. ❑ Explore past successes. ❑ Imagine the future. ❑ Summarize ambivalence. ❑ Ask about next step—go to goal setting in Level 3, if appropriate	*Summarize benefits of physical activity.* *What are some reasons you would like things to stay just like they are?* *What concerns do you have about not increasing your activity?* *What good things would happen if you were more physically active?* *Have you ever been physically active?* *Have you ever enjoyed a physical activity?* *What would your life be like if . . .? What is the first thing you notice? How do you feel?*

Level 3—Ready (8–10 on LMF 4.1 Assessment Graphic)	

❑ Praise positive behaviors. ❑ Review current activity program. ❑ Explore change options to develop a broadly stated goal. ○ Elicit client's thoughts. ○ Look to the past. ○ Go over list of possibilities. (See LMF 8.3 Physical Activity Options.) ○ Validate physician approval. ❑ Client selects an appropriate activity goal. Do not set a goal if physician approval was not obtained. ❑ Explain goal setting basics, if appropriate. ❑ Identify a specific physical activity goal. ○ Small talk. ○ Look to the past. ○ Stated goal is achievable, measurable, totally under client's control, and stated positively. ❑ Develop an action plan. ○ Discuss target heart rate zone. ○ Investigate physical environment. ○ Examine social support. ○ Review cognitive environment. ○ Explain positive coping talk, if necessary. ○ Discuss target heart rate zone. ❑ Select tracking technique—chart, journal, etc. ❑ Ask your client to verbalize the goal. ❑ Write down goal on a card and give it to your client.	*Do you have an idea of what will work for you?* *What activities have you enjoyed in the past?* *Here are some options.* *This seems to be the best choice, but will it work for you?* *What is the smallest physical activity goal you believe is worth pursing?* *When have you exercised before?* *Would this goal work for someone else?* *Do you have everything you need? Do you have walking shoes?* *Is there anyone who could help you achieve your goal?* *What will you say to yourself if you miss a day that you planned to walk?* *Just to be sure we are both clear about your goal, could you please restate the goal?*

Resolving Phase (25 minutes) *(continued)*

Level 4—Active

❏ Praise positive behaviors.
❏ Review current activity program.
❏ Review sport-specific nutrient needs.
❏ Relapse prevention.
  ○ Explain relapse prevention.
  ○ Use Assessment Graphic, LMF 4.1, to assess confidence to continue.
  ○ Identify barriers.
  ○ Explore solutions to barriers.
 	
  ○ Explain that setbacks are common.
  ○ Identify social supporters.
❏ Identify a specific goal and action plan, if appropriate.
❏ Go to goal setting in Level 3, if appropriate.
	It is wonderful that you have such a physically active lifestyle.
	
	Let's look at how confident you are that you can maintain your level of activity.
	How come you chose 9 instead of 10?
	Do you have any ideas of how to overcome difficulties with . . .?
	Setbacks are to be expected. It is important to just start up again.

Closing Phase (5 minutes)

❏ Support self-efficacy.
❏ Review issues and strengths.
❏ Use a relationship-building response (respect).
❏ Restate goals:
  ○ Food goal
  ○ Food diary
  ○ Exercise goal
❏ Review next meeting time.
❏ Discuss videotaping or audiotaping session 3 or 4.
❏ Shake hands.
❏ Express appreciation for participation.
  ○ Use support and partnership statement.
	I am very impressed with . . .
	
	I have really enjoyed working with you. If you have any questions or concerns, do not hesitate to call me.

POST-SESSION 2

EVALUATE AND DOCUMENT SESSION 2

❏ Complete LMF 7.5 Interview Assessment Form.
❏ Document session using ADIME format.

PLAN FOR SESSION 3

❏ Make adjustments to your client's food monitoring tool, if necessary.
❏ Prepare an education intervention as indicated in the Client Assessment Questionnaire and your discussions.

PREPARE FOR SESSION 3

❏ Review the following procedures/guidelines:
  ○ Analysis and flow of counseling interview/counseling session (Chapter 4).
  ○ Nutrition Counseling Motivational Algorithm (Figure 4.2 in Chapter 4).
  ○ Goal setting process (Chapter 5).
  ○ Session 3 guidelines and checklist (Chapter 14).
  ○ Cue management, countering, and barriers counseling (Chapter 6).
  ○ Social support and social disclosure (Chapter 7).
❏ Review your notes from Sessions 1 and 2.
❏ Bring copies of the following forms:
  ○ LMF 4.1 Assessment Graphic.
  ○ LMF 5.3 24-Hour Recall/Usual Diet Form—This form will be used to complete a 24-hour recall if your client does not bring three-day food records.
❏ Bring visuals to estimate portion size.
❏ Bring visuals for tailored education intervention.
❏ Bring audio- or videotape recorder if taping session 3.
❏ Minimize distractions.
❏ Remind yourself of the six relationship-building responses: attending, reflection, legitimation, support, partnership, and respect.

SESSION 3 CHECKLIST	
TASK	**POSSIBLE DIALOGUE**
Involving Phase (5 minutes)	

❏ Greeting
   ○ Verbal greeting.
   ○ Shake hands.
❏ Set agenda.
	Good morning. It is nice to see you again.
	
	In the session today, I thought we would review how your goal/s worked out this week, go over [client's requested educational need], go over food record keeping for computer analysis of your food intake, discuss some behavior approaches to modifying food intake, and talk about social support.

Transition ❏ Transition statement	*I'd like to start by asking about your week.*
Exploration-Education Phase (25 minutes)	

❏ Investigate client thoughts about the week
❏ *If appropriate, use:*
   ○ Reflection statements.
   ○ Legitimation statements.
   ○ Respect statements.
	How did your week go?

❏ Assess adherence to goals using LMF 4.1.
   ○ Food goal.
   ○ Exercise goal.
   ○ Identify barriers to goal achievement.
   ○ Clarify client strengths and weaknesses.
   ○ Summarize—keep present goals or modify.
❏ Review food diary:
   ○ Portion sizes.
   ○ Preparation methods.
   ○ Condiments.
❏ Tailor education intervention.
❏ Review food management tool, if appropriate.
❏ Discuss role of behavior in food choices:
   ○ Cue management.
   ○ Countering.
   ○ Barriers management.
❏ Discuss importance of support system and possible supports:
   ○ Support buddies.
   ○ Organizations.
   ○ Self-help groups.
   ○ Classes.
❏ Social disclosure.
	Let's look at the assessment graphic again. What number would you pick to describe how closely you have been following your plan?
	
	So you feel good about your diet, but you are not sure if you can keep it going.
	So overall this week you . . .
	
	A variety of behavior change methods have been used successfully to help people change their habits. I thought we could talk about some of them and choose one to help you with your desired changes.
	
	Another topic I'd like to address today is the importance of a support system and various ways to build one.

Exploration-Education Phase (25 minutes) *(continued)*	
Transition to Resolving Phase ❏ Transition statement.	*Now, I would like to talk about any specific changes you would like to make.*
Resolving Phase (25 minutes)	

❏ Identify a specific goal.
   ○ Select a new or modify a previous goal.
   ○ Review options tool, if appropriate.
   ○ Engage in small talk.
   ○ Look to the past.
   ○ Build on past.
   ○ The stated goal is achievable, measurable, totally under client's control, and stated positively.
	What is the smallest goal you believe is worth pursing?
	When have you eaten . . . before?
	Explain how it happened that you ate . . . before. Would this work for someone else?

❏ Develop an action plan.
   ○ Investigate physical environment.
   ○ Examine social support.
   ○ Review cognitive environment. Explain positive coping talk, if necessary.
❏ Select a tracking technique—chart, journal, etc.
❏ Do a microanalysis of scenario; review the plan step by step.
❏ Ask your client to verbalize the goal.
❏ Write down goal on a card and give it to your client.	

Closing Phase (5 minutes)	
❏ Support self-efficacy. ❏ Review issues and strengths. ❏ Use a relationship-building response (respect). ❏ Restate food, exercise and behavior goals. ❏ Review next meeting time. ❏ Set date/time to call with reminder. ❏ Shake hands. ❏ Express appreciation for participation. ❏ Use support and partnership statement.	*I am very impressed with . . .*

POST-SESSION 3

EVALUATE AND DOCUMENT SESSION 3

❑ Complete the following forms:
   ○ Complete LMF 7.5 Interview Assessment Form.
   ○ Document session using ADIME format.

PLAN FOR SESSION 4

❑ Make adjustments on your client's food monitoring tool, if necessary.
❑ Prepare an education intervention as indicated in the Client Assessment Questionnaire and your discussions.
   ○ Complete a computer analysis of the food diary data.

PREPARE FOR SESSION 4

❑ Review the following procedures/guidelines:
   ○ Analysis and flow of counseling interview/counseling session (Chapter 4).
   ○ Nutrition Counseling Motivational Algorithm (Figure 4.2 in Chapter 4).
   ○ Session 4 guidelines and checklist (Chapter 14).
   ○ Modifying Cognitions and Problem Solving (Chapter 6).
   ○ Relapse Prevention (Chapter 7).
   ○ Ending the Counseling Relationship (Chapter 7).
   ○ Goal setting process (Chapter 5).
❑ Review your notes from Sessions 1, 2, and 3.
❑ Bring copies of the following forms:
   ○ LMF 7.4, Prochaska's and DiClemente's Spiral of Change.
❑ Bring visuals to estimate portion size.
❑ Bring audio- or videotape recorder if taping session 4.
❑ Minimize distractions.
❑ Remind yourself of the six relationship-building responses: attending, reflection, legitimation, support, partnership, and respect.

SESSION 4 CHECKLIST	
TASK	**POSSIBLE DIALOGUE**
Involving Phase (5 minutes)	

❏ Greeting
   ○ Verbal greeting
   ○ Shake hands.
	Good morning. It is nice to see you again.

❏ Set agenda.	*Let's discuss your concerns about this past week, how the goals and action plan worked out, your computerized diet analysis, relapse prevention and cognitive restructuring.*
Transition ❏ Transition statement.	*So let's move on and take a look at your week.*
Exploration-Education Phase (5 minutes)	

❏ Investigate client thoughts about the week.
 ❏ Summarize.
 ❏ Assess present regimen:
   ○ Food goals.
   ○ Physical activity.
   ○ Behavior strategy.
   ○ Continue or change.
 ❏ Review computerized analysis:
   ○ Compare to standard.
   ○ Highlight areas of concern.
   ○ Provide feedback, education.
   ○ Ask opinion of comparison.
   ○ Give your opinion, if appropriate.
 ❏ Summarize.
 ❏ Tailor education intervention.
 ❏ Relapse prevention.
   ○ Describe behavior change and relapse—show LMF 7.4 Prochaska and DiClemente's Spiral of Change.
   ○ Identify high-risk situations.
 	
   ○ Determine apparent irrelevant decisions.
	So overall this week you . . .
	
	What do you think about the analysis?
	
	Behavior change can be compared to a journey. . . .
	
	When are you likely to have the greatest difficulty keeping your goals?
	Sometimes problems occur because of mini decisions that seem harmless on the surface.

Transition to Resolving Phase ❏ Transition statement.	

Resolving Phase (25 minutes)

- ❏ Coping strategies:
  - ○ Urge management: contract, countering, urge surfing.
- ❏ Cognitive restructuring counseling:
  - ○ Educate about the process.
  - ○ Investigate dysfunctional thinking patterns.
  - ○ Explore validity of self-destructive statements.
  - ○ Explain thought stopping.
  - ○ Prepare constructive responses.
  - ○ Use imagery.

Ending the Counseling Relationship

- ❏ Review the issues that brought your client to you in the first place.
- ❏ Identify goals and progress in meeting them.
- ❏ Emphasize success.
- ❏ Summarize current status.
- ❏ Explore the future.
- ❏ Provide and elicit feedback regarding the significance of the relationship.
- ❏ Summarize.

Some people are surprised to learn that what they are thinking can influence their ability to make lifestyle changes. Many people are not aware that thinking patterns can be changed. We are not obligated to keep a thought in our head.

Closing Phase (5 minutes)

- ❏ Support self-efficacy.
- ❏ Use a relationship-building response (respect).
- ❏ Shake hands.
- ❏ Express appreciation for participation.
- ❏ Give a certificate of appreciation.

POST-SESSION 4

Now that you have completed the four-session program, take the time to reflect and evaluate the experience.

EVALUATE AND DOCUMENT SESSION 4

❑ Complete the following forms:
- ○ Complete LMF 7.5 Interview Assessment Form.
- ○ Document session using ADIME format.
- ○ Complete LMF 7.6 Counseling Responses Competency Assessment.

EVALUATE THE TOTAL COUNSELING INTEREVENTION

❑ Complete the Chapter 13 assignment, "Evaluate Your Counseling Effectiveness."

❑ Reflect on the entire four-session program and answer the following questions:

1. Describe your best counseling experience. What did you specifically say or do to facilitate the positive interaction? Give examples.

2. Describe any behavioral strategies you used during the intervention. Were they effective? Explain.

3. Describe any difficulties you encountered. What do you believe caused them? Is there anything you could have done differently to prevent or alleviate the impact of the difficulties?

4. What did you learn from this experience?

Appendices

DASH Food Plan

Dietary Approaches to Stop Hypertension (DASH) is a heart-healthy dietary regimen rich in fruits, vegetables, fiber, and low-fat dairy foods and low in saturated and total fat. Although the DASH food plan was developed to address high blood pressure, the plan has been found to be useful for everyone to guide healthful eating. The booklet, *In Brief: Your Guide to Lowering Your Blood Pressure with DASH*, is available from the National Heart Lung and Blood Institute website.

The recommended servings of food groups under the DASH food plan are adjusted according to calorie intake. See Table A.1.

The food management tool, *What's on Your Plate*, can be used by clients to both record food intake and define food group servings. The number of servings indicated on the following form contains food group guidelines for a 2,000 kcalorie intake. You will need to adjust the servings and the calories, as indicated in Table A.1 for clients who need a different level of kcalorie intake.

Table A.1 DASH Eating Plan: Daily Servings Based on Calorie Level

Food Groups	Servings/Day		
	1,600 Calories/Day	2,600 Calories/Day	3,100 Calories/Day
Grains*	6	10–11	12–13
Vegetables	3–4	5–6	6
Fruits	4	5–6	6
Fat-free or low-fat milk and milk products	2–3	3	3–4
Lean meats, poultry, and fish	3–6	6	6–9
Nuts, seeds, and legumes	3/week	1	1
Fats and oils	2	3	4
Sweets and added sugars	0	≤2	≤2

* Whole grains are recommended for most gram servings as a good source of fiber and nutrients.

Table A.2   Following the DASH Diet

Food Group	Daily Servings (except as noted)	Serving Sizes	Examples and Notes	Significance of Each Food Group to the Dash Eating Plan
Grains*	6–8	1 slice bread 1 oz dry cereal[†] ½ cup cooked rice, pasta, or cereal	whole wheat bread and rolls, whole wheat pasta, English muffin, pita bread, bagel, cereals, grits, oatmeal, brown rice, unsalted pretzels and popcorn	major sources of energy and fiber
Vegetables	4–5	1 medium fruit ¼ cup dried fruit ½ cup fresh, frozen, or canned fruit ½ cup fruit juice	broccoli, carrots, collards, green beans, green peas, kale, lima beans, potatoes, spinach, squash, sweet potatoes, tomatoes	rich sources of potassium, magnesium, and fiber
Fruits	4–5	1 medium fruit ¼ cup dried fruit ½ cup fresh, frozen, or canned fruit ½ cup fruit juice	apples, apricots, bananas, dates, grapes, oranges, grapefruit, grapefruit juice, mangoes, melons, peaches, pineapples, raisins, strawberries, tangerines	important sources of potassium, magnesium, and fiber
Fat-free or low-fat milk and milk products	2–3	1 cup milk or yogurt 1½ oz cheese	fat-free (skim) or low-fat (1%) milk or buttermilk, fat-free, low-fat, or reduced-fat cheese, fat-free or low-fat regular or frozen yogurt	major sources of calcium and protein
Lean meats, poultry, and fish	6 or less	1 oz cooked meats, poultry, or fish 1 egg[‡]	select only lean; trim away visible fats; broil, roast, or poach; remove skin from poultry	Rich sources of protein and magnesium
Nuts, seeds, and legumes	4–5 per week	$1/3$ cup or 1½ oz nuts 2 Tbsp peanut butter 2 Tbsp or ½ oz seeds ½ cup cooked legumes (dry beans and peas)	almonds, hazelnuts, mixed nuts, peanuts, walnuts, sunflower seeds, peanut butter, kidney beans, lentils, split peas	rich sources of energy, magnesium, protein, and fiber
Fats and oils[§]	2–3	1 tsp soft margarine 1 tsp vegetable oil 1 Tbsp mayonnaise 2 Tbsp salad dressing	soft margarine, vegetable oil (such as canola, corn, olive, or safflower), low-fat mayonnaise, light salad dressing	the DASH study had 27 percent of calories as fat, including fat in or added to foods
Sweets and added sugars	5 or less per week	1 Tbsp sugar 1 Tbsp jelly or jam ½ cup sorbet, gelatin 1 cup lemonade	fruit-flavored gelatin, fruit punch, hard candy, jelly, maple syrup, sorbet and ices, sugar	sweets should be low in fat

Whole grains are recommended for most grain servings as a good source of fiber and nutrients.

[†] *Serving sizes vary between 1/2 cup and 11/4 cups, depending on cereal type. Check the product's Nutrition Facts label.*

[‡] *Since eggs are high in cholesterol, limit egg yolk intake to no more than four per week; two egg whites have the same protein content as 1 oz of meat.*

[§] *Fat content changes serving amount for fats and oils. For example, 1 Tbsp of regular salad dressing equals one serving; 1 Tbsp of a low-fat dressing equals one-half serving; 1 Tbsp of a fat-free dressing equals zero servings.*

Whats on Your Plate? How Much Are You Moving

Date:		Number of Servings by DASH Food Group									
Food	Amount (Serving Size)	Sodium (mg)	Grains	Vegetables	Fruits	Milk Products	Meats, Fish, and Poultry	Nuts, Seeds, and Legumes	Fats and Oils	Sweets and Added Sugars
Example: whole wheat bread, with soft (tub) margarine	2 slices 2 tsp	299 52	2						2	
Breakfast										
Lunch										
Dinner										
Snacks										
Day's Totals										
Compare yours with the DASH eating plan at 2,000 calories.		2,300 or 1,500 mg per day	6–8 per day	4–5 per day	4–5 per day	2–3 per day	6 or less per day	4–5 per week	2–3 per day	5 or less per week
Physical Activity Log Record your minutes per day for each activity. Aim for at least 30 minutes of moderate-intensity physical activity on most days of the week.										

Body Mass Index

Body mass index (BMI) is the preferred weight-for-height standard and is used as a determinant of health risk and a predictor of mortality. You can use the chart to determine health risk as well as for identifying a desirable weight for your clients. See the guidelines provided under the chart.

Table B.1 Body Mass Index (BMI) Chart*

	18	19	20	21	22	23	24	25	26	27	28	29	30	31	32	33	34	35	36	37	38	39	40
Height	Body Weight (Pounds)																						
4'10"	86	91	96	100	105	110	115	119	124	129	134	138	143	148	153	158	162	167	172	177	181	186	191
4'11"	89	94	99	104	109	114	119	124	128	133	138	143	148	153	158	163	168	173	178	183	188	193	198
5'0"	92	97	102	107	112	118	123	128	133	138	143	148	153	158	163	168	174	179	184	189	194	199	204
5'1"	95	100	106	111	116	122	127	132	137	143	148	153	158	164	169	174	180	185	190	195	201	206	211
5'2"	98	104	109	115	120	126	131	136	142	147	153	158	164	169	175	180	186	191	196	202	207	213	218
5'3"	102	107	113	118	124	130	135	141	146	152	158	163	169	175	180	186	191	197	203	208	214	220	225
5'4"	105	110	116	122	128	134	140	145	151	157	163	169	174	180	186	192	197	204	209	215	221	227	232
5'5"	108	114	120	126	132	138	144	150	156	162	168	174	180	186	192	198	204	210	216	222	228	234	240
5'6"	112	118	124	130	136	142	148	155	161	167	173	179	186	192	198	204	210	216	223	229	235	241	247
5'7"	115	121	127	134	140	146	153	159	166	172	178	185	191	198	204	211	217	223	230	236	242	249	255
5'8"	118	125	131	138	144	151	158	164	171	177	184	190	197	203	210	216	223	230	236	243	249	256	262
5'9"	122	128	135	142	149	155	162	169	176	182	189	196	203	209	216	223	230	236	243	250	257	263	270
5'10"	126	132	139	146	153	160	167	174	181	188	195	202	209	216	222	229	236	243	250	257	264	271	278
5'11"	129	136	143	150	157	165	172	179	186	193	200	208	215	222	229	236	243	250	257	265	272	279	286
6'0"	132	140	147	154	162	169	177	184	191	199	206	213	221	228	235	242	250	258	265	272	279	287	294
6'1"	136	144	151	159	166	174	182	189	197	204	212	219	227	235	242	250	257	265	272	280	288	295	302
6'2"	141	148	155	163	171	179	186	194	202	210	218	225	233	241	249	256	264	272	280	287	295	303	311
6'3"	144	152	160	168	176	184	192	200	208	216	224	232	240	248	256	264	272	279	287	295	303	311	319
6'4"	148	156	164	172	180	189	197	205	213	221	230	238	246	254	263	271	279	287	295	304	312	320	328
6'5"	151	160	168	176	185	193	202	210	218	227	235	244	252	261	269	277	286	294	303	311	319	328	336
6'6"	155	164	172	181	190	198	207	216	224	233	241	250	259	267	276	284	293	302	310	319	328	336	345
	Under Weight (<18.5)	Healthy Weight (18.5–24.9)						Overweight (25–29.9)					Obese (≥30)										

*Find your height along the left-hand column and look across the row until you find the number that is closest to your weight. The number at the top of that column identifies your BMI. **Find your client's height and run your finger along the corresponding horizontal line until you come to the weight that matches the desired BMI, such as 24.**

Exchange Lists for Weight Management

Chapter 6 introduced the topic of exchange list food plans. This appendix provides details from the *2008 Choose Your Foods: Exchange Lists for Weight Management*, a booklet designed by the American Dietetic Association and the American Diabetes Association. This planning guide can be used to design a food group plan based on carbohydrate, fat and protein content of food. Each food item in a group has a defined serving size and similar macronutrient distribution and kcalorie level. The nutrition counselor designs a food plan providing the number of servings of each food group to meet a desired kcalorie level. Because macronutrients and kcalories for each food item in the food group are similar, any item can be "exchanged" for another food item in the food group. For example, an exchange food plan may state that 3 servings from the fruit group should be consumed each day. A client can look in the booklet under fruits and decide to eat any of the food items listed in that category, such as, one kiwi, three dates, or one cup cubed papaya. See Table C.5.

DETERMINE DESIRED CALORIE LEVEL. CALCULATE TOTAL ENERGY EXPENDITURE (TEE)

When you use the *Exchange Lists for Weight Management* to design a food plan for a client, you need to begin with a kcalorie level. The following provides one way of estimating calorie needs of an individual, TEE. We will begin by using the Harris-Benedict equation to calculate resting energy expenditure (REE).

Step 1: Calculate Resting Energy Expenditure (REE)

FORMULAS

Harris-Benedict Equations

Women: REE (kcal) = 655 + 9.56 wt (kg) + 1.85 ht. (cm)—4.68 (age)

Men: REE (kcal) = 66.5 + 13.75 wt (kg) + 5.0 ht. (cm)—6.78 (age)

Use the following website to check your calculations: http://www-users.med.cornell.edu/~spon/picu/calc/beecalc.htm

Step 2: Use Table C.1 to determine a Physical Activity Level (PA)

Step 3: Multiply REE Times PA to Obtain the Estimated TEE (kcal/day) to Maintain Weight

$$REE \times PA = TEE$$

Step 4: If Weight Loss Is Desired, Subtract 500 kcal/day to Obtain Adjusted Caloric Intake Required to Achieve Weight Loss of Approximately One Pound Per Week

Table C.1 Physical Activity (PA) Level Factors

Activity Level	PA Men	PA Women	Typical Daily Living Activities
Sedentary	1.00	1.00	Only physical activities typical of daily living
Low active	1.11	1.12	30–60 minutes of moderate activity
Active	1.25	1.27	≥60 minutes of moderate activity
Very active	1.48	1.45	≥60 minutes of moderate activity plus 60 min. vigorous or 120 min. Moderate activity

Note: Moderate activity is equivalent to walking at 3 to 4½ mph.

DETERMINE SERVINGS OF FOOD GROUPS (EXCHANGES)

We will use the Exchange Lists for Weight Management found in this appendix to calculate the number of servings (exchanges) for each food group. In order to do these calculations, you need to know the grams of carbohydrate, protein, and fat for each food group. Table C.2 lists each of these categories.

- Choose a desired intake of carbohydrate, protein, fat, and alcohol.
- Start your calculations with your desired kcalories. Use your calculated total energy expenditure (TEE).
- If alcohol intake is desired, use 100 kcalories for an estimate of each serving, and subtract the alcohol kcalories from your TEE. For example, if your TEE is 1900 kcalories and one serving of alcohol is planned, then reduce your desired calorie intake to 1800 kcalories.
- Calculate the grams of carbohydrate, protein, and fat the eating plan should contain. See Exhibit C.1 for an example, and use the template in Exhibit C.2 to guide the calculations.
- Calculate numbers of food groups (exchanges) of each macronutrient.
- Calculate carbohydrate exchanges first. Select numbers of servings of milk, fruit, and vegetable exchanges based on what is a reasonable number for your client. Calculate total grams of carbohydrate accounted for in these

three groups, and then divide the remaining grams of carbohydrate by 15, the number of grams of carbohydrate in one exchange of the starch group. Then total the grams of protein accounted for in the carbohydrate exchanges.

- Determine the number of meat and meat alternate exchanges by dividing the remaining grams of protein not accounted for in the carbohydrate group seven, the grams of protein in one protein exchange.
- Determine the number of fat exchanges by adding up the fat already accounted for in the carbohydrate and meat and meat alternate exchanges. Remaining fat grams should be divided by 5, the number of grams of fat in one fat exchange.
- See Table C.3 for samples of distributions of food groups for various calorie levels.

DESIGN A SAMPLE MEAL PLAN

In order to understand how to distribute the food groups throughout the day, design a sample meal plan. See Exhibit C.3 for a sample 2000 kcalorie meal plan and use the template in Exhibit C.4 to design your own sample meal plan based on food preferences. In order to accomplish this task, you need to know how the foods are distributed throughout the exchange food groups and the serving sizes of individual foods. This information can be found in Tables C.4 through C.14. Table C.3 provides sample food group plans for various kcalorie intakes.

Table C.2 The Food Lists

Lists	Typical Item/ Portion Size	Carbohydrate (g)	Protein (g)	Fat (g)	Energy[a] (kcal)
Carbohydrates					
Starch[b]	1 slice bread	15	0–3	0–1	80
Fruits	1 small apple	15	—	—	60
Milk					
Fat-free, low-fat, 1%	1 c fat-free milk	12	8	0–3	100
Reduced-fat, 2%	1 c reduced-fat milk	12	8	5	120
Whole	1 c whole milk	12	8	8	160
Sweets, desserts, and other carbohydrates[c]	2 small cookies	15	varies	varies	varies
Nonstarchy vegetables	½ c cooked carrots	5	2	—	25
Meat and Meat Substitutes					
Lean	1 oz chicken (no skin)	—	7	0–3	45
Medium-fat	1 oz ground beef	—	7	4–7	75
High-fat	1 oz pork sausage	—	7	8+	100
Plant-based proteins	½ c tofu	varies	7	varies	varies
Fats	1 tsp butter	—	—	5	45
Alcohol	12 oz beer	varies	—	—	100

[a]The energy value for each exchange list represents an approximate average for the group and does not reflect the precise number of grams of carbohydrate, protein, and fat. For example, a slice of bread contains 15 grams of carbohydrate (60 kcalories), 3 grams protein (12 kcalories), and a little fat—rounded to 80 kcalories for ease in calculating. A ½ cup of vegetables (not including starchy vegetables) contains 5 grams carbohydrate (20 kcalories) and 2 grams protein (8 more), which has been rounded down to 25 kcalories.
[b]The Starch list includes cereals, grains, breads, crackers, snacks, starchy vegetables (such as corn, peas, and potatoes), and legumes (dried beans, peas, and lentils).
[c]The Sweets, Desserts, and Other Carbohydrates list includes foods that contain added sugars and fats such as sodas, candy, cakes, cookies, doughnuts, ice cream, pudding, syrup, and frozen yogurt.

EXHIBIT C.1   Example of Calculations for Development of an Eating Plan Based on Exchange Lists for Weight Management

1800 kcal desired = roughly 50% carbohydrate, 20% protein, 30% fat, no alcohol

If alcohol is planned, reduce the number of desired calories by the amount of alcohol calories desired. For example, one glass of white wine (5 fl oz) = 100 kcal. Then, begin calculations with 1700 kcal, rather than 1800.

Carbohydrate (CHO): 1800 kcal $\times$ .50 = 900 kcal $\div$ 4 kcal/gm = 225 gms

Protein (PRO):   1800 kcal $\times$ .20 = 360 kcal $\div$ 4 kcal/gm = 90 gms

Fat (FAT):   1800 kcal $\times$ .30 = 540 kcal $\div$ 9 kcal/gm = 60 gms

FOOD GROUP	SERVING NUMBER	CHO 225 gms	PRO 90 gms	FAT 60 gms	KCAL 1800
MILK					
Fat-free, low-fat, 1%	2	24	16	6	200
Reduced fat, 2%	0				
Whole	0				
VEGETABLES (nonstarchy)	4	20	8	—	100
FRUIT	4	60	—	—	240

CHO SUM 104 gms

225 gms total CHO–104 gms CHO used (SUM) = 121 gms CHO left

121 CHO gms left $\div$ 15 gms CHO/starch exchange = ~8 servings of starch

FOOD GROUP	SERVING NUMBER	CHO 225 gms	PRO 90 gms	FAT 60 gms	KCAL 1800
STARCH	8	120	24	—	640
OTHER CARBOHYDRATES	0				

PRO SUM 48 gms

90 gms total PRO–48 gms PRO used (SUM) = 42 gms PRO left

42 gms PRO left $\div$ 7 gms PRO/meat exchange = 6 meat exchanges

FOOD GROUP	SERVING NUMBER	CHO 225 gms	PRO 90 gms	FAT 60 gms	KCAL 1800
MEAT					
Lean	6	—	42	18	270
Medium-fat	0				
High-fat	0				
Plant-based proteins	0				

FAT SUM 24 gms

60 gms total FAT–24 gms FAT used (SUM) = 36 gms FAT left

36 gms FAT left $\div$ 5 gms FAT/FAT exchange = ~7 FAT exchanges

FOOD GROUP	SERVING NUMBER	CHO 225 gms	PRO 90 gms	FAT 60 gms	KCAL 1800
FAT	7	—	—	35	315
TOTALS		224	90	59	1765

EXHIBIT C.2   Template for Calculations to Develop an Eating Plan Based on Exchange Lists for Weight Management

____ kcal—alcohol calories = ____ kcal

____ kcal—roughly ____ % carbohydrate, ____ % protein, ____ % fat

Carbohydrate (CHO): ____ kcal × ____ = ____ kcal ÷ 4 kcal/gm = ____ gms

Protein (PRO):   ____ kcal × ____ = ____ kcal ÷ 4 kcal/gm = ____ gms

Fat (FAT):   ____ kcal × ____ = ____ kcal ÷ 9 kcal/gm = ____ gms

FOOD GROUP	SERVING NUMBER	CHO gms	PRO gms	FAT gms	KCAL
MILK					
Fat-free, low-fat					
Reduced fat					
Whole					
VEGETABLES (nonstarchy)					
FRUIT					

CHO SUM ___ gms

___ gms total CHO– ___ gms CHO used (SUM) = ___ gms CHO left

___ CHO gms left ÷ 15 gms CHO/starch exchange = _____ servings of starch

STARCH					
OTHER CARBOHYDRATES					

PRO SUM ___ gms

___ gms total PRO– ___ gms PRO used (SUM) = ___ gms PRO left

___ gms PRO left ÷ 7 gms PRO/meat exchange = _____ meat exchanges

MEAT					
Lean					
Medium-fat					
High-fat					
Plant-based proteins					

FAT SUM ___ gms

___ gms total FAT– ___ gms FAT used (SUM) = ___ gms FAT left

___ gms FAT left ÷ 5 gms FAT/FAT exchange = ___ FAT exchanges

FAT					
ALCOHOL					
TOTALS					

Table C.3 Sample Food Group Servings for Exchange Patterns

kcal/day	1200*	1600*	1800*	2000**	2400**	2800**	3200**
Exchange List							
Milk, low fat	2	2	2	2	2	3	3
Vegetable	3	3	4	4	5	6	6
Fruit	3	4	4	5	5	6	7
Starch	5	8	8	10	14	15	18
Meat, lean	4	4	6	5	5	6	7
Fat	2	4	7	9	12	13	15

*Food group distribution is based on 50% carbohydrate, 20% protein, 30% fat, no alcohol.
**Food group distribution is based on 55% carbohydrate, 15% protein, 30% fat, no alcohol.

Note: Protein is calculated at 20% of calorie intake for lower calorie food patterns in order to offer more flexibility. For high calorie food patterns, 15% is used to keep protein intake within reasonable limits.

EXHIBIT C.3   Sample Meal Plan for 2000 kcal

Total Food Groups

Starch _____ 10 _____

Fruit _____ 5 _____

Milk

Fat-free _____ 2 _____

Reduced fat _____

Whole _____

Vegetables _____ 4 _____

**Sweets, Desserts, and
Other Carbohydrates** _____

Meat and Meat Substitutes

Lean _____ 5 _____

Med-fat _____

High-fat _____

Plant-based _____

Fats _____ 9 _____

Alcohol _____

B/L/D/S*	NUMBER AND TYPES OF FOOD GROUP SERVINGS	SAMPLE FOOD SUGGESTIONS
B	_2_ Starch	1 cup cooked oatmeal
	_2_ Fruit	2 tbsp craisins
	_1_ Milk _____ (type)	1 cup fat-free milk
	___ Vegetables	1 oz Canadian bacon
	_1_ Meat _____ (type)	¾ cup blueberries
	___ Fats	
	___ Sweets, Alcohol	

(continued)

EXHIBIT C.3   Sample Meal Plan for 2000 kcal *(continued)*

B/L/D/S*	NUMBER AND TYPES OF FOOD GROUP SERVINGS	SAMPLE FOOD SUGGESTIONS
L	__3__ Starch	2 slices whole wheat bread
	__1__ Fruit	3 prunes
	__1__ Milk _____ *(type)*	1 cup lentil soup (counts as 1 starch & 1 lean meat)
	__2__ Vegetables	½ cup raw carrots
	__1__ Meat _____ *(type)*	½ cup tomatoes
	__2__ Fats	1 cup fat-free milk
	_____ Sweets, Alcohol	2 tbsp light plant stanol ester spread
D	__4__ Starch	⅔ cup brown rice
	__1__ Fruit	1 corn on the cob (counts as 2 starches)
	_____ Milk _____ *(type)*	17 small grapes
	__2__ Vegetables	1 cup cooked broccoli
	__3__ Meat _____ *(type)*	3 ounces trout
	__5__ Fats	1 tbsp light plant stanol ester spread
	_____ Sweets, Alcohol	4 tsp olive oil
S	__1__ Starch	3 cups lower fat popcorn (counts as 1 starch)
	__1__ Fruit	8 dried apricots
	_____ Milk _____ *(type)*	6 almonds
	_____ Vegetables	10 peanuts
	_____ Meat _____ *(type)*	
	__2__ Fats	
	_____ Sweets, Alcohol	

*Breakfast, Lunch, Dinner, Snack.

EXHIBIT C.4   Sample Meal Plan for _____

Total Food Groups

Starch _____

Fruit _____

Milk

  Fat-free _____

  Reduced fat _____

  Whole _____

Vegetables _____

Sweets, Desserts, and Other Carbohydrates _____

Meat and Meat Substitutes

  Lean _____

  Med-fat _____

  High-fat _____

  Plant-based _____

Fats _____

Alcohol _____

B/L/D/S*	NUMBER AND TYPES OF FOOD GROUP SERVINGS	SAMPLE FOOD SUGGESTIONS
	____ Starch ____ Fruit ____ Milk ____ (type) ____ Vegetables ____ Meat ____ (type) ____ Fats ____ Sweets, Alcohol	
	____ Starch ____ Fruit ____ Milk ____ (type) ____ Vegetables ____ Meat ____ (type) ____ Fats ____ Sweets, Alcohol	
	____ Starch ____ Fruit ____ Milk ____ (type) ____ Vegetables ____ Meat ____ (type) ____ Fats ____ Sweets, Alcohol	
	____ Starch ____ Fruit ____ Milk ____ (type) ____ Vegetables ____ Meat ____ (type) ____ Fats ____ Sweets, Alcohol	

*Breakfast, Lunch, Dinner, Snack.

Table C.4 Starch

The Starch list includes bread, cereals and grains, starchy vegetables, crackers and snacks, and legumes (dried beans, peas, and lentils). 1 starch choice = 15 grams carbohydrate, 0–3 grams protein, 0–1 grams fat, and 80 kcalories.

Note: In general, one starch exchange is ½ cup cooked cereal, grain, or starchy vegetable; ⅓ cup cooked rice or pasta; 1 ounce of bread product; ¾ ounce to 1 ounce of most snack foods.

FOOD	SERVING SIZE
Bread	
Bagel, large (about 4 oz)	¼ (1 oz)
▽ Biscuit, 2½ inches across	1
☺ reduced-kcalorie	2 slices (1½ oz)
white, whole-grain, pumpernickel, rye, unfrosted raisin	1 slice (1 oz)
Chapatti, small, 6 inches across	1
▽ Cornbread, 1¾ inch cube	1 (1½ oz)
English muffin	½
Hot dog bun or hamburger bun	½ (1 oz)
Naan, 8 inches by 2 inches	¼
Pancake, 4 inches across, ¼ inch thick	1
Pita, 6 inches across	½
Roll, plain, small	1 (1 oz)
▽ Stuffing, bread	⅓ cup
▽ Taco shell, 5 inches across	2
Tortilla, corn, 6 inches across	1
Tortilla, flour, 6 inches across	1
Tortilla, flour, 10 inches across	⅓
▽ Waffle, 4-inch square or 4 inches across	1
Cereals and Grains	
Barley, cooked	⅓ cup
Bran, dry	
☺ oat	¼ cup
☺ wheat	½ cup
☺ Bulgur (cooked)	½ cup
Cereals	
☺ bran	½ cup
cooked (oats, oatmeal)	½ cup
puffed	1½ cups
shredded wheat, plain	½ cup
sugar-coated	½ cup
unsweetened, ready-to-eat	¾ cup

FOOD	SERVING SIZE
Cereals and Grains *(continued)*	
Couscous	⅓ cup
Granola	
low-fat	¼ cup
▽ regular	¼ cup
Grits, cooked	½ cup
Kasha	½ cup
Millet, cooked	⅓ cup
Muesli	¼ cup
Pasta, cooked	⅓ cup
Polenta, cooked	⅓ cup
Quinoa, cooked	⅓ cup
Rice, white or brown, cooked	⅓ cup
Tabbouleh (tabouli), prepared	½ cup
Wheat germ, dry	3 Tbsp
Wild rice, cooked	½ cup
Starchy Vegetables	
Cassava	⅓ cup
Corn	½ cup
on cob, large	½ cob (5 oz)
☺ Hominy, canned	¾ cup
☺ Mixed vegetables with corn, peas, or pasta	1 cup
☺ Parsnips	½ cup
☺ Peas, green	½ cup
Plantain, ripe	⅓ cup
Potato	
baked with skin	¼ large (3 oz)
boiled, all kinds	½ cup or ½ medium (3 oz)
▽ mashed, with milk and fat	½ cup
french fried (oven-baked)[a]	1 cup (2 oz)
☺ Pumpkin, canned, no sugar added	1 cup
Spaghetti/pasta sauce	½ cup
☺ Squash, winter (acorn, butternut)	1 cup
☺ Succotash	½ cup
Yam, sweet potato, plain	½ cup

(continued)

Table C.4 Starch *(continued)*

FOOD	SERVING SIZE
Crackers and Snacks[b]	
Animal crackers	8
Crackers	
▽ round-butter type	6
saltine-type	6
▽ sandwich-style, cheese or peanut butter filling	3
▽ whole-wheat regular	2–5 (¾ oz)
☺ whole-wheat lower fat or crispbreads	2–5 (¾ oz)
Graham cracker, 2½-inch square	3
Matzoh	¾ oz
Melba toast, about 2-inch by 4-inch piece	4
Oyster crackers	20
Popcorn	3 cups
▽ ☺ with butter	3 cups
☺ no fat added	3 cups
☺ lower fat	3 cups

FOOD	SERVING SIZE
Crackers and Snacks[b] *(continued)*	
Pretzels	¾ oz
Rice cakes, 4 inches across	2
Snack chips	
fat-free or baked (tortilla, potato), baked pita chips	15–20 (¾ oz)
▽ regular (tortilla, potato)	9–13 (¾ oz)
Beans, Peas, and Lentils[c]	

The choices on this list count as 1 starch + 1 lean meat.

☺ Baked beans	⅓ cup
☺ Beans, cooked (black, garbanzo, kidney, lima, navy, pinto, white)	½ cup
☺ Lentils, cooked (brown, green, yellow)	½ cup
☺ Peas, cooked (black-eyed, split)	½ cup
🧂 ☺ Refried beans, canned	½ cup

KEY

☺ = More than 3 grams of dietary fiber per serving.

▽ = Extra fat, or prepared with added fat. (Count as 1 starch + 1 fat.)

🧂 = 480 milligrams or more of sodium per serving.

[a]Restaurant-style french fries are on the Fast Foods list.
[b]For other snacks, see the Sweets, Desserts, and Other Carbohydrates list. For a quick estimate of serving size, an open handful is equal to about 1 cup or 1 to 2 ounces of snack food.
[c]Beans, peas, and lentils are also found on the Meat and Meat Substitutes list.

Table C.5 Fruits

Fruit[a]

The Fruits list includes fresh, frozen, canned, and dried fruits and fruit juices. 1 fruit choice = 15 grams carbohydrate, 0 grams protein, 0 grams fat, and 60 kcalories.

Note: In general, one fruit exchange is ½ cup canned or fresh fruit or unsweetened fruit juice; 1 small fresh fruit (4 ounces); 2 tablespoons dried fruit.

FOOD	SERVING SIZE	FOOD	SERVING SIZE
Apple, unpeeled, small	1 (4 oz)	Peaches	
Apples, dried	4 rings	canned	½ cup
Applesauce, unsweetened	½ cup	fresh, medium	1 (6 oz)
Apricots		Pears	
canned	½ cup	canned	½ cup
dried	8 halves	fresh, large	½ (4 oz)
☺ fresh	4 whole (5½ oz)	Pineapple	
Banana, extra small	1 (4 oz)	canned	½ cup
J Blackberries	¾ cup	fresh	¾ cup
Blueberries	¾ cup	Plums	
Cantaloupe, small	⅓ melon or 1 cup cubed (11 oz)	canned	½ cup
		dried (prunes)	3
Cherries		small	2 (5 oz)
sweet, canned	½ cup	☺ Raspberries	1 cup
sweet fresh	12 (3 oz)	☺ Strawberries	1¼ cup whole berries
Dates	3		
Dried fruits (blueberries, cherries, cranberries, mixed fruit, raisins)	2 Tbsp	☺ Tangerines, small	2 (8 oz)
		Watermelon	1 slice or 1¼ cups cubes (13½ oz)
Figs			
dried	1½		
☺ fresh	1½ large or 2 medium (3½ oz)	**Fruit Juice**	
Fruit cocktail	½ cup		
Grapefruit		Apple juice/cider	½ cup
large	½ (11 oz)	Fruit juice blends, 100% juice	⅓ cup
sections, canned	¾ cup	Grape juice	⅓ cup
Grapes, small	17 (3 oz)	Grapefruit juice	½ cup
Honeydew melon	1 slice or 1 cup cubed (10 oz)	Orange juice	½ cup
		Pineapple juice	½ cup
☺ Kiwi	1 (3½ oz)	Prune juice	⅓ cup
Mandarin oranges, canned	¾ cup		
Mango, small	½ (5½ oz) or ½ cup		
Nectarine, small	1 (5 oz)		
☺ Orange, small	1 (6½ oz)		
Papaya	½ or 1 cup cubed (8 oz)		

KEY

☺ = More than 3 grams of dietary fiber per serving.

▽ = Extra fat, or prepared with added fat. (Count as 1 starch + 1 fat.)

▤ = 480 milligrams or more of sodium per serving.

[a]The weight listed includes skin, core, seeds, and rind.

Table C.6 Milk

The milk list groups milks and yogurts based on the amount of fat they have (fat-free/low fat, reduced fat, and whole). Cheeses are found on the Meat and Meat Substitutes list and cream and other dairy fats are found on the Fats list.

Note: In general, one milk choice is 1 cup (8 fluid ounces or ½ pint) milk or yogurt.

FOOD	SERVING SIZE
Milk and Yogurts	
Fat-free or low-fat (1%)	
1 fat-free/low-fat milk choice = 12 g carbohydrate, 8 g protein, 0–3 g fat, and 100 kcal.	
Milk, buttermilk, acidophilus milk, Lactaid	1 cup
Evaporated milk	½ cup
Yogurt, plain or flavored with an artificial sweetener	⅔ cup (6 oz)
Reduced-fat (2%)	
1 reduced-fat milk choice = 12 g carbohydrate, 8 g protein, 5 g fat, and 120 kcal.	
Milk, acidophilus milk, kefir, Lactaid	1 cup
Yogurt, plain	⅔ cup (6 oz)
Whole	
1 whole milk choice = 12 g carbohydrate, 8 g protein, 8 g fat, and 160 kcal.	
Milk, buttermilk, goat's milk	1 cup
Evaporated milk	½ cup
Yogurt, plain	8 oz

Dairy-Like Foods

FOOD	SERVING SIZE	COUNT AS
Chocolate milk		
fat-free	1 cup	1 fat-free milk + 1 carbohydrate
whole	1 cup	1 whole milk + 1 carbohydrate
Eggnog, whole milk	½ cup	1 carbohydrate + 2 fats
Rice drink		
flavored, low fat	1 cup	2 carbohydrates
plain, fat-free	1 cup	1 carbohydrate
Smoothies, flavored, regular	10 oz	1 fat-free milk + 2½ carbohydrates
Soy milk		
light	1 cup	1 carbohydrate + ½ fat
regular, plain	1 cup	1 carbohydrate + 1 fat
Yogurt		
and juice blends	1 cup	1 fat-free milk + 1 carbohydrate
low carbohydrate (less than 6 grams carbohydrate per choice)	⅔ cup (6 oz)	½ fat-free milk
with fruit, low-fat	⅔ cup (6 oz)	1 fat-free milk + 1 carbohydrate

Table C.7 Sweets, Desserts, and Other Carbohydrates

1 other carbohydrate choice = 15 grams carbohydrate, variable grams protein, variable grams fat, and variable kcalories.

Note: In general, one choice from this list can substitute for foods on the Starch, Fruits, or Milk lists.

FOOD	SERVING SIZE	COUNT AS
Beverages, Soda, and Energy/Sports Drinks		
Cranberry juice cocktail	½ cup	1 carbohydrate
Energy drink	1 can (8.3 oz)	2 carbohydrates
Fruit drink or lemonade	1 cup (8 oz)	2 carbohydrates
Hot chocolate		
regular	1 envelope added to 8 oz water	1 carbohydrate + 1 fat
sugar-free or light	1 envelope added to 8 oz water	1 carbohydrate
Soft drink (soda), regular	1 can (12 oz)	2½ carbohydrates
Sports drink	1 cup (8 oz)	1 carbohydrate
Brownies, Cake, Cookies, Gelatin, Pie, and Pudding		
Brownie, small, unfrosted	1¼-inch square, ⅞ inch high (about 1 oz)	1 carbohydrate + 1 fat
Cake		
angel food, unfrosted	1/12 of cake (about 2 oz)	2 carbohydrates
frosted	2-inch square (about 2 oz)	2 carbohydrates + 1 fat
unfrosted	2-inch square (about 2 oz)	1 carbohydrate + 1 fat
Cookies		
chocolate chip	2 cookies (2¼ inches across)	1 carbohydrate + 2 fats
gingersnap	3 cookies	1 carbohydrate
sandwich, with crème filling	2 small (about ⅔ oz)	1 carbohydrate + 1 fat
sugar-free	3 small or 1 large (¾–1 oz)	1 carbohydrate + 1–2 fats
vanilla wafer	5 cookies	1 carbohydrate + 1 fat
Cupcake, frosted	1 small (about 1¾ oz)	2 carbohydrates + 1–1½ fats
Fruit cobbler	½ cup (3½ oz)	3 carbohydrates + 1 fat
Gelatin, regular	½ cup	1 carbohydrate
Pie		
commercially prepared fruit, 2 crusts	⅙ of 8-inch pie	3 carbohydrates + 2 fats
pumpkin or custard	⅛ of 8-inch pie	1½ carbohydrates + 1½ fats
Pudding		
regular (made with reduced-fat milk)	½ cup	2 carbohydrates
sugar-free or sugar- and fat-free (made with fat-free milk)	½ cup	1 carbohydrate
Candy, Spreads, Sweets, Sweeteners, Syrups, and Toppings		
Candy bar, chocolate/peanut	2 "fun size" bars (1 oz)	1½ carbohydrates + 1½ fats
Candy, hard	3 pieces	1 carbohydrate
Chocolate "kisses"	5 pieces	1 carbohydrate + 1 fat
Coffee creamer		
dry, flavored	4 tsp	½ carbohydrate + ½ fat
liquid, flavored	2 Tbsp	1 carbohydrate
Fruit snacks, chewy (pureed fruit concentrate)	1 roll (¾ oz)	1 carbohydrate
Fruit spreads, 100% fruit	1½ Tbsp	1 carbohydrate
Honey	1 Tbsp	1 carbohydrate
Jam or jelly, regular	1 Tbsp	1 carbohydrate
Sugar	1 Tbsp	1 carbohydrate

(continued)

Table C.7 Sweets, Desserts, and Other Carbohydrates *(continued)*

FOOD	SERVING SIZE	COUNT AS
Candy, Spreads, Sweets, Sweeteners, Syrups, and Toppings *(continued)*		
Syrup		
chocolate	2 Tbsp	2 carbohydrates
light (pancake type)	2 Tbsp	1 carbohydrate
regular (pancake type)	1 Tbsp	1 carbohydrate
Barbeque sauce	3 Tbsp	1 carbohydrate
Cranberry sauce, jellied	¼ cup	1½ carbohydrates
Gravy, canned or bottled	½ cup	½ carbohydrate + ½ fat
Salad dressing, fat-free, low-fat, cream-based	3 Tbsp	1 carbohydrate
Sweet and sour sauce	3 Tbsp	1 carbohydrate
Doughnuts, Muffins, Pastries, and Sweet Breads		
Banana nut bread	1-inch slice (1 oz)	2 carbohydrates + 1 fat
Doughnut		
cake, plain	1 medium (1½ oz)	1½ carbohydrates + 2 fats
yeast type, glazed	3¾ inches across (2 oz)	2 carbohydrates + 2 fats
Muffin (4 oz)	¼ muffin (1 oz)	1 carbohydrate + ½ fat
Sweet roll or Danish	1 (2½ oz)	2½ carbohydrates + 2 fats
Frozen Bars, Frozen Desserts, Frozen Yogurt, and Ice Cream		
Frozen pops	1	½ carbohydrate
Fruit juice bars, frozen, 100% juice	1 bar (3 oz)	1 carbohydrate
Ice cream		
fat-free	½ cup	1½ carbohydrates
light	½ cup	1 carbohydrate + 1 fat
no sugar added	½ cup	1 carbohydrate + 1 fat
regular	½ cup	1 carbohydrate + 2 fats
Sherbet, sorbet	½ cup	2 carbohydrates
Yogurt, frozen		
fat-free	⅓ cup	1 carbohydrate
regular	½ cup	1 carbohydrate + 0–1 fat
Granola Bars, Meal Replacement Bars/Shakes, and Trail Mix		
Granola or snack bar, regular or low-fat	1 bar (1 oz)	1½ carbohydrates
Meal replacement bar	1 bar (1⅓ oz)	1½ carbohydrates + 0–1 fat
Meal replacement bar	1 bar (2 oz)	2 carbohydrates + 1 fat
Meal replacement shake, reduced kcalorie	1 can (10–11 oz)	1½ carbohydrates + 0–1 fat
Trail mix		
candy/nut-based	1 oz	1 carbohydrate + 2 fats
dried fruit-based	1 oz	1 carbohydrate + fat

KEY

= 480 milligrams or more of sodium per serving.

[a]You can also check the Fats list and Free Foods list for other condiments.

Table C.8   Nonstarchy Vegetables

The Nonstarchy Vegetables list includes vegetables that have few grams of carbohydrates or kcalories; starchy vegetables are found on the Starch list. 1 nonstarchy vegetable choice = 5 grams carbohydrate, 2 grams protein, 0 grams fat, and 25 kcalories.

Note: In general, one nonstarchy vegetable choice is ½ cup cooked vegetables or vegetable juice or 1 cup raw vegetables. Count 3 cups of raw vegetables or 1½ cups of cooked vegetables as one carbohydrate choice.

Nonstarchy Vegetables[a]	Nonstarchy Vegetables[a]
Amaranth or Chinese spinach	Kohlrabi
Artichoke	Leeks
Artichoke hearts	Mixed vegetables (without corn, peas, or pasta)
Asparagus	Mung bean sprouts
Baby corn	Mushrooms, all kinds, fresh
Bamboo shoots	Okra
Beans (green, wax, Italian)	Onions
Bean sprouts	Oriental radish or daikon
Beets	Pea pods
☺ Borscht	☺ Peppers (all varieties)
Broccoli	Radishes
☺ Brussels sprouts	Rutabaga
Cabbage (green, bok choy, Chinese)	☺ Sauerkraut
☺ Carrots	Soybean sprouts
Cauliflower	Spinach
Celery	Squash (summer, crookneck, zucchini)
☺ Chayote	Sugar pea snaps
Coleslaw, packaged, no dressing	☺ Swiss chard
Cucumber	Tomato
Eggplant	Tomatoes, canned
Gourds (bitter, bottle, luffa, bitter melon)	☺ Tomato sauce
Green onions or scallions	☺ Tomato/vegetable juice
Greens (collard, kale, mustard, turnip)	Turnips
Hearts of palm	Water chestnuts
Jicama	Yard-long beans

KEY

☺ = More than 3 grams of dietary fiber per serving.

🗂 = 480 milligrams or more of sodium per serving.

[a]Salad greens (like chicory, endive, escarole, lettuce, romaine, spinach, arugula, radicchio, watercress) are on the Free Foods list.

Table C.9 Meat and Meat Substitutes

The Meat and Meat Substitutes list groups foods based on the amount of fat they have (lean meat, medium-fat meat, high-fat meat, and plant-based proteins).

FOOD	AMOUNT
Lean Meats and Meat Substitutes	

1 lean meat choice = 0 grams carbohydrate, 7 grams protein, 0–3 grams fat, and 100 kcalories.

FOOD	AMOUNT
Beef: Select or Choice grades trimmed of fat: ground round, roast (chuck, rib, rump), round, sirloin, steak (cubed, flank, porterhouse, T-bone), tenderloin	1 oz
ⓘ Beef jerky	1 oz
Cheeses with 3 grams of fat or less per oz	1 oz
Cottage cheese	¼ cup
Egg substitutes, plain	¼ cup
Egg whites	2
Fish, fresh or frozen, plain: catfish, cod, flounder, haddock, halibut, orange roughy, salmon, tilapia, trout, tuna	1 oz
ⓘ Fish, smoked: herring or salmon (lox)	1 oz
Game: buffalo, ostrich, rabbit, venison	1 oz
ⓘ Hot dog with 3 grams of fat or less per oz (8 dogs per 14 oz package) *Note: May be high in carbohydrate.*	1
Lamb: chop, leg, or roast	1 oz
Organ meats: heart, kidney, liver *Note: May be high in cholesterol.*	1 oz
Oysters, fresh or frozen	6 medium
Pork, lean	
ⓘ Canadian bacon	1 oz
rib or loin chop/roast, ham, tenderloin	1 oz
Poultry, without skin: Cornish hen, chicken, domestic duck or goose (well-drained of fat), turkey	1 oz
Processed sandwich meats with 3 grams of fat or less per oz: chipped beef, deli thin-sliced meats, turkey ham, turkey kielbasa, turkey pastrami	1 oz
Salmon, canned	1 oz
Sardines, canned	2 medium
ⓘ Sausage with 3 grams of fat or less per oz	1 oz
Shellfish: clams, crab, imitation shellfish, lobster, scallops, shrimp	1 oz
Tuna, canned in water or oil, drained	1 oz
Veal, lean chop, roast	1 oz

Medium-Fat Meat and Meat Substitutes

1 medium-fat meat choice = 0 grams carbohydrate, 7 grams protein, 4–7 grams fat, and 130 kcalories.

FOOD	AMOUNT
Medium-Fat Meat and Meat Substitutes *(continued)*	
Beef: corned beef, ground beef, meatloaf, Prime grades trimmed of fat (prime rib), short ribs, tongue	1 oz
Cheeses with 4–7 grams of fat per oz: feta, mozzarella, pasteurized processed cheese spread, reduced-fat cheeses, string	1 oz
Egg *Note: High in cholesterol, so limit to 3 per week.*	1
Fish, any fried product	1 oz
Lamb: ground, rib roast	1 oz
Pork: cutlet, shoulder roast	1 oz
Poultry: chicken with skin; dove, pheasant, wild duck, or goose; fried chicken; ground turkey	1 oz
Ricotta cheese	2 oz or ¼ cup
ⓘ Sausage with 4–7 grams of fat per oz	1 oz
Veal, cutlet (no breading)	1 oz

High-Fat Meat and Meat Substitutes

1 high-fat meat choice = 0 grams carbohydrate, 7 grams protein, 8+ grams fat, and 150 kcalories. These foods are high in saturated fat, cholesterol, and kcalories and may raise blood cholesterol levels if eaten on a regular basis. Try to eat 3 or fewer servings from this group per week.

FOOD	AMOUNT
Bacon	
ⓘ pork	2 slices (16 slices per lb or 1 oz each, before cooking)
ⓘ turkey	3 slices (½ oz each before cooking)
Cheese, regular: American, bleu, brie, cheddar, hard goat, Monterey jack, queso, and Swiss	1 oz
▽ ⓘ Hot dog: beef, pork, or combination (10 per lb-sized package)	1
ⓘ Hot dog: turkey or chicken (10 per lb-sized package)	1
Pork: ground, sausage, spareribs	1 oz
Processed sandwich meats with 8 grams of fat or more per oz: bologna, pastrami, hard salami	1 oz
ⓘ Sausage with 8 grams fat or more per oz: bratwurst, chorizo, Italian, knockwurst, Polish, smoked, summer	1 oz

(continued)

Table C.9 Meat and Meat Substitutes *(continued)*

Plant-Based Proteins

1 plant-based protein choice = variable grams carbohydrate, 7 grams protein, variable grams fat, and variable kcalories. Because carbohydrate content varies among plant-based proteins, you should read the food label.

FOOD	SERVING SIZE	COUNT AS
"Bacon" strips, soy-based	3 strips	1 medium-fat meat
☺ Baked beans	⅓ cup	1 starch + 1 lean meat
☺ Beans, cooked: black, garbanzo, kidney, lima, navy, pinto, white[a]	½ cup	1 starch + 1 lean meat
☺ "Beef" or "sausage" crumbles, soy-based	2 oz	½ carbohydrate + 1 lean meat
"Chicken" nuggets, soy-based	2 nuggets (1½ oz)	½ carbohydrate + 1 medium-fat meat
☺ Edamame	½ cup	½ carbohydrate + 1 lean meat
Falafel (spiced chickpea and wheat patties)	3 patties (about 2 inches across)	1 carbohydrate + 1 high-fat meat
Hot dog, soy-based	1 (1½ oz)	½ carbohydrate + 1 lean meat
☺ Hummus	⅓ cup	1 carbohydrate + 1 high-fat meat
☺ Lentils, brown, green, or yellow	½ cup	1 carbohydrate + 1 lean meat
☺ Meatless burger, soy-based	3 oz	½ carbohydrate + 2 lean meats
☺ Meatless burger, vegetable- and starch-based	1 patty (about 2½ oz)	1 carbohydrate + 2 lean meats
Nut spreads: almond butter, cashew butter, peanut butter, soy nut butter	1 Tbsp	1 high-fat meat
☺ Peas, cooked: black-eyed and split peas	½ cup	1 starch + 1 lean meat
🖭 ☺ Refried beans, canned	½ cup	1 starch + 1 lean meat
"Sausage" patties, soy-based	1 (1½ oz)	1 medium-fat meat
Soy nuts, unsalted	¾ oz	½ carbohydrate + 1 medium-fat meat
Tempeh	¼ cup	1 medium-fat meat
Tofu	4 oz (½ cup)	1 medium-fat meat
Tofu, light	4 oz (½ cup)	1 lean meat

KEY

☺ = More than 3 grams of dietary fiber per serving.

▽ = Extra fat, or prepared with added fat. (Add an additional fat choice to this food.)

🖭 = 480 milligrams or more of sodium per serving (based on the sodium content of a typical 3-oz serving of meat, unless 1 or 2 oz is the normal serving size).

[a]Beans, peas, and lentils are also found on the Starch list; nut butters in smaller amounts are found in the Fats list.

Table C.10 Fats

Fats and oils have mixtures of unsaturated (polyunsaturated and monounsaturated) and saturated fats. Foods on the Fats list are grouped together based on the major type of fat they contain. 1 fat choice = 0 grams carbohydrate, 0 grams protein, 5 grams fat, and 45 kcalories.

Note: In general, one fat exchange is 1 teaspoon of regular margarine, vegetable oil, or butter; 1 tablespoon of regular salad dressing.
    When used in large amounts, bacon and peanut butter are counted as high-fat meat choices (see Meat and Meat Substitutes list).
Fat-free salad dressings are found on the Sweets, Desserts, and Other Carbohydrates list. Fat-free products such as margarines, salad dressing, mayonnaise, sour cream, and cream cheese are found on the Free Foods list.

FOOD	SERVING SIZE
Monounsaturated Fats	
Avocado, medium	2 Tbsp (1 oz)
Nut butters (*trans* fat-free): almond butter, cashew butter, peanut butter (smooth or crunchy)	1½ tsp
Nuts	
almonds	6 nuts
Brazil	2 nuts
cashews	6 nuts
filberts (hazelnuts)	5 nuts
macadamia	3 nuts
mixed (50% peanuts)	6 nuts
peanuts	10 nuts
pecans	4 halves
pistachios	16 nuts
Oil: canola, olive, peanut	1 tsp
Olives	
black (ripe)	8 large
green, stuffed	10 large
Polyunsaturated Fats	
Margarine: lower-fat spread (30%–50% vegetable oil, *trans* fat-free)	1 Tbsp
Margarine: stick, tub (*trans* fat-free) or squeeze (*trans* fat-free)	1 tsp
Mayonnaise	
reduced-fat	1 Tbsp
regular	1 tsp
Mayonnaise-style salad dressing	
reduced-fat	1 Tbsp
regular	2 tsp
Nuts	
Pignolia (pine nuts)	1 Tbsp
walnuts, English	4 halves
Oil: corn, cottonseed, flaxseed, grape seed, safflower, soybean, sunflower	1 tsp
Oil: made from soybean and canola oil—Enova	1 tsp
Plant stanol esters	
light	1 Tbsp
regular	2 tsp

FOOD	SERVING SIZE
Polyunsaturated Fats *(continued)*	
Salad dressing	
🅢 reduced-fat	2 Tbsp
Note: May be high in carbohydrate.	
🅢 regular	1 Tbsp
Seeds	
flaxseed, whole	1 Tbsp
pumpkin, sunflower	1 Tbsp
sesame seeds	1 Tbsp
Tahini or sesame paste	2 tsp
Saturated Fats	
Bacon, cooked, regular or turkey	1 slice
Butter	
reduced-fat	1 Tbsp
stick	1 tsp
whipped	2 tsp
Butter blends made with oil	
reduced-fat or light	1 Tbsp
regular	1½ tsp
Chitterlings, boiled	2 Tbsp (½ oz)
Coconut, sweetened, shredded	2 Tbsp
Coconut milk	
light	⅓ cup
regular	1½ Tbsp
Cream	
half and half	2 Tbsp
heavy	1 Tbsp
light	1½ Tbsp
whipped	2 Tbsp
whipped, pressurized	¼ cup
Cream cheese	
reduced-fat	1½ Tbsp (¾ oz)
regular	1 Tbsp (½ oz)
Lard	1 tsp
Oil: coconut, palm, palm kernel	1 tsp
Salt pork	¼ oz
Shortening, solid	1 tsp
Sour cream	
reduced-fat or light	3 Tbsp
regular	2 Tbsp

KEY

🅢 = 480 milligrams or more of sodium per serving.

Table C.11 Free Foods

A "free" food is any food or drink choice that has less than 20 kcalories and 5 grams or less of carbohydrate per serving.

- Most foods on this list should be limited to 3 servings (as listed here) per day. Spread out the servings throughout the day. If you eat all 3 servings at once, it could raise your blood glucose level.
- Food and drink choices listed here without a serving size can be eaten whenever you like.

FOOD	SERVING SIZE
Low Carbohydrate Foods	
Cabbage, raw	½ cup
Candy, hard (regular or sugar-free)	1 piece
Carrots, cauliflower, or green beans, cooked	¼ cup
Cranberries, sweetened with sugar substitute	½ cup
Cucumber, sliced	½ cup
Gelatin	
dessert, sugar-free	
unflavored	
Gum	
Jam or jelly, light or no sugar added	2 tsp
Rhubarb, sweetened with sugar substitute	½ cup
Salad greens	
Sugar substitutes (artificial sweeteners)	
Syrup, sugar-free	2 Tbsp
Modified Fat Foods with Carbohydrate	
Cream cheese, fat-free	1 Tbsp (½ oz)
Creamers	
nondairy, liquid	1 Tbsp
nondairy, powdered	2 tsp
Margarine spread	
fat-free	1 Tbsp
reduced-fat	1 tsp
Mayonnaise	
fat-free	1 Tbsp
reduced-fat	1 tsp
Mayonnaise-style salad dressing	
fat-free	1 Tbsp
reduced-fat	1 tsp
Salad dressing	
fat-free or low-fat	1 Tbsp
fat-free, Italian	2 Tbsp
Sour cream, fat-free or reduced-fat	1 Tbsp
Whipped topping	
light or fat-free	2 Tbsp
regular	1 Tbsp
Condiments	
Barbecue sauce	2 tsp
Catsup (ketchup)	1 Tbsp
Honey mustard	1 Tbsp
Horseradish	

FOOD	SERVING SIZE
Condiments (continued)	
Lemon juice	
Miso	1½ tsp
Mustard	
Parmesan cheese, freshly grated	1 Tbsp
Pickle relish	1 Tbsp
Pickles	
🗟 dill	1½ medium
sweet, bread and butter	2 slices
sweet, gherkin	¾ oz
Salsa	¼ cup
🗟 Soy sauce, light or regular	1 Tbsp
Sweet and sour sauce	2 tsp
Sweet chili sauce	2 tsp
Taco sauce	1 Tbsp
Vinegar	
Yogurt, any type	2 Tbsp

Drinks/Mixes

Any food on the list—without a serving size listed—can be consumed in any moderate amount.

- 🗟 Bouillon, broth, consommé
- Bouillon or broth, low-sodium
- Carbonated or mineral water
- Club soda
- Cocoa powder, unsweetened (1 Tbsp)
- Coffee, unsweetened or with sugar substitute
- Diet soft drinks, sugar-free
- Drink mixes, sugar-free
- Tea, unsweetened or with sugar substitute
- Tonic water, diet
- Water
- Water, flavored, carbohydrate free

Seasonings

Any food on this list can be consumed in any moderate amount.
Flavoring extracts (for example, vanilla, almond, peppermint)
Garlic
Herbs, fresh or dried
Nonstick cooking spray
Pimento
Spices
Hot pepper sauce
Wine, used in cooking
Worcestershire sauce

KEY

🗟 = 480 milligrams or more of sodium per serving.

Table C.12   Combination Foods

Many foods are eaten in various combinations, such as casseroles. Because "combination" foods do not fit into any one choice list, this list of choices provides some typical combination foods.

FOOD	SERVING SIZE	COUNT AS
Entrees		
📦 Casserole type (tuna noodle, lasagna, spaghetti with meatballs, chili with beans, macaroni and cheese)	1 cup (8 oz)	2 carbohydrates + 2 medium-fat meats
📦 Stews (beef/other meats and vegetables)	1 cup (8 oz)	1 carbohydrate + 1 medium-fat meat + 0–3 fats
Tuna salad or chicken salad	½ cup (3½ oz)	½ carbohydrate + 2 lean meats + 1 fat
Frozen Meals/Entrees		
📦😊 Burrito (beef and bean)	1 (5 oz)	3 carbohydrates + 1 lean meat + 2 fats
📦 Dinner-type meal	generally 14–17 oz	3 carbohydrates + 3 medium-fat meats + 3 fats
📦 Entrée or meal with less than 340 kcalories	about 8–11 oz	2–3 carbohydrates + 1–2 lean meats
Pizza		
📦 cheese/vegetarian, thin crust	¼ of a 12 inch (4½–5 oz)	2 carbohydrates + 2 medium-fat meats
📦 meat topping, thin crust	¼ of a 12 inch (5 oz)	2 carbohydrates + 2 medium-fat meats + 1½ fats
📦 Pocket sandwich	1 (4½ oz)	3 carbohydrates + 1 lean meat + 1–2 fats
📦 Pot pie	1 (7 oz)	2½ carbohydrates + 1 medium-fat meat + 3 fats
Salads (Deli-Style)		
Coleslaw	½ cup	1 carbohydrate + 1½ fats
Macaroni/pasta salad	½ cup	2 carbohydrates + 3 fats
📦 Potato salad	½ cup	1½–2 carbohydrates + 1–2 fats
Soups		
📦 Bean, lentil, or split pea	1 cup	1 carbohydrate + 1 lean meat
📦 Chowder (made with milk)	1 cup (8 oz)	1 carbohydrate + 1 lean meat + 1½ fats
📦 Cream (made with water)	1 cup (8 oz)	1 carbohydrate + 1 fat
📦 Instant	6 oz prepared	1 carbohydrate
📦 with beans or lentils	8 oz prepared	2½ carbohydrates + 1 lean meat
📦 Miso soup	1 cup	½ carbohydrate + 1 fat
📦 Oriental noodle	1 cup	2 carbohydrates + 2 fats
Rice (congee)	1 cup	1 carbohydrate
📦 Tomato (made with water)	1 cup (8 oz)	1 carbohydrate
📦 Vegetable, beef, chicken noodle, or other broth-type	1 cup (8 oz)	1 carbohydrate

KEY

😊 = More than 3 grams of dietary fiber per serving.

▽ = Extra fat, or prepared with added fat.

📦 = 600 milligrams or more of sodium per serving (for combination food main dishes/meals).

Table C.13 Fast Foods

The choices in the Fast Foods list are not specific fast-food meals or items, but are estimates based on popular foods. Ask the restaurant or check its website for nutrition information about your favorite fast foods.

FOOD	SERVING SIZE	COUNT AS
Breakfast Sandwiches		
🗋 Egg, cheese, meat, English muffin	1 sandwich	2 carbohydrates + 2 medium-fat meats
🗋 Sausage biscuit sandwich	1 sandwich	2 carbohydrates + 2 high-fat meats + 3½ fats
Main Dishes/Entrees		
🗋 ☺ Burrito (beef and beans)	1 (about 8 oz)	3 carbohydrates + 3 medium-fat meats + 3 fats
🗋 Chicken breast, breaded and fried	1 (about 5 oz)	1 carbohydrate + 4 medium-fat meats
Chicken drumstick, breaded and fried	1 (about 2 oz)	2 medium-fat meats
🗋 Chicken nuggets	6 (about 3½ oz)	1 carbohydrate + 2 medium-fat meats + 1 fat
🗋 Chicken thigh, breaded and fried	1 (about 4 oz)	½ carbohydrate + 3 medium-fat meats + 1½ fats
🗋 Chicken wings, hot	6 (5 oz)	5 medium-fat meats + 1½ fats
Oriental		
🗋 Beef/chicken/shrimp with vegetables in sauce	1 cup (about 5 oz)	1 carbohydrate + 1 lean meat + 1 fat
🗋 Egg roll, meat	1 (about 3 oz)	1 carbohydrate + 1 lean meat + 1 fat
Fried rice, meatless	½ cup	1½ carbohydrates + 1½ fats
🗋 Meat and sweet sauce (orange chicken)	1 cup	3 carbohydrates + 3 medium-fat meats + 2 fats
🗋 ☺ Noodles and vegetables in sauce (chow mein, lo mein)	1 cup	2 carbohydrates + 1 fat
Pizza		
Pizza		
🗋 cheese, pepperoni, regular crust	⅛ of a 14 inch (about 4 oz)	2½ carbohydrates + 1 medium-fat meat + 1½ fats
🗋 cheese/vegetarian, thin crust	¼ of a 12 inch (about 6 oz)	2½ carbohydrates + 2 medium-fat meats + 1½ fats
Sandwiches		
🗋 Chicken sandwich, grilled	1	3 carbohydrates + 4 lean meats
🗋 Chicken sandwich, crispy	1	3½ carbohydrates + 3 medium-fat meats + 1 fat
Fish sandwich with tartar sauce	1	2½ carbohydrates + 2 medium-fat meats + 2 fats
Hamburger		
🗋 large with cheese	1	2½ carbohydrates + 4 medium-fat meats + 1 fat
regular	1	2 carbohydrates + 1 medium-fat meat + 1 fat
🗋 Hot dog with bun	1	1 carbohydrate + 1 high-fat meat + 1 fat
Submarine sandwich		
🗋 less than 6 grams fat	6-inch sub	3 carbohydrates + 2 lean meats
🗋 regular	6-inch sub	3½ carbohydrates + 2 medium-fat meats + 1 fat
Taco, hard or soft shell (meat and cheese)	1 small	1 carbohydrate + 1 medium-fat meat + 1½ fats
🗋 ☺ Salad, main dish (grilled chicken type, no dressing or croutons)		1 carbohydrate + 4 lean meats
Salad, side, no dressing or cheese	Small (about 5 oz)	1 vegetable

(continued)

Table C.13 Fast Foods *(continued)*

FOOD	SERVING SIZE	COUNT AS
Sides/Appetizers		
▽ French fries, restaurant style	small	3 carbohydrates + 3 fats
	medium	4 carbohydrates + 4 fats
	large	5 carbohydrates + 6 fats
🖻 Nachos with cheese	small (about 4½ oz)	2½ carbohydrates + 4 fats
🖻 Onion rings	1 serving (about 3 oz)	2½ carbohydrates + 3 fats
Desserts		
Milkshake, any flavor	12 oz	6 carbohydrates + 2 fats
Soft-serve ice cream cone	1 small	2½ carbohydrates + 1 fat

KEY

☺ = More than 3 grams of dietary fiber per serving.

▽ = Extra fat, or prepared with added fat.

🖻 = 600 milligrams or more of sodium per serving (for fast-food main dishes/meals).

Table C.14 Alcohol

1 alcohol equivalent = variable grams carbohydrate, 0 grams protein, 0 grams fat, and 100 kcalories.

Note: In general, one alcohol choice (½ ounce absolute alcohol) has about 100 kcalories. For those who choose to drink alcohol, guidelines suggest limiting alcohol intake to 1 drink or less per day for women, and 2 drinks or less per day for men. To reduce your risk of low blood glucose (hypoglycemia), especially if you take insulin or a diabetes pill that increases insulin, always drink alcohol with food. While alcohol, by itself, does not directly affect blood glucose, be aware of the carbohydrate (for example, in mixed drinks, beer, and wine) that may raise your blood glucose.

ALCOHOLIC BEVERAGE	SERVING SIZE	COUNT AS
Beer		
light (4.2%)	12 fl oz	1 alcohol equivalent + ½ carbohydrate
regular (4.9%)	12 fl oz	1 alcohol equivalent + 1 carbohydrate
Distilled spirits: vodka, rum, gin,	1½ fl oz	1 alcohol equivalent
whiskey, 80 or 86 proof		
Liqueur, coffee (53 proof)	1 fl oz	1 alcohol equivalent + 1 carbohydrate
Sake	1 fl oz	½ alcohol equivalent
Wine		
dessert (sherry)	3½ fl oz	1 alcohol equivalent + 1 carbohydrate
dry, red or white (10%)	5 fl oz	1 alcohol equivalent

D

Lifestyle Management Forms

This appendix contains a variety of forms that nutrition counselors are likely to find useful during nutrition counseling interventions.

- LMF 4.1 Assessment Graphic
- LMF 5.1 Client Assessment Questionnaire
- LMF 5.2 Food Record
- LMF 5.3 24-Hour Recall, Usual Diet
- LMF 5.4 Food Frequency Questionnaire
- LMF 5.5 Feedback—Anthropometric
- LMF 5.6 Client Concerns and Strengths
- LMF 5.7 Student Nutrition Interview Agreement
- LMF 6.1 Eating Behavior Journal
- LMF 6.2 Counseling Agreement
- LMF 7.1 Symptoms of Stress
- LMF 7.2 Stress Journal
- LMF 7.3 Tips to Reduce Stress
- LMF 7.4 Prochaska and DiClemente's Spiral of Change
- LMF 7.5 Interview Assessment Form
- LMF 7.6 Counseling Responses Competency
- LMF 7.7 The CARE Assessment
- LMF 8.1 Benefits of Regular Moderate Physical Activity
- LMF 8.2 Physical Activity Log
- LMF 8.3 Physical Activity Options
- LMF 8.4 Physical Activity Readiness Questionnaire, PAR_Q
- LMF 8.5 Medical Release
- LMF 8.6 Physical Activity Assessment and Feedback Form
- LMF 14.1 Registration for Nutrition Clinic
- LMF 14.2 Student Nutrition Counseling Agreement

Assessment Graphic*

NOT READY				NOT SURE	NOT SURE				
1	2	3	4	5	6	7	8	9	10

*For readiness to change 1 = not ready; 10 = very ready
For adherence to dietary goals 1 = never; 10 = always
For confidence in making a lifestyle change: 1 = not ready; 10 = very ready
For degree of importance for making a lifestyle change: 1 = not ready; 10 = very ready

Client Assessment Questionnaire

DEMOGRAPHIC DATA

Name _____ Date: _____

Address _____ Home telephone: _____

_____ Cell telephone: _____

Fax: _____ Email _____

Sex: M F Age: _____ Birthdate _____ Height _____ Weight _____

HEALTH HISTORY

1. What medical concerns (e.g., pregnancy), if any, do you have at the present time?

_____

2. Indicate if you have had blood relatives with any of the following problems:

Cancer   ☐ yes ☐ no   High blood pressure ☐ yes ☐ no
Diabetes   ☐ yes ☐ no   Osteoporosis   ☐ yes ☐ no
Heart disease   ☐ yes ☐ no   Thyroid disorder   ☐ yes ☐ no
High cholesterol   ☐ yes ☐ no

3. Do you have complaints about any of the following?

_____ Appetite   _____ Constipation   _____ Menstrual difficulties
_____ Bleeding gums   _____ Diarrhea   _____ Seeing in dim light
_____ Bruising   _____ Edema   _____ Sudden weight change
_____ Chewing or swallowing _____ Indigestion   _____ Stress

4. Do you use tobacco in any way? ☐ yes ☐ no
How much? _____

Did you recently stop smoking? ☐ yes ☐ no

5. Do you enjoy physical activity? ☐ yes ☐ no Explain _____

6. List any food allergies or intolerances.

_____

_____

Lifestyle Management Form 5.1

List any prescribed, over-the-counter, herbal, or vitamin/mineral supplements you take.

_____

_____

_____

DIET HISTORY

1. Do you follow a special dietary plan, such as, low cholesterol, kosher, vegetarian?

    _____

2. Have you ever followed a special diet? _____ Explain _____

3. Do you have any problems purchasing foods that you want to buy? _____

4. Are there certain foods that you do not eat? _____

5. Do you eat at regular times each day? ☐ yes ☐ no How often? _____

6. Identify any foods you particularly like. _____

7. Do you drink alcohol? ☐ yes ☐ no How often? _____

8. What change would you like to make?

    ☐ Improve my eating habits   ☐ Improve my activity level
    ☐ Learn to manage my weight   ☐ Improve my cholesterol/triglyceride levels
    ☐ Other _____

9. Please add any additional information you feel may be relevant to understanding your nutritional health. _____

10. In order to tailor your counseling experience to your needs, it would be useful to know your expectations. Please check one of the following to indicate the amount of structure you believe meets your needs:

    ☐ *Tell me exactly what to eat for all my meals and snacks. I want a detailed food plan.* Example: ½ cup oatmeal, 1 cup skim milk, 6 oz. orange juice, 1 slice whole wheat toast, 1 teaspoon margarine

    ☐ *I want a lot of structure but freedom to select foods. I want to use the exchange system.* Example: 1 milk, 2 starch, 1 fruit, and 1 fat exchange

    ☐ *I want some structure and freedom to select foods. I want to use a food group plan.* Example: 1 serving of dairy foods, fruits, and fat and oil group; 2 servings of grains

    ☐ *I don't want a diet. I just want to eat better. I will just set food goals.*

Lifestyle Management Form 5.1

SOCIOECONOMIC HISTORY

1. What is the highest level of formal education you received? _____

   Other type of school _____

2. Are you employed? _____ Occupation _____

3. How many people in your household? _____ Ages? _____

4. Present marital status (circle one):

   Single   Married   Divorced   Widowed   Separated   Engaged

5. Do you have a refrigerator? _____ Stove? _____

6. Who prepares most of the meals in your home? _____ Who shops? _____

7. Do you use convenience foods daily? ☐ yes ☐ no

8. How often do you eat out?_____ Where? _____

9. Have you made any food changes in your life you feel good about? ☐ yes ☐ no

10. Who could support and encourage you to make these changes? _____

EDUCATION INTERESTS

What information would you like from your counselor?

☐ Supermarket shopping tour   ☐ Eating out   ☐ Exercise
☐ Weight management   ☐ Portion size   ☐ Alcohol calories
☐ Healthy food preparation   ☐ Eating less fat   ☐ Meal planning
☐ Fiber   ☐ Walking program   ☐ Snack foods
☐ Food labels   ☐ Other _____

Thank you for your willingness to share this information and to take part in the Nutrition Clinic. We look forward to working with you to make lifestyle changes in order to meet your food and fitness objectives.

Food Record

Name: _____ Date: _____

- Complete this form as accurately as possible, using the examples as a guide.
- Use only one form per day. Do not put any thing on this form which pertains to another day.
- Record all foods and beverages, including water, you consumed from the time you woke up to the time you went to bed.

TIME	FOOD / DRINK	TYPE	PREPARATION	AMOUNT
8:00 AM	Bagel	Cinnamon Raisin	Toasted	one half
8:00 AM	Milk	1% fat	Fresh	8 ounces
NOON	Chicken	leg and thigh	Fried	1 each

24-Hour Recall/Usual Diet Form

Date _____ Day of the Week _____

Food and Drink Consumed	Amount from Each Group						
	Milk	Meat	Fruit	Veggie	Grain	Oil	Disc[a]
Name & Type							
TOTALS							
MyPyramid Recommendations for 2000 Calories[b]	3 c	5 ½ oz	2 c	2 ½ c	6 oz[c]	6 tsp	≤ 267 calories
EVALUATION[c]							

[a]Discretionary calories
[b]These are approximations for a general evaluation of food intake for an adult. Exact amounts of food groups vary according to gender, age, and activity level. See www.mypyramid.gov for a customized food guide.
[c]At least 3 servings of whole grains should be eaten each day.
[d]Evaluation: **L** = low   **A** = adequate   **E** = excessive

Food Group Serving Sizes
Serving Sizes and MyPyramid Recommendations

BREADS, CEREALS, AND OTHER GRAIN PRODUCTS

What counts as 1 ounce of grains?

1 slice bread	½ bun, bagel, or English muffin
½ c cooked cereal, rice, or pasta	1 small roll, biscuit, or muffin
1 c ready-to-eat cereal	3 to 4 small or 2 large crackers

VEGETABLES

What counts as 1 cup of vegetables?

1 cup of raw or cooked vegetables or vegetable juice	1 cup tofu
2 cups of raw leafy greens	1 medium baked potato, 20 French fries
1 cup cooked dry beans and peas (such as pinto beans or split peas)	

FRUITS

What counts as 1 cup of fruit?

1 cup of fruit or 100% fruit juice	1 small apple
½ cup of dried fruit	1 medium pear, grapefruit
1 large banana, orange, peach	32 seedless grapes

MEAT, POULTRY, FISH, AND ALTERNATES

What counts as 1 ounce of meat or meat equivalent?

1 ounce of meat, poultry or fish	1 tablespoon of peanut butter
¼ cup cooked dry beans, 1 falafel patty (2 ¼", 4 oz)	½ ounce of nuts or seeds (12 almonds, 24 pistachios, 7 walnut halves)
1 egg	¼ cup (about 2 ounces) of tofu,
12 Tbsp. hummus	1 oz tempeh, cooked

MILK, YOGURT, AND, CHEESE

What counts as 1 cup of milk?

1 c milk or yogurt	$1/3$ c shredded cheese
2 oz process cheese food	2 slices Swiss cheese
1½ oz cheese	

OILS

What counts as 1 teaspoon of oil?

1 teaspoon vegetable oil (soy, corn, peanut, and sesame)	1 tablespoon mayonnaise type dressing, Italian dressing
1¼ teaspoon mayonnaise	8 large canned olives

FATS, SWEETS, AND ALCOHOLIC BEVERAGES

- Foods high in fat include margarine, salad dressing, oils, mayonnaise, sour cream, cream cheese, butter, gravy, sauces, potato chips, chocolate bars.
- Foods high in sugar include cakes, pies, cookies, doughnuts, sweet rolls, candy, soft drinks, fruit drinks, jelly, syrup, gelatin, desserts, sugar, and honey.
- Alcoholic beverages include wine, beer, and liquor.

Food Frequency Questionnaire

SERVING SIZES	FOOD GROUP	SERVINGS PER DAY	SERVINGS PER WEEK	NEVER or RARELY
1 slice bread 1 cup dry cereal ½ cup cooked rice, pasta, or cereal ½ bun, bagel, or English muffin 1 small roll, biscuit, or muffin	**Refined Grains**–white bread, pasta, cereals			❑
	Whole Grains–whole wheat bread, brown rice, oatmeal, bran cereal			❑
1 cup raw leafy vegetable ½ cup cooked or raw vegetables 6 oz vegetable juice	**Vegetables**			❑
6 oz fruit juice 1 medium fruit ¼ cup dried fruit ½ cup fresh, frozen, or canned fruit	**Fruits**			❑
8 oz milk 1 cup yogurt 1½ oz cheese 2 oz process cheese	**Dairy**—low-fat or fat-free ice cream, milk, cheese, yogurt; frozen yogurt			❑
	Dairy—whole milk, regular cheese, regular ice cream			❑
3 oz cooked meats, poultry, or fish	**Meats, Poultry, Fish**—lean			❑
	Meats, Poultry, Fish—high-fat: sausage, cold cuts, spareribs, hot dogs, eggs, bacon			❑
1/3 cup or 1½ oz nuts 2 Tbsp or ½ oz seeds ½ cup cooked dry beans 4 oz tofu, 1 cup soy milk 2 Tbsp peanut butter	**Nuts, Seeds and Dry Beans**			❑
1 Tbsp regular dressing 2 Tbsp light salad dressing 1 tsp oil 1 Tbsp low-fat mayonnaise 1 tsp margarine, butter	**Fats and Oils**			❑
8 oz lemonade 1½ oz candy 8 oz. soda	**Sweets**			❑
12 oz beer, 4 oz wine 1 shot hard liquor	**Alcohol**			❑

Anthropometric Feedback Form

Volunteer's Measurements	Standard
Actual weight =	
Body Mass Index =	Desirable = 19–25
Waist circumference =	High risk = men > 35", women > 40"

Client Concerns and Strengths Log

1. List all concerns expressed by your client or identified by you.

2. Write NC (no control) next to all concerns in which you or your client has no control.

3. Categorize in the following chart the remaining concerns in which there is some degree of control and as a result could be addressed by a goal:

Nutritional	Behavioral	Exercise

4. List strengths and skills.

5. Categorize the strengths and skills in the following chart:

Nutritional	Behavioral	Exercise

6. What strengths and skills can be used to address the concerns? List them in the following chart.

Strengths and Skills	Concerns	Possible Intervention Strategies

Student Nutrition Interview Agreement

Thank you for your willingness to participate in the nutrition counseling clinic offered by _____. This interview is designed to provide interviewing experience for nutrition counseling students. The objective is for the student to work on counseling skills, gather information about a health problem, and learn something about your health issues. While discussing your situation, you may receive some benefit by clarifying your health concerns and possibly formulating a decision to make a behavior change. However, this experience is not designed to be an intervention.

After this meeting, students will be required to write a report about their findings. This report is shared only with the course instructor. Information in the report may be shared with other students during classroom discussions. However, at no time will your name be used in those discussions. In all other respects, the information you give will be held in absolute and strictest confidence.

We thank you very sincerely for your willingness to participate and for your help in the education of future nutrition counselors. If you have any questions or problems during this project, please call the course instructor, _____ at _____.

I, _____, **have read and understood the above statement.**

Print your name here

_____   _____
Your signature here   *Today's date*

_____   _____
Counselor signature here   *Today's date*

Lifestyle Management Form 6.1

Eating Behavior Journal

Name _____   Day/Date _____   Physical Activities[1] _____

Time	Location/ Place	Foods and Beverages Consumed Amounts/Description	Degree of Hunger[A]	Social Situation[2]	Comments[3]

[1]Include type of activities and minutes engaged in the activities
[2]Use the following rating scale: 0 = not hungry, 1 = hungry, 2 = very hungry
[3]Indicate activities and who you were with, if anyone
[4]Record significant thoughts (I'm doing fine. I am a loser.); feelings (angry, happy, worried); concerns (Maybe I should have had the turkey sandwich.)

Source: Adapted from Pastors et al. *Facilitating Lifestyle Change A Resource Manual*. Chicago, IL: American Dietetic Association; © 1996. Reprinted with permission.

Counseling Agreement

Name _____ Date: _____

My plan is to do the following:

_____

_____

This activity will be accomplished on _____

My reward will be (specify when, where and what) _____

_____

_____

_____   _____
Client signature   Date

_____   _____
Counselor signature   Date

Symptoms of Stress

Physical Symptoms

- Muscular tension
- Headaches
- Insomnia
- Twitching eyelid
- Fatigue
- Backaches
- Neck/shoulder pain
- Digestive disorders
- Teeth grinding
- Changes in eating/sleep patterns
- Sweaty palms

Emotional Symptoms

- Anxiety
- Frequent crying
- Irritability
- Frustration
- Depression
- Worrying
- Nervousness
- Moodiness
- Anger
- Self-doubt
- Resentment

Mental Symptoms

- Short concentration
- Forgetfulness
- Lethargy
- Pessimism
- Low productivity
- Confusion

Social Symptoms

- Loneliness
- Nagging
- Withdrawal from social contact
- Isolation
- Yelling at others
- Reduced sex drive

Source: Adapted from Goliszek A. 60 *Second Stress Management*, 2nd ed. Far Hills, New Jersey: New Horizon Press; 2004.

Stress Awareness Journal

Name: _____ Date: _____

Time	Symptom of Stress	Activities*	Internal Self-talk

*List any eating activities before, during symptom, or after experiencing the symptom.

Tips for Reducing Stress

- ❏ Learn to say "no." Don't over-commit. Delegate work at home and work.

- ❏ Organize your time. Use a daily planner. Prioritize your tasks. Make a list and a realistic timetable. Check off tasks as they are completed. This gives you a sense of control of overwhelming demands and reduces anxiety.

- ❏ Be physically active. Big muscle activities, such as walking, are the best for relieving tension.

- ❏ Develop a positive attitude. Surround yourself with positive quotes, soothing music, and affirming people.

- ❏ Relax or meditate. Schedule regular massages, use guided imagery tapes or just take ten minutes for quiet reflection time in a park.

- ❏ Get enough sleep. Small problems can seem overwhelming when you are tired.

- ❏ Eat properly. Be sure to eat at least five servings of fruits and vegetables and three servings of whole grains every day. Limit intake of alcohol and caffeine.

- ❏ To err is human. Don't treat a mistake as a catastrophe. Ask yourself what will be the worst thing that will happen.

- ❏ Work at making friends and being a friend. Close relationships don't just happen. Compliment three people today. Send notes to those who did a good job.

- ❏ Accept yourself. Appreciate your talents and your limitations. Everyone has them.

- ❏ Laugh. Look at the irony of a difficult situation. Watch movies and plays and read stories that are humorous.

- ❏ Take three deep breaths.

- ❏ Forgive. Holding onto grudges only causes you more stress and pain.

Prochaska's and DiClemente's Spiral of Change

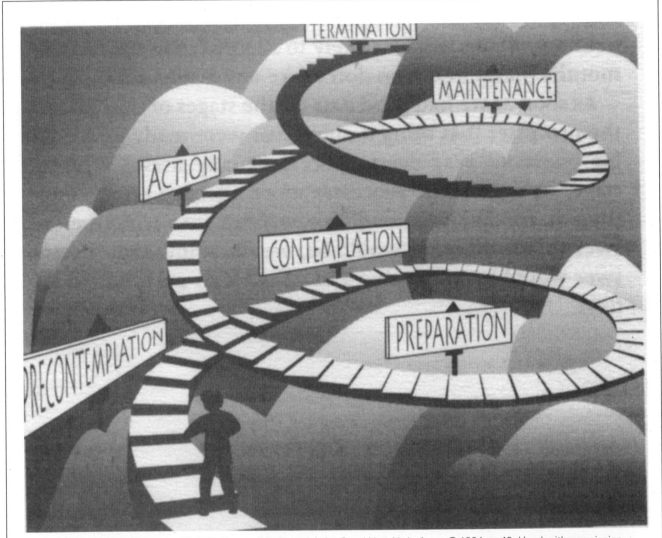

Source: Prochaska JO, Norcross JC, DiClemente CC, Changing for Good.New York: Avon; © 1994, p. 49. Used with permission.

Interview Checklist[1]

Interviewer _____ Observer _____ Date _____

Goal of the Interview: _____

I. FLOW OF THE INITIAL INTERVIEW
A. Involving Phase

1. Greeting	Yes ❑	No ❑
a) Verbal greeting	Yes ❑	No ❑
b) Shakes hands	Yes ❑	No ❑
2. Introduction of self	Yes ❑	No ❑
3. Attention to self comfort—Other obligations finished or planned for a later time; attention focused. (Self-evaluation only)	Yes ❑	No ❑
4. Attention to client's comfort—Physical comfort, noise and visual distractions minimized	Yes ❑	No ❑
5. Small talk, if appropriate	Yes ❑	No ❑
6. Establishes counseling objectives	Yes ❑	No ❑
a) Opening question—*What brings you here today?*	Yes ❑	No ❑
b) Establishes client's long-term objectives	Yes ❑	No ❑
c) Explains counseling process	Yes ❑	No ❑
d) Discusses weight monitoring, if appropriate	Yes ❑	No ❑
7. Establishes agenda	Yes ❑	No ❑
8. Transition Statement—*Now that we have gone over the basics of the program, we can explore your needs in greater detail.*	Yes ❑	No ❑

B. Exploration-Education Phase

1. Reviews completed assessment forms	Yes ❑	No ❑
2. Compares assessment to a standard, point-by-point, non-judgmental	Yes ❑	No ❑
3. Asks client thoughts about comparison	Yes ❑	No ❑
4. Segment summary—identifies problems, reiterates self-motivational statements, checks accuracy	Yes ❑	No ❑
5. Asks client if s/he would like to make changes	Yes ❑	No ❑
6. Assesses motivation—use a ruler to determine readiness to change	Yes ❑	No ❑
7. Tailors educational experiences to client needs	Yes ❑	No ❑

[1]This evaluation form is based on the Brown Interview Checklist, Brown University School of Medicine, Novack, DH, Goldstein, MG, Dubé CE, 1986.

C. Resolving Phase

Level 1 (numbers 1 to 3 on assessment graphic)

1. Raises awareness—Discusses benefits of change	Yes ❑	No ❑
2. Raises awareness—Personalizes benefits	Yes ❑	No ❑
3. Asks open-ended questions regarding importance of change	Yes ❑	No ❑
4. Provides summary	Yes ❑	No ❑
5. Offers advice, if appropriate	Yes ❑	No ❑
6. Expresses support	Yes ❑	No ❑

Level 2 (numbers 4 to 7 on assessment graphic)

1. Raises awareness—Discusses benefits of change and diet options	Yes ❑	No ❑
2. Asks open-ended questions regarding confidence in ability to change	Yes ❑	No ❑
3. Asks open-ended questions to identify barriers	Yes ❑	No ❑
4. Examines pros and cons	Yes ❑	No ❑
5. Imagines the future	Yes ❑	No ❑
6. Explores past successes	Yes ❑	No ❑
7. Explores support networks	Yes ❑	No ❑
8. Summarizes ambivalence	Yes ❑	No ❑

Level 3 (numbers 8 to 10 on assessment graphic)

1. Praises positive behaviors	Yes ❑	No ❑
2. Explores change options	Yes ❑	No ❑
a) Asks client's ideas for change	Yes ❑	No ❑
b) Uses an options tool, if appropriate	Yes ❑	No ❑
c) Explores concerns regarding selected option	Yes ❑	No ❑
3. Explains goal setting process	Yes ❑	No ❑
4. Identifies a specific goal from a broad goal—uses small talk, explores past experiences, builds on past	Yes ❑	No ❑
5. Goal is SMART: specific, measurable, attainable (client has control), rewarding (stated positively), time-bound	Yes ❑	No ❑
6. Designs a plan of action	Yes ❑	No ❑
a) Investigates physical environment	Yes ❑	No ❑
b) Examines social support	Yes ❑	No ❑
c) Examines cognitive environment, explains coping talk, if needed	Yes ❑	No ❑
d) Defines a tracking technique	Yes ❑	No ❑
e) Client verbalizes goal	Yes ❑	No ❑
7. Writes down goal	Yes ❑	No ❑

D. Closing Phase

1. Supports self-efficacy	Yes ❑	No ❑
2. Reviews issues and strengths	Yes ❑	No ❑
3. Uses "respect" relationship building response	Yes ❑	No ❑
4. Restates goal or goals	Yes ❑	No ❑
5. Reviews next meeting time	Yes ❑	No ❑
6. Shakes hands	Yes ❑	No ❑
7. Expresses appreciation for participation	Yes ❑	No ❑
8. Uses "support and partnership" relationship building responses—	Yes ❑	No ❑

II. INTERPERSONAL SKILLS

A. Facilitation (Attending) Skills

1. Eye contact—Appropriate length to enhance client comfort	Yes ❑	No ❑
2. Uses silences to facilitate client's expression of thoughts and feelings	Yes ❑	No ❑
3. Open posture—Arms uncrossed, facing client	F* ❑ P ❑	No ❑
4. Head nod, *mm-hm*, repeats client's last statement	F* ❑ P ❑	No ❑

*F = Frequently, P = Partially

B. Relationship Skills (Conveying Empathy)

1. Reflection—Restates the client's expressed emotion or inquires about emotions	F* ❑ P ❑	No ❑
2. Legitimation—Expresses understandability of client's emotions	Yes ❑	No ❑
3. Respect—Expresses respect for the client's coping efforts or makes a statement of praise	Yes ❑	No ❑
4. Support—Expresses willingness to be helpful to client addressing his/her concerns	Yes ❑	No ❑
5. Partnership—Expresses willingness to work with client	Yes ❑	No ❑

*F = Frequently, P = Partially

III. PATIENT RESPONSES

	Often	Sometimes	Seldom
A. Client freely discusses his/her concerns.	|_____	_____|
B. Client appears comforted and relaxed.	|_____	_____|
C. Client appears engaged in the counseling session.	|_____	_____|
D. Client freely offers information about his/her condition and life context.	|_____	_____|

IV. GENERAL COMMENTS

Counseling Responses Competency Assessment

Audio- or video-tape a counseling session and listen to the tape several times to complete the following assessment:

- Track the number of times <u>you</u> made each response by placing slash marks next to the name of the response. Note that this is an evaluation of your responses, not your client responses.
- For each category of responses, give an example from the tape. In cases where the particular response category was not demonstrated on the tape, write an example that may have been effective with your client and then complete the category evaluation.
- Select an intent and focus of the response. You may wish to review a discussion of these topics in Chapter 3.
- Indicate the effectiveness of your particular response, and explain why it was or was not effective. For responses that do not receive the most effective rating, write alternative responses that you believe would have worked better.
- Some of your responses may not fit any of the categories. This assessment covers many basic counseling responses but it is possible that some of your statements do not appear to fit into any of the categories. If that is the case, such material would not be evaluated. The following is an example of a competency evaluation for one response:

Example

Questions   _/ / /_

Example   What brings you here? Are you looking to lower your blood pressure?

Intent (circle one):   *To acknowledge*   (*To explore*)   *To challenge*

Focus (circle one):   *information*   (*experience*)   *feelings*   *thoughts*   *behaviors*

❑ Effective   ☒ Somewhat Effective   ❑ Not Effective   Explain I asked

two questions at the same time. I made an assumption that the main issue was blood pressure.

Alternative Response   What brings you here today?

---

1. ***Attending***   _____

   Example _____

   Intent (circle one): *To acknowledge*   *To explore*   *To challenge*

   Focus (circle one): *information*   *experience*   *feelings*   *thoughts*   *behaviors*

   ❑ Effective   ❑ Somewhat Effective   ❑ Not Effective   Explain _____

   Alternative Response_____

2. **Empathizing (Reflecting)** _____

Example _____

Intent (circle one): *To acknowledge   To explore   To challenge*

Focus (circle one): *information   experience   feelings   thoughts   behaviors*

❏ Effective   ❏ Somewhat Effective   ❏ Not Effective   Explain _____

Alternative Response _____

3. **Legitimation** _____

Example _____

Intent (circle one): *To acknowledge   To explore   To challenge*

Focus (circle one): *information   experience   feelings   thoughts   behaviors*

❏ Effective   ❏ Somewhat Effective   ❏ Not Effective   Explain _____

Alternative Response _____

4. **Respect** _____

Example _____

Intent (circle one): *To acknowledge   To explore   To challenge*

Focus (circle one): *information   experience   feelings   thoughts   behaviors*

❏ Effective   ❏ Somewhat Effective   ❏ Not Effective   Explain _____

Alternative Response _____

5. **Personal Support** _____

Example _____

Intent (circle one): *To acknowledge   To explore   To challenge*

Focus (circle one): *information   experience   feelings   thoughts   behaviors*

❏ Effective   ❏ Somewhat Effective   ❏ Not Effective   Explain _____

Alternative Response _____

6. **Partnership** _____

Example _____

Intent (circle one): *To acknowledge   To explore   To challenge*

Focus (circle one): *information   experience   feelings   thoughts   behaviors*

❏ Effective   ❏ Somewhat Effective   ❏ Not Effective   Explain _____

Alternative Response _____

7. Mirroring (Parroting) _____

Example _____

Intent (circle one): *To acknowledge   To explore   To challenge*

Focus (circle one): *information   experience   feelings   thoughts   behaviors*

❏ Effective   ❏ Somewhat Effective   ❏ Not Effective   Explain _____

Alternative Response _____

8. Paraphrasing _____

Example _____

Intent (circle one): *To acknowledge   To explore   To challenge*

Focus (circle one): *information   experience   feelings   thoughts   behaviors*

❏ Effective   ❏ Somewhat Effective   ❏ Not Effective   Explain _____

Alternative Response _____

9. Giving feedback (Immediacy) _____

Example _____

Intent (circle one): *To acknowledge   To explore   To challenge*

Focus (circle one): *information   experience   feelings   thoughts   behaviors*

❏ Effective   ❏ Somewhat Effective   ❏ Not Effective   Explain _____

Alternative Response _____

10. Questioning _____

Example _____

Intent (circle one): *To acknowledge   To explore   To challenge*

Focus (circle one): *information   experience   feelings   thoughts   behaviors*

❏ Effective   ❏ Somewhat Effective   ❏ Not Effective   Explain _____

Alternative Response _____

11. Clarifying (Probing, Prompting) _____

Example _____

Intent (circle one): *To acknowledge   To explore   To challenge*

Focus (circle one): *information   experience   feelings   thoughts   behaviors*

❏ Effective   ❏ Somewhat Effective   ❏ Not Effective   Explain _____

Alternative Response _____

12. Noting a Discrepancy (Confrontation, Challenging) _____

Example _____

Intent (circle one): *To acknowledge   To explore   To challenge*

Focus (circle one): *information   experience   feelings   thoughts   behaviors*

❑ Effective   ❑ Somewhat Effective   ❑ Not Effective   Explain _____

Alternative Response _____

13. Directing (Instructions) _____

Example _____

Intent (circle one): *To acknowledge   To explore   To challenge*

Focus (circle one): *information   experience   feelings   thoughts   behaviors*

❑ Effective   ❑ Somewhat Effective   ❑ Not Effective   Explain _____

Alternative Response _____

14. Advice _____

Example _____

Intent (circle one): *To acknowledge   To explore   To challenge*

Focus (circle one): *information   experience   feelings   thoughts   behaviors*

❑ Effective   ❑ Somewhat Effective   ❑ Not Effective   Explain _____

Alternative Response _____

15. Allowing Silence _____

Example _____

Intent (circle one): *To acknowledge   To explore   To challenge*

Focus (circle one): *information   experience   feelings   thoughts   behaviors*

❑ Effective   ❑ Somewhat Effective   ❑ Not Effective   Explain _____

Alternative Response _____

16. Self-Referent _____

Example _____

Intent (circle one): *To acknowledge   To explore   To challenge*

Focus (circle one): *information   experience   feelings   thoughts   behaviors*

❑ Effective   ❑ Somewhat Effective   ❑ Not Effective   Explain _____

Alternative Response _____

The CARE Measure

© Stewart W Mercer 2004

1. Please rate the following statements about today's consultation. Please tick one box for each statement and *answer every statement.*

How was the counsellor at ...	Poor	Fair	Good	Very Good	Excellent	Does Not Apply
1. Making you feel at ease... *(being friendly and warm towards you, treating you with respect; not cold or abrupt)*	☐	☐	☐	☐	☐	☐
2. Letting you tell your "story"... *(giving you time to fully describe your illness in your own words; not interrupting or diverting you)*	☐	☐	☐	☐	☐	☐
3. Really listening... *(paying close attention to what you were sayings; not looking at the notes or computer as you were talking)*	☐	☐	☐	☐	☐	☐
4. Being interested in you as a whole person ... *(asking/knowing relevant details about your life, your situation; not treating you as "just a number")*	☐	☐	☐	☐	☐	☐
5. Fully understanding your concerns... *(communicating that he/she had accurately understood your concerns; not overlooking or dismissing anything)*	☐	☐	☐	☐	☐	☐
6. Showing care and compassion... *(seeming genuinely concerned, connecting with you on a human level; not being indifferent or detached)*	☐	☐	☐	☐	☐	☐
7. Being Positive... *(having a positive approach and a positive attitude; being honest but not negative about your problems)*	☐	☐	☐	☐	☐	☐
8. Explaining things clearly... *(fully answering your questions, explaining clearly, giving you adequate information; not being vague*	☐	☐	☐	☐	☐	☐
9. Helping you to take control... *(exploring with you what you can do to improve your health yourself; encouraging rather than "lecturing" you)*	☐	☐	☐	☐	☐	☐
10. Making a plan of action with you ... *(discussing the options; involving you in decisions as much as you want to be involved; not ignoring your views)*	☐	☐	☐	☐	☐	☐

Potential Benefits of Regular Moderate Physical Activity

There are many potential benefits to becoming physically active. Review this list to identify what is most important to you.

➡ Reduces risk of dying prematurely

➡ Reduces risk or aids in the management of:
- heart disease
- diabetes
- high blood pressure
- cancer
- falls

➡ Aids in the support of strong bones

➡ Improves mood, self-esteem and self-image

➡ Reduces feelings of depression and anxiety

➡ Lowers cholesterol

➡ Lowers triglycerides

➡ Controls of blood sugar levels

➡ Strengthens heart and lungs

➡ Decreases stress

➡ Improves sleep, reduces risk of sleep apnea

➡ Improves productivity

➡ Increases stamina and energy

➡ Makes you feel better

➡ Maintains weight or aids loss of weight

➡ Maintains ability to function and preserves independence in older adults

Physical Activity Log

- Record all physical activity for a week. Remember to include regular daily activities such as climbing stairs, gardening, and walking to the office from a parking lot.
- Include all forms of physical fitness activities including stretching, weight lifting, balancing, and aerobic movement.

Day of the Week	Type of Activity	Amount of Time
Sunday		
Monday		
Tuesday		
Wednesday		
Thursday		
Friday		
Saturday		

Physical Activity Options

➤ Look for Everyday Opportunities
Short bursts of activity throughout the day make a difference.

- Use steps instead of elevators or escalators.
- Park your car in a distant section of the parking lot.
- Leave work five minutes later. Take a walk around the building.
- Get off the train or bus one stop early and walk the rest of the way.
- Take a walk during lunch.
- March, stretch, or do squats while brushing your teeth.
- Pace around the house or do arm curls with a can of food while talking on the phone.
- Jump rope, stretch, jog in place, or lift weights while watching TV.
- Be prepared. Keep walking shoes in your car or in your desk.
- Take your bike with you to a conference and explore the local scenery before driving home.

➤ Plan a Daily Routine
Think about cost, convenience, and bad weather options when planning a program. Look for creative ways to keep the activities enjoyable.

- Schedule time for physical activity. Write it in your calendar.
- Vary the physical activities. Plan to bike one day a week, jog two days a week, and go to the gym three days a week.
- Join a walking club, a biking club, etc.
- Add variety to the activity. Have several walking trails; ask a friend to join you in your walks; or listen to music or recorded books during your walks.

➤ Plan Physically Active Leisure Time Events
Look for activities the whole family can enjoy.

- Have a family baseball or soccer game.
- Plan a bike tour, mountain hike, or canoe trip.
- Explore a cave.

Physical Activity Readiness Questionnaire, PAR-Q*

(A Questionnaire for People Aged 15 to 69)

Regular physical activity is fun and healthy, and increasingly more people are starting to become more active every day. Being more active is very safe for most people. However, some people should check with their doctor before they start becoming much more physically active.

If you are planning to become much more physically active than you are now, start by answering the seven questions in the box below. If you are between ages of 15 and 69, the PAR-Q will tell you if you should check with your doctor before you start. If you are over 69 years of age, and you are not used to being very active, check with your doctor.

Common sense is your best guide when you answer these questions. Please read the questions carefully and answer each one honestly: check YES or NO.

YES	NO	
❏	❏	1. Has your doctor ever said that you have a heart condition *and* that you should only do physical activity recommended by a doctor?
❏	❏	2. Do you feel pain in your chest when you do physical activity?
❏	❏	3. In the past month, have you had chest pain when you were not doing physical activity?
❏	❏	4. Do you lose your balance because of dizziness or do you ever lose consciousness?
❏	❏	5. Do you have a bone or joint problem that could be made worse by a change in your physical activity?
❏	❏	6. Is your doctor currently prescribing drugs (for example, water pills) for your blood pressure or heart condition?
❏	❏	7. Do you know of any other reason why you should not do physical activity?

*Reference: PAR-Q & You, Physical Activity Readiness Questionnaire - PAR-Q (revised 1994), Canadian Society for Exercise Physiology. Reprinted with permission from American College of Sports Medicine. *Guidelines for Exercise Testing and Prescription*, 5th Ed. Philadelphia, PA: Williams & Wilkins, 1995, p. 14–16.

If you answered YES to one or more questions:

Talk with your doctor by phone or in person BEFORE you start becoming much more physically active or BEFORE you have a fitness appraisal. Tell your doctor about the PAR-Q and which questions you answered YES.

- You may be able to do any activity you want—as long as you start slowly and build up gradually. Or, you may need to restrict your activities to those that are safe for you. Talk with your doctor about the kinds of activities you wish to participate in and follow his/her advice.
- Find out which community programs are safe and helpful for you.
- Develop an exercise plan with the aid of an exercise specialist.

If you answered NO honestly to *all* PAR-Q questions, you can be reasonably sure that you can:

- Start becoming much more physically active—begin slowly and build up gradually. This is the safest and easiest way to go.
- Take part in a fitness appraisal—this is an excellent way to determine your basic fitness so that you can plan the best way for you to live actively.

DELAY BECOMING MUCH MORE ACTIVE:

- If you are not feeling well because of a temporary illness such as a cold or a fever—wait until you feel better; or
- If you are or may be pregnant—talk to your doctor before you start becoming more active.

Please note: If your health changes so that you then answer YES to any of the above questions, tell your fitness or health professional. Ask whether you should change your physical activity plan.

Informed Use of the PAR-Q: The Canadian Society for Exercise Physiology, Health Canada, and their agents assume no liability for persons who undertake physical activity, and if in doubt after completing this questionnaire, consult your doctor prior to physical activity.

I have read, understood and completed this questionnaire. Any questions I had were answered to my full satisfaction.

Name _____

Signature _____ Date _____

Signature
of parent or guardian_____ Witness_____

© Canadian Society for Exercise Physiology   Supported by Health Canada

Medical Release

Your patient has enrolled in our nutrition counseling lifestyle management program. We have asked this person to seek medical consultation to evaluate if there should be any limitations to his or her involvement in our clinic. If a client wishes to lose weight, a program is designed allowing for a modest weight loss of 1 to 2 pounds per week. Students counsel clients under the supervision of food and nutrition faculty. Please completely read the following statements and sign the form if you believe your client can safely participate in a lifestyle management program to alter eating and exercise behaviors.

Date: _____

This is to certify that I have examined the person named below:

NAME: _____

ADDRESS: _____

_____

CITY, STATE, ZIP _____

This person was found to be in satisfactory health. There are no reasons to prohibit this person from participating in a lifestyle management program that advocates changes in eating behaviors and modest exercise goals tailored to the client's level of readiness.

Health Practitioner _____

Address _____

_____

For further information, please contact _____ at _____

Physical Activity Assessment and Feedback Form

The following contains your evaluation of the physical activity assessment. Do not be surprised if you do not meet all the standards set by national organizations, most North Americans do not. One consequence of recent technological advances has been to decrease the need to move. This is a serious concern for our health. As evidence has been accumulating about the benefits of regular physical activity, several governmental and health agencies have issued official statements and/or instituted national programs to combat this problem. These include:

- American Medical Association
- American Heart Association
- Centers for Disease Control
- American College of Sports Medicine
- National Institutes of Health
- Office of the Surgeon General and Health Canada.

> *Many Americans may be surprised at the extent and strength of the evidence linking physical activity to numerous health improvements.*
> *– David Satcher, Former Director of the Centers for Disease Control and Prevention[1]*

Benefits of regular *moderate* exercise:

➤ Reduces your risk or aids in the management of
- heart disease
- diabetes
- high blood pressure
- cancer
➤ Aids in the support of strong bones
➤ Improves your mood, self-esteem and self-image
➤ Increases energy
➤ Maintains or aids in loss of weight
➤ Maintains function and preserves independence in older adults.

[1] Foreword, *Physical Activity and Health a Report of the Surgeon General*, Atlanta, GA: Department of Health and Human Services, 1996.

Lifestyle Management Form 8.6

Physical Activity Standard[1]	Standard Met	Standard Not Met
Muscular Strength: Engage in muscle strengthening activities that are moderate or high intensity and involve all muscle groups on 2 or more days a week.	❑	❑
Flexibility: Engage in activities that stretch major muscle groups at least 2 times per week.	❑	❑
Endurance (Minimum): Engage in at least 150 minutes of moderate or 75 minutes of vigorous aerobic activity a week.	❑	❑
Endurance (Additional Benefits): Engage in at least 300 minutes of moderate or 150 minutes of vigorous aerobic activity a week.	❑	❑

Motivation Level	Implication
Level 1, Not ready	❑ Would you consider learning more about how moderate physical activity could help your health?
Level 2, Unsure	❑ For some reason you are not sure that you are ready to begin a physical activity program. Your counselor will explore your ambivalence with you to see if you are ready to make plans to increase your physical activity level.
Level 3, Ready	❑ Great, you are ready to begin or increase your activity level. Your counselor can provide you with resources to aid in developing a plan.
Level 4, Action	❑ Congratulations, you are already actively involved in a physical activity program. Your counselor will review with you the standards set by authorities. If you do not meet all of them, you may wish to make some alterations.

Physical Activity Readiness:

❑ Talk to your doctor before becoming much more physically active or having a fitness appraisal as indicated by the following:

    ○ PAR-Q Readiness Questions (LMF 8.4) ○ Woman over age 50 ○ Man over age 40

❑ Delay an increase in physical activity due to pregnancy or illness.

[1]Standards are based on American College of Sports Medicine Position Standards, 1998 and 2008 Physical Activity Guidelines for Americans.
Note: Reevaluate readiness if you experience dizziness, chest pain, undue shortness of breath, difficulty breathing or unusual discomfort after beginning an exercise program.

Registration for Nutrition Clinic

Counselor	Participant
_____	_____
Name	*Name*
_____	_____
Cell Telephone: _____	***Cell Telephone:*** _____
Best times to call: _____	Best times to call: _____
Home Telephone: _____	***Home Telephone:*** _____
Best times to call: _____	Best times to call: _____
Email: _____	**Email:** _____

Your meeting day is: _____   **Location of meetings:** _____

Your meeting time is: _____   **Room number:** _____

Length of meetings is approximately one hour. If welcome packet forms have not been completed previous to the first session, the first counseling session may take an extra 20 minutes.

The dates of your 4 meetings are as follows: _____   _____

_____   _____

- Please complete 2 copies of this agreement form. The client copy should be given to the participant and the clinic copy should be given to the counselor.

- Thank you for your interest in our program. Please note that any cancellations of meetings should be made directly between each participant and counselor.

- If you have any questions, please contact the instructor, _____
  Phone number _____
  Email _____

Nutrition Counseling Agreement

Thank you for your interest in the nutrition counseling clinic offered by _____. This program is designed to provide a mutually beneficial experience for both students and volunteer adult clients. You will work one on one with an advanced nutrition counseling student for _____ sessions, each one lasting approximately one hour. During the registration process, clients are assigned a counselor, a counseling room, and meeting times. The counseling sessions provide clients an opportunity to explore and find solutions for nutrition and weight issues. At the same time, students will be working on their nutrition counseling skills. Although students will be following a well defined counseling guideline, each session will be tailored to their client's needs. Students can only assist clients in achieving weight loss if the client is overweight by National Institutes of Health Standards. Normal and underweight clients can still take part in the program with the goal of improving the quality of their diets.

Your student counselor will use a client-centered, motivational approach during sessions with you. This means your counselor will work collaboratively with you to explore your nutrition and weight issues, brainstorm resources and solutions, and help you to set achievable goals each week. Students will ask you questions about your health and family history as well as present day food habits. Two of the nutrition assessment forms will be given to you at registration. You can look at them before signing this form. Students will have a variety of tools at their disposal including food models and educational handouts. Students are encouraged to engage their clients in hands-on experiences. Therefore, at times your counseling session may take place in a grocery store, the student cafeteria, or the gym.

Physical activity is an important part of fitness and weight management. Experience has shown that our clients have a variety of orientations to this topic. If you are already very active in this area, you will be encouraged to continue your program. However, if exercise has not been an enjoyable experience, you will be invited to explore this issue. As long as there is no medical problem and you are ready to take action, weekly activity goals will be developed with you. For certain clients, we have a structured walking protocol that can be followed.

During the course of the counseling program, your student counselor may discuss his or her counseling interactions with the course instructor. The student will write a report about the counseling experience. This report is only shared with the course instructor. Your counselor may give a case study presentation concerning this experience to the nutrition counseling class, but at no time in these presentations will your name be used. In all other respects, information you give the student will be held in absolute and strictest confidence.

We thank you sincerely for your willingness to participate and for your help in the education of future nutrition counselors. If you have any questions or problems during this project, please call the course instructor, _____, at _____.

I, _____, have read and understand the above statement and agree to
       Print your name here
meet with _____ at agreed times and places on the registration form.

_____   _____
   Your signature here   *Today's date*

_____   _____
   Counselor signature here   *Today's date*

Index

diagnosis domains of, 129
diagram of, 126
documentation and, 134
examples of, 129
intervention and, 131–133
model, 126
monitoring and evaluation and, 133–134
nutritional assessment and, 127–128
nutritional diagnosis and, 128–131
objectives of, 126
PES statements, 129–131
standardized language and, 127
Nutrition counseling/counselor
algorithm for, 225
characteristics of effective, 5–7
defined, 2
foundation of, 2
goals in, 47
motivational algorithm for, 77–79
physical activity, role in guidance of, 192
program models for, 77
protocols for, 81
psychotherapy and, boundary between, 333
Nutrition counseling protocols
closing phase, 94
exploration-education phase, 85–88
greeting, 81–82
involving phase, 81–85
resolving phase, 88–89, 202–208
Nutrition education, 2
Nutrition education interventions, 275–297
for congregate meal program, 280–281
data collection methods for, 279–280
during nutrition counseling, 156–160
generalization statements for, 290–293
instructional plan for, 293–297
keys to successful
educational philosophy, 281–283
evaluations, 319–320
goals, 287–290
instructional planning, 301–311
learning strategies, 301–311
mass media materials, 311–319
needs assessment, 277–279
objectives, 287–290
theory-based interventions, 283–285
setting for, 276–277
social marketing and, 285–287
Nutrition evaluation, 133
"Nutrition Jeopardy," 309
Nutrition monitoring, 133
Nutrition Screening Initiative checklist, 234

O
OARS, 34–35
Obama, Michelle, 191

Obesity, 119
adult, 218
bias free care for, 237
childhood, 191
nutritional risks of, 230
in older people, 232
preschool, 218
qualitative research on, 279
solutions to, 228
Objectives
ABC's of, 288
classifying, 292
defined, 276, 288
establishment of, 287–290
evaluating, 289
types of, 288–289
Observation evaluations, 185
Obstacle thinking, 153. *See also* Cognitive restructuring
Older adults, 231–236
developmental factors of, 232
food behavior of, 231–232
intervention strategies for, 232–236
MyPyramid for, 233
national programs for, 233
nutritional risks of, 232
obesity in, 232
150 kilocalorie activities, 197
On-line courses, 309
Open-ended questions, 64
culturally sensitive, 224
exploring, 90–92
for promoting talk change, 91–92
Open groups, 260–261
Opening counseling sessions, 83
Operant conditioning approach to behavioral therapy, 29
Opportunity thinking, 153
Options
change, 206–207
diet, 89–90, 91
for physical activity, 27, 205
Options tool, 103–104
ORID (Objective, Reflective, Interpretive, and Decisional) discussion method, 256–257
Outcomes, measuring of, 133
Overhead transparency projector, 306–308
Overweight. *See* Obesity

P
Pair-Share, 256
Palm OS, 119
Paraphrasing, 63
Parroting, 62–63
Partnership, 62
Pavlov, Ivan, 29
Pawtucket Heart Health Program, 287
PBS Kids, 309